Health and Safety at Work

Studies in Employment and Social Policy

Volume 42

Series Editors

Prof. Dr Alan C. Neal
Prof. Dr Manfred Weiss
Prof. Dr Birgitta Nyström

Launched in 1997, the Studies in Employment and Social Policy series now boasts over thirty titles, addressing key policy and development issues in the fields of Industrial Relations, Labour Law, Social Security, and International Labour Regulation.

Under the direction of its General Editors, Professor Alan Neal (Founding Editor of the International Journal of Labour Law and Industrial Relations, and Convenor of the European Association of Labour Law Judges), Professor Manfred Weiss (past-President of the International Industrial Relations Association) and Professor Birgitta Nyström (Professor of Private Law at the Law Faculty, University of Lund, Sweden), Studies in Employment and Social Policy seeks to provide a forum for highlighting international and comparative research on contemporary areas of significance for evaluation and regulation of the world of work.

With contributors from leading figures in the field, Studies in Employment and Social Policy brings together key policy-makers, academics, and regulators, providing a unique context in which to analyse and evaluate the rapid and dramatic work and social policy developments taking place across the globe.

Health and Safety at Work

European and Comparative Perspective

Edited by
Edoardo Ales

Wolters Kluwer
Law & Business

Published by:
Kluwer Law International
PO Box 316
2400 AH Alphen aan den Rijn
The Netherlands
Website: www.kluwerlaw.com

Sold and distributed in North, Central and South America by:
Aspen Publishers, Inc.
7201 McKinney Circle
Frederick, MD 21704
United States of America
Email: customer.service@aspenpublishers.com

Sold and distributed in all other countries by:
Turpin Distribution Services Ltd
Stratton Business Park
Pegasus Drive, Biggleswade
Bedfordshire SG18 8TQ
United Kingdom
Email: kluwerlaw@turpin-distribution.com

Printed on acid-free paper.

ISBN 978-90-411-4661-8

Printed and Bound by CPI Group (UK) Ltd, Croydon, CR0 4YY.

List of Editor and Contributors

Edoardo Ales is Professor of Labour Law, Università degli studi di Cassino e del Lazio Meridionale (Italy). He also teaches EU Labour Law and Social Policies at the Libera Università degli studi sociali LUISS – G. Carli in Rome and Labour Law at the Pontificial Lateran University in Rome. He has been visiting professor at the Johannes Wolfgang Goethe Universität in Frankfurt a. Main (Germany). He is an active participant in the European Working Group on Labour Law (EWL). He is also member of the Scientific Committee and national expert for Italy in the European Labour Law Network (ELLN).

Joaquin Aparicio Tovar is Professor of Labour Law and Social Security Law, University of Castilla-la Mancha (Spain). He is Dean of the Faculty of Industrial Relations and Human Resources of the same University. He is an active participant in the EWL.

Mark Bell is Professor of Law and Head of the School of Law, University of Leicester (United Kingdom). He conducts research in the areas of national and European anti-discrimination law and employment law. He is the author of Racism and Equality in the European Union (OUP, 2008) and Anti-Discrimination law and the European Union (OUP, 2002). He is an active participant in the European Working Group on Labour Law and a member of the Executive Committee of the Society of Legal Scholars. He was a member of the European Commission's Network of Legal Experts in the Non-Discrimination Field (2004-2010) and has worked with the ILO to develop training on the legal protection of precarious workers.

Olaf Deinert is Professor of Civil Law, Labour Law and Social Security Law, University of Göttingen (Germany), honorary judge at the Federal Labour Court of Germany. His research includes all fields of labour law, especially with a focus on relations to social security law and private international law.

Filip Dorssemont is Professor of Labour Law, Université Catholique de Louvain (Belgium). His main subjects of research deal with legal issues of industrial relations from a Belgian, comparative and European perspective, fundamental workers' rights and human rights at work and EU labour law. He graduated in Law (1993) and in Philosophy (1994) and wrote a PhD Thesis related to the legal status of trade unions in a comparative perspective (Antwerp University, 2001). After a period as post-doc researcher lecturer at Utrecht University, the author was able to return to his home country due to a Ulysses grant of the FNRS (2008).

Alessia Giurini is PhD at the Università degli studi di Cassino e del Lazio Meridionale. Her field of research is the protection of occupational health and safety, with specific reference to the gender dimension and new psychosocial risks.

Elisabeth Kohlbacher is Research Assistant, Institute for Austrian and European Labour Law and Social Security Law at the Vienna University of Economics and Business (Austria). Her main research interests lie in the interface of European Law and national Law with a specific interest in questions related to the right to take collective action. She has researched and written on European Social Security Coordination and is co-publisher of the Austrian Yearbook on Social Security Law.

Teun Jaspers is Emeritus Professor of Labour Law and Social Policy, Utrecht University (The Netherlands). He is an active participant in the EWL.

Erika Kovács is Assistant Professor, Institute of Labour Law and Social Law at the Vienna University of Economics and Business (Austria). She wrote her doctoral thesis at the University of Trier (Germany) on trade union recognition and was the director of the Institute of Comparative and European Labour Law at the University of Pécs (Hungary). Her research interests lie in European labour law, Hungarian, Austrian and comparative labour law.

Franz Marhold is Professor of Labour Law and Social Security Law, Vienna University of Economics and Business (Austria). His current key areas of interest and research are industrial relations and social dialogue on the one hand and cross border health care provisions on the other. He is vice president of the European Institute for Social Security in Leuven and serves as member of the scientific advisory board of the Max Planck Institute for international social law and social policy in Munich. In addition, he is listed as an attorney with the Superior Court of Braunschweig as well as with the Viennese chamber of attorneys.

Luca Miranda is PhD, Università degli studi di Cassino e del Lazio Meridionale. His main area of study and research is the protection of the health and safety in the working environment, with specific reference to the collective dimension and the role of social partners. He is the author of a book and some articles in Spanish and Italian about the regulation of health and safety in a comparative perspective.

Koen Naert, is Group Environment Health and Safety Manager at Bekaert.

Frans Pennings is Professor of Labour Law and Social Security Law, Utrecht University (The Netherlands), Tilburg University (The Netherlands) and Gothenburg University (Sweden). He is an active participant in the EWL.

Anne Van Regenmortel is Professor of Social Security Law, University of Antwerp (Belgium) and Member of the Antwerp Bar. She has published numerous articles and papers in the field of labour law and social security law. She is also a highly esteemed expert in the domain of well-being at work. For thirty years she has been involved in the training of prevention advisers.

Corinne Sachs-Durand is Professor Emeritus of Labour Law, University of Strasbourg (France). Her main research interests are in the field of French labour law in general and more specifically in workers' representation. Moreover, she belongs to several European working groups, dealing with European Union law and comparative law. She is an active participant in the EWL.

Maria Steinberg is Doctor of Law and Senior Lecturer, University of Örebro (Sweden). Her teaching is in health and safety laws and in labour law. Her main research is in the field of democracy at the workplace, safety representatives, psychosocial issues and gender in relation with the Swedish Work Environment Act. She has a background working at the Swedish Work Environment Authority.

Berta Valdés de la Vega is Professor of Labour Law and Social Security Law, University of Castilla-la Mancha (Spain). She is an active participant in the EWL.

Summary of Contents

Table of Contents

Preface

This book is the product of the common effort of a group of labour lawyers involved in the Pontignano International Seminars and in the European Working Group on Labour Law (EWL). In 2010, the 27th edition of Pontignano took place in Gaeta (Italy) on the subject 'Health and safety at work'. The seminar was, as usual, based on national reports discussed in working groups within a comparative perspective. From the very beginning, the group got the impression that national reports could be re-elaborated into national chapters to be collected in a book, opening with a European Law chapter and closing with a comparative chapter. A template with four broad subjects to be developed in the national chapters was elaborated in order to facilitate comparative analysis. Edoardo Ales took the responsibility of being the editor of the book. He also committed himself to drafting the comparative chapter. Berta Valdés de la Vega committed herself to drafting the European chapter.

The deadline for the delivery of the first draft of national chapters was fixed on March 2011. Drafts were presented and discussed in a workshop organized in Cassino in May 2011. That is when Austria was included in the project. As a result of the workshop, a new template was elaborated. The new template added further and more specific subjects to be developed within each national chapter. However, the template did not contain punctual instructions thus leaving the authors free to follow the approach to the subject matter chosen by each national system.

On the suggestion of the Dutch members of the group, the editor got in touch with Kluwer Law International in order to propose the book for publication. The book was accepted for publication in the series 'Studies in Employment and Social Policy', directed by Alan Neal. Following a meeting of the group during the 28th edition of Pontignano in Gaeta, it was decided to include in the project also a country of new accession. Hungary was chosen for this purpose. In November 2011, on the suggestion of the Director of the Series, a Scandinavian country – Sweden, was also included int the project.

The deadline for the delivery of the final version of the national chapters was fixed on March 2012. In July 2012, during the 29th edition of Pontignano in Modena (Italy) the editor presented a first draft of the comparative chapter which was discussed and approved by the authors of the national chapters. At the same time, a glossary of common technical terms to be used in the book was agreed upon by the group. As the most important result, ‘health and safety at work’ was substituted by ‘occupational health and safety’. In December 2012, the European chapter and the comparative chapter were delivered. The comparative chapter was submitted to a final check and approved by the authors of the national chapters.

Acknowledgements

Further to the authors of the chapters, other people have effectively contributed to the realization of this project and the publishing of this book. A warm thanks goes to Lorenzo Gaeta (University of Siena) who, since 2003, has enthusiastically taken charge of the organization of Pontignano seminars. A crucial contribution to the comparative analysis has been provided by Paolo Pascucci (University of Urbino) in the 2011 workshop in Cassino. Decisive indications on the relationship between work organization and occupational health and safety law have been offered by Luigi Enrico Golzio and Tommaso Maria Fabbri (University of Modena and Reggio Emilia – Marco Biagi Foundation) in the 2012 meeting in Modena. References to Swiss Law are due to Camilla Giudici (University of Zurich). The burden of the organization of the meetings has been borne by Antonio Riccio (University of Cassino) and Iacopo Senatori (Marco Biagi Foundation).

Last but not least, the group and the editor would like to thank Alan Neal as Director of the Series 'Studies in Employment and Social Policy' and Ewa Szkatula as Acquisition Editor of Wolters Kluwer Law & Business.

Edoardo Ales
Mainz, 25 January 2013

Chapter 1
Occupational Health and Safety: An EU Law Perspective

Berta Valdés de la Vega

§1.01 INTRODUCTION

Law-making activity concerning occupational health and safety has being going on practically since the beginning of the creation of the common market. The first directives on this subject formed part of the provisions that were to contribute to the functioning and creation of this market, although they also had the additional effect of performing the function of safeguarding occupational health and safety. The improvement in living standards and working conditions would be the result of the functioning of the common market and the approximation of the laws, regulations and administrative provisions, according to the wording of Article 117 of the Treaty establishing the European Economic Community (EEC) signed in 1957. The aim was primarily of an economic nature, to be achieved through functionalist integration and to the detriment of the pro-European body of opinion which sought political integration and federalism. However, twenty years from the launch of the fundamental freedoms of movement, the number of work-related accidents and illnesses remained high, the 'human and social consequences [of which] are incalculable'. This fact already appeared as the justification for proposing raising the level of protection in matters of health and safety within the First Action Programme of the European Communities on Safety and Health at Work in 1978.[1] But the EC Treaty did not confer any powers on social policy. This imbalance between social and economic matters has been and continues to be a source of tension, which is undoubtedly increased by the tension generated between the supranational and the national level of the Member States.

1. Council Resolution of 29 Jun. 1978. OJ C 165/1.

Since then, the European Union's (EU's) Social Policy has seen many changes, and so have the aims and the intensity of European normative intervention with regard to the health and safety of workers. This evolution is in part a result of the formation process of the European Union's own legal system, the complexity and content of which derive from the progressive acquisition of powers by the European institutions, and the increase in the declared aims of the various Treaties. But the other aspect of this evolution is conceptual, and may be identified with an important change in the way in which occupational health and safety is understood. What began with safety in a purely industrial setting as a 'concept of ensuring workplace safety through technical labour standards' evolved into a concept of the safety of working conditions understood in a non-technical way.[2] As from the mid-1980s, a different approach to the treatment of occupational health and safety emerged, specifically as a citizen's right, especially following the Community Charter of Fundamental Social Rights of Workers, adopted (with the exception of the United Kingdom) in 1989.[3] This recognizes that the protection of health is a right linked to the value of human dignity, which necessarily entails that it 'becomes a part of the ultimate axiological element that legitimises the Democratic State' becoming constitutional in nature.[4] According to this premise, it is hardly acceptable that the protection of people's health and personal safety should be subordinate to economic activities and business interests, as was subsequently acknowledged by Directive 89/391/EEC.[5] This was the concept that would subsequently emerge as European legislation and case law evolved.

The construction of the EU legal system has been complicated, as has also been the case with the supranational reality with which the citizens of the Member States ought to identify themselves as an existential condition of the supranational body itself. The current situation of occupational health and safety is due to the formation process of the EU, with all the complexity that this entails. For this reason, any research into this subject requires identifying and selecting those elements which have, to a greater or lesser extent, constituted milestones in the development of the protection of workplace health and safety. As has already been stated, the legal system has been built up through the aims or objectives established in those Treaties that have allocated or conferred powers. The evolution of both elements has defined the progress of protecting the health and safety of employees, but legislative activity can also be explained by political advances. Thus it is also appropriate to identify the most significant aspects of the action programmes or of the European strategies for occupational health and safety, and to observe the interaction between political advances and legislative activities.

---

2. L. Miranda, *La dimensión colectiva de la tutela del entorno de trabajo en Europa: perfiles comparativos entre Italia, Francia y Eslovenia* (Albacete: Bomarzo, 2012), 32.
3. A. Neal, *Regulating health and safety at work: developing European Union policy for the millennium*, 14 Intl. J. Comp. Labour L. & Indus. Rel. 236 (1998): 'This Community Charter "marked an early stage" in the treatment of the protection of health and safety as a "human right".'
4. J. Aparicio Tovar, 'Sicurezza sul lavoro', in *Dizionario di diritto del lavoro comunitario*, eds. A. Baylos Grau, B. Caruso, M. D'Antona & S. Sciarra (Bologna: Monduzzi, 1996), 568.
5. Of 12 Jun. 1989, on the introduction of measures to encourage improvements in the safety and health of workers at work.

We will also see how the technique of legislative intervention will evolve from safeguards based on detailed and highly-technical normative acts, to a different kind of regulation which is more based on general protection duties and principles. From another perspective, it is also possible to speak of changes in the forms of intervention, which have gone from traditional binding regulations using a variety of instruments to the use of other techniques that fall outside of the legal system as such. In fact, the regulations that have been introduced in matters of safety and health in the European legal system constitute a typical manifestation of legislative harmonization. This technique, based on binding normative instruments with the aim of standardizing national laws, was widely used until the 1990s. As from then, the inadequacy of this method of legislative harmonization started to become apparent and the strategy changed direction. There is greater use of soft law instruments, together with the greater role of private individuals in the drafting of those instruments. This is particularly apparent in relation to the technical harmonization of the rules governing the production, marketing and use of products (which include safety conditions). In these cases, the harmonization technique is maintained for the regulation of the principles, whilst the specification of the technical details is referred to private bodies, and their application becomes voluntary.[6]

§1.02 LEGAL BASES THAT JUSTIFY NORMATIVE ACTION REGARDING OCCUPATIONAL HEALTH AND SAFETY

The path that has led to the current development in occupational health and safety has not been swift or straightforward. A detailed description of this journey would probably result in various stages being identified, marked out by both the successive action programmes and by legislative initiatives. However, to contemplate this path from the legal bases of the legislative acts makes it possible to see how the conferral of powers by the Member States to what is nowadays the EU has evolved. This shows how the social Europe has grown, by incorporating social aims into the Treaties and setting up shared powers, the exercise of which has not always been without dispute,[7] and requires the application of the principle of subsidiarity.

[A] The Approximation of Laws and the Functioning of the Common Market

Until the 1980s, intervention in occupational health and safety was carried out under the provisions of the Treaty establishing the EEC in its original version. This intervention was from the start conditioned by one essential objective, viz. the mission of the

6. B. Caruso, 'L'Europa, il diritto alla salute e l'ambiente di lavoro', in *Ambiente, salute e sicurezza*, ed. L. Montuschi (Turin: Giappichelli, 1998), 7.
7. For example, the UK applied to the Court of Justice for the annulment of Directive 93/104/EC on the grounds that the legal basis for the said Directive (Article 118 A EEC Treaty) did not confer powers over the organization of working time (Case C-84/94, *United Kingdom of Great Britain and Northern Ireland v. Council of the European Union*).

European Community as stated at Article 2 of the EC Treaty: 'to promote a harmonious development of economic activities (...) accelerated raising of the standard of living (...)', which objective would be achieved by way of 'the establishment of a common market (...)'. The attainment of these Community objectives within the scope of social policy was specified at Article 117 of the EC Treaty with the need to 'promote improved working conditions and an improved standard of living for workers, so as to make possible their harmonisation while the improvement is being maintained', although it was indicated that this improvement would derive from the functioning of the common market. As a result, powers over social matters were not conferred on the EC Institutions. Powers related to occupational health and safety ('prevention of occupational accidents and diseases, and occupational hygiene') were the preserve of the Member States, as was stated at Article 118 of the EC Treaty, which only conferred on the Commission the task of 'promoting close cooperation between Member States' through research, reports, or consultations.[8]

As a result of the above, the directives concerning occupational safety and hygiene from this period had their legal basis in Article 100 of the EC Treaty,[9] and as such their main justification was to harmonize and improve national laws or remove the obstacles, in order to ensure, in both cases, the functioning of the common market.[10] On this point, the direction of the EC Treaty was clear and decisive in the mechanism for the approximation of national laws through Article 100 of the EC Treaty, by requiring that normative intervention should arise where it 'directly affects' the establishment or functioning of the common market. Therefore, these directives did not seek to create a body of occupational health and safety rights and duties, but rather sought uniformity in the mechanisms for controlling certain risks, and therefore in material working conditions. The regulatory method used and the type of duties contained in these directives were 'of a technical nature and subordinate to scientific imperatives', with the right being defined in accordance with technical values.[11] The majority of them corresponded to a limited notion of occupational hygiene[12] which had the aim of preventing certain professional illnesses caused by exposure to hazardous

8. G. Arrigo, 'La politica sociale comunitaria: il quadro evolutivo', in *Percorsi di diritto del lavoro*, eds. D. Garofalo & M. Ricci (Bari: Cacucci, 2006), 60
9. Also at Article 235 EEC Treaty for Directive 82/501/EEC given that this directive does not only seek the approximation of laws, but also lays down specific provisions relating to environmental-protection objectives and workplace health and safety where the powers to intervene required for this purpose are not envisaged in the Treaty.
10. F. Wright, *The developments of occupational health and safety regulation in the European Community*, 8 Intl. J. Comp. Labour L. & Indus. Rel. 33 (1992): 'Other reasons have been put forward to explain the importance of this area, such as the human, social and economic costs of accidents and ill-health or that more efficient work practices will bring increased productivity'.
11. F. Camas Roda, *La normativa Internacional y Comunitaria de Seguridad y Salud en el Trabajo* (Valencia: Tirant lo Blanch, 2003), 134.
12. Except for Directive 77/576/EEC relating to the provision of safety signs at places of work, aimed at unifying the types of signs for all places of work except for the three cases stated at Article 1.

agents or substances, and as such the material scope of their application was limited to certain businesses or industries.[13]

However, they were also based on the aims set out for the most part in the Workplace Health & Safety Action Programmes of 1978 and 1984, where raising the level of protection against occupational risks through prevention and control was an essential aim. The influence of these Programmes meant that subsequent directives would begin to address certain aspects that were distinct from the exclusively economic aims. Some provisions were aimed at employers as the party with direct responsibility for specific prevention measures, and not just at the Member States;[14] others featured the right of workers to information or to have medical check-ups. Overall, Directive 80/1107/EEC, with its four specific directives, would be the most relevant at this first stage, and where a certain tension may be observed between provisions that are overtly technical and of a scientific nature, and others which have the protection of the health and safety of employees as their main purpose. This introduced the need to avoid or reduce risks as may be 'reasonably practicable' (Article 3.1 of Directive 80/1107/EEC, or Article 6 of Directive 83/477/EEC), which went beyond the simple duty to apply technical measures. Other measures would also start to appear, placing duties on employers to monitor health, individual protection or information, without this resulting in a general duty to protect the health and safety of employees.[15]

Ultimately, during this period, Community occupational health and safety policy was marked by the Commission Recommendations to the Member States, and in particular, with the first two European Community Workplace Health & Safety Action Programmes.[16] However, the limitations to normative action were considerable given the legal basis on which it was operating, but this was to change in 1986. We may conclude by stating that this stage is characterized by the tension deriving from two dominant objectives: market integration, and the improvement of protection levels for workers. This duality between 'fairness and effectiveness'[17] would also be reflected in the allocation of powers and in the legal bases used as the grounds for normative acts, and this was particularly notable beginning with the Single European Act, as we shall now see.

---

13. Specifically, Directives 78/610/EEC, 80/1107/EEC, 82/501/EEC, 82/605/EEC, and 83/477/EEC. In contrast to previous directives, Directive 86/188/EEC on the protection of workers from the risks related to exposure to noise at work (the third specific Directive under Directive 80/1107/EEC) is to be applied to all workers with two exceptions, subsequently subject to review.
14. In the case of Directive 77/576/EEC, the Member States are the parties responsible for ensuring the uniformity of safety signage.
15. F. Camas Roda, *La normativa Internacional y Comunitaria de Seguridad y Salud en el Trabajo* (Valencia: Tirant lo Blanch, 2003), 143.
16. G. Arrigo, 'La politica sociale comunitaria: il quadro evolutivo', in *Percorsi di diritto del lavoro*, eds. D. Garofalo & M. Ricci (Bari: Cacucci, 2006), 61: 'In fact, the social policy became a positive integration factor as it contributed to reach the Treaty objectives: its evolution was not any more dependent from the "functioning of the common market".'
17. R. Baldwin & T. Daintith, *Harmonization and hazard: regulating workplace health and safety in the European Community* 1 (Graham & Trotman 1991).

[B] The Conferral of a Specific Power for the Protection of the Health and Safety of Employees

The amendment of the EC Treaty by the Single European Act (SEA) in 1986 brought with it an important change as for the first time, pursuant to Article 118A of the EC Treaty, the Community was conferred 'internal legislative competence in the area of social policy'.[18] Although the precept was included in the chapter on 'social provisions', it was not a wide power for implementing measures on workers' rights, for which the legal basis continued to be Article 100 of the EC Treaty, but rather a specific power on the protection of the health and safety of employees. In other words, the possibility of harmonizing provisions on occupational safety by way of a qualified majority under Article 118A EC Treaty repealed the general rule contained at Article 100A.2 EC Treaty which required normative acts relating to the rights and interests of dependent workers to be adopted unanimously.

Meanwhile, the new Article 100A of the EC Treaty was to introduce qualified majority voting for the adoption of measures aimed at the approximation of the provisions of the Member States, although with the known economic aim of allowing the establishment and functioning of the internal market. The rights and interests of workers could not have Article 100A of the EC Treaty as their legal basis, but this precept could be used for measures aimed at the approximation of laws governing health, safety, environmental protection and consumer protection, which would naturally include occupational health and safety. This entails the existence of two different precepts under which Community institutions would be able to find the legal basis for acting in connection with the health and safety of employees. The difference between them, or to be more precise, the choice of one particular legal basis for adopting a normative act would depend in particular on the aim and content of the act, and in any event, on objective elements open to jurisdictional control. Where the primordial aim is the functioning of the internal market, the proper legal basis would be Article 100A of the EC Treaty, and should be based on an 'appropriate level of protection', close to that attained in the most advanced Member States in this matter.[19]

However, normative acts that have the aim of improving the protection of the health and safety of employees should be based on Article 118A of the EC Treaty, even where this act should also have an effect on the establishment and functioning of the internal market. The wording of Article 118A of the EC Treaty, as opposed to Article 100A of the EC Treaty, means that the improvement of workers' safety, hygiene and health at work 'is an objective which should not be subordinated to purely economic considerations'.[20] Harmonization through this precept should be carried out by way of minimum requirements, with directives being adopted by a qualified majority and in

---

18. Opinion 2/91 issued by the Court of Justice on 19 Mar. 1993 (Rec. P I-1061).
19. F. Camas Roda, *La normativa Internacional y Comunitaria de Seguridad y Salud en el Trabajo* (Valencia: Tirant lo Blanch, 2003), 160.
20. Fifteenth recital of Directive 89/391/EEC, on the introduction of measures to encourage improvements in the safety and health of workers at work, later repeated in other directives, e.g., in the fifth recital of Directive 93/104/EC concerning certain aspects of the organization of working time.

cooperation with the European Parliament. The definition of 'minimum requirements' does not pre-judge the intensity of the action considered necessary in order to meet the objectives, but only indicated the possibility that Member States might adopt stricter protection rules.[21]

This change, as with many changes throughout the construction of the EU, was a result of intense negotiations between competing interests, and this was reflected in the complex wording of Article 118A of the EC Treaty. The interpretation of 'working environment' was particularly divisive, and in this regard the judgment of the Court of Justice, Case C-84/94 on Directive 93/104/EC concerning certain aspects of the organization of working time,[22] was important. In this case, the United Kingdom challenged the use of Article 118A as the legal basis for the adoption of this directive. This precept stated that 'Member States shall pay particular attention to encouraging improvements, especially in the working environment, as regards the health and safety of workers ... '. The Government of the United Kingdom considered that the concept of 'working environment' was only related to 'the physical conditions and risks at the place of work', and that Article 118A only constituted legal basis for directives the aim of which was 'objectively and genuinely related to the safety and the health of workers'.

However, the Court considered that 'working environment' should be interpreted broadly, and that therefore the power conferred on the Council must also be broad. The reference to international sources in order to back its argument on the concept of health and safety was based on the Constitution of the World Health Organization. In its Preamble, health is defined as 'a state of complete physical, mental and social well-being that does not consist only in the absence of illness or infirmity'. Ultimately, it should be understood that the power conferred on the Council was aimed at promoting the improvement, as a part of overall progress, of the conditions that could affect the health and safety of workers within the working environment, and these factors could be physical or of any other kind, including certain aspects relating to the organization of working time.[23]

Another result of the tensions that existed when agreeing the text of the SEA was the apparent introduction of a limit on the powers envisaged by Article 118A of the EC Treaty. The aim was to avoid imposing administrative, financial and legal constraints that would hinder the creation and development of small and medium-sized undertakings. However, this precept is to be interpreted by the Court of Justice to mean that 'such undertakings may be the subject of special economic measures', but that does not mean that they cannot 'be subject to binding measures'.[24]

---

21. E. Ales, 'Lo sviluppo della dimensione sociale comunitaria: un'analisi "genealógica"', in *Diritto del lavoro dell'Unione Europea*, eds. F. Carinci & A. Pizzoferrato (Milano: Wolters Kluwer, 2010), 137.
22. Judgment of the Court of 12 Nov. 1996, *United Kingdom of Great Britain and Northern Ireland v. Council of the European Union.*
23. Case C-84/94, *United Kingdom of Great Britain and Northern Ireland v. Council of the European Union*, para. 15.
24. Case C-84/94, *United Kingdom of Great Britain and Northern Ireland v. Council of the European Union*, para. 44.

The next important change took place with the signing of the Maastricht Treaty in 1992 and the inclusion of the Social Policy Protocol, especially the procedure for social dialogue established at Articles 3 and 4. Once again, there are two legal bases capable of providing legitimacy to normative action on the health and safety of employees: Article 118A of the EC Treaty, and Article 2 of the Social Policy Protocol. However, in practice this never actually happened given that the refusal of the United Kingdom to sign the Protocol resulted in the Commission continuing to use Article 118A of the EC Treaty as the legal basis in this area. The social dialogue procedure thus 'remained without application in the field of occupational health and safety'.[25]

This situation changed with the Amsterdam Treaty signed in 1997, which included the Agreement on social policy under Title XI of the Treaty, applicable now to all Member States. As from this time, there would be one single legal basis – that envisaged at Article 137 of the EC Treaty – pursuant to which 'the Community shall support and complement the activities of the Member States', specifically in the improvement of the working environment, in order to protect workers' health and safety. When using this channel for actions concerning health and safety, the Commission must consult with social partners and follow the social dialogue procedure laid down at Articles 138 and 139 of the EC Treaty. In addition, the Advisory Committee on Safety, Hygiene and Health Protection at Work[26] should be consulted, which has, since the 1970s, assisted the Commission in its activities in this area.

The current Treaty on the Functioning of the European Union (TFEU) does not include any innovations at Title X on Social Policy with regard to the legal bases that would provide legitimacy for legislative acts on occupational health and safety. The same may be said of the Charter of Fundamental Rights of the European Union, which currently has the same legally-binding force as the Treaties. Article 31 provides that 'every worker has the right to working conditions which respect his or her health, safety and dignity', but the application of this precept shall be carried out 'respecting the limits of the powers of the Union as conferred on it in the Treaties', as is stated at Article 51 of the Charter. This acknowledgement of rights does not establish any new power or task, or to put it more clearly, 'the provisions of the Charter shall not extend in any way the competences of the Union as defined in the Treaties' (Article 6 of the Treaty on European Union). Ultimately, the legal basis for workplace health and safety shall continue to be Article 153 of the TFEU. However, it remains to be seen how the Court of Justice will interpret Article 31 of the Charter in relation to normative acts adopted under the legal basis of Article 153 TFEU.

As we shall see below, from another perspective, the evolution of the legislative output in this area will be linked to the trends in legislative policy, which may be extracted from a variety of documents, with the Programmes and Strategies regularly produced on this subject being particularly relevant.

25. S. Smismans, *Towards a new community strategy on health and safety at work? Caught in the institutional web of soft procedures*, 19 Intl. J. Comp. Labour L. & Indus. Rel. 72 (2003).
26. Subsequently, two different advisory committees would be merged: the Advisory Committee for Safety, Hygiene and Health Protection at Work, and the Safety and Health Commission for the Mining and Other Extractive Industries. The Advisory Committee on Safety and Health at Work was set up by way of the Council Decision of 22 Jul. 2003 (2003/C 218/01).

[C] Technical Harmonization and New Approach Directives

The rules governing the production, marketing and use of products include the safety conditions for products that are to be handled by workers, be they instruments (machines, tools, working equipment) or physical, chemical or biological substances. The goal of free movement of goods means that the technical characteristics of the products of the different Member States must have a certain uniformity, meeting equivalent standards of protection that allow them to be mutually accepted.[27] The technical harmonization of the safety conditions of products has precisely this aim, to allow the free movement of goods which have been produced under a variety of national specifications. But, as hazards can be 'designed out of productions processes', the effect 'may be to produce not merely safe products but to some extent a homogeneous degree of hazard avoidance in the production process itself'.[28]

Now then, these technical safety conditions are constantly evolving given that scientific progress and research provide newer and better protective elements to avoid hazards. The main way in which technical specifications intervene is 'by way of the creation of the standardization plan and the production of technical standards', which advance in tandem with scientific advances and 'represent the level of knowledge existing on a subject and at a particular time'.[29] These technical standards may refer to the level of concentration of a substance or the time to which exposure is limited, but may also consist in rules or instructions for the use of technical equipment or protective elements. In any event, they must be kept continually up to date in accordance with the industrial materials, substances, or innovations incorporated into the productive processes. Technical standards have a standardizing effect and give products, equipment or materials a level of uniformity and, importantly for present purposes, they also have the effect of protecting the health and safety of both consumers and workers. This leads us to the relationship between legal and technical rules, and this is precisely the harmonization problem that the implementation of the New Approach seeks to tackle, as we shall now see.

The free movement of goods, as one of the fundamental elements of the single market, needed to be able to overcome obstacles to trade resulting from the introduction of divergent national technical rules and regulations. For this purpose, an ambitious process of technical harmonization was commenced in the 1960s at a Community-wide level, which had its legal basis in Article 100 of the EC Treaty.[30] However, the pace of harmonization was quite slow, due first to the procedure for the drafting of the normative act based on the unanimity required at the Council. But the

27. Following the Judgment of the Court of Justice, Case C-120/78 ('Cassis de Dijon') *Rewe-Zentral v. Bundesmonopolverwaltung fuer Branntwein*, obstacles to trade resulting from the different regulations of the Member States shall only be valid, in the absence of Community measures, if they are justified on the basis of essential health and safety requirements.
28. R. Baldwin & T. Daintith, *Harmonization and hazard: regulating workplace health and safety in the European Community* 7 (Graham & Trotman 1991).
29. J. Sancho Cuesta, *La seguridad e higiene en el ordenamiento jurídico comunitario* (Madrid: Ministry of Employment and Social Security, 1993), 114.
30. A. Salerno *Diritto comunitario della sicurezza del lavoro* (Padova: Cedam, 1990), 162.

second significant issue was the incorporation of the technical rules into the legal text, and as such the content of the directives was highly-technical, which meant they needed to be amended when advances were made or technical changes arose. In the 1980s, the need to establish another more flexible system was put forward, a different regulation technique to attain the goal of creating the Single Market in 1992. This is the reason for the incorporation of Article 100A of the EC Treaty into the Single European Act, allowing the adoption of provisions (directives or regulations) on the basis of a qualified majority. A second response is the 1985 Council Resolution on a New Approach to harmonization and standardization,[31] which referred the task of defining the technical characteristics of products to harmonized standards. The aim of the New Approach would be twofold: the abolition of technical barriers to trade within the EU, and the development of a 'European doctrine in the field of safety, health, protection of consumers, workers and environment'.[32]

It may be said that with this Resolution, two original and innovative instruments were developed, the New Approach to product regulation, and the Global Approach to conformity assessment. In both of these, public intervention is limited to just the essential, i.e., legislative harmonization is limited, by way of directives, to the adoption of the essential safety requirements in accordance with the types of risk (safety, health, environment, consumer protection, etc.) and the circumstances (in the home, in the workplace, in movement, etc.). The drafting of the technical specifications required in order to produce and market products which meet these safety requirements is not to be incorporated into the directive or performed by Community institutions.

This task is assigned to the private bodies in charge of industrial standardization. The drafting of these harmonized rules or technical standards within the scope of the EU and in relation to health and safety corresponds to the European standardization organizations – the European Committee for Standardization (CEN), the European Committee for Electrotechnical Standardization (CENELEC), and the European Telecommunications Standards Institute (ETSI). These organizations set and update European standards on a regular basis.[33] The technical specifications they draft must be in compliance with the essential requirements laid down by the directives, but shall not be binding. However, there is a presumption that the manufacture of products in compliance with the technical specifications issued by these bodies shall be in compliance with the essential requirements laid down by the directives.

Currently, the technical directives under the so-called New Approach relating to matters of safety and health are based on Article 114 of the TFEU. The aim of the New Approach Directives is to approximate the laws of the Member States in order to remove obstacles to free trade, but take as their starting-point a high level of protection. However, as a general rule, Member States are not allowed to maintain or introduce

31. OJ L 136/01.
32. J. Pelkmans *The New Approach to Technical Harmonization and Standardization*, XXV J. Com. Mkt. Stud. 254 (1987).
33. In which process the procedures applicable shall be those envisaged in the Directive of 28 Mar. 1983, laying down a procedure for the provision of information in the field of technical standards and regulations (83/189/EEC), subsequently amended by Directive 88/182/EEC.

measures that are stricter than those envisaged in the directives.[34] This is an important difference as compared to directives adopted under Article 153 TFEU, i.e., those with the aim of improving occupational health and safety, which provisions are defined as 'minimum requirements' and where national legislation can impose stricter (higher) conditions.

The 'New Approach' Directives do not lay down duties for employers (except for those relating to implementation), although numerous products covered by these directives are used in the workplace. Employer duties relating to the use of work equipment at work are governed by a variety of directives[35] based on Article 153 TFEU. This means that employers may only acquire work equipment or personal protective equipment that meets the technical standards on design or manufacture laid down in the Directives based on the 'New Approach' of Article 114 TFEU. Member States' powers to transpose Directives based on Article 153 TFEU introducing stricter provisions or a higher degree of protection for the health and safety of workers has a limit. These provisions may not affect the marketing conditions of the products, i.e., they may not affect the provisions within the scope of application of a 'New Approach' Directive.[36] It seems then that the free movement of goods would be imposed on the pretensions of those Member States that wanted to have higher standards of protection. The only possibility of doing so would be to resort to the justification that can offer 'mandatory requirements' to protect workers and citizens' health (Article 36 TFUE) with support in the Community strategy on health and safety at work (2007–2012), but this is something that must be solved, in case of conflict, by the Court of Justice.

§1.03 ACTION PROGRAMMES AND STRATEGIES FOR OCCUPATIONAL HEALTH AND SAFETY

The development of the regulations governing occupational health and safety represents one of the most advanced aspects of the EU's Social Policy. However, the European institutions have not limited themselves to passing legislation, they also engage in awareness and promotion campaigns for a safe and healthy working environment in cooperation with other bodies, especially the European Agency for Safety and Health at Work[37] and the European Foundation for the Improvement of

34. Guide to the implementation of Directives based on the New Approach and the Global Approach (Luxembourg: European Communities, 2000), 11.
35. For example, the following: Directive 2009/104/EC of the European Parliament and of the Council of 16 Sep. 2009 concerning the minimum safety and health requirements for the use of work equipment by workers at work (second individual Directive within the meaning of Article 16(1) of Directive 89/391/EEC), Council Directive 89/656/EEC of 30 Nov. 1989 on the minimum health and safety requirements for the use by workers of personal protective equipment at the workplace (third individual directive within the meaning of Article 16 (1) of Directive 89/391/EEC).
36. Guide to the implementation of Directives based on the New Approach and the Global Approach (Luxembourg: European Communities, 2000), 25.
37. A critical comment about the efficiency of the Agency in S. Smismans, *Towards a new community strategy on health and safety at work? Caught in the institutional web of soft procedures*, 19 Intl. J. Comp. Labour L. & Indus. Rel. 62–65 (2003).

Living and Working Conditions. The EU's occupational health and safety policy is set out fundamentally in the various Programmes that have been approved since the 1970s, starting with the Social Action Programme of 1974.[38] This Programme set up the Advisory Committee on Safety, Hygiene and Health Protection at Work,[39] which had the task of assisting the Commission in the preparation and implementation of activities in the fields of safety, hygiene and health protection at work and, specifically, it was to make a major contribution to the preparation of the First Action Programme on safety and health at work adopted by resolution in 1978.[40]

From then to the present day, with the current Community strategy on health and safety at work (2007–2012),[41] the approach and the policy on this matter have seen significant changes. For example, there has been a shift from the general aims of this First Action Programme, focussing on increasing occupational safety (especially with regard to machines, premises, and installations), limiting the exposure of workers to toxic substances and the examination of the causes of accidents and illnesses linked to work, to objectives such as 'aiming for a situation in which work enhances a person's health and well-being and in which being able to find and retain a job improves the general health of the population', defined in the current 2007–2012 strategy. However, there is an aim that is common to all programmes and strategies, which has at various times been defined as being a priority, which consists in reducing occupational accidents and diseases to a minimum. Specifically, the Commission has set a global target to reduce by 25% the rate of occupational accidents within the EU-27 for the current period. This goal has been described as ambitious in the current Community strategy on health and safety at work (2007–2012).

The first three action programmes on occupational safety and hygiene had legislative work as the keystone of Community action. This is logical, given that in the formation of the common market it was essential to put into place harmonized minimum health and safety requirements. The purpose of acting through normative intervention was reinforced in the third 'Programme concerning safety, hygiene and health at work',[42] which was the most significant from the point of view of legislative policy aimed specifically at defining the objectives of the normative actions. Specifically, directives were to cover the maximum number of risks possible, provide adequate protection for sectors with a high level of risk and certain categories of workers who were especially vulnerable, and finally, ensure that Community provisions adopted on the basis of Articles 100A of the EC Treaty (creation of the internal market) and 118A of the EC Treaty (social policy) were consistent. Through this programme, the Commission defined the basic principles that are to oversee safety, hygiene, and health in the workplace, and this culminated with the approval of the Framework Directive in 1989.

---

38. Approved by the Resolution of 24 Jan. 1974, OJ C 13 of 12 Feb. 1974.
39. Council Decision of 27 Jun. 1974 on the setting up of an Advisory Committee on Safety, Hygiene and Health Protection at Work (74/325/EEC). OJ 1974 L 185. See *supra* n. 26.
40. OJ C 165/1 of 11 Jul. 1978.
41. OJ C 145 of 30 Jun. 2007.
42. OJ C 28 of 3 Feb. 1988.

Beginning with the fourth Community programme on this subject covering the period 1996–2000, there was a slight change in political focus, which subsequently became more notable in successive strategies (2001–2006 and 2007–2012). Certain common elements appear in all of them and the strategies are designed to act in basically two different fields.

The common elements are, first, the identification of the economic and social/labour context – which is subject to rapid technological and organizational changes– as a starting-point for new approaches, which are sometimes termed as innovative in the programmes or strategies in this subject. In accordance with the foregoing, the second element is the need to constantly adapt and update the measures adopted in order to protect workers, in the face of new risks and hazards in a changing working environment. Therefore, this is a subject that is constantly evolving, both scientifically and technically, without forgetting that science describes the reality but the decisions on risk management are basically political.[43] The third is the principle that legislative action is not the only method, and most likely not even the main method at the current time, for taking action in the field of health and safety.

With regard to this last observation, the strategies distinguish between two areas of activity. One is formed by non-legislative measures aimed at increasing the levels of occupational health and safety, from which we may highlight awareness, i.e., the accurate and efficient communication of the existing body of law in matters of health and safety. These non-legislative accompanying measures' would be to 'promote and supplement' legislative objectives.[44] The second area of activity is related to work of a legislative nature, including the need to simplify the normative framework and adapt it to technical and labour changes, oversee its enforcement, or act on new risks, all in close cooperation with social partners.

Within this evolution, the new Community strategy on health and safety at work for the period 2001–2006 represents an important milestone by introducing an innovative approach based on the need to adapt to the changes that arise at work and in society through a global perspective. This strategy arrives, as is stated by the Economic and Social Committee, 'after a period of several years without much strategic thinking on health and safety at work'[45] during which important changes had taken place in the world of work and with the challenges concerning health and safety that could arise in connection with the expansion of the EU on the nearby horizon. It also owes a lot to the European Social Agenda[46] adopted at the European Council in Nice in December 2000, which promotes the development of this strategy and sets out its main aspects, which basically consist of the following: (a) to consolidate or adapt existing standards; (b) to

43. A.B. Muñoz Ruiz, *El Sistema Normativo de prevención de riesgos laborales* (Valladolid: Lex Nova, 2009), 102: 'In the risk management the use of value judgements on issues like "the acceptability of the risk" are required.'
44. A. Neal, *Regulating health and safety at work: developing European Union policy for the millennium*, 14 Intl. J. Comp. Labour L. & Indus. Rel. 244 (1998).
45. Opinion of the Economic and Social Committee on the 'Communication from the Commission – Adapting to change in work and society: a new Community strategy on health and safety at work 2002–2006' (COM (2002) 118 final).
46. OJ C 157 of 30 May 2001.

respond to new risks; (c) to promote the application of legislation in SMEs; (d) to develop exchange of good practice and collaboration between labour inspection institutions. The Social Agenda also expresses an opinion on the types of measures to be adopted in this matter, stating that a variety of different pre-existing Community instruments should be used, especially the open method of co-ordination, whilst at the same time space is reserved for the initiative of the social partners.

The strategy for the 2001–2006 period is based on a novel concept within the scope of the European Community, that of occupational health and safety as an essential component of the quality of the work, which is also in line with the aim of 'creating more and better jobs' set down by the Lisbon European Council of March 2000.[47] The Commission expresses, both in this strategy and the following one for the period 2007–2012, its deep concern over the number of occupational accidents, with particular emphasis, 'over and above these human tragedies', on the loss of productive capacity and the economic cost that results from the payment of indemnities and benefits. The 'non-quality' of work having as a result accidents or health problems is a 'waste of resources', which demonstrates the complex relationship between occupational health and competitiveness. Part of this reasoning leads the Commission to opt for a Community policy in matters of health and safety that promotes a real 'well-being at work', this being taken to mean physical, moral and social well-being that cannot be measured only by an absence of accidents or occupational illnesses. The concept of 'well-being at work' is the one developed by the ILO.

The current strategy presented by the Commission under the title 'Improving quality and productivity at work: Community strategy 2007–2012 on health and safety at work'[48] follows the broad lines established in the previous strategy. It identifies the major health and safety challenges and highlights the new trends in employment with more sub-contracting, increased numbers of self-employed workers, and an increase in small and medium-sized enterprises. Likewise, it identifies a change in the nature of occupational risks deriving from new risk factors (violence at work, addictions), and the transformation of work. Facing up to the new risks by drawing up methods for their identification and evaluation is one of the aims of this strategy. For this purpose, the Risk Observatory[49] of the European Agency for Safety and Health at Work needs to improve risk anticipation. However, no new legislative policy goals have been established to face up to these challenges. The view is that the legislative framework merely needs to be adapted to technical progress, along with greater simplification as the task of codifying the directives continues. The number of occupational accidents and diseases remains unacceptably high, but this is considered an enforcement problem. For this reason, two of the aims will be to 'ensure the proper application of European Union law and support small and medium-sized enterprises in the enforcement of the current legislation'.

47. O. Bonardi, 'La sicurezza del lavoro nella Comunità Europea, nella Costituzione e nella legge di semplificazione', *Rivista giuridica del lavoro*, no. 3 (2004): 241.
48. COM (2007) 62 final.
49. The creation of the Observatory was envisaged in the 2002–2006 Strategy, and was based on examples of good practice by businesses or specific sectors of activity, and its aims include the organization of exchanging experience and information. It began work in 2003.

However, the aim is not to resolve the proper application of the normative framework using the enforcement resources of the Member States, but rather through other means, actions or instruments.[50] Some of these include spreading good practice, increasing training and awareness, using economic incentives at a European and national level, or resorting to inspectors in order to promote a higher standard of compliance with the law through 'persuasion and encouragement, then, where necessary, through coercive measures'.[51] Co-ordination is also perceived as being an essential element given that 'the degree to which the Community strategy is successful will depend on the Member States being committed to adopting coherent national strategies' that are consistent with it. Cooperation is also tackled from another perspective, as was already established in 2002, through the role played by the Senior Labour Inspectors Committee (SLIC), encouraging the exchange of information and experience, and organizing mutual cooperation and assistance.[52]

As may be seen, this latest strategy consolidates the change of direction and ostensibly abandons legislative harmonization as the key to the EU's policy on safety and health in the workplace. This does not mean that there are no legislative initiatives at all on this subject, but rather these have been finalized in the form of Framework Agreements (on work-related stress and on harassment and violence at work), with a significant weakness in their content and tremendous complexity in their application, as we shall see below.

§1.04 THE FRAMEWORK DIRECTIVE

Probably the most significant occupational health and safety legislative act adopted since the Treaty establishing the EEC was Directive 89/391/EEC of 12 June 1989 on the introduction of measures to encourage improvements in the safety and health of workers at work (hereinafter 'Framework Directive' or 'FD').[53] This Directive marked a substantial change to the legislative policy of the Institutions so far by opting to establish a protection model based on the 'principle of absolute prevention', distilled into the general duty of care envisaged at Article 5.[54] This was possible given the incorporation, through the SEA, of a specific legal basis under the social policy chapter. This duty is developed by way of setting out a series of principles, and the organization of prevention is described with instruments and the rights and duties of the parties involved. The aim is to reduce the risks to a minimum within a context of permanent updating of prevention. Ultimately, the general framework for the protection of health and safety defined in the Framework Directive has the following main planks: (a) a notion of safety and health based on constant evolution; (b) the creation of a

50. O. Bonardi, 'La sicurezza del lavoro nella Comunità Europea, nella Costituzione e nella legge di semplificazione', *Rivista giuridica del lavoro*, no. 3 (2004): 243
51. COM (2007) 62 final, 6.
52. Commission Decision 95/319/EC setting up a Committee of Senior Labour Inspectors.
53. Amended by various acts: Regulation (EC) No 1882/2003 of 29 Sep. 2003, Directive 2007/30/EC of 20 Jun. 2007, Regulation (EC) No 1137/2008 of 22 Oct. 2008.
54. B. Caruso, 'L'Europa, il diritto alla salute e l'ambiente di lavoro', in *Ambiente, salute e sicurezza*, ed. L. Montuschi (Turin: Giappichelli, 1998), 4.

prevention culture which covers all kinds of risks; (c) the definition of the employer as a safety debtor, especially through a general duty; (d) management of prevention at the company through a model based on dialogue and the balanced participation of employers and workers, and (e) the widespread application of the directive to all production sectors.

[A] The Improvement in Occupational Health and Safety: General Application

The aim of the Framework Directive is the application of measures for the improvement of occupational health and safety. Article 3, which is dedicated to defining the basic concepts that are going to be used in the Directive, does not include those of 'safety' or 'health'. Safety is linked to technical progress which ought to be taken into account by employers as one of the guiding principles for the performance of their duties. The concept of health has evolved with social progress and the scientific advances regarding people's health, as may be seen from the new Community strategy on health and safety at work for the period 2001–2006. In any event, the Framework Directive refers to certain aspects which are necessarily linked to the concepts of 'safety and health'. Specifically, Article 5.1 FD provides that safety and health should be protected in broad terms given that this should refer to 'every aspect related to the work'.

Amongst the principles that ought to be applied in order to define the limits of this duty, we may find the duty to adapt the work to the individual, which includes the design of work places, working and production methods, and that monotonous and repetitive work ought to be avoided (Article 6.2 (d) FD). Thus it may be interpreted that this duty to adapt to the individual entails an integral view of health, taking into account all aspects that form a part of human beings and which includes physical integrity, but also moral integrity.[55] Article 6.2 (g) FD also supports this view where it states that avoiding harm requires preventive action which includes 'technology, organization of work, working conditions, social relationships and the influence of factors related to the working environment'. Such a broad definition of the factors that have an effect on health is consistent with the notion of understanding health within its dual physical and moral dimensions, and therefore including risks of a psycho-social nature.[56]

With the Framework Directive, the European legislator has designed a normative framework of general application to all sectors of activities – public or private, with the exception of certain specific activities in the public sector, such as the armed forces, the police or civil protection brigades – and with regard to those risks that may arise in the workplace (Article 2 FD). The scope of this exception has been the subject of interpretation by the European Court of Justice (ECJ) in a number of judgments. In all of them, the Court has reiterated that the scope of application of the Framework

55. M. Peruzzi, *The prevention of psychosocial risks in European Union law*, I Working Papers di Olympus no. 14, 23 (2012): www.uniurb.it/olympus.
56. This view was upheld in the judgment of the Court of Justice in Case C-84/94, already cited.

Directive has been defined broadly, and from this it may be deduced that the exceptions to this scope should be interpreted narrowly. The aim of this exception is:

> ensuring the proper operation of services essential for the protection of public health, safety and order in cases, such as a catastrophe, the gravity and scale of which are exceptional and a characteristic of which is the fact that, by their nature, they do not lend themselves to planning as regards the working time of teams of emergency workers.[57]

The principle used by the legislator when establishing the exception to the scope of application of the Framework Directive is not based on the sectors of activity, such as the armed forces, the police, or civil protection brigades, 'but rather is based exclusively on the specific nature of certain special tasks carried out by the workers in these sectors', and this on the basis of the absolute need to ensure proper protection of the whole. The activities of assisting persons who are injured or ill organized by associations such as the Deutsches Rotes Kreuz, the doctors of Primary Care Teams at public hospitals,[58] or the activity performed by the front-line teams of a public fire-fighting service,[59] do not have these characteristics, and as such Directives 89/391 and 93/104 are applicable to them. From the content of Article 2 FD - scope of application - and from its interpretation by the Court of Justice, a 'compulsory generalization principle of the prevention strategy'[60] may be deduced.

[B] The General Duty on the Employer and the Standard of the Maximum Safety Technologically Possible

The integral protection model designed by the Framework Directive has as its essential aim the formulation of a general duty on the employer to 'ensure the safety and health of workers in every aspect related to the work' (Article 5.1 of the FD). This duty constitutes the basis of the protection system envisaged by this Directive, and converts the employer into the 'safety debtor' as a consequence of its position in the management and organization of production.[61] Naturally, the employees are the 'beneficiaries

57. Judgment in Joined Cases C-397/01 to C-403/01 *Bernhard Pfeiffer and others v. Deutsches Rotes Kreuz*, Kreisverband Waldshut eV, 55th recital.
58. Case C-303/98 *Sindicato de Médicos de Asistencia Pública (Simap) v. Conselleria de Sanidad y Consumo de la Generalidad Valenciana*. Also Case C-241/99 *Confederación Intersindical Galega (CIG) v. Servicio Galego de Saúde* (Sergas) in relation to the activities of medical and nursing staff at continuous care services and non-hospital emergencies.
59. Case C-52/04 *Personalrat der Feuerwehr Hamburg v. Leiter der Feuerwehr Hamburg*. In all of these situations, the aim is to determine the scope of application of Directive 93/104 concerning certain aspects of the organization of working time. This directive defines its scope of application by expressly referring to Article 2 of the Framework Directive, and therefore to the exceptions envisaged there. Thus it is necessary to define the scope of application of the FD which then in turn defines the scope of application of Directive 93/104.
60. B. Caruso, 'L'Europa, il diritto alla salute e l'ambiente di lavoro', in *Ambiente, salute e sicurezza*, ed. L. Montuschi (Turin: Giappichelli, 1998), 4.
61. J. Aparicio Tovar, 'Sicurezza sul lavoro', in *Dizionario di diritto del lavoro comunitario*, eds. A. Baylos Grau, B. Caruso, M. D'Antona & S. Sciarra (Bologna: Monduzzi, 1996), 12. The employer is always the safety debtor, irrespective of the size of the company, and as such a Member State may not release companies employing ten or fewer employees from the duties

of the rights corresponding to the employer's duties, which makes them the creditors of the safety debtor who is in turn the employer'.[62] However, the problem of interpretation which has been generated by this precept is the determination of the content of the duty. Part of the discussion by learned opinion has been about the nature of the duty, i.e., if the duty is one of means or outcomes, given that the compulsory intensity of the safety duty is very high.[63] The general principles of prevention under Article 6.2 FD allow the establishment of certain guidelines in order to organize the content of this duty. To consider them jointly would lead one to believe that the duty to ensure safety implies, first of all, a duty to avoid health hazards by eliminating the risks that have been detected. Second, a risk may be accepted 'exceptionally, on the condition that it is properly assessed, under control, combated, and is as small as possible'. In other words, in the evaluation of the risks and in the decision on the prevention measures, the employer should exercise all possible care (Article 5.4 FD). Third, the measures adopted in order to eliminate or mitigate the risk 'should take into account the best possibilities and technical means', as well as adaptation to the personal conditions of the worker.[64] Therefore, the standard of a prudent employer involves diligent conduct, as high as possible, for which the only limit is the possibilities offered by the state of the technology.[65] These principles of prevention reflect the choice made by the legislator for a particular model of risk management which gives priority to the elimination of risk at source, as opposed to the adoption of protection measures, and in respect of the latter, to favour protection of the group over that of the individual.

In the general duty of safety, there is no specific and defined conduct required, and instead prevention becomes a constant objective. The duty of safety is presented as an open and flexible duty within the logic of continuous improvement, which confirms the principle of 'maximum safety technologically possible'.[66] As is stated by the Court of Justice in its judgment of 15 November 2001, the occupational risks that are to be

---

envisaged at Articles 9.1 (a) and 10.3 (a), Case C-5/00, *Commission of the European Communities v. Federal Republic of Germany*, para. 35.

62. J. Aparicio Tovar, 'Sicurezza sul lavoro', in *Dizionario di diritto del lavoro comunitario*, eds. A. Baylos Grau, B. Caruso, M. D'Antona & S. Sciarra (Bologna: Monduzzi, 1996), 577. Domestic servants are excluded from the concept of workers under Article 3 (a) FD. The definition of this precept protects workers employed by an employer, and self-employed workers are excluded from the scope of application of the FD. However, Council Recommendation 2003/134/EC concerning the improvement of the protection of the health and safety at work of self-employed workers urges Member States to apply the national laws on health and safety on a voluntary basis, or that certain protection measures should be established for this group.
63. P. Rodière, *Droit social de L'Union Européenne* (Paris: LGDJ, 2002), 454; S. González Ortega & J. Aparicio Tovar, *Comentarios a la Ley 31/1995 de Prevención de Riesgos Laborales* (Madrid: Trotta, 1996), 107.
64. J. Aparicio Tovar & S. González Ortega, 'La Ley de prevención de riesgos laborales', *Revista de Relaciones Laborales*, Tomo I (1996): 329.
65. B. Rodríguez Sanz de Galdeano, 'El alcance de la obligación de seguridad del empresario en la directiva marco: la disyuntiva entre la seguridad tecnológicamente factible y la razonablemente factible', *Revista de Derecho Social*, no. 39 (2007): 122.
66. M. Lai, 'Salute e sicurezza sul lavoro', in *Diritto del lavoro dell'Unione Europea*, eds. F. Carinci & A. Pizzoferrato (Milano: Wolters Kluwer, 2010), 665. In this same regard, J. Aparicio Tovar, 'Sicurezza sul lavoro', in *Dizionario di diritto del lavoro comunitario*, eds. A. Baylos Grau, B. Caruso, M. D'Antona & S. Sciarra (Bologna: Monduzzi, 1996), 582.

evaluated are not conclusively defined, 'but are continually changing in relation, particularly, to the progressive development of working conditions and scientific research concerning such risks'.[67] Therefore, with the use of standards in order to define the proper conduct of employers in matters of safety and health in the workplace, it is possible to avoid the rules from becoming rapidly out of date because of scientific progress or changes in the working environment. Now then, the content of the employer's duty may be very different in accordance with the type of standard used.

During the processing of the Framework Directive, the adoption of the principle of maximum safety technologically possible (common in continental Europe), or the standard of 'reasonably practicable' (traditional in the United Kingdom) was discussed. The latter allows the assessment of another kind of factors, for the most part those that support the interests of the company. What is reasonably practicable would derive from the balance between the cost, particularly the economic cost, of applying the technological means required and the results that would be obtained. Ultimately, the relationship between the cost and the probability of the accident and the extent of the injuries caused would be weighed up in order to determine the measure to be adopted in order to deem the duty to have been discharged. The generalized application of this kind of standard opens the door to the possibility of lowering protection, which would be contrary to 'harmonization with progress'.[68] Finally, the Framework Directive did not contain what the United Kingdom was seeking, but it did include the possible exclusion or limitation of the employer's liability that all States may establish pursuant to Article 5.4 FD.

With regard to the reasonably practicable standard, the Commission considered that the United Kingdom was in breach of the duties deriving from Article 5 of the Framework Directive by limiting the duty of employers contained in the said precept to a duty to ensure safety and health only 'in so far as was reasonably practicable'. However, the Court of Justice, in its judgment of 14 June 2007,[69] considered that the Commission had not shown that limiting the employer's duties to what was reasonably practicable amounted to a breach of the Directive. This judgment might have dispelled any doubts about the content and nature of the duty of safety and the system of liabilities that derives from it. However, it 'does not convincingly resolve the main issue and adds new elements of confusion' to these matters, given that in all cases the Framework Directive creates a duty to ensure safety 'irrespective of the economic cost of the measures'.[70]

The Framework Directive dedicates just a handful of provisions to the liabilities that may arise in the event of non-compliance with health and safety requirements. As

67. Case C-49/00, *Commission of the European Communities v. Italian Republic*, para. 13.
68. J. Aparicio Tovar, 'Sicurezza sul lavoro', in *Dizionario di diritto del lavoro comunitario*, eds. A. Baylos Grau, B. Caruso, M. D'Antona & S. Sciarra (Bologna: Monduzzi, 1996), 582.
69. Case C-127/05, *Commission of the European Communities v. United Kingdom of Great Britain and Northern Ireland*, para. 58.
70. B. Rodríguez Sanz de Galdeano, 'El alcance de la obligación de seguridad del empresario en la directiva marco: la disyuntiva entre la seguridad tecnológicamente factible y la razonablemente factible', *Revista de Derecho Social*, no. 39 (2007): 117.

a result, the system of liabilities of the Member States is not uniform and significant differences are apparent, even though they are all required to abide by the provisions of Articles 5.2 (no exclusion of liability in the event the employer procures external services for prevention tasks) and 5.3 (the duties of the workers in respect of safety do not affect the liability of the employer). Obviously the exercise of the power envisaged at Article 5.4 FD on the exclusion or limitation of employers' liability where occurrences are due to unusual and unforeseeable circumstances 'despite the exercise of all due care' will also result in differences between the national systems.

[C] Balanced Participation in the Management of Health and Safety

The Framework Directive does not have the sole aim of increasing the protection of workers, but rather it also seeks to establish specific measures, and it is not indifferent, as is indicated in the Judgment of the Court of Justice of 22 May 2003,[71] to the manner in which prevention and protection are managed. In effect, it clearly defines the 'background philosophy that supports it', given that the choice of legislative policy is in favour of the adoption of participating strategies in order to tackle the problem of the management of safety in the workplace.[72] For this purpose, Article 7 establishes the basic elements of the protection and prevention services at companies, adding in the last two sections those aspects that the Member States must necessarily or voluntarily define in their internal legislation. The margin allowed in the design of these services has given rise to the commencement by the Commission of proceedings for breaches by the Member States of their duty to transpose. Specifically it has been necessary to clarify that the possibility of outsourcing the services for the organization of prevention activities (Article 7.3 FD) is a subsidiary duty for the employer, which should in the first instance designate its own workers, pursuant to the provisions of Article 7.1 FD.[73] The way in which this designation is to be carried out is formulated in the directive in broad terms, allowing for selection, appointment or designation, and as such without imposing a duty to indicate a specific selection procedure.[74]

Articles 10 and 11 FD establish the workers' (and/or their representatives') rights to information, consultation, and participation, which are all relevant for completing the management model for prevention designed in the Framework Directive. However, the legislator's decision to establish shared management of prevention through a balanced participation model (Article 11.1. FD) is noteworthy. It is true that, as recital nine of the Framework Directive indicates, the legislative systems in matters of safety and health are very different, but this does not mean the Framework Directive seeks to

71. Case C-441/01, *Commission of the European Communities v. Kingdom of the Netherlands*, para. 38. This view was upheld in the Judgment of the Court of Justice of 6 Apr. 2006, Case C-428/04, *Commission of the European Communities v. Republic of Austria*
72. A. Lo Faro, 'Azione collettiva e tutela dell'ambiente di lavoro in Europa', *Giornale di diritto del lavoro e di relazioni industriali*, no. 1 (1991): 186.
73. Case C-441/01, *Commission of the European Communities v. Kingdom of the Netherlands*, para. 20.
74. Judgment of the Court of Justice of 12 Jun. 2003, Case C 425/01, *Commission of the European Communities v. Portuguese Republic*, para. 20.

construct a single model. The lack of powers to act in this matter, where trade-union freedom is present, would make this difficult. The Framework Directive does not alter the established national systems, but it does establish certain guidelines that must necessarily be included and which are in line with the configuration of the 'truly innovative profile' of the Framework Directive represented by balanced participation.[75]

The type of participation instituted by the FD presupposes 'a community of interests' between the parties overseeing the safety measures at work, which requires reciprocal cooperation as an ideal component of management that ensures the highest level of protection.[76] However, the Member States must ensure that the workers or their representatives are able to appeal to the relevant authorities where they consider that the measures adopted by the employer are not sufficient to ensure safety and health in the workplace. This is a model with a system of shared powers, where the collective participation of the workers should be specialized and for all matters relating to safety and the protection of health. The matters to which this participation applies, listed at Article 11 FD, are broad, especially given the general power of participation over 'any measure which may substantially affect safety and health'.

The effectiveness of this model of balanced participation in the prevention of risks requires certain additional measures. Specifically these refer to the workers' right to 'adequate time off work, without loss of pay' and to provide them with the necessary means in order to exercise their rights (Article 11.5 FD). Representative immunity must also be assured such that workers may not be placed at a disadvantage because of their activities.

Now then, the entire system for the collective management of prevention is based on an indispensable assumption: joint management does not lead to joint liability. The legislator has been clear on this point by specifying that the participation of the workers in the area of safety and health in the workplace does not affect the principle of the employer's liability, and the employer continues to have the general duty. This is a necessary condition for the effectiveness of this model, given that 'the definite liability of the employer represents fertile ground for the powers of [collective] intervention [in the balanced participation] to take root'.[77]

§1.05 THE LEGISLATIVE DEVELOPMENT AFTER THE FRAMEWORK DIRECTIVE AND THE AUTONOMOUS FRAMEWORK AGREEMENTS

Directive 89/391/EEC establishes a framework that is generally applicable, and refers complementary regulation for certain sectors or activities (Article 16 FD) to subsequent specific directives. For this purpose, the Annex to the directive includes a list of areas

75. L. Miranda, *La dimensión colectiva de la tutela del entorno de trabajo en Europa: perfiles comparativos entre Italia, Francia y Eslovenia* (Albacete: Bomarzo, 2012), 61.
76. L. Miranda, *La dimensión colectiva de la tutela del entorno de trabajo en Europa: perfiles comparativos entre Italia, Francia y Eslovenia* (Albacete: Bomarzo, 2012), 62.
77. L. Miranda, *La dimensión colectiva de la tutela del entorno de trabajo en Europa: perfiles comparativos entre Italia, Francia y Eslovenia* (Albacete: Bomarzo, 2012), 54.

in which specific directives may be adopted, without this list being exhaustive. The relationship between the Framework Directive and the specific directives is based on the full application of the Framework Directive to those areas regulated by the specific directives, which may lay down more rigorous or specific provisions. So far the number of specific directives adopted has been high and they affect a variety of areas such as: workplaces, equipment, signs, personal protective equipment;[78] exposure to chemical agents and chemical safety;[79] exposure to physical hazards or to biological agents, workload, ergonomic and psycho-social risks and specific Sector and worker related provisions; exposure to physical hazards;[80] exposure to biological agents;[81] provisions on workload, ergonomic and psycho-social risks;[82] and, finally sector specific and worker related provisions.[83]

78. Directive 2009/104/EC of 16 Sep. 2009 concerning the minimum safety and health requirements for the use of work equipment by workers at work (second individual Directive within the meaning of Article 16(1) of Directive 89/391/EEC); Directive 1999/92/EC of 16 Dec. 1999 on minimum requirements for improving the safety and health protection of workers potentially at risk from explosive atmospheres (15th individual Directive within the meaning of Article 16(1) of Directive 89/391/EEC); Directive 92/58/EEC of 24 Jun. 1992 on the minimum requirements for the provision of safety and/or health signs at work (ninth individual Directive within the meaning of Article 16(1) of Directive 89/391/EEC); Directive 89/656/EEC of 30 Nov. 1989 on the minimum health and safety requirements for the use by workers of personal protective equipment at the workplace (Third individual Directive within the meaning of Article 16(1) of Directive 89/391/EEC); Directive 89/654/EEC of 30 Nov. 1989 concerning the minimum safety and health requirements for the workplace (First individual Directive within the meaning of Article 16(1) of Directive 89/391/EEC).

79. Directive 2004/37/EC of 29 Apr. 2004 on the protection of workers from the risks related to exposure to carcinogens or mutagens at work (Sixth individual Directive within the meaning of Article 16(1) Directive 89/391/EEC); Directive 98/24/EC of 7 Apr. 1998 on the protection of the health and safety of workers from the risks related to chemical agents at work (Fourteenth individual Directive within the meaning of Article 16(1) of Directive 89/391/EEC).

80. Directive 2006/25/EC on the minimum health and safety requirements regarding the exposure of the workers to risks arising from physical agents (artificial optical radiation, 19th individual Directive within the meaning of Article 16(1) of Directive 89/391/EEC); Directive 2004/40/EC of 29 Apr. 2004 on the minimum health and safety requirements regarding the exposure of the workers to risks arising from electromagnetic fields and waves (18th individual Directive within the meaning of Article 16(1) of Directive 89/391/EEC), amended by directive 2012/11/EU of 19 Apr. 2012; Directive 2003/10/EC of 6 Feb. 2003 on the minimum health and safety requirements regarding the exposure of workers to the risks arising from physical agents (noise) (Seventeenth individual Directive within the meaning of Article 16(1) of Directive 89/391/EEC); Directive 2002/44/EC of the of 25 Jun. 2002 on the minimum health and safety requirements regarding the exposure of workers to the risks arising from physical agents (vibration) (Sixteenth individual Directive within the meaning of Article 16(1) of Directive 89/391/EEC).

81. Directive 2000/54/EC of the European Parliament and of the Council of 18 Sep. 2000 on the protection of workers from risks related to exposure to biological agents at work (Seventh individual Directive within the meaning of Article 16(1) of Directive 89/391/EEC).

82. Directive 90/270/EEC of 29 May 1990 on the minimum safety and health requirements for work with display screen equipment (Fifth individual Directive within the meaning of Article 16(1) of Directive 89/391/EEC); Directive 90/269/EEC of 29 May 1990 on the minimum health and safety requirements for the manual handling of loads where there is a risk particularly of back injury to workers (Fourth individual Directive within the meaning of Article 16(1) of Directive 89/391/EEC).

83. Directive 93/103/EC of 23 Nov. 1993 concerning the minimum safety and health requirements for work on board fishing vessels (Thirteenth individual Directive within the meaning of Article 16(1) of Directive 89/391/EEC); Directive 92/104/EEC of 3 Dec. 1992 on the minimum

In addition to the collection of directives formed by the Framework Directive and the specific directives, various other directives have been adopted which also have their legal basis on Article 153 TFEU or on the corresponding previous numberings of the same precept (previously Article 137 TEU). These are directives which, whether directly or indirectly, have the purpose of regulating the health and safety of workers, and some of them are a result of the process of social dialogue envisaged at current Articles 154 and 155 TFEU (previously Articles 138 and 139 TEU).[84] A start has also been made on codifying the directives adopted prior to the Framework Directive, in accordance with the guidelines envisaged in the 2007–2012 Strategy.

Directive 2003/88/EC, concerning certain aspects of the organization of working time, is one of the directives aimed at simplifying the legislative framework by way of codification.[85] The purpose is to provide clarity to the regulation of this subject previously covered by Directive 93/104/EC and significantly amended by Directive 2000/34/EC. There is no doubt that work and rest times have an effect on the health of workers, and for this reason the legal basis chosen at the pertinent time for the drafting of Directive 93/104/EC was Article 118A of the EC Treaty. Currently, the third recital of Directive 2003/88/EC acknowledges Directive 89/391/EEC as its base directive, without prejudice to the more rigorous or specific provisions that may be stipulated at Directive 2003/88/EC. However, this directive has been the subject of much debate,

requirements for improving the safety and health protection of workers in surface and underground mineral-extracting industries (Twelfth individual Directive within the meaning of Article 16(1) of Directive 89/391/EEC); Directive 92/91/EEC of 3 Nov. 1992 concerning the minimum requirements for improving the safety and health protection of workers in the mineral-extracting industries through drilling (Eleventh individual Directive within the meaning of Article 16(1) of Directive 89/391/EEC); Directive 92/85/EEC of 19 Oct. 1992 on the introduction of measures to encourage improvements in the safety and health at work of pregnant workers and workers who have recently given birth or are breastfeeding (Tenth individual Directive within the meaning of Article 16(1) of Directive 89/391/EEC); Directive 92/57/EEC of 24 Jun. 1992 on the implementation of minimum safety and health requirements at temporary or mobile construction sites E individual Directive within the meaning of Article 16(1) of Directive 89/391/EEC)

84. Directive 2003/88/EC of the European Parliament and of the Council of 4 Nov. 2003 concerning certain aspects of the organization of working time; Directive 2002/15/EC of the European Parliament and of the Council of 11 March 2002 on the organization of the working time of persons performing mobile road transport activities; Council Directive 2000/79/EC of 27 Nov. 2000 concerning the European Agreement on the Organisation of Working Time of Mobile Workers in Civil Aviation concluded by the Association of European Airlines (AEA), the European Transport Workers' Federation (ETF), the European Cockpit Association (ECA), the European Regions Airline Association (ERA) and the International Air Carrier Association (IACA); Council Directive 2010/32/EU of 10 May 2010 implementing the Framework Agreement on prevention from sharp injuries in the hospital and healthcare sector concluded by HOSPEEM and EPSU; Council Directive 94/33/EC of 22 Jun. 1994 on the protection of young people at work; Council Directive 92/29/EEC of 31 Mar. 1992 on the minimum safety and health requirements for improved medical treatment on board vessels; Council Directive 91/383/EEC of 25 Jun. 1991 supplementing the measures to encourage improvements in the safety and health at work of workers with a fixed- duration employment relationship or a temporary employment relationship.

85. With the same aim, e.g., Directive 2009/148/EC of the European Parliament and of the Council of 30 Nov. 2009 on the protection of workers from the risks related to exposure to asbestos at work (codified version).

amongst other reasons because of the definition of working time[86] and rest time and the numerous exceptions allowed for, without forgetting the opt-out clause envisaged at Article 22. Furthermore, the effect of standardizing on the subject has been weak given that 'a significant part of their content was left to the will of the States themselves or to the social partners'.[87] It has been stated that in the Directive, priority is given to 'considering working time as a matter of business organization and as an instrument of employment policy, rather than considering it as an instrument for the protection of the health and safety of the workers'.[88] There is no doubt that these were the debating points that led to the rejection in February 2009 of the approval of a new directive, as the need for revision of Directive 2003/88/EC had arisen shortly after it was adopted, in reform proposals as far back as 2004, and subsequently in 2005 and 2008.

[A] Psycho-Social Risks and the Impact of the Framework Agreements on Work-Related Stress and on Harassment and Violence at Work

Dialogue between social partners has also resulted in Framework Agreements for which the social parties have not requested the Council to adopt a decision on their implementation pursuant to the procedure envisaged at Article 155.2 TFEU. From all of these agreements, two in particular stand out since they tackle, on an inter-sectoral basis, occupational health and safety, both of them with reference to psycho-social risks. They are the Framework Agreement on work-related stress (2004), and the subsequent Framework Agreement on harassment and violence at work (2007).[89] The first, from 2004, was agreed during the timeframe of the fifth programme on safety and health, the Community strategy on health and safety at work for the period 2001–2006. This Strategy has as one of its central aims that of promoting genuine 'well-being at work', especially including risks of a moral and social nature, which are precisely the sorts of risks covered by these two agreements.

The importance of mental health at work is stated in the European Community Strategy for Health and Safety at Work (2007–2012): 'problems associated with poor mental health constitute the fourth most frequent cause of incapacity for work'.[90] This statement was made in 2007 and as such these are confirmed figures. Forecasts made by the World Health Organization worsen this situation by estimating that by 2020,

86. The Court of Justice has considered the definition of 'working time' in relation to the periods when doctors and other medical staff are 'on call'. The most significant judgments have been Case C-303/98 *Sindicato de Médicos de Asistencia Pública (Simap) v. Conselleria de Sanidad y Consumo de la Generalidad Valenciana*, Order of the Court of Justice of 3 Jul., Case C-241/99 *Confederación Intersindical Galega (CIG) v. Servicio Galego de Saúde (Sergas)*, and Case C-151/2002 *Landeshauptstadt Kiel v. Norbert Jaeger*.
87. J. Rentero Jover, 'La regulación de la Jornada en el ordenamiento comunitario. Los cambios en curso' in *Derecho Social europeo* (Madrid: Consejo General del Poder Judicial 2007), 352.
88. N. Martínez Yáñez, 'Tiempo de trabajo y periodo de descanso en la Directiva 03/88/CE y en la jurisprudencia del Tribunal de Justicia de las Comunidades Europeas', *Revista de Derecho Social*, no. 25 (2004): 150.
89. Within this same category of 'Autonomous Agreements' executed by social partners, we find the Framework Agreement on inclusive labour markets (2010), and the Framework Agreement on telework (2002).
90. COM (2007) 62 final, 14.

depression will be the main cause of incapacity for work. This does not mean that the Framework Directive or the specific directives[91] are not applicable to psych-social risks. Directive 89/391/EEC sets a general framework, for which reason it does not specify any risk in particular, or the risk factors or the measures to be taken within the general prevention policy. But both from its wording and from the interpretation made by the EU Court of Justice case law of 15 November 2001[92] it may be seen that it applies to all kinds of risks, therefore including psycho-social risks, along with the duty to avoid them, or where this is impossible, to assess them and prevent them.

The problem arises from the very nature of the risks and the difficulty in identifying them and establishing the measures necessary in order to prevent them. As a result, the need for effective complementary action by the Member States, or in default of them, by the EU in this matter, may be confirmed. The regulation of specific risks may be one of the aims of the individual directives adopted in accordance with Article 16 FD, and for this purpose the Commission launched a first round of consultations with the social partners in 2002 on work-related stress, and another one subsequently on harassment and violence at work. However, the adoption of these two Framework Agreements as a solution to these problems is not truly satisfactory, whether from the point of view of the contents[93] or their efficacy and enforceability. However in both cases, it is stated explicitly that they are work-related risks, and this is undoubtedly relevant.[94]

Their efficacy and enforceability is the most interesting at the current time, given that the impact of the Agreements on the legal systems of the Member States is totally different from that of the directives. The decision of the social partners to opt for an autonomous agreement means that it has to be implemented by the members of the EU signatory parties in accordance with national procedures and practices. This means that the transposition process cannot be considered to be equivalent in the different Member States, and nor will the results be comparable. This has already been reflected in the 'Report on the implementation of the European social partners' Framework Agreement on Work-related Stress'.[95] This report points out that there are 'persistent discrepancies in the levels of protection' and concludes that 'implementation of the Agreement has not yet ensured a minimum degree of effective protection for workers from work-related stress throughout the EU'. Similar problems and deficiencies may be

91. For example, there are explicit references to mental stress, mental fatigue or psycho-social factors in Directives 90/270/EEC, 92/85/EEC and 2010/32/EU.
92. Case C-49/00, *Commission of the European Communities v. Italian Republic.*
93. For example, the aim of the Framework Agreement on work-related stress is not directly the improvement of health, but rather 'increasing awareness and understanding' of this problem. As such, the agreement does not provide any definition of stress, but only describes it in a way that is 'intentionally vague, scientifically inaccurate, besides being individually focussed and subjective', according to M. Peruzzi, 'The prevention of psychosocial risks in European Union law', *I Working Papers di Olympus*, no. 14 (2012): www.uniurb.it/olympus, 27.
94. A.B. Muñoz Ruiz, *El Sistema Normativo de prevención de riesgos laborales* (Valladolid: Lex Nova, 2009), 121
95. Commission Staff Working Paper 'Report on the implementation of the European social partners' Framework Agreement on Work-related Stress' SEC (2011) 241 final.

expected in relation to the application of the Framework Agreement on harassment and violence at work.

It could be said that the European social dialogue may provide some 'added value to occupational health and safety' – in particular at the sectoral level – but it would be wrong to believe 'that one can do without new legislative intervention because (autonomous) social dialogue at several levels could compensate for such intervention'.[96] Therefore, it is necessary to monitor the application of the agreements, and it will probably be advisable to adopt additional measures. It is true that the collective autonomy of the social partners provides full backing for its decisions, but the difficulty inherent to acting in relation to these kinds of risks would have made it advisable to opt for the channel envisaged at Article 155.2 TFEU. This assertion is notably not in accordance with the opinions of the European institutions, which consider social dialogue 'to be an essential instrument of governance given the proximity of the social partners to the requirements of the reality of working environments'.[97]

BIBLIOGRAPHY

Ales, E. 'Lo sviluppo della dimensione sociale comunitaria: un'analisi "genealogica"'. In *Diritto del lavoro dell'Unione Europea*, edited by F. Carinci & A. Pizzoferrato. Milanofiori Assago: Wolters Kluwer, 2010.

Aparicio Tovar, J. 'Sicurezza sul lavoro'. In *Dizionario di diritto del lavoro comunitario*, edited by A. Baylos Grau, B. Caruso, M. D'Antona & S. Sciarra. Bologna: Monduzzi, 1996.

Aparicio Tovar, J. & S. González Ortega. *Comentarios a la Ley 31/1995 de Prevención de Riesgos Laborales*. Madrid: Trotta, 1996.

Aparicio Tovar, J. & S. González Ortega. 'La Ley de prevención de riesgos laborales'. *Revista de Relaciones Laborales*, 1, (1996).

Arrigo, G. 'La politica sociale comunitaria: il quadro evolutivo'. In *Percorsi di diritto del lavoro*, edited by Garofalo Domenico & Ricci Maurizio. Bari: Cacucci, 2006.

Baldwin, R. & T. Daintith. *Harmonization and hazard: regulating workplace health and safety in the European Community*. London: Graham & Trotman, 1991.

Bonardi, O. 'La sicurezza del lavoro nella Comunità Europea, nella Costituzione e nella legge di semplificazione'. *Rivista giuridica del lavoro*, 3, (2004).

Camas Roda, F. *La normativa Internacional y Comunitaria de Seguridad y Salud en el Trabajo*. Valencia: Tirant lo Blanch, 2003.

Caruso, B. 'L'Europa, il diritto alla salute e l'ambiente di lavoro'. In *Ambiente, salute e sicurezza*, edited by L. Montuschi. Torino: Giappichelli, 1998.

Lai, M. 'Salute e sicurezza sul lavoro'. In *Diritto del lavoro dell'Unione Europea*, edited by F. Carinci, & A. Pizzoferrato. Milanofiori Assago: Wolters Kluwer, 2010.

96. S. Smismans, *Towards a new community strategy on health and safety at work? Caught in the institutional web of soft procedures*, 19 Intl. J. Comp. Labour L. & Indus. Rel. 77 (2003).
97. L. Miranda, *La dimensión colectiva de la tutela del entorno de trabajo en Europa: perfiles comparativos entre Italia, Francia y Eslovenia* (Albacete: Bomarzo, 2012), 85.

Lo Faro, A. 'Azione collettiva e tutela dell'ambiente di lavoro in Europa'. *Giornale di diritto del lavoro e relazioni industriali*, 1, (1991).

Lozano Lares, F. 'El marco jurídico comunitario de la seguridad y salud laboral'. In *Una aproximación al derecho social comunitario*, edited by J. Cruz Villalón & T. Pérez del Rio. Madrid: Tecnos, 2000.

Martínez Yáñez, N. 'Tiempo de trabajo y periodo de descanso en la Directiva 03/88/CE y en la jurisprudencia del Tribunal de Justicia de las Comunidades Europeas'. *Revista de Derecho Social*, 25, 2004.

Miranda, L. *La dimensión colectiva de la tutela del entorno de trabajo en Europa*. Albacete: Bomarzo, 2012.

Muñoz Ruiz, A.B. *El Sistema Normativo de prevención de riesgos laborales*. Valladolid: Lex Nova, 2009.

Neal, A. 'Regulating health and safety at work: developing European Union policy for the millennium'. *The International Journal of Comparative Labour Law and Industrial Relations* 14, no. 3 (1998).

Pelkmans, J. 'The New Approach to Technical Harmonization and Standardization'. *Journal of Common Market Studies* XXV, no. 3 (1987).

Peruzzi, M. 'The prevention of psychosocial risks in European Union law' *I Working Papers di Olympus* 14, (2012), www.uniurb.it/olympus.

Rentero Jover, J. 'La regulación de la Jornada en el ordenamiento comunitario. Los cambios en curso.' In *Derecho Social europeo*, Madrid: Consejo General del Poder Judicial, 2007.

Rodière, P. *Droit social de l'Union Européenne*. Paris: LGDJ, 2002.

Rodríguez Sanz de Galdeano, B. 'El alcance de la obligación de seguridad del empresario en la directiva marco: la disyuntiva entre la seguridad tecnológicamente factible y la razonablemente factible'. *Revista de Derecho Social*, 39, (2007).

Salerno, A. *Diritto comunitario della sicurezza del lavoro*. Padova: Cedam, 1990.

Sancho Cuesta, J. *La seguridad e higiene en el ordenamiento jurídico comunitario*. Madrid: Ministerio de Trabajo y Seguridad Social, 1993.

Smismans, S. 'Towards a new community strategy on health and safety at work? Caught in the institutional web of soft procedures'. *The International Journal of Comparative Labour Law and Industrial Relations* 19, no. 3 (2003).

Wright, F. 'The developments of occupational health and safety regulation in the European Communities'. *The International Journal of Comparative Labour Law and Industrial Relations* 8, no. 1 (1992).

CHAPTER 2

Occupational Health and Safety in Austria: 'The Labour Inspectorate Is Watching You'

Elisabeth Kohlbacher and Franz Marhold

§2.01 INTRODUCTION

As early as the nineteenth century, provisions concerning the protection of employees, especially their health and safety, were introduced to several laws. Health and safety provisions from these days can be found in the Industrial Code (*Gewerbeordnung*-GewO)[1] especially. In 1951, the general regulation for the protection of employees was adopted, based on the GewO. Yet, only in 1972, occupational safety provisions were separated from the GewO and an independent Employee Protection Act (*ArbeitnehmerInnenschutzgesetz*-ASchG)[2] was adopted.[3]

Laws that provide health and safety measures at work are necessary because of the state's duty to protect the peoples' most important objects of legal protection,[4] i.e., life and limb, health and property. The individual work contract often does not contain any measure aimed at protecting these objects. Thus, and due to the fact that employees are in a weaker position than their employer, laws on health and safety try to improve the employees' situation by providing the necessary measures.[5] As a consequence, the existence of an actual work contract is not necessarily a prerequisite for the application of the ASchG on the employment relationship. The Act itself states mere employment as a prerequisite.

1. GewO 1859, RGBl 227/1859.
2. ASchG, BGBl 234/1972.
3. Heinz Mayer, 'Der Arbeitnehmerschutz im öffentlichen Dienst', in *Rechtsfragen des technischen Arbeitnehmerschutzes*, ed. Theodor Tomandl (Wien: Braumüller, 1999), 75.
4. OGH 27/03/1995, 1 Ob 544/95 ZIK 1995, 124 = JBl 1995, 651 = EvBl 1995/156.
5. Reinhard Resch, 'Zur Rechtsstellung der Arbeitnehmer nach dem ArbeitnehmerInnenschutzgesetz', in *Rechtsfragen des technischen Arbeitnehmerschutzes*, ed. Theodor Tomandl (Wien: Braumüller, 1999), 4.

The actual ASchG[6] was adopted by legislature in 1994, shortly before Austria's accession to the European Union (EU) in 1995. Most of the standards provided by the main directive concerning health and safety at work, Directive 89/391/EEC (hereinafter 'Framework Directive'), were implemented through the ASchG.

The provisions of the ASchG have to be classified as provisions of public law. However, they often appear as protective laws in the fields of private law.[7]

Besides the ASchG, there exist numerous regulations that specify its provisions. The most important ones are:

- *Bauarbeiterschutzverordnung* (BauV; regulation on the protection of workers at construction sites).
- *Verordnung über die Fachausbildung der Sicherheitsfachkräfte und die Besonderheiten der sicherheitstechnischen Betreuung für den untertägigen Bergbau* (SFK-VO; regulation on the education of safety representatives and on the peculiarities concerning safety related technical supervision for underground mining).
- *Elektroschutzverordnung* 2003 (ESV 2003; regulation on electrical safety).
- *Verordnung über die Sicherheitsvertrauenspersonen* (SVP-VO; regulation on safety representatives).
- *Verordnung über die Sicherheits- und Gesundheitsschutzdokumente* (DOK-VO; regulation on health and safety related documents).
- *Verordnung über arbeitsmedizinische Zentren* (AMZ-VO; regulation on occupational health centres).
- *Kennzeichnungsverordnung* (KennV; labelling regulation).
- *Bildschirmarbeitsverordnung* (BS-V; regulation on work with monitors).
- *Verordnung biologische Arbeitsstoffe* (VbA; regulation on biological working materials).
- *Arbeitsstättenverordnung* (AStV; work site regulation).
- *Verordnung über sicherheitstechnische Zentren* (STZ-VO; regulation on centres for safety related issues).
- *Arbeitsmittelverordnung* (AM-VO; work equipment regulation).
- *Verordnung über Beschäftigungsverbote und -beschränkungen für Arbeitnehmerinnen* (regulation on employment bans and restrictions for female employees).
- *Bühnen-Fachkenntnisse-Verordnung* (Bühnen-FK-V; regulation on stage skills).
- *Verordnung explosionsfähige Atmosphären* (VEXAT; regulation on explosible atmospheres) with appendix.
- *Sprengarbeitenverordnung* (SprengV; blasting regulation).
- *Verordnung Gleichstellung von Bewilligungsverfahren* (regulation on coordination of authorization procedures).

6. BGBl 450/1994.
7. Reinhard Resch, 'Zur Rechtsstellung der Arbeitnehmer nach dem ArbeitnehmerInnenschutzgesetz', in *Rechtsfragen des technischen Arbeitnehmerschutzes*, ed. Theodor Tomandl (Wien: Braumüller, 1999), 12.

- *Bohrarbeitenverordnung* (BohrarbV; drilling regulation).
- *Verordnung Lärm und Vibrationen* (VOLV; regulation on noise and vibrations).
- *Fachkenntnisnachweis-Verordnung* (FK-V; regulation on the evidence of skills).
- *Grenzwerteverordnung* 2007 (GKV 2007; regulation on limits).
- *Verordnung über die Gesundheitsüberwachung am Arbeitsplatz 2008* (VGÜ 2008; regulation on health surveillance at the work place).
- *Verordnung optische Strahlung* (VOPST; regulation on optical radiation).

§2.02 BASIC CONCEPTS

[A] Typologies of Risks

According to Nöstlinger,[8] risk has to be defined as an 'event that threatens the employee's health and/or safety'. On a worst-case basis, the non-implementation of the provisions of the ASchG can result either in an occupational accident and/or disease. As a consequence the employee who is not healthy anymore [according to the WHO's definition, health can be defined as 'physical, psychic(al), mental and social state of full well-being which goes beyond the mere absence of diseases and which enables and motivates you to lead an economically as well as socially active life'] suffers from reduced or eliminated work capacity. Thus, both phenomena entitle the employee (or his or her relatives) to benefits granted by the occupational accident insurance institution (hereinafter 'the Insurance'). Consequently, occupational accidents as well as occupational diseases shall be discussed in this section.

[1] Occupational Accidents

According to § 175/1 General Social Insurance Act (*Allgemeines Sozialversicherungsgesetz*-ASVG)[9] occupational accidents are accidents that are spatially, temporarily and causally linked to the employment which is in turn causal for insurance. Jurisdiction and doctrine define accident as a temporarily limited event that leads to an injury.[10]

In order to substantiate this general rule, paragraph two of § 175 ASVG enlists various examples of accidents that do not necessarily happen at the work site but which are connected to the occupation, e.g., accidents that happen on the way to or from work. Apart from these concretizations, paragraph two also enlists examples which go beyond the paragraph one-definition of an occupational accident.

8. Walter Nöstlinger, *Handbuch Arbeitnehmerschutz* (Wien: Manz, 2006), 9.
9. ASVG, BGBl 1955/189.
10. Theodor Tomandl, *Grundriss des österreichischen Sozialrechts*, 6th ed. (Wien: Manz, 2009), 122.

Yet, not only do actual occupational accidents entitle the employee to claims against the Insurance, but also accidents that are equal to the aforementioned accidents and that happen during specific activities, e.g., the participation in an employee's meeting or during the rescue of a person who is in danger of life (for further examples, see § 176/1 ASVG).

The causality of the occupation for the accident can pose difficulties. Accidents often do not only happen as a result of the occupation but are also caused by other circumstances, as e.g., in case of inherent defects that only appear due to the occupational accident. Contrary to the two theories of general civil law (theory of equivalence and theory of adequacy[11]), the 'theory of the constitutive prerequisite' (*Theorie der wesentlichen Bedingung*) is applied in practice. According to this theory, only prerequisites that are constitutive for the accident, especially when compared to other possible prerequisites, are causal for the accident.[12]

In principle, actual fault of the employee does not necessarily imply the loss of the employee's possibility to file a claim for disability pension. Yet, in case of extraordinarily dangerous behaviour, the employee might lose the benefits the Insurance provides in principle, e.g., when s/he works under the influence of alcohol, although jurisdiction does concede the possibility to file a claim for disability pension as long as the dangerous behaviour was not the only reason for the accident.[13]

[2] Occupational Diseases

Contrary to accidents at work, which are temporarily limited events, occupational diseases pose a temporarily longer impairment to the employee's work capacity. Yet, not every disease is an occupational disease. In principle, according to § 177 ASVG, only those diseases enlisted in annex 1 to the ASVG which occur due to an occupation also enlisted in the annex are accredited as occupational diseases that entitle the employee to claims against the Insurance. However, § 177/2 ASVG stipulates the possibility of a case by case-recognition of occupational diseases by the Insurance. Apart from the prerequisite of verified scientific knowledge about the outbreak of the disease due to the usage of dangerous substances at work, the case-by-case recognition needs to be approved by the minister of labour, social affairs and consumer protection.[14]

The occupational diseases enlisted in annex 1 to § 177 ASVG have one common feature: they are all physical diseases and not mental or psychophysical risks. Yet, more and more employees suffer from mental problems which are often stress related. Another form of psychic risk is mobbing, either exerted by the employer him or herself or by other employees.

11. Helmut Koziol & Rudolf Welser, *Bürgerliches Recht II*, 13th ed. (Wien: Manz, 2006), 309 et seq.
12. Theodor Tomandl, *Grundriss des österreichischen Sozialrechts*, 6th ed. (Wien: Manz, 2009), 137 et seq., however criticizes the application of this theory.
13. OGH 27/03/1990, 10 ObS 52/90 SSV-NF 4/49; Konrad Grillberger, *Österreichisches Sozialrecht*, 8th ed. (Wien/New York: Springer, 2010), 65.
14. Konrad Grillberger, *Österreichisches Sozialrecht*, 8th ed. (Wien/New York: Springer, 2010), 66 et seq.

[3] Mobbing: An Occupational Disease?

With respect to employment relationships, mobbing was first defined by the Swede Heinz Leymann. According to Leymann, mobbing is a:

> hostile and unethical communication, which is directed in a systematic way by one or a few individuals mainly towards one individual who, due to mobbing, is pushed into a helpless and defenceless position, being held there by means of continuing mobbing activities.[15]

Recent surveys show that in 2008 in Austria 2.2% of employed men and 2.5% of employed women were suffering from mobbing situations at work. Altogether, 93,000 employees reported that they were morally harassed.[16]

In Austria, the phenomenon 'mobbing' does not exist as a legal expression, nor does a provision in the ASchG or its regulations deal with it. Yet, as recorded at the very beginning of this article, the employer has a duty to take care of his or her employees. According to Schnorr,[17] the employer is obliged to contribute to the solution of conflicts occurring due to mobbing by other employees. Jurisdiction recorded an infringement of the employer's duty of care in the following situations:

- Lack of support to an employee who was morally harassed by colleagues.[18]
- Repeated moral harassment of an employee by his or her direct superior; furthermore, the other superiors did not take any remedy despite numerous requests for help.[19]
- Conversation and discussion with the mobbing employees as only remedy which, as a result, did not prevent the harassers from repeating their attacks.[20]
- Request for statements from the mobbing employees and the offer to be transferred as only remedies.[21]

Employees who morally harass their working colleagues infringe their duty of good faith towards the employer. Consequently, the employer disposes of various possibilities to sanction this violation, particularly depending on the employee's level of fault. Besides an informal admonition, the employer can transfer the mobbing employee, though s/he has to be aware of the existence of a possible protection from being transferred according to § 101 Labour Relations Act (*Arbeitsverfassungsgesetz*-AVG).[22]

15. For example. Heinz Leymann, 'Mobbing and Psychological Terror at Workplaces', *Violence and Victims* 5, 120.
16. http://www.statistik.at/web_de /presse/032606.
17. Gerhard Schnorr, 'Probleme des allgemeinen sozialen Kontaktes zwischen Arbeitnehmern eines Betriebes', in *Innerbetriebliche Konflikte aus rechtlicher Sicht*, ed. Theodor Tomandl (Wien: Braumüller, 1977), 32.
18. OGH 26/08/2004, 8 ObA 3/04f Arb 12.459 = ZAS 2005, 241 (Posch) = RdA 2005/37 (Resch) = ARD 6508/5/2005.
19. LG St. Pölten 02/07/1998, 30 Cga 83/97m Arb 11.751 = ARD 5114/2/2000.
20. OLG Wien 28/02/2000, 8 Ra 349/99y ARD 5223/11/2001.
21. OLG Wien 13/03/2009, 8 Ra 80/08f ARD 5959/5/2009.
22. ArbVG, BGBl 22/1974.

Furthermore, the employer has the possibility to terminate the work contract prematurely due to person-related reasons. The last and also the most serious remedy is the summary dismissal. However, the mobbing has to be intense and of a kind that has been going on for a long period in order to justify the summary dismissal.[23]

[4] Sexual Harassment and Improper Sexual Behaviour as Risks at the Work Place

The ASchG as well as General Civil Code (*Allgemeines Bürgerliches Gesetzbuch*-ABGB)[24] and the Employee Code (*Angestelltengesetz*-AngG)[25] stipulate the employer's duty of care. Although the prevention of sexual harassment and improper sexual behaviour at work are part of the employer's duty of care these phenomena are governed by the Act on equal treatment (*Gleichbehandlungsgesetz*-GlBG).[26]

According to § 6/2 GlBG, sexual harassment has to be defined as any sexually related conduct affecting a person's dignity which is undesirable, inappropriate or indecent for the female/male employee concerned and which creates an intimidating or humiliating working environment or subsequently results in disadvantages at the job as a result of the victim's reaction. In the context of the GlBG, it is irrelevant whether the employer him or herself is the harasser (§ 6/1/1) or whether s/he acts culpably by denying the employee concerned protection against harassment by third parties, either colleagues or customers (§ 6/1/2-4).

Section 7/2 GlBG defines the phenomenon of improper sexual behaviour. Again, it is irrelevant whether the employer him or herself is the 'molestator' or whether s/he acts culpably by denying the employee concerned protection against molestation by third parties.

Any violation of the equal treatment principle, i.e., any sexual harassment or improper sexual behaviour, entitles the employee suffering discrimination to file for injunctive relief (§ 12/11 GlBG), either against the perpetrator or, in case of denied protection, against the employer. If the employee concerned has suffered other damages than pecuniary loss, s/he is also entitled to claim compensation for personal damage; in case of sexual harassment, s/he is entitled to lodge a claim for a sum of a minimum of EUR 720, in all other cases a maximum of EUR 400. Furthermore, the employee may resign from his or her contract due to a legitimate reason (§ 26 Z 4 AngG or § 82a lett. b GewO).

23. Christa Kolodej & Thomas Majoros, 'Mobbing und die Fürsorgepflicht des/der Arbeitgebers/in', *Das Recht der Arbeit* 2010, 162.
24. ABGB, JGS 946/1811.
25. AngG, BGBl 292/1921.
26. GlBG, BGBl 100/1993.

§2.03 PERSONAL SCOPE AND DUTIES OF THE PARTIES INVOLVED

[A] Personal Scope

[1] Employees

According to § 2/1 ASchG the term 'employee' comprises not only workers under the regime of a work contract, but also volunteers or apprentices. The ASchG applies to all these workers, also in case of nullity of the work contract.[27] Furthermore, the ASchG enlists in its § 1 groups of people who are actually employed under the regime of a work contract but who do not benefit from the provisions of the Act. Thus, the ASchG does not apply to:

- People employed at government agencies or administrative offices.
- People employed in the fields of agriculture and forestry [to these people, §§ 77–94 of the Act on work in agriculture (*Landarbeitsgesetz*)[28] applies].
- Servants and domestic personnel; however, § 8 of the Law on servants and domestic personnel (*Hausgehilfen- und Hausangestelltengesetz*) und[29] stipulates the duty of care of the employer towards his or her employees. The focus lies on the protection of the employees' personal rights, e.g., life, health and property (as enlisted in §§ 16 and 17 ABGB).
- Home workers, although the Act on work at home (*Heimarbeitsgesetz*)[30] provides in its § 16 the obligation that work sites where people carry out their home work must not be a risk to the employees' life, health or morality.
- Clergymen.

The non-application of the ASchG to these categories of employees does not imply that there exist no health and safety provisions at all for these groups. The protection of employees at government agencies or administrative offices, for example, is provided by the B-BSG which contains provisions that follow the pattern of the ASchG.

According to Winkler,[31] home workers are excluded from the ASchG because of the lack of a specific area of responsibility the employer has towards the home workers. Melzer-Azodanloo,[32] however, comes to a different conclusion. In principle, the ASchG (apart from the aforementioned exceptions) applies to all people that are in an employment relationship to the employer. Yet, numerous people are in an employment relationship though they do not work at the work site of the employer, e.g., telecommuters who work from home or from another place dislocated from the company's

---

27. Hans Floretta & Karl Spielbüchler & Rudolf Strasser, *Arbeitsrecht I*, 4th ed. (Wien: Manz, 1998), 435.
28. LandarbG, BGBl 287/1984.
29. HausgG, BGBl Nr 235/1962.
30. HeimAG, BGBl 1961/105.
31. Gottfried Winkler, 'Europa- und verfassungsrechtliche Aspekte des ArbeitnehmerInnenschutzgesetzes', in *Rechtsfragen des technischen Arbeitnehmerschutzes*, ed. Theodor Tomandl (Wien: Braumüller, 1999), 64.
32. Nora Melzer-Azodanloo, *Tele-Arbeitsrecht* (Wien: Linde, 2001), 62 et seq.

work site. With regard to this group of people, the provisions of the ASchG only apply to those who work at offices dislocated from the actual work site but which are nevertheless qualified as work site according to § 2/3. If those people work at their home place, Melzer-Azodanloo records that in principle the provisions of the ASchG do not apply.

Yet, to specific groups of employees, some or all health and safety related provisions apply if only the law stipulates the application explicitly, e.g., to people who work at home with help of computers and monitors. §§ 67 and 68 of the ASchG as well as the Regulation on work with monitors stipulate special provisions for these tele-home-workers, especially with regard to ergonomic work places and eye-sight examinations.

[2] Subcontractors and Customers

The provisions regarding health and safety at work primarily apply to employees working for one specific employer. Yet, there are two other groups of people to whom the ASchG and its regulations apply – first to seconded employees, second to customers.

Entrepreneurs may ask another employer (contractor) to second some employees to work at their company for a limited period of time. In order to guarantee the employees' health and safety at their temporary work site, § 9 ASchG [as well as § 6/1 Act on (temporary) provision of labour (*Arbeitskräfteüberlassungsgesetz-* AÜG)[33]] provides that for as long as the seconded employees work in that company, the user shall be considered as their 'employer' with regard to the ASchG. As a result, for the time the employees are made available, the user is liable for the compliance with health and safety provisions, as well as with to the duty of care principle towards the seconded employees.

According to § 9/2 ASchG, prior to the sub-contracting, the user is obliged to inform the contractor about the necessary qualifications for the job as well as about possible special physical abilities that might be required. In addition, the user has to concede the contractor access to all documents related to health and safety at work. The contractor, on the other hand, has the duty to inform the user about all relevant provisions concerning personal safety at work, particularly with regard to the Act on working time (*Arbeitszeitgesetz*-AZG).[34] Yet, the contractor remains liable for the employee with respect to social insurance obligations (§ 5/1 AÜG).

If the contractor's business consists of providing employees to one or various users in a way that the contractor him or herself actually does not need the workforce of the employees (temporary agency work), AÜG is to be applied. Yet, according to § 2/2 AÜG, an employee can only be seconded with his or her prior consent to the sub-contracting.

33. AÜG, BGBl 196/1988 as amended by BGBl I 68/2002.
34. AZG, BGBl 461/1969; Günther Löschnigg, *Arbeitsrecht*, 11th ed. (Wien: ÖGB, 2011), 209.

With regard to the employees' health and safety at work, § 6/2 AÜG stipulates temporary work agency's duty to inform and instruct the user about the personal health and safety instructions concerning the employees. According to § 6/3 AÜG, for the time the employees are working for the user, the duty of care is also part of the user's obligations towards the employees. In case of non-implementation of health and safety provisions towards the temporary agency workers, the temporary works agency has to terminate immediately the contract with the user (§ 6/4 AÜG).

At work sites, not only the employees' but also the customers' and other third parties' health and safety are at a potential risk. Yet, the provisions of the ASchG only apply to employees. If third parties suffer from damages that result from the non-implementation of health and safety provisions, the employer's and/or employee's liability follows in principle the scheme of the ABGB. More detailed information on the liability, its links to the Insurance as well as on the internal regress will be given *infra.*

[B] Duties of the Parties Involved

[1] The Employer as the Main Responsible for Health and Safety at the Workplace

The employer is the main responsible for health and safety at work. His or her responsibility is, on the one hand, based on the duty of care principle, provided by § 1157 ABGB and § 18 AngG, on the other hand, on the provisions of the ASchG. It is his or her duty to provide a safe workplace and to ensure the implementation of health and safety instructions at work.[35] As a principle, only the employer is liable for the compliance with safety measures set up by law (§ 15 Abs 8 ASchG).

The duty of care for the employee, which arises from the individual contract of employment, implies that the employer is not only under an obligation of public law, but also under private law to comply with health and safety provisions. In this way, the compliance with health and safety legislation becomes part of the obligations of the individual work contract.[36]

[a] General Duties of the Employer

Section 3 ASchG provides general duties of the employer which are then specified in the *supra* paragraphs of the Act as well as in numerous regulations based on the ASchG and/or on EU-regulations and directives. Thus, the employer is obliged to e.g:

- Hold a register of health and safety-documents.
- Take the necessary measures to ensure protection of life, health and safety at work, including prevention.
- Provide information and instruction of the employees.

35. Franz Marhold & Michael Friedrich, *Österreichisches Arbeitsrecht*, 2nd ed. (Wien/New York: Springer, 2012), 238.
36. Günther Löschnigg, *Arbeitsrecht*, 11th ed. (Wien: ÖGB, 2011), 391.

- Provide organization and other measures.
- Provide occupational health care analyses for specific groups of workers.

According to § 3 ASchG the employer has to take all these measures at his or her expenses. Not only is the employer liable for the implementation of the health and safety provisions, but also for the employees' compliance with these provisions. Yet, if the employer him or herself cannot attend to particular work places or construction sites, s/he has to appoint a capable substitute who will be liable for the employees' compliance with health and safety measures.[37]

[b] *Information, Evaluation and Monitoring: Crucial Obligations of the Employer*

First of all, the employer is obliged to provide an appropriate signposting for health and safety protection. Furthermore, it is the employer's duty to identify possible dangers and threats to the employees' health and safety at work as well as to evaluate them. In order to comply with this duty, the employer can also consult qualified persons, e.g., occupational health and safety specialists (*Sicherheitsfachkräfte*) or occupational health doctors (*Arbeitsmediziner*). The employer is obliged to take into consideration especially those employees who are exposed to specific risks or who need special protection.[38]

The next step is the definition of the necessary safety measures, based on the results of the evaluation process. The main aim is that health risks are not only shifted but eliminated. To make sure that health and safety measures are actually effective, the employer is obliged to monitor their implementation and effectiveness. Finally, it is the employer's obligation to adapt health and safety measures to the outcome of the evaluation and monitoring process.[39] Section 5 ASchG obliges the employer to record the results of the identification as well as of the evaluation of the risks in specific health and safety-documents.[40] In case of occupational accident the employer may be obliged to recheck the results of the evaluation process. The same measure may be necessary when labour inspectorate insists on it due to well-founded reasons.[41]

Apart from risk evaluation, one of the main duties of the employer is to inform the employees on risks for their health and safety and to make sure that measures of risk prevention are actually implemented. Yet, the employer is not only obliged to provide the employees with necessary information, but also to listen to their health and safety related questions as well as to instruct them before they start working in the company.

37. Günther Löschnigg, *Arbeitsrecht*, 11th ed. (Wien: ÖGB, 2011), 899.
38. Günther Löschnigg, *Arbeitsrecht*, 11th ed. (Wien: ÖGB, 2011), 899.
39. Lukas Stärker, 'Die Evaluierungspflicht des Arbeitgebers' in *Rechtsfragen des technischen Arbeitnehmerschutzes*, ed. Theodor Tomandl (Wien: Braumüller, 1999), 40 et seq.
40. Gottfried Winkler, 'Europa- und verfassungsrechtliche Aspekte des ArbeitnehmerInnens-chutzgesetzes', in *Rechtsfragen des technischen Arbeitnehmerschutzes*, ed. Theodor Tomandl (Wien: Braumüller, 1999), 68.
41. Günther Löschnigg, *Arbeitsrecht*, 11th ed. (Wien: ÖGB, 2011), 899.

Further instructions may follow, e.g., when risk prevention measures are set up for the first time or after accidents.

Finally, accidents as well as 'almost-accidents' have to be recorded and the records have to be stored for at least five years (§§ 12–14 ASchG).

[c] Preventive and Protective Services: Occupational Safety Advisors, Occupational Health Doctors and the Occupational Safety Committee

Second, the employer is obliged to nominate occupational safety advisors (*Sicherheitsfachkräften*). Their main field of activity is to control the employer and to advise employers, employees, safety representatives and the works council as well as to support the employer at implementing health and safety provisions.

Medical care for the employees is carried out by occupational health doctors. Before the appointment of these preventive services, the employer is obliged to consult the Occupational Safety Committee.

Safety advisors as well as occupational health doctors are obliged to:

- record their activities;
- participate in the occupational safety committee's meetings;
- issue an annual report on their activities if there is not occupational safety committee;
- inspect work places;
- inform the responsible for health and safety related issues on deficits, possible dangers and risks.

In companies with work sites of less than fifty employees, preventive service care has to be carried out through work site inspection. It is up to the employer to choose whether safety service care shall be realized by safety advisors or by a Centre for safety related issues. According to § 75 ASchG and the Regulation on centres for safety related issues,[42] the head of such a centre has to be a safety advisor who works on a full-time basis in the centre.

Centres for safety related issues have to provide their services to companies for at least seventy hours per week. In order to be able to comply with this requirement, other safety advisors have to be employed by the centre.[43] If less than fifty employees work at a single work site, the employer him or herself can carry out the tasks of preventive service if s/he is as qualified as a safety advisor (see *infra* for the prerequisites). If up to fifty employees work at a single work site and if the company employs on the whole not more than 250 people, the company can use preventive services provided by a Centre for prevention, according to § 78 ASchG. These Centres have to be set up by the Insurance which has to provide for the necessary work capacity. When commissioning the required safety advisors, priority has to be given to external preventive services and centres for safety related issues. If the employer decides to use the services of a Centre

42. STZ-VO, BGBl II 1998/450.
43. Günther Löschnigg, *Arbeitsrecht*, 11th ed. (Wien: ÖGB, 2011), 906.

for prevention (for safety service and/or for occupational health service) or if s/he wants to carry out the safety service him or herself is obliged to inform the safety representatives and the works council about it.[44]

If safety advisors or occupational health doctors do not fulfil their obligations, the employer is allowed to recall them, but only after having forwarded the issue to the Occupational Safety Committee. If safety advisors or occupational health doctors are dismissed or if their work contract is terminated prematurely due to their activities as prevention task force, they have the same protection as safety representatives, which allows them to challenge termination or dismissal at court.[45]

Safety advisors can be appointed only after having completed a special vocational training acknowledged by the Federal Ministry of Labour, Social Affairs and Consumer Protection. The employer has the possibility to either appoint an in-house safety advisor or an external qualified person or to use the services either of a specialized centre for prevention or of a Centre for safety related issues (§§ 73, 78a ASchG). The prerequisites for running one of these centres are enlisted in § 75 ASchG. In order to guarantee regularly available support in safety topics to the extent of a minimum of seventy hours per week, further safety specialists apart from the executive safety specialist have to be employed.[46]

When counselling and supporting the aforementioned group of people, safety advisors are not bound to any instructions of the employer. The latter is obliged to provide them with the necessary rooms and equipment and to consult them in various fields (§ 73 ASchG).

In order to become an occupational health doctor, physicians have to complete a special vocational training on occupational health care. Employers can, again, appoint either in-house physicians or qualified external occupational health doctors or they can use the services of an occupational health care centre. In order to be allowed to run one of these occupational health care centres, prerequisites that are similar to those mentioned before for centres for safety related issues have to be complied with (§ 80 ASchG).

Occupational health doctors are, as well as safety advisors, obliged to provide employees, safety representatives and the works council with required information. Moreover, the employer has to make sure that occupational health doctors advise the aforementioned group of people. However, occupational health doctors must not infringe their obligation of secrecy.[47]

According to § 88 ASchG, in companies with more than 100 employees the employer is obliged to set up an Occupational Safety Committee. However, if at least three quarters of the places of work are 'desk jobs' or places of work similar to 'desk jobs' it is only from 250 employees or more the employer's duty to set up an Occupational Safety Committee. Members of the committee are the employer or his or her representative, the person liable for compliance with health and safety provisions,

44. Günther Löschnigg, *Arbeitsrecht*, 11th ed. (Wien: ÖGB, 2011), 910.
45. Günther Löschnigg, *Arbeitsrecht*, 11th ed. (Wien: ÖGB, 2011), 908.
46. Günther Löschnigg, *Arbeitsrecht*, 11th ed. (Wien: ÖGB, 2011), 906.
47. Günther Löschnigg, *Arbeitsrecht*, 11th ed. (Wien: ÖGB, 2011), 909.

health and safety advisors, occupational health doctors, safety representatives and one competent delegate of the works council. The committee has to meet at least twice a year, though it may also come together when one third of its members insist on it. If the employer runs more than one work site, a central occupational safety committee, which has to be called at least once a year, has to be set up.

The committee's main duty is to guarantee vice versa information and co-ordination among the committee's members and the groups they represent, as well as to work for better conditions of employment and improved protection of the employees' health and safety.[48]

[d] Health Surveillance and Medical Examinations

The ASchG provides numerous obligations the employer has to fulfil concerning medical examinations of the employees, though it is not only the employer but also the Labour Inspectorate who is responsible for appropriate health surveillance at work. However, the employer exclusively has to pay the expenses of the examinations. Only in case of special examinations related to activities that could be causal for occupational diseases is the employer entitled to refund of expenses against the Insurance.[49]

With certain activities there comes along a high risk of occupational diseases. In these cases it is crucial that preventive occupational health examinations take place before the possibly affected employees start to work as well as in periodical intervals in case of continued high risk activities.[50]

In order to provide efficient medical examination, the employer has to make sure that occupational health doctors can enter the employees' work places. Furthermore, the employer is obliged to render any necessary information concerning the employees' health to the examiner. The results have to be recorded by the employer as long as the employee belongs to his or her company. After termination of the employment relation, records have to be forwarded to the Insurance which has to keep it for at least forty years (§ 58 ASchG).

The results of the examinations have to be recorded by occupational health doctors as well, yet it rests with the Labour Inspectorate to decide whether the employees are physically apt for their respective activities. The inspectorate has to issue an official notice and it is important to emphasize that the employer as well as the employee is party to the administrative proceeding.

48. Günther Löschnigg, *Arbeitsrecht*, 11th ed. (Wien: ÖGB, 2011), 911.
49. Günther Löschnigg, *Arbeitsrecht*, 11th ed. (Wien: ÖGB, 2011), 902 et seq.
50. Günther Löschnigg, *Arbeitsrecht*, 11th ed. (Wien: ÖGB, 2011), 901.

[e] The Spatial Dimension of Employer's Responsibility

The place of employment is defined in § 2/3 ASchG. Buildings as well as outdoor work places come within the scope of application of this provision. According to § 2/4 ASchG work place is defined as the area where the employee carries out his or her work.[51]

The employer is obliged to set up and run work places in compliance with health and safety provisions. In order to provide safe work places, the employer e.g., has to ensure that:

- Work places in buildings are safely built, with sufficient daylight or electric lighting.
- Exits and transport routes can be used easily and safely.
- Outdoor work places can be left quickly and easily.
- Work places are provided with enough air for breathing and that humidity, bad odour etc are avoided.
- Adequate measures in case of fire and explosion are provided, especially concerning the evacuation of employees.
- First aid facilities are available as well as first aid wards in work places where more than 250 employees work; in addition, the employer has to ensure that a certain number of employees is instructed in first aid measures.
- Washing and changing rooms are installed, as well as restrooms, lockers and recreation rooms.

Yet, the employee is not only protected at work but also on his or her way to and from work. Although the ASchG does not provide any distinct measures regarding the employee's health and safety on his or her way to/from work, the ASVG defines accidents that happen on this way as occupational accidents (§ 175/2/1). As a result, accidents that happen on the way to/from work are also covered by benefits of the Insurance.

In order to be able to define the 'way to and from work' a starting point has to be found. According to jurisdiction, the way to work starts at the exterior front of the place where the employee is a permanent resident, e.g., the door of the house or the garage.[52] The permanent living place is the place where the employee's main focus of private life is situated, e.g., flat of his or her family.[53]

[f] The Temporal Dimension of Employer's Responsibility

Occupational diseases often only appear (long) after termination of the employment relationship. Yet, in many cases it is most difficult to ascertain whether the illness is related to the employment and caused by it. Due to the fact that in principle only occupational diseases enlisted in annex 1 to § 177 ASVG entitle the (former) employee

51. Walter Schrammel, *Arbeitsrecht II*, 6th ed. (Wien: Braumüller, 2008), 46.
52. OGH 09/02/1988, 10 ObS 25/88 SZ 61/16 = EvBl 1988/98 = SSV-NF 2/17.
53. Konrad Grillberger, *Österreichisches Sozialrecht*, 8th ed. (Wien/New York: Springer, 2010), 61.

to monetary claims against the Insurance, the assessment of the actual causation of the employment for the occupational disease is crucial. In order to better illustrate the difficulties of the proof of causation, a short example may be cited:[54]

> Franz S. worked for the E-U-company for fifteen years (from 1956 until 1971). In 1983, lung cancer was diagnosed and in 1984, Franz S. died. Consequently, according to § 215 ASVG, his wife filed a claim for a widow's pension. However, the Insurance denied her claim due to the fact that according to the institution's opinion Franz S. had not suffered from an occupational disease (Franz S. had worked under the influence of asbestos) but from a constitutional disease.
> The main question the Supreme Court of Justice had to deal with in the end was whether the widow of Franz S. had to prove that the lung cancer was probably caused by her husband's work under the influence of asbestos and whether the Insurance on the other hand had to prove that the widow was not entitled to file a claim for the pension.

The Supreme Court of Justice came to the conclusion that in fact, the widow had to prove that her husband's disease was caused by his work under the influence of asbestos. Yet, she only had to prove that the causation was highly probable (prima facie-proof). The Insurance, on the other side, would have had to prove that Franz S.' lung cancer was the atypical result of circumstances that were not related to his work under the influence of asbestos. Due to the fact that the Insurance was not able to prove this atypical development, the widow's claim entitled her to the pension.

In conclusion, we can state the following:

- The causation of the occupation at work for the occupational disease has to be proved; however, a prima facie-proof is sufficient.
- A claim for pension due to an occupational disease can also be justified after the termination of an employment relationship.
- With regard to the limitation of the claim, according to § 337/2 ASVG, § 1489 ABGB has to be applied. Thus, the claim can be filed in principle within three years after knowledge of the damage and its cause or the person who caused it. However, the lodging of the claim is limited to the absolute period of thirty years after the damaging event.[55]

[2] The Liable Authorized Representatives

In companies that are legal persons, the responsibility for the undertaking lies within those persons who represent the company. With respect to the responsibility for compliance with health and safety provisions in the undertaking, these representatives

54. OGH 14/06/1988, 10 ObS 23/88 SSV-NF 2/65.
55. See Franz-Stefan Meissel & Wilma Dehn, '§ 1489' in *Kommentar ABGB* 3rd ed., eds Helmut Koziol & Peter Bydlinski & Raimund Bollenberger (Wien/New York: Springer, 2010).

are allowed to appoint executive employees as so-called liable authorized representatives [§ 23 Labour Inspection Act (*Arbeitsinspektionsgesetz*-ArbIG),[56] § 9/2 Administrative Criminal Code (*Verwaltungsstrafgesetz*-VStG)[57]]. Yet, according to § 23 ArbIG and § 9/3 VStG, not only representatives of legal persons but also sole proprietors are allowed to appoint liable authorized representatives.[58] Furthermore, the relevant administrative authority has the right to commit the employer to such an appointment (§ 9/2 VStG).

In case of a legally effective appointment, the liable authorized representative will be held liable for not having taken appropriate and sufficient measures according to the provisions of the ASchG. The employer him or herself, however, will still be held liable for wilful inducement or non-prevention of infringement.[59]

Due to the fact that the liable authorized representative will be charged for all infringements connected to the non-implementation of health and safety measures at work, there are numerous prerequisites to be complied with when appointing a liable authorized representative. If the company is a legal person, the liable authorized representative can either be one of the representatives or a simple employee.

Yet, with respect to the appointment of simple employees as liable authorized representatives, § 23 ArbIG stipulates (for legal persons as well as for sole proprietor companies) that only executive employees can be appointed legally effectively. Szymanski/Oberhauser/Marx[60] record that whether a salaried employee is actually an executive employee or not, has to be proved on an individual basis. According to jurisdiction, employees are executive employees in terms of § 23 ArbIG when they have been awarded certain responsibilities by their employer. On the one hand, these responsibilities enable the executive employee to take specific measures and to prevent infringements of the ASchG; on the other hand, the executive employee will be held liable for any violation.[61]

According to § 23/1 ArbIG and § 9/4 VStG, all of the following prerequisites have to be fulfilled for a lawful appointment:

- The person has to have its principal residence in Austria (however, there are exceptions for persons who have their principal residence within the EEA).
- The appointee must not benefit from any immunity or extraterritoriality (criminal traceability).
- The employee or representative has to accept the appointment in a way that can be proved.
- The relevant labour inspectorate has to be informed about the appointment itself in writing, as well as about the appointee's consent.

56. ArbIG, BGBl 27/1993.
57. VStG, BGBl Nr 52/1991.
58. Rudolf Thienel & Eva Schulev-Steindl, *Verwaltungsverfahrensrecht*, 5th ed. (Wien: Verlag Österreich, 2009), 429.
59. Walter Schrammel, *Arbeitsrecht II*, 6th ed. (Wien: Braumüller, 2008), 42 et seq.
60. Eva-Elisabeth Szymanski & Helga Oberhauser & Alexandra Marx, *Kommentar Arbeitnehmerinnenschutzgesetz* (Wien/Graz: NWV, 2004), 270.
61. VwGH 07/04/1995, 94/02/0470 VwSlg 14.236 A/1995.

As already mentioned before, in general the employer him or herself is only liable in case of wilful inducement or non-prevention of infringement. However, if the employer issues a 'special order' which contains the instruction not to act conform to the ASchG, the representative is exempt from his or her liability.

Finally, with regard to the actual monetary liability of liable authorized representatives, § 9/7 VStG stipulates that legal persons as well as sole proprietors are liable, i.e., the appointees can file for regress against the legal person as well as against the sole proprietor. Consequently, also the legal person/sole proprietor is party to the administrative procedure.[62]

[3] The Main Responsible for Compliance with Health and Safety Measures in the Building Sector

According to § 1 of the Act on co-ordination at construction sites (*Bauarbeitenkoordinationsgesetz*-BauKG),[63] co-ordination is the major measure in order to achieve compliance with health and safety provisions at construction sites. Preparation and realization of the project especially have to be well co-ordinated. The main responsibility lies with the builder who has to appoint a special co-ordinator for each construction site where employees of more than one employer work together, one for the stadium of planning, one for the stadium of actual realization of the project (§ 2/6 and 7 BauKG). However, under certain circumstances, the builder him or herself can also act as co-ordinator (§ 3/1 and 3 BauKG). Finally, the builder is also allowed to delegate the responsibility to the project manager.[64]

[4] The Employee as an Individual and His or Her Responsibility for Occupational Health and Safety

Not only the employer but also the employee has to comply with health and safety provisions set up in the ASchG. However, the employer is obliged to take care of the employees' compliance with health and safety provisions.[65] Furthermore, the employer must not tolerate any behaviour of the employees contrary to health and safety provisions.

The employees themselves are obliged to participate in the evaluation process and to inform the employer immediately when insufficiencies concerning health and safety measures occur (duty of alert according to § 15/5 ASchG). If the employer does not comply with this duty of information, s/he may be charged an administrative fine (§ 130/4/6 ASchG; for further information see *infra*). In case of imminent danger, the

62. Rudolf Thienel & Eva Schulev-Steindl, *Verwaltungsverfahrensrecht*, 5th ed. (Wien: Verlag Österreich, 2009), 429.
63. BauKG, BGBl I 37/1999.
64. Günther Löschnigg, *Arbeitsrecht*, 11th ed. (Wien: ÖGB, 2011), 888.
65. Hans Floretta & Karl Spielbüchler & Rudolf Strasser, *Arbeitsrecht I*, 4th ed. (Wien: Manz, 1998), 439.

employee him or herself has to take necessary and reasonable measures in order to prevent any harm to other employees' health and life (§ 15/6 ASchG).

Together with the employer, the safety representatives and with persons liable for the prevention of occupational accidents at work, employees have to work towards compliance with health and safety measures (§ 15/7 ASchG). According to § 13 ASchG, employers are obliged to hear the employees on all issues related to health and safety at work. In case there are no safety representatives appointed and no works council exists, all employees are to be heard about and have to participate in the tasks of safety representatives stipulated in § 11/5 and 6 ASchG.

[5] The AUVA: The General Accident Insurance Institution

The General Accident Insurance Institution for all dependent employees, self-employed persons and freelancers (*Allgemeine Unfallversicherungsanstalt*-AUVA), which is also known as the Austrian Workers' Compensation Board, is the Austrian social insurance for occupational risks for more than 3 million of persons (of which 0.42 million are self-employed persons, 1.57 million are white collar workers and 1.21 million are blue collar workers) and 1.3 million school children and students. They care for about 170,000 victims of occupational accidents and diseases annually. About 350,000 persons involved in occupational accidents of all kinds are treated in their trauma and rehabilitation centres every year. About 73,000 compensations per year are paid to victims of occupational accidents and diseases or their families. The insured persons' interests are represented by delegates of the employers' and employees' organizations (Social Partners).

In the administrative bodies (Boards and Committees), the members of the autonomous administration decide the policy for the insured persons according to the General Social Insurance Act and other relevant laws.

The AUVA's legal duties comprise the prevention of occupational accidents and diseases, occupational medical care, first aid for occupational accidents, post traumatic treatment, rehabilitation, compensation and research. Occupational accidents and diseases are required to be reported to the AUVA, employers or educational authorities are even bound by law to inform the AUVA.

With respect to health and safety at work, the AUVA offers advice, training, expert opinion and information. Furthermore, the AUVA provides employers with information on adapting the working process as well as insured persons with all information concerning their needs. The Safety Technology Testing and Research Centre undertakes as part of the AUVA fundamental research and provides expert opinion. Furthermore, the Technical Safety Laboratory, a certified and officially licensed testing facility, offers European Conformity testing for manufacturers and producers. With regard to the occupational medical services offered by the AUVA, insured persons threatened by occupational diseases receive prophylactic therapy; examinations and vaccinations help maintain health. Yet, the AUVA does not only help maintain health, it also helps disabled persons regain the highest possible degree of physical, mental, occupational and social ability by means of rehabilitation.[66]

§2.04 WORKERS REPRESENTATIVES AND THEIR OBLIGATIONS WITH REGARD TO HEALTH AND SAFETY AT WORK

In every company of at least five employees older than the age of eighteen it is, according to § 40/1 ArbVG, the employees' duty to establish a works council, i.e., that the employees and not the employer are obliged to establish a works council.[67] The main obligation of the works council, which represents employees on health and safety issues, is to control and to survey the implementation of the provisions of the ASchG as well as the respective regulations.

Yet, the works council's obligations with reference to health and safety at work are not provided in the ASchG but in the ArbVG. According to § 89 Z 3 ArbVG, which provides the works council's duty to monitor the implementation and compliance with health and safety provisions, the works council is allowed to inspect the work site. The employer, on the other hand, is obliged to inform the works council immediately about any occupational accident as well as on any trial related to issues concerning the employees.

Section 92a/1 ArbVG stipulates the works council's right to be heard by the employer concerning all health and safety matters as well as the employer's obligation to consult the works council on these issues. According to § 92a/2 ArbVG the employer is obliged to grant the works council access to health and safety-documents, to place documents and findings concerning e.g., labour organization or specific working materials at the works council's disposal, and to provide information about regulatory requirements, orders and authorizations as well as to hear the works council on e.g., health and safety related issues and measures or on first aid measures.

With regard to the appointment or the dismissal of e.g., safety advisors and occupational health doctors, the employer is obliged to consult the works council before the appointment or the dismissal. If the employer does not comply with this obligation, the act of appointment or dismissal is void (§ 92a/3 ArbVG). Only if the intended appointment or dismissal is dealt with in the Health and Safety Committee can the consultation with the works council be omitted.

Finally, the works council can delegate its responsibilities concerning health and safety at work enlisted in § 92a/1/1-3 ArbVG to the safety representatives (§ 92a/4 ArbVG). The safety representatives as well as the employer have to be informed about the delegation; the information is prerequisite for the legal effectiveness of the delegation.

---

66. http://www.auva.at/portal27/portal/auvaportal/channel_content/cmsWindow?action=2&p_menuid=3410&p_t abid=6.
67. Thomas Rauch, 'Die Rolle des Arbeitgebers bei der Errichtung eines Betriebsrats', *Arbeits- und Sozialrechtskartei* (2010): 318.

[A] Safety Representatives

In companies with more than ten employees who work there regularly, the employer is obliged to appoint so-called safety representatives (§ 10 ASchG). Resch[68] points out that safety representatives are, also according to the Framework Directive, representatives of the employees. As a result, in case of appointment of these safety representatives, the employer has to comply with his or her obligations of information, instruction etc. towards them and not towards the employees.

The Regulation on safety representatives[69] stipulates the following minimum numbers of safety representatives (Table 2.1).

Table 2.1 Safety Representatives at Workplace

Number of Employees		*Number of Safety Representatives*
From	*To*	
11	50	1
51	100	2
101	300	3
301	500	4
501	700	5
701	900	6
901	1,400	7
1,401	2,200	8
2,201	3,000	9
3,001	3,800	10
3,801	4,600	11
4,601	5,400	12
5,401	6,200	13
6,201	7,000	14
7,001	7,800	15
7,801	8,600	16
8,601	9,400	17
9,401	10,200	18

If there are more than 10,200 employees, one safety representative for another 800 employees has to be appointed. If there exists a workers representatives' organ, a member of the works council can also be appointed as safety representative. In principle, the appointment of any safety representative has to be approved by the

68. Reinhard Resch, 'Zur Rechtsstellung der Arbeitnehmer nach dem ArbeitnehmerInnenschutzgesetz', in *Rechtsfragen des technischen Arbeitnehmerschutzes*, ed. Theodor Tomandl (Wien: Braumüller, 1999), 21.
69. BGBl 1996/172.

works council. Only if no works council is established do all employees have to be informed in writing on the appointment. When at least one third of the employees object the employers' choice (again in writing) the employer is obliged to select another person.[70]

Generally speaking, safety representatives are appointed for four years, although there is the possibility of prior revocation. In order to be able to accomplish their tasks without any drawbacks, safety representatives are entitled to paid leave. Yet, it is crucial to note that the appointment of safety representatives does not diminish the employer's responsibility for the implementation of health and safety measures. Any transfer of responsibilities towards safety representatives is void.[71]

[B] Duties of the Safety Representatives

According to § 11/1 ASchG the safety representatives' main duties are to:

- provide information, support and advice for the workers and workers representatives as provided in collective labour law;
- represent workers;
- advise/support employers in implementing health and safety provisions;
- take care of the actual implementation of health and safety measures;
- inform the employer about any potential risks;
- co-operate with occupational health and safety advisors and occupational health doctors.

Safety representatives are not bound to any instructions issued by the employer and they have the right to intervene by the employers as well as by the competent authorities concerning all matters of health and safety at work. Furthermore, they have access to all health and safety related documents, accident records and test results.

In case of notice of termination or dismissal because of their activities, safety representatives have special protection. First, if the employer wants to give notice to the safety representative or if s/he wants to dismiss him or her, the employer is obliged to inform the statutorily established employees' interest group (Chamber of Labour). According to § 105 ArbVG and § 9 Act on the adjustment of the provisions concerning employment relationships (*Arbeitsvertragsrechtsanpassungsgesetz*-AVRAG)[72] the safety representative him or herself can then challenge this termination or dismissal at court, but only if s/he was given notice or dismissed due to his or her activity as safety representative. If the court decides that the notice/dismissal was unlawful, the employment relationship continues to exist.[73]

---

70. Franz Marhold & Michael Friedrich, *Österreichisches Arbeitsrecht*, 2nd ed. (Wien/New York: Springer, 2012), 241 et seq.
71. Günther Löschnigg, *Arbeitsrecht*, 11th ed. (Wien: ÖGB, 2011), 904 et seq.
72. AVRAG, BGBl 459/1993 as amended by BGBl 450/1994 due to Art. 8/4 Directive 89/391.
73. Franz Marhold & Michael Friedrich, *Österreichisches Arbeitsrecht*, 2nd ed. (Wien/New York: Springer, 2012), 242.

§2.05 CONSEQUENCES ON THE PARTIES AND REMEDIES IN CASE OF THE VIOLATION OF HEALTH AND SAFETY PROVISIONS

As we have already stated before, provisions for health and safety at work are not only part of the individual work contract but also provisions of public law, first of all. In order to ensure the employer's compliance with these provisions, labour inspection has to supervise their implementation. In case of non-compliance, labour inspection is obliged to enforce them. The bodies that carry out these obligations/tasks are the labour inspectorates. The main source regulating labour inspection obligations is the Labour Inspection Act.

[A] Compliance and Control: The Labour Inspectorates

Labour inspection is part of the Federal Ministry of Labour, Social Affairs and Consumer Protection. The Central Labour Inspectorate is the supreme body of supervision monitoring all matters related to the tasks of nineteen labour inspectorates. These nineteen regional labour inspectorates (each of the federal provinces disposes of at least one labour inspectorate) and one special inspectorate for construction work which respond directly to the Central Labour Inspectorate which is attached to the ministry (§ 16 ArbIG). For all matters related to traffic and similar issues, there exists a special traffic labour inspectorate.[74] In 2009, 297 labour inspectors and 106 administrative experts carried out the labour inspectorate's tasks. The labour inspectorate's expenses amounted to approximately EUR 26,75 million in 2009, of which EUR 21,45 million were spent on personnel, EUR 0,02 million on expenditures for legal obligations and EUR 5,28 million on the cost of material.

Labour inspection in the fields of agriculture and forestry is carried out by specialized agricultural and forestry labour inspection bodies, according to provisions of the federal states corresponding to the LAG.

[1] Agenda of the Labour Inspectorates

As already mentioned before, the labour inspectorate's main obligation is to verify the employer's adherence to the provisions of the ASchG which cater for the protection of life and health of the working population. Furthermore, it informs and advises in all matters of occupational health and safety at work. Within the scope of its activities, labour inspection mediates between conflicting parties in the world of work. In addition, it investigates occupational accidents and matters concerning complaints about a nuisance or abuse. Apart from their obligations according to the ArbIG, labour inspectorates are allowed to issue special certificates of exemption, e.g., in matters concerning working time according to the AZG. Finally, labour inspection is involved in lectures, training and discussions concerning vocational and further training of those

74. Günther Löschnigg, *Arbeitsrecht*, 11th ed. (Wien: ÖGB, 2011), 970 et seq.

liable for protection at work and it creates awareness for aspects of health and safety at work.

Labour inspection is entitled and/or obliged to the following tasks according to the ArbIG:

- Assist and offer consultation to employers as well as employees in all matters of occupational health and safety (§ 3/1).
- Enter and inspect plants, work places and construction sites at any time with or without prior notice (§ 4).
- Interview employers (or their representatives) and employees (§ 7) as well as ask employers for written information (§ 7/2); all interviewed persons are legally bound to respond to the inspectorate's questions.
- Inspect documents referring to safety at work or employment (§ 8).
- Take pictures (§ 8 and others) and take measurements (§ 5); in case of lack of qualification of the labour inspectors concerning special issues, they are allowed to call experts (§ 5/2).
- Take samples and arrange for analyses (§ 5/3).
- Obtain information on material and machines from producers and distributors (§ 6/2).
- Request the competent authorities to enforce measures for the protection of employees (§ 9/3-5; § 10).
- By all means attend to complaints without disclosing the source.

In addition, according to the ASchG, labour inspection, i.e., its occupational health doctors, is obliged to issue an official administrative notice about the employees' physical aptitude. However, in case of the employees' aptitude, e.g., the labour inspectorate can refrain from issuing the official notice (§ 53 ASchG). The employer as well as the employee is party to the respective administrative procedure.

As mentioned at the very beginning of this section, labour inspection's main duty is to monitor the employer's adherence to health and safety provisions. If shortcomings are ascertained, labour inspection has numerous rights and/or obligations to remedy the employer's infringement.

Where provisions for the protection of employees are not complied with, labour inspection first of all enters into consultation with those liable and asks them in writing to remedy the situation within a fixed period (§ 9 Abs 1 ArbIG). If the ascertained shortcomings are not remedied within a fixed or extended period, labour inspection lays an information with the competent authority (§ 9 Abs 2 ArbIG) which is the office for district administration according to § 26 VStG, § 1 Administrative Enforcement Act (*Verwaltungsvollstreckungsgesetz*).[75] Only in case of minor deviations from technical measurements can labour inspection refrain from laying the information.[76] If the labour inspector discovers a serious infringement s/he is obliged to lay an information

75. VVG, BGBl 53/1991.
76. Günther Löschnigg, *Arbeitsrecht*, 11th ed. (Wien: ÖGB, 2011), 972 et seq.

immediately, without any prior notice or request to take remedies necessary (§ 9 Abs 3 ArbIG).

When labour inspection comes to the conclusion that according to the relevant legal provisions measures concerning the protection of life, health and morals should have been taken by the employer, it is obliged to request the competent authority to take appropriate remedial actions (§ 10/1 ArbIG).

Only in case of imminent danger to the employee's life or health is labour inspection obliged to enforce measures without delay, such as e.g., prohibit any continuation of work prior to the elimination of the danger. When doing so, the inspectorate has to issue an official notice although the enforced measures are only valid for fourteen days unless an official notice in writing confirms the measures within this period of time.[77] Yet, labour inspection is also allowed to enforce measures before issuing an official notice as well as to arrange for its implementation.[78]

According to the ASchG, employers are subject to administrative criminal liability in case of infringement to health and safety provisions. In general, it is labour inspection which detects the shortcomings. Yet, the labour inspectorate is not allowed to issue the official criminal-administrative notice itself. It is the duty of the administrative criminal authority to prosecute the infringement, although the labour inspectorate has to claim a certain degree of penalty. Furthermore, the labour inspectorate is party of the following criminal-administrative procedure (§ 11 ArbIG). As party of the procedure the labour inspectorate is allowed to give notice of appeal as well as to raise an objection (§ 11/3 ArbIG). The same is valid for administrative procedures held in case of infringement concerning health and safety instructions (§ 12 ArbIG).[79] However, in both criminal administrative and administrative procedures only the minister of labour, social affairs and consumer protection is allowed to file administrative appeal at the Supreme Court of Administration (§ 13 ArbIG).[80]

[2] Data on the Labour Inspectorates' Measures, Efficiency and Penalties

In 2009, the labour inspectorates carried out the various aforementioned measures at 66.240 places of employment, i.e., in 12,803 companies. In 44% of the cases (all in all 145,786 single measures), labour inspectors visited places of employment in order to monitor the employers' adherence to health and safety provisions. In addition, labour inspectors controlled 376,699 working days of drivers, participated in 17,148 magisterial hearings and delivered 10,434 occupational health care analyses. 62,194 employees were examined by occupational health doctors on their physical aptitude for the respective activities of their place of employment. Only thirty-two were detected as not physically apt.

77. Hans Floretta & Karl Spielbüchler & Rudolf Strasser, *Arbeitsrecht I*, 4th ed. (Wien: Manz, 1998), 438.
78. Günther Löschnigg, *Arbeitsrecht*, 11th ed. (Wien: ÖGB, 2011), 973.
79. Hans Floretta & Karl Spielbüchler & Rudolf Strasser, *Arbeitsrecht I*, 4th ed. (Wien: Manz, 1998), 436 et seq.
80. http://www.arbeitsinspektion.gv.at/NR/rdonlyres/F5C28318-D67E-4B52-AF95-12DAB1DF5125/0/arbig_d_gb_2009_Broschuere.pdf.

In 32% of all construction sites visited (i.e., 19,952 sites), labour inspectorate detected infringements to health and safety provisions. On the whole (all places of employment visited), 68,927 infringements were detected and 2,202 official criminal-administrative notices issued.

According to data provided by the AUVA, the number of employees affected by an occupational accident decreased from 116,407 in 2008 to 99,052 in 2009, as well as the number of deadly injured persons (115 in 2008, 98 in 2009). However, the number of occupational diseases listed in the annex to the ASVG increased from 1,477 to 1,589, and eighty people died as a result of suffering from an occupational disease.[81]

[B] Typologies of Liabilities in Case of Infringement of Provisions Concerning Health and Safety at Work

With respect to the employer's duty of care and his or her obligations according to the ASchG, but also with respect to the legal system as a whole, we have to distinguish between criminal, administrative and civil liability. Special focus has to be put on the employer's civil liability and the role of the occupational accident insurance institution.

[1] Criminal Liability of the Employer

The ASchG does not include any provisions concerning the employer's criminal liability. Thus, we have to resort to the crimes and offences stipulated in the Austrian Criminal Code (*Strafgesetzbuch*).[82] With regard to the employer's duties concerning health and safety at work (protection of the employee's life, health and morality), the commitment of offences like homicide, bodily injury caused by negligence and other common crimes with a harmful result to the employee's health may be reproached to the employer in case of non-implementation of health and safety related provisions.

[2] Administrative Liability

The sanctions provided in the ASchG are administrative sanctions, thus administrative offices and not courts have to deal with infringements committed by the employer as well as the employee. Unique prerequisite for the condemnation to pay an administrative fine is the non-implementation of the provisions of the ASchG, not an actual personal injury or property damage.

According to § 130/1 ASchG, the employer has to pay administrative fines from EUR 145 up to EUR 7,260, in case of repetitive offence from EUR 290 up to EUR 14,530 if s/he infringes specific provisions of the ASchG or its regulations, specified in the aforementioned paragraph. Paragraph two stipulates the employer's obligation to pay

81. http://www.arbeitsinspektion.gv.at/NR/rdonlyres/D40D1716-C0E8-4F64-B648-552086A1873A/0/JB_2009.pdf.
82. StGB, BGBl 60/1974.

an administrative fine (of the same as amount of the aforementioned fines) in case of non-implementation of duties regulated by official notice.

The amount of the fines may seem quite elevated at first sight. However, the government proposal[83] to the ASchG states that employers often save extensive investments by non-implementation of health and safety measures. Yet, these infringements lead to a significant increase of the risks for the employees' lives or safety. Thus, according to the materials, only high administrative fines can prevent employers from not implementing the provisions of the ASchG.

Apart from the user also the contractor may be subject to administrative liability. According to § 130/3 ASchG the contractor may also be charged an administrative fine from EUR 145 up to EUR 7,260, in case of repetitive offence from EUR 290 up to EUR 14,530 if s/he infringes the duties related to the sub-contracting of employees. The same is valid for the user.

As already mentioned, apart from the employer also the employee has to respect and adhere to the provisions of the ASchG. As a result, the employee may be charged administrative fines as well.

Employees who, despite information and proved written order by the employer or the labour inspectorate on how to act according to health and safety provisions, do not act according to their obligations recorded in the ASchG or its regulations are subject to administrative fines of up to EUR 218, in case of repetitive offence up to EUR 360. However, there is no administrative liability for employees who incite colleagues to infringe provisions of the ASchG.[84]

In comparison to the fines, the employer can be condemned to pay, EUR 360 as the maximum admissible fine seems to be rather low. However, much higher fines would violate the criteria developed by the Constitutional Court with respect to minimum penalties.[85] According to the court, special attention has to be paid to the economic interest of the offender. The higher the economic advantage of the infringement, the higher the administrative fine can be. Yet, if an employee offends his or her obligations according to the ASchG, the employee's economic advantage due to the infringement is rather low. As a result, the administrative fine must not be too high.

Besides the liabilities of employers and employees, § 130/6 ASchG stipulates sanctions for those who run a centre for safety related issues or a centre for occupational medical care without fulfilling the prerequisites specified in § 75/1 and § 80/1 ASchG. The fine's amount is again from EUR 145 up to EUR 7,260 and in case of repetitive offence from EUR 290 up to EUR 14,530.

83. 1590 BlgNR 18.GP.
84. Reinhard Resch, 'Zur Rechtsstellung der Arbeitnehmer nach dem ArbeitnehmerInnenschutzgesetz', in *Rechtsfragen des technischen Arbeitnehmerschutzes*, ed. Theodor Tomandl (Wien: Braumüller, 1999), 20.
85. VfGH 20/06/1994, B 1908/93, B 1971/93 VfSlg 13.790.

[3] Civil Liability: The Role of the Occupational Accident Insurance

With respect to the civil liability (of the employer) for personal injuries resulting from occupational accidents or damages due to occupational diseases, the Austrian legal system provides numerous provisions which shall be displayed in the following chapters.

The first provision that has to be outlined is the employer's exemption from tortuous liability (apart from cases where the employer has to be reproached with wilful and malicious injury) in combination with the employee's right to claim benefits from the occupational accident insurance institution. Thus, first of all a short overview of the Austrian system of occupational accident insurance shall be given.

Main task of the Insurance is to prevent occupational accidents and diseases (the occupational accident insurance is the supportive organization of the centres for prevention that have to be set up by the accident insurance institution; see also *infra*). Benefits of the occupational accident insurance are only granted to people who suffer from personal injuries due to occupational accidents or diseases. Main goal is to enable these victims to work again; in case of general invalidity, the loss of income shall be compensated by invalidity pensions.[86]

The historical intention of the legislator with regard to the occupational accident insurance was to protect employees from negative consequences of an accident at work (e.g., risk of dismissal). Due to the fact that the employee was often not able to prove the employer's fault and causal action, civil tortuous liability was an inefficient remedy.[87] As a result, civil tortuous liability of the employer was substituted by benefits of the occupational accident insurance which are granted independent of any fault. Consequently, employees can claim benefits even if the occupational accident or disease result from their deliberate non-application of health and safety provisions.[88] Thus occupational accident insurance has a function of liability insurance for the individual employment contract parties as well. Basis of this concept is the idea that the employer shall be exempt from liability for risks that come along with the management of a company.[89]

Yet, it is not the Federal state that has to bear the costs of work-related injuries. As a consequence of the employer's exemption from tortuous liability, accident insurance is funded by employer contributions exclusively.[90]

Apart from e.g., therapeutic treatment after accidents or remedies and aids such as crutches and other orthopaedic aids, the occupational accident insurance has to bear the expenses for occupational disablement pensions. Yet, not every victim of an

86. Theodor Tomandl, *Grundriss des österreichischen Sozialrechts*, 6th ed. (Wien: Manz, 2009), 120.
87. Theodor Tomandl, *Grundriss des österreichischen Sozialrechts*, 6th ed. (Wien: Manz, 2009), 121.
88. Konrad Grillberger, *Österreichisches Sozialrecht*, 8th ed. (Wien/New York: Springer, 2010), 55.
89. Franz Marhold, 'Quale tutela per il danno alla persona del lavoratore nell'ordinamento austriaco? Tutele civilistiche, lavoristiche e di sicurezza sociale', in *Öffnung und Wandel – Die internazionale Dimension des rechts II – FS für Willibald Posch*, eds Tomislav Borić et al. (Wien: LexisNexis ARD Orac, 2011), 487.
90. Konrad Grillberger, *Österreichisches Sozialrecht*, 8th ed. (Wien/New York: Springer, 2010), 55.

occupational accident or an disease is entitled to compensation. In order to file a claim for this pension, according to § 203 ASVG the degree of the reduction of the workforce has to be at least 20% for more than three months. Furthermore, the disablement pension will only be granted for the period the employee suffers from an impairment of his or her work capacity of at least 20%.

In case of a parallel right to compensation (sick pay) against the health insurance institution, the disablement pension will only be granted after the discontinuation of sick pay, yet not later than twenty-seven weeks after the occurrence of the insured event (§ 203/2 ASVG).

The amount of compensation depends on the one hand on the grade of the reduction of the work capacity, on the other hand on the assessment basis (§ 205 ASVG). In case of general invalidity, the pension will be 66.66% of the assessment basis, in case of the reduction of work capacity the employee can claim the corresponding percentage. Thus, if the employee suffers from a 50%-reduction of his or her work capacity, the compensation will be 50% of the 66.66% of the assessment basis.[91]

With respect to non-material damages, the ASVG also provides a so-called surrogate for compensation. Due to the fact that the grant of this surrogate for compensation is closely related to the employer's exemption from liability, we will deal with it *infra*.

According to § 333 ASVG, the employer is exempt from any tortuous liability for personal injuries of the employee that result from an occupational accident or disease. In addition, the employer's exemption from liability also comprises compensation for immaterial damages, although, apart from the only seldom granted surrogate for compensation of immaterial damages, the general accident insurance institution (AUVA) does not pay the costs for these damages.[92]

One of the reasons for the employer's exemption from liability in case of personal injuries (and immaterial damage) is that the employer alone has to pay accident insurance contributions. Another motive for exempting the employer from any tortuous liability is to avoid internal conflicts between employers and employees.[93] The exemption is not only valid concerning claims of the employee but also concerning claims of bereaved family members in case of death of the employee due to occupational accident or disease.

Exempt from tortuous liability is not only the employer, but also the legal or appointed representatives of a company or the supervisor of the undertaking. Furthermore, also the sub-contractor benefits from this exemption from liability.[94]

Apart from the aforementioned group of people, due to § 1313a ABGB, which stipulates the employer's liability for damages caused by his or her assistants in

91. Konrad Grillberger, *Österreichisches Sozialrecht*, 8th ed. (Wien/New York: Springer, 2010), 55.
92. Franz Marhold, 'Quale tutela per il danno alla persona del lavoratore nell'ordinamento austriaco? Tutele civilistiche, lavoristiche e di sicurezza sociale', in *Öffnung und Wandel – Die internazionale Dimension des rechts II – FS für Willibald Posch*, eds Tomislav Borić et al. (Wien: LexisNexis ARD Orac, 2011), 490.
93. Günther Löschnigg, *Arbeitsrecht*, 11th ed. (Wien: ÖGB, 2011), 526.
94. Gert-Peter Reissner, *Lern- und Übungsbuch Arbeitsrecht*, 4th ed. (Wien: Manz, 2011), 335 et seq.

compliance with the contract, also these assistants are exempted from any liability to other employees.[95]

At the very beginning of this chapter we pointed out that health and safety provisions for employees in the public sector follow the pattern of the ASchG. However, with respect to the employer's exemption from tortuous liability, the Act on health and accident insurance for civil servants (*Beamten Kranken- und Unfallversicherungsgesetz*)[96] does not provide any exemption similar to the one stipulated in § 333 ASVG. According to jurisdiction,[97] § 333 ASVG cannot be applied analogously.

Benefits paid by accident insurance are granted independent from any fault of the employee. As a result, the employee is entitled to these benefits even in case of non-implementation of health and safety measures. The employer, on the other hand, is not exempt from his or her liability in case of wilful and malicious injury. Yet, the employee's claim for compensation is, different to tort law according to ABGB, reduced to the amount that exceeds the pension granted by the occupational accident insurance (§ 333/2 ASVG).

At work, employees often have to deal with means of transport that are subject to special liability insurance according to legal provisions. If an occupational accident happens in connection with one of these means of transport, the employer is not exempt from his or her liability. The amount of compensation the employee can claim is limited to the amount of the insurance benefit, though. However, if the employer acts with deliberate intention, this limit is void and the employer has to compensate also the amount that exceeds the insurance benefit.[98]

In case of an occupational accident or disease, employees often suffer from immaterial damages in addition to personal injuries. In principle, the exemption from tortuous liability also comprises compensation for immaterial damage as well as compensation for deformation notwithstanding a possible employer's gross fault. Moreover, as already mentioned before, according to §§ 203 et seq. ASVG, accident insurance only grants pensions for general invalidity or reduction of work capacity.

The 48th amendment to the ASVG introduced the so-called surrogate for compensation of immaterial damages (*Integritätsabgeltung*) in § 213a ASVG in 1990[99] (for employees in the agricultural sector, § 149m Act on social insurance for farmers (*Bauern-Sozialversicherungsgesetz*)[100] provides a surrogate for compensation of immaterial damages as well). This surrogate payment can be compared to ABGB's compensation for immaterial damage (§ 1325) or its compensation for disfigurement

95. Franz Marhold, 'Quale tutela per il danno alla persona del lavoratore nell'ordinamento austriaco? Tutele civilistiche, lavoristiche e di sicurezza sociale', in *Öffnung und Wandel – Die internazionale Dimension des rechts II – FS für Willibald Posch*, eds Tomislav Borić et al. (Wien: LexisNexis ARD Orac, 2011), 489.
96. B-KUVG, BGBl 200/1967.
97. OGH 23/03/1977, 8 Ob 218/76 SZ 50/45 = EvBl 1977/251 = JBl 1978, 536.
98. Konrad Grillberger, *Österreichisches Sozialrecht*, 8th ed. (Wien/New York: Springer, 2010), 76.
99. BGBl 264/1989.
100. BSVG, BGBl 559/1978.

(§ 1326).[101] Yet, in order to be entitled to claims according to § 213a ASVG, the following indispensable prerequisites have to be fulfilled:

- The employer caused the damage by disregarding health and safety provisions (which comprises also provisions concerning working time or concerning protection for special groups of people, e.g., pregnant women) by gross negligence.[102]
- Grave and lasting damage to the employee's physical or mental integrity.
- The employee is entitled to a pension for general invalidity or for reduction of work capacity according to §§ 203 ff ASVG.

According to Grillberger,[103] the employee is also entitled to claims for this surrogate payment in case of non-implementation of health and safety measures by other employees who acted by gross negligence.

The amount of the surrogate payment is not a fixed sum nor can it be calculated according to legal prerequisites. In fact, § 213a/1 ASVG only stipulates that the surrogate has to be 'adequate', though the maximum payment is limited to the double annual maximum assessment basis.[104] Further provisions that substantiate the surrogate payment can be found in regulations issued by the AUVA. According to these regulations, the employee loses his or her right to file a claim for surrogate payment if health and safety provisions are disregarded by gross negligence.[105] In addition, the employee is not entitled to this payment if a reasonable possibility to enforce a civil claim for compensation exists.[106]

Doctrine criticizes the quite restrictive prerequisites in order to be entitled to file a claim for surrogate payment.[107] Due to the fact that jurisdiction does not admit the provisions of the road traffic regulations as provisions that cater for the employees' health and safety,[108] there is hardly any possibility to claim surrogate payment in case of traffic accidents. Furthermore, the criterion 'gross negligence' is scarcely fulfilled.[109] As a result, courts rather seldom decide in favour of the employee.

101. Walter Pöltner, 'Die Integritätsabgeltung in der gesetzlichen Unfallversicherung', *Das Recht der Arbeit* 1990, 152
102. According to § 1324 ABGB; OGH 19/07/1994, 10 ObS 76/94 SSV-NF 8/64.
103. Konrad Grillberger, *Österreichisches Sozialrecht*, 8th ed. (Wien/New York: Springer, 2010), 72.
104. Section 178/2 in conjunction with § 213a/3 ASVG; OGH 04/03/2008, 10 Ob S 145/07g ASoK 2009, 40 (*Marhold-Weinmeier*).
105. Reinhard Resch, *Sozialrecht* 5th ed. (Wien: Manz, 2012), 108 et seq. Compare as well OGH 08.06.1995, 10 ObS 84/95 SSV-NF 9/51.
106. Konrad Grillberger, *Österreichisches Sozialrecht*, 8th ed. (Wien/New York: Springer, 2010), 73; however, Rudolf Reischauer, 'Neuerungen im Bereich des Arbeitgeber-Haftungsprivilegs im Zusammenhang mit Kfz-Verkehr und Integritätsabgeltung', *Das Recht der Arbeit* 1992, 326 e.g., criticizes these restrictions.
107. Franz Marhold, 'Quale tutela per il danno alla persona del lavoratore nell'ordinamento austriaco? Tutele civilistiche, lavoristiche e di sicurezza sociale', in *Öffnung und Wandel – Die internazionale Dimension des rechts II – FS für Willibald Posch*, eds Tomislav Borić et al. (Wien: LexisNexis ARD Orac, 2011), 492 et seq.
108. OGH 15/09/1992, 10 ObS 169/92 SSV-NF 6/89 = JBl 1993, 257 = DRdA 1993/31.
109. See Andreas Gerhartl, 'Einschränkung der Schadenersatzpflicht bei Arbeitsunfällen', *Zeitschrift für Arbeits- und Sozialrecht* (2009): 253.

As we have already stated, according to § 333/1 ASVG the employer or his or her representatives are not exempt from their tortuous liability in case of wilful and malicious injury. Moreover, the occupational accident insurance can claim regress against employers and people on a par with employers who caused damage by gross negligence or deliberately. Yet, accident insurance is only entitled to claim its expenses related to general invalidity or reduction of workforce pension or to other benefits granted to the employees, e.g., remedies and aids. Surrogate payment for compensation of immaterial damages is explicitly not included in the accident insurance institution's right to claim regress (§ 334/1 ASVG).

Goal of the accident insurance's right to claim regress is to enforce the implementation of health and safety provisions by the employer.[110] If the employer knew that the single sanction for not implementing health and safety provisions was administrative liability only, the probability of an actual implementation would be rather low. The Austrian legislator hoped that by establishing the occupational accident insurance's right to regress, more employers would fulfil their obligations according to the ASchG and its regulations. Yet, Grillberger[111] claims that it is a mystery why the Insurance's right of regress does not comprise the regress for surrogate for compensation of immaterial damages.

The occupational accident insurance's right to claim regress is independent from claims the employee is entitled to. As a result, the employer cannot object to a possible contributory negligence of the employee (§ 334/4 ASVG). Furthermore, not only can the accident insurance institution but also health insurance institutions and pension insurance institutions claim regress if they are subject to remedies and aids claims or pension claims due to an occupational accident or disease.[112]

At work, not only employers are liable for personal injuries that result from occupational accidents or diseases, but also employees when harming other employees. In principle, employees are liable as soon as prerequisites stated in ABGB are complied with (damage, causal action that led to the damage, illegal conduct and fault). Due to the fact that the injured employee is a third party according to §§ 3, 4 Employees' liability act (*Dienstnehmerhaftpflichtgesetz*-DNHG),[113] the employee liable for damage can file for regress against the employer if the prerequisites of §§ 3, 4 DNHG are complied with. On the other hand, in case of claims against the employer due to his or her liability for his or her employees according to § 1313a ABGB, the employer can file for regress against the employee liable for the damage.

With regard to personal injuries, the employee has to compensate costs for curative measures in principle, furthermore for loss of earnings as well as for damages for pain and suffering (§ 1325 ABGB) and finally for possible claims for deformation (§ 1326 ABGB). However, in case of a personal injury that results from an occupational accident, the Insurance will pay for the costs of treatment, remedies and aids. In case of fulfilment of the prerequisites stipulated in §§ 203 ff ASVG, pension for general

110. Günther Löschnigg, *Arbeitsrecht*, 11th ed. (Wien: ÖGB, 2011), 526.
111. Konrad Grillberger, *Österreichisches Sozialrecht*, 8th ed. (Wien/New York: Springer, 2010), 76.
112. Konrad Grillberger, *Österreichisches Sozialrecht*, 8th ed. (Wien/New York: Springer, 2010), 77.
113. DNHG, BGBl 80/1965.

invalidity or for reduction of workforce will be granted. With regard to these payments, consequently the injured employee does not suffer from any more damages. Yet, the Insurance will not compensate immaterial damages.

Due to the fact that in principle, the Insurance shall not discharge the employee from his or her liabilities according to ABGB, it can file for regress against the liable employee (§ 332/1 ASVG). However, if the employee liable for the damage works in the same company as the injured employee, the Insurance can only file for regress if the liable employee caused the occupational accident or disease by gross negligence or deliberately or if the accident results from handling a vehicle subject to special legal liability (§ 332/5 ASVG).

In case of a claim for immaterial damage against the liable employee, the latter is liable even when damage is caused by slight negligence. Although § 3 DNHG provides the possibility for the employee to file for regress against the employer in case of damaging a third party (e.g., a customer), the employee cannot file for regress when damaging another employee. This apparent inequity can easily be explained: The employer alone has to pay for all contributions to the Insurance. Consequently, § 333 ASVG stipulates the employer's exemption from liability in case of personal injuries resulting from occupational accidents. A possible regress of the liable employee against the employer in case of damaging another employee would thwart the system of the employer's exemption from liability.[114]

With regard to non-implementation of health and safety provisions by employers, also third parties, e.g., customers, can be victims of personal injuries. Furthermore, third parties' damages can result from employees' non-compliance with health and safety measures provided in § 15 ASchG (e.g., unjustified removal of safety devices).

In case of personal injury of the third party due to an employee's fault, the third party can claim compensation against the employee as soon as the prerequisites according to §§ 1293 ff ABGB are fulfilled. If the third party is either in a contractual or legal relationship with the employer or if the employer consciously uses the services of an incapable employee, the third party can file for compensation against the employer (§§ 1313a, 1315 ABGB). If both employer and employee are subject to possible claims of the third party, the provisions of the DNHG (§§ 3 and 4) stipulate the prerequisites of the employer's or employee's right to file for regress against the one claimed by the third party.[115]

Employees in the public sector who act on behalf of the agency or office and who damage third parties while applying the law are not subject to the third party's claims. In fact, according to § 1 Office liability Act (*Amtshaftungsgesetz*),[116] the respective legal entity (e.g., the Federal or the Regional government) has to compensate the third party's (personal) injuries. However, when certain prerequisites are fulfilled, the legal entity can file for regress against the employee.

114. Franz Marhold & Michael Friedrich, *Österreichisches Arbeitsrecht*, 2nd ed. (Wien/New York: Springer, 2012), 275 et seq.
115. Franz Marhold & Michael Friedrich, *Österreichisches Arbeitsrecht*, 2nd ed. (Wien/New York: Springer, 2012), 265 et seq.
116. AHG, BGBl 20/1949.

[C] Other Sanctions in Case of Non-implementation of Health and Safety Provisions

We have already point out the employer's administrative liability according to ASchG. In order to prevent possible dangers to employees' occupational health and safety, other practical sanctions can be imposed on the employer by official notice. According to § 96 ASchG the competent administrative authority can prohibit further employment at the work site. In addition, it has the authority to impose adequate interim measures of protection, e.g., complete or partial close-down of the work site. Possible appeals do not dispose of a suspension effect. The official notice expires after one year, unless it is limited to a shorter period (§ 96/3).

Apart from sanctions imposed by the office of district administration, according to § 10/3 ArbIG the labour inspectorate can also issue official decrees in case of imminent danger for the employee's health and safety.[117] In addition, § 10/4 ArbIG stipulates labour inspection's right to issue immediate measures without even enacting an official notice. If the employer does not follow the labour inspection's instruction or if s/he resists the imposed measures, labour inspection is allowed to enlist the public security agency's assistance.

The non-compliance with health and safety measures can also lead to the employee's premature resignation from the employment relationship. To a certain degree, this termination of the contract also constitutes a sanction for the employer, given that the employee has the right to a severance pay.

In case of danger for the employees' life, health or 'morality', the employee is allowed to refuse work until the employer has complied with his or her obligations. Indispensable prerequisite for the employee's right to refuse work is that the continued work performance is unacceptable. The right of the employee to refuse work is independent from any fault of the employer. According to § 8 AVRAG the employee must not suffer from any consequence due to his or her refusal to work (except for cases where the employee acts by gross negligence). If, however, the employer wants to terminate the employment relationship prematurely due to the employee's refusal to work, the employee has the right to appeal against the dismissal at court within one week after the dismissal has reached him or her.[118] In case of further non-compliance with health and safety provisions which leads to the impossibility to work without endangering the employee's health and safety, the employee has the right to resign prematurely for just cause due to the employer's breach of contract.[119]

The endangering of employees' health is a physical and/or mental ground of premature resignation from the employment relationship according to § 26/1 AngG[120]

117. Eva-Elisabeth Szymanski & Helga Oberhauser & Alexandra Marx, *Kommentar Arbeitnehmerinnenschutzgesetz* (Wien/Graz: NWV, 2004), 230.
118. Reinhard Resch, 'Zur Rechtsstellung der Arbeitnehmer nach dem ArbeitnehmerInnenschutzgesetz', in *Rechtsfragen des technischen Arbeitnehmerschutzes*, ed. Theodor Tomandl (Wien: Braumüller, 1999), 22.
119. Günther Löschnigg, *Arbeitsrecht*, 11th ed. (Wien: ÖGB, 2011), 392.
120. OGH 17/12/1997, 9 ObA 196/97g DRdA 1998, 359 = RdW 1998, 638.

(for white collars) or § 82 GewO 1859 (for blue collars).[121] Unique prerequisite that has to be fulfilled is the imminent danger for the employee's health in case of further employment. At the time of resignation, a third party has to be able to determine actual danger for the employee's health.[122] Furthermore, information of the employer by the employee about the danger for the employee's health is mandatory. If the employee fails to inform the employer, the resignation will not be justified.

Apart from the possibility to resign prematurely according to § 26/1 AngG, employees can also resign from the employment relationship according to § 26/3 AngG. If the employer fails to implement health and safety provisions, i.e., s/he fails to comply with his or her duty of care for the employee, the latter can resign prematurely for just cause due to inacceptable continuation of the employment (relationship). Yet, indispensable prerequisite is that the employer fails to comply with the employee's 'request' to implement the health and safety-obligations within an adequate period of time.[123] According to prevailing case law,[124] only if the employer actually fails to comply with the request or if s/he declares from the very beginning on that s/he will not comply with these provisions, can the employee be sure about the employer's refusal to comply with his or her duty of care.

After handing in a justified resignation, the employee will generally be entitled to all those claims s/he would be entitled to if the employment relationship had been terminated by the employer. In case of premature resignation due to the employer's fault, the employee is entitled to file a claim for payment during the notice period according to § 29 AngG, § 84 GewO 1859 and § 1162b ABGB. Furthermore, the employee is entitled to compensation for further damages that result from the justified premature resignation.[125]

[D] Workplace Health Promotion

In Austria, workplace health promotion is a rather recent phenomenon. In 1994 only, when adopting the actual ASchG, the legislator introduced provisions to the Act (§§ 81/1, 88/2) that enable health and safety specialists to enforce occupational health promotion and to combine it with traditional health and safety measures. In fact, occupational health promotion is carried out by these health and safety specialists most of the time. Yet, since 2005, efforts have been made in order to interconnect measures stipulated in the ASchG and occupational health promotion in form of workplace health management.[126] In 1998, the Act on occupational health promotion

121. Michael Friedrich, '§ 26-13' in *Kommentar AngG*, eds Franz Marhold & Gerald Burgstaller & Helmut Preyer (Wien: Manz, 2005).
122. Franz Marhold & Michael Friedrich, *Österreichisches Arbeitsrecht*, 2nd ed. (Wien/New York: Springer, 2012), 351.
123. Michael Friedrich, '§ 26-132' in *Kommentar AngG*, eds. Franz Marhold & Gerald Burgstaller & Helmut Preyer (Wien: Manz, 2005).
124. OGH 08/05/1956, 4 Ob 15/56 Arb 6436.
125. Franz Marhold & Michael Friedrich, *Österreichisches Arbeitsrecht*, 2nd ed. (Wien/New York: Springer, 2012), 362.
126. Walter Nöstlinger, *Handbuch Arbeitnehmerschutz* (Wien: Manz, 2006), 298.

(*Gesundheitsförderungsgesetz*-GfG)[127] was adopted and in 2006, the implementation of the measures provided in this Act was delegated to *Gesundheit Österreich GmbH* (Health Austria Company Ltd.).

With respect to traditional health and safety measures provided in the ASchG, only those occupational health promotion-measures can be classified as health and safety measures which are aimed at improving concrete working conditions that vary from company to company. As a result and also due to the fact that these measures are carried out by occupational health and safety specialists according to the aforementioned provisions of the ASchG, they are not subject to the provisions of the GfG. With regard to the occupational health and safety specialists' tasks, only those occupational health promotion-measures which are related to the improvement of concrete working conditions can be included in their compulsory minimum operating time.

§2.06 CONCLUSIONS

The aim of health and safety provisions is to protect employees' most important objects of legal protection, i.e., life and limb, health and property. Due to the duty of care principle, which arises from the individual contract of employment, the employer is the main responsible for health and safety at work. It is his or her duty to provide a safe workplace and to ensure the implementation of occupational health and safety. However, there is the possibility to appoint liable authorized representatives, i.e., to delegate the main responsibility for the employees' health and safety to these representatives. In case of a legally effective appointment, the liable authorized representative will be held liable for not having taken appropriate and sufficient measures according to the provisions of the ASchG. The employer him or herself, however, will still be held liable for wilful inducement or non-prevention of infringement. Apart from its obligation to verify the employer's adherence to health and safety provisions, the Labour Inspectorate informs and advises in all matters of occupational health and safety, it mediates between conflicting parties and investigates occupational accidents and matters concerning complaints about a nuisance or abuse. Furthermore, labour inspectorates are allowed to issue special certificates of exemption and are involved in lectures, training and discussions concerning vocational and further training of those responsible for protection at work. The nineteen regional labour inspectorates' expenses amounted to approximately EUR 26,75 million in 2009.

BIBLIOGRAPHY

Floretta, Hans, Karl Spielbüchler & Rudolf Strasser. *Arbeitsrecht I*. 4th edn. Wien: Manz, 1998.

Friedrich, Michael '§ 26'. In *Kommentar AngG*, edited by Franz Marhold, Gerald Burgstaller, & Helmut Preyer, Wien: Manz, 2005.

127. GfG, BGBl I 51/1998.

Gerhartl, Andreas. 'Einschränkung der Schadenersatzpflicht bei Arbeitsunfällen'. *Zeitschrift für Arbeits- und Sozialrecht* (2009): 252.

Grillberger, Konrad. *Österreichisches Sozialrecht*. 8th edn. Wien/New York: Springer, 2010.

Kolodej, Christa & Majoros, Thomas. 'Mobbing und die Fürsorgepflicht des/der Arbeitgebers/in'. *Das Recht der Arbeit* (2010): 157.

Koziol, Helmut & Welser Rudolf. *Bürgerliches Recht II*. 13th edn. Wien: Manz, 2006.

Leymann, Heinz. 'Mobbing and Psychological Terror at Workplaces'. *Violence and Victims* 5, 119-126 (1990).

Löschnigg, Günther. *Arbeitsrecht*. 11th edn. Wien: ÖGB, 2011.

Marhold, Franz & Friedrich, Michael. *Österreichisches Arbeitsrecht*. 2nd edn. Wien/New York: Springer, 2012.

Marhold, Franz. 'Quale tutela per il danno alla persona del lavoratore nell'ordinamento austriaco? Tutele civilistiche, lavoristiche e di sicurezza sociale'. In *Öffnung und Wandel – Die internazionale Dimension des rechts II – FS für Willibald Posch*, edited by Borić, Tomislav et al. Wien: LexisNexis ARD Orac, 2011, 487.

Mayer, Heinz. 'Der Arbeitnehmerschutz im öffentlichen Dienst'. In *Rechtsfragen des technischen Arbeitnehmerschutzes*, edited by Tomandl, Theodor. Wien: Braumüller, 1999.

Meissel, Franz-Stefan & Dehn, Wilma '§ 1489'. In *Kommentar ABGB* 3rd edn. edited by Koziol, Helmut & Bydlinski, Peter & Bollenberger, Raimund. Wien/New York: Springer, 2010.

Melzer-Azodanloo, Nora. *Tele-Arbeitsrecht*. Wien: Linde, 2001.

Nöstlinger, Walter. *Handbuch Arbeitnehmerschutz*. Wien: Manz, 2006.

Pöltner, Walter 'Die Integritätsabgeltung in der gesetzlichen Unfallversicherung'. *Das Recht der Arbeit* (1990): 152.

Rauch, Thomas 'Die Rolle des Arbeitgebers bei der Errichtung eines Betriebsrats'. *Arbeits-und Sozialrechtskartei* (2010): 318.

Reischauer, Rudolf 'Neuerungen im Bereich des Arbeitgeber-Haftungsprivilegs im Zusammenhang mit Kfz-Verkehr und Integritätsabgeltung'. *Das Recht der Arbeit* (1992): 317.

Reissner, Gert-Peter. *Lern- und Übungsbuch Arbeitsrecht*. 4th edn. Wien: Manz, 2011.

Resch, Reinhard 'Zur Rechtsstellung der Arbeitnehmer nach dem ArbeitnehmerInnenschutzgesetz'. In *Rechtsfragen des technischen Arbeitnehmerschutzes*, edited by Tomandl, Theodor. Wien: Braumüller, 1999.

Resch, Reinhard. *Sozialrecht* 5th edn. Wien: Manz, 2012.

Schnorr, Gerhard. 'Probleme des allgemeinen sozialen Kontaktes zwischen Arbeitnehmern eines Betriebes'. In *Innerbetriebliche Konflikte aus rechtlicher Sicht*, edited by Tomandl, Theodor. Wien: Braumüller, 1977.

Schrammel, Walter. *Arbeitsrecht II*, 6th edn. Wien: Braumüller, 2008.

Stärker, Lukas. 'Die Evaluierungspflicht des Arbeitgebers'. In *Rechtsfragen des technischen Arbeitnehmerschutzes*, edited by Tomandl, Theodor. Wien: Braumüller, 1999.

Szymanski, Eva-Elisabeth & Oberhauser, Helga & Marx, Alexandra. *Kommentar Arbeitnehmerinnenschutzgesetz*. Wien/Graz: NWV, 2004.

Thienel, Rudolf & Schulev-Steindl, Eva. *Verwaltungsverfahrensrecht*. 5th edn. Wien: Verlag Österreich, 2009.

Tomandl, Theodor. *Grundriss des österreichischen Sozialrechts*. 6th edn. Wien: Manz, 2009.

Winkler, Gottfried. 'Europa- und verfassungsrechtliche Aspekte des ArbeitnehmerInnenschutzgesetzes'. In *Rechtsfragen des technischen Arbeitnehmerschutzes*, edited by Tomandl, Theodor. Wien: Braumüller, 1999.

CHAPTER 3

Well-Being at Work in Belgium: A Matter of Co-operation

F. Dorssemont, K. Naert and A. Van Regenmortel

§3.01 INTRODUCTION

The Belgian regulations regarding health and safety have a long history. The first norms with regard to safety and hygiene were established in the regulations concerning dangerous, unhealthy or unsuitable workplaces. The first impulse was the Imperial Decree of 15 October 1810[1] which gave the administrative authorities permission to set up *unhealthy* workplaces, to monitor these and to take all the necessary measures with regard to *public* health and safety. The control and supervision of these workplaces was realized by the establishment of the Labour Inspectorate under the Act of 5 May 1888.[2] This regulation breathed the spirit of the French Revolution with the freedom of labour and industry as its most important *leitmotiv*. There certainly was no such thing as a social protection concern yet. The employee was not protected because his or her special qualification required it but as a member of the population who just like any other citizen had to be protected against the risks of industrial production.[3]

The Act of 2 July 1899[4] was the first general employment safety act with a clearly protective character. The focus shifted to the 'salubrity' of the workplaces and the safety of the workers within the enterprise itself. Characteristic of this act was the fact that it delegated a broad mandate to the Government to take all appropriate measures

1. Decree of 15 Oct. 1810 regarding the establishments spreading unhealthy or objectionable odours, *Pasin.* 1788–1814, 179.
2. Law of 5 May 1888 regarding the control of dangerous, unhealthy or unsuitable establishments and the steam engines and steam boilers, *Belgian Official Journal*, 13 May 1888.
3. For a more detailed analysis, see especially: M. Sojcher-Rousselle, *Droit de la sécurité et de la santé de l'homme au travail* (Bruylant 1979), 54–55.
4. Law of 2 Jul. 1899 regarding the health and safety of workers employed in industrial and trade companies, *Belgisch Staatsblad* (hereinafter '*Belgian Official Journal*'), 1 Dec. 1899.

to guarantee healthy workplaces and employment and the safety of the workers. This led to a pre-eminently *regulatory* approach of the safety issue. There was an unshakeable belief in technically detailed rules promoted by a legalistic and positivistic approach. Each risk was considered to require instructions which clearly indicated how the employer was expected to behave and which were subject to criminal sanctions. These detailed rules were also considered necessary to fill the gap that had been created by the establishment of the first occupational accident insurance in 1903 which introduced the principle of civil immunity for employers, their agents and/or appointees (colleagues/employees). This had the effect that from now on a civil liability claim was only allowed in cases of intent and only for damage not covered by the occupational accident insurance. Thus the general safety obligations of the first Employment Contract Act of 1900 were largely invalidated. The former Article 11 – now Article 20, 2° – of the Employment Contract Act obliged the employer to 'make sure with due diligence that the work was done in decent circumstances regarding the health and safety of the workers'. Case law interpreted these obligations so broadly and flexibly that quite far-reaching safety duties were imposed on the employer. These dynamics were largely lost by the introduction of the occupational accident insurance. As a result of civil immunity, the civil courts no longer had to speak out on the contents and range of the employer's duty of safety in the case of damage. The general safety duties were replaced by a set of technical and detailed rules that henceforth were to be subject to criminal sanctions. As a matter of fact, this detailed regulation scheme developed at top speed and on a double legal basis, notably the system concerning dangerous, unhealthy and unsuitable establishments on the one hand and the Act of 2 July 1899 on the other. The result was rigid protection with outmoded and obscure texts.

Not until 1946 were attempts made to assemble all these regulatory provisions in an orderly way into one document, notably the General Rules on Employment Safety (ARAB/RGPT).[5] However, this attempt was not successful. These General Rules did not produce co-ordination, but at best nothing more than a compilation of the various implementing orders. The technically detailed regulation was maintained as a result of which there was an increasing risk of employers taking refuge to the strict observance of these rules in order to avoid the responsibility for organizing prevention. Not until the Prevention Decree of 1975 did a real change in approach occur, namely by abandoning the regulatory approach and opting for a more self-regulating system. From now on the employer was considered to be responsible for the organization of the safety policy within his or her enterprise by obliging him or her to 'take the indispensable material measures to protect the employees against the *demonstrable* risks inherent to the work'.[6]

Moreover, the employer was expected to adapt the work to the human being, with which the duty to consider ergonomics made its first appearance. Based on the conviction that a solid and dynamic prevention policy also requires competent experts,

5. Approved by the Regent Decrees of 11 Feb. 1946 (*Belgian Official Journal*, 3 April 1946) and 27 Sep. 1947 (*Belgian Official Journal*, 3 Oct. 1947).
6. Article 54 *quater* 2 ARAB/RGPT.

all employers were obliged to appoint safety managers whose expertise had to be confirmed by compulsory additional training. Only in the case of less than twenty employees this obligation did not apply, meaning that the employer him or herself could act as a safety manager. By putting the safety manager or one of his or her assistants in charge of the secretariat of the Committee for Prevention and Protection at Work (hereinafter 'the Committee') the expertise became also available within the Committee. To avoid the danger that the presence of experts would lead to an unresponsive attitude of the employer, the idea of a *hierarchical structure* was introduced. This did not only emphasize the necessity of an integrated safety policy but also provided a hierarchical liability. The employer was entitled to train the members of the hierarchical structure although this was limited to the development of a training programme. The concrete elaboration was passed on to the Committee.

In 1996 – partly due to the influence of the Directive 89/391/EEC of 12 June 1989[7] (hereinafter 'Framework Directive') – a restructuring operation of the Belgian safety regulation was launched. The former Safety Act of 10 June 1952 was replaced by the new Well-Being Act.[8] Similarly the replacement of ARAB/RGPT by the new Code on Well-Being at Work was proposed. This process of codification has not yet been completed although it has reached its ultimate stage so that very soon ARAB/RGPT will be a thing of the past.

In this contribution, we wish to outline the present Belgian framework on the basis of the following points of interest: (1) an analysis of the basic concepts regarding health and safety; (2) a survey of the various actors involved in well-being, with special attention for the particular situation of working with third parties; (3) the involvement of employees and/or their representatives and (4) the available sanctions in the case of violations regarding health and safety and. On this basis, we will try to point out the specificity of the Belgian safety regulations and to assess the impact of the Framework Directive in this field.

§3.02 BASIC CONCEPTS

For over forty years, the Act of 10 June 1952 regarding 'the workers' health and safety as well as the salubrity of work and workplaces' set the framework for employment protection in the Belgian companies. In 1996, this framework act was exchanged for a new basic Act 'concerning the well-being of workers in the performance of their work'. Belgium clearly distanced itself from the terms 'health and safety' as they had been used for a long time in Convention no. 155 of 1981 and later in the Framework Directive. The lively parliamentary debate accompanying this choice of words illustrates the disputed yet well-considered character of the option taken. Indeed, a 'broader' interpretation and a 'wider dimension' of health and safety had to be advocated at all costs. Only by using the new term of 'well-being' new fields – such as

7. Council Directive 89/391 of 12 Jun. 1989 on the introduction of measures to encourage improvements in the health and safety of workers at work (OJ 1989 L183, 1).
8. Law of 4 Aug. 1996 on the well-being of workers in the performance of their work, *Belgian Official Journal*, 18 Sep. 1996 (hereinafter 'Well-Being Act').

psychosocial pressure and the effects of the environment on the working conditions – and important developments in the more traditional fields – safety at work, occupational medicine and hygiene at work[9] – could be sufficiently emphasized. More than before the focus was on 'the quality of labour'. Moreover, the terms 'health and safety' held the danger of reflecting inadequately the typicality of the new measures – such as the use of general principles of prevention that take the level of technology and science into consideration as well as a stronger focus on the worker as a whole person.[10]

Whether or not these new points of emphasis would not have been just as possible by sticking to the terms 'health and safety' is a different question. To many people the concepts of health and safety already imply quality of work.[11] As demonstrated by the ruling of the European Court of Justice (ECJ),[12] European Union (EU) Law also advocates a broad interpretation of the terms 'health and safety'. Therefore, certain measures with regard to working time, psychological factors, employment, training in hygiene and safety, the protection of young workers and the protection of employee representatives against dismissal or any other infringement of their terms of employment can be considered as part of the 'working environment', meaning that it is not conceived as a definite norm, but rather as a reflection of social and technological developments.[13]

So it was to be expected that in Belgium too the new term 'well-being' would not be welcomed with enthusiasm by everyone. Through several amendments the return to the old wordings 'health and safety'[14] was advocated – sometimes with the addition of 'well-being'[15] – in order to be more in line with the international and European repertory of concepts. Furthermore the concept of 'well-being' was considered to be too broad. Indeed, well-being is understood to mean 'the totality of factors regarding the circumstances in which work is done' which – initially as an example – refers to the following areas: (1) safety at work; (2) protection of the employee's health at work; (3) psychosocial pressure caused by work, in particular violence, bullying and sexual harassment at work; (4) ergonomics; (5) hygiene; (6) decoration of the workplace and (7) the undertaking's measures with regard to the effects of the environment on the aforementioned fields. Not only is this definition too vague, it does not contain any

9. Precisely these three disciplines underwent a clear evolution as a result of which ergonomics, the science of adapting work to the human being, is a point of special interest. The term 'industrial medicine' was exchanged for the description 'protection of the employee's health' so as to emphasize the growing attention for preventive measures (Explanatory Memorandum to the Bill regarding the well-being of employees in the execution of their work. *Parliamentary Documents*, Chamber of Deputies, Extraordinary Session 1995, nos. 71/1, 17–18.)
10. *Parliamentary Documents*, Chamber of Deputies, Extraordinary Session 1996, nos. 71/1, 40 and 71.
11. W. Donceel, *Praktijkgids bij de Welzijnswet* (Intersentia Rechtswetenschappen 1999), 22–23.
12. ECJ 12 Nov. 1996, C-84/94, *United Kingdom of Great Britain and Northern Ireland v. the Council of the European Union* [1996] ECR I-5755, with the opinion of Adv. Gen. P. Léger.
13. P. Léger, opinion at the ECJ 12 Nov. 1996, C-84/94, *United Kingdom of Great Britain and Northern Ireland v. Council of the European Union* [1996] ECR I-5766–5767.
14. Amendment 38 by Bacquelaine, *Parliamentary Documents*, Chamber of Deputies, Extraordinary Session 1995, no. 71/4; Amendment 60 by Wauters and Detienne, *Parliamentary Documents*, Chamber of Deputies, Extraordinary Session 1995, no. 71/5.
15. Amendment 1 by Nelis-Vanliedekerke and De Coene, *Parliamentary Documents*, Senate, no. 1-318/2.

objectives either.[16] This criticism was only partly met. The definition of the concept of 'well-being' was maintained but for the sake of legal certainty, and partly due to the National Labour Council's recommendation,[17] a clear exhaustive list was opted for. Whether this brought about the desired legal certainty is doubtful. Although the various fields are listed exhaustively, they are still defined rather broadly without any substantive limitation.[18]

But even apart from the Belgian context one could ask whether the choice for the concept of well-being is a good choice. The concept of well-being refers to *feeling well physically and mentally* and therefore contains a subjective dimension which is hard to express legally.[19] However, the term does indicate clearly that in an employment relationship as well a respectful treatment of the worker as a person is required. Thus the concept of 'well-being' bears close resemblance to the concept of 'dignity' given the fact that both concepts do not approach the employee as a worker but rather as a human being.[20] That way the term 'well-being' is more in keeping with a fundamental rights approach of the employment relation in which the employee is recognized as a citizen and a bearer of fundamental rights – *in casu* the right to 'decent' treatment. Incidentally, Belgium has been making the link between 'well-being' and 'dignity' for quite some time. A case in point is the Belgian social assistance system which guarantees all citizens a minimum protection. The ultimate safety net in this context is that each person is entitled to social services whose aim it is to enable everyone to live a life of *human dignity*.[21]

It is no coincidence that the institutions whose task it is to guarantee this right are called 'Public Centres for Social *Well-Being*'. It is precisely this service to those in need that is so characteristic of the welfare services. Accordingly, well-being as a separate branch of law is more associated with the law concerning the functioning of welfare services. One could hold that with the new Well-Being Act the link between 'well-being' and 'dignity' is extended to the area of labour relations for the first time. However, the fact that initially social well-being law did in no way govern the labour relations, could possibly explain why the term 'well-being at work' is not easily accepted and lacks the necessary visibility. Actually the legislators themselves could contribute to a better understanding by using these terms more consistently. For instance, although the term 'well-being' has been chosen deliberately, the old terminology of 'health and safety' is reverted to[22] in the description of the employees' obligations – perhaps in servile imitation of the Framework Directive. Likewise, the new terminology was 'forgotten' in the recently introduced Social Criminal Code[23] just

16. Amendment 38 by Bacquelaine, *Parliamentary Documents*, Chamber of Deputies, Extraordinary Session 1995, no. 71/4.
17. National Labour Council, recommendation 1.110 of 1 Dec. 1994 concerning the preliminary draft law on the well-being of employees in the execution of their work, 4.
18. W. Donceel, *Praktijkgids bij de Welzijnswet* (Intersentia Rechtswetenschappen 1999), 27–28.
19. For a more in-depth analysis of the idea of 'well-being at work', see F. Héas, *Le bien-être au travail*, JCP S (2010): 14–15.
20. F. Héas, *Le bien-être au travail*, JCP S (2010): 17.
21. Article 1 Organic Law of 8 Jul. 1976 concerning the Public Centres for Social Well-Being.
22. Article 6 Well-Being Act.
23. Law of 6 Jun. 2010 establishing the Social Criminal Code.

as (certain) breaches of the Well-Being Act were lumped together[24] under 'Health and Safety at Work'. Moreover, no assessment was made as to the necessity of a terminological adaptation of the obligations referring to 'safety obligations' that existed prior to the Well-Being Act. With a few exceptions[25] not a single change was made. Article 20, 2° of the Belgian Employment Contract Act for instance still obliges the employer to see to it with due diligence that the work is done in decent circumstances with regard to 'the health and safety' of his or her employees.

The above analysis clearly demonstrates that an actual definition of the concept of well-being itself is lacking. Well-being is more formulated as an objective to be pursued by the employer through measures in the seven previously listed fields, the totality of these working conditions constituting the employees' well-being. But two other crucial concepts, notably 'risk' and 'prevention' are not described more specifically in the Well-Being Act either. Yet the multidisciplinary approach of prevention, which is the basis of the Belgian well-being regulations, does focus on the concept of 'risk'. In line with the Framework Directive, the Belgian legislator[26] obliges the employer to pursue a policy of prevention, in which a number of general principles of prevention should be taken into account, such as preventing risks, assessment of risks that cannot be avoided, risk control at the source... The term 'risk' is used several times without there being a general definition of this concept in the framework act. This was not considered expedient for two reasons. First of all, it is said that both at the national and European level the ambiguities in this respect have not yet been clarified and that because of this it was not considered appropriate to run ahead of things.[27] In addition, the concept is also said to differ too strongly per industry. This also explains why the various implementing orders concerning the concrete fields of risk do include such a description.[28]

For the definition of the concept of 'prevention' we are also referred to the implementing orders. In the Royal Decree of 27 March 1998[29] which provides for a further elaboration of the principle of 'handling prevention structurally and systematically' prevention is understood to mean: 'the whole of provisions or measures taken or laid down at all stages of the activity of the enterprise or institution, and at all levels, in order to prevent or reduce occupational hazards.' Except for the specification that the measures of prevention need to be taken at 'all levels', this is a literal rendering of the concept of 'prevention' in the Framework Directive.[30] In doing so, the concept of

24. Articles 123 through 133 Law of 6 Jun. 2010.
25. See for instance Art. 46 § 1, 7° Occupational Accidents Act where the reference to 'legislative and statutory provisions regarding safety and hygiene at work' – albeit only in 2004 – was changed to 'legislative and statutory obligations regarding the well-being of employees in the execution of their work'.
26. Article 5 Well-Being Act.
27. *Parliamentary Documents*, Senate, Z. 1995–96, no. I-318/3.
28. For instance, Article 2, 6° Royal Decree of 28 May 2003 concerning the health supervision of employees (*Belgian Official Journal*, 16 Jun. 2003) which with a view to the application of this Royal Decree describes the concept of risk as 'the probability that the possible damage to a job or activity occurs, in circumstances of use or exposure, as this job or activity is carried out'.
29. Article 2, 5 ° Royal Decree of 27 Mar. 1998 concerning the policy with regard to the well-being of employees in the execution of their work, *Belgian Official Journal*, 31 Mar. 1998.
30. Article 3(d) Framework Directive 89/391 of 12 Jun. 1989.

prevention is given an instrumental meaning. Prevention is not an aim as such but 'a set of measures' with the aim of 'preventing and reducing risks'. In line with this, Article 9 of the implementing order obliges the employer, on the basis of a risk analysis, to establish three levels of prevention measures in the following order: (a) prevention measures aimed at avoiding risks; (b) prevention measures aimed at preventing damage and (c) prevention measures aimed at limiting the damage. In doing so, measures 'for the prevention and restriction of damage' are also regarded as prevention measures which is hard to reconcile with the earlier narrower description in Article 2, 5° which only considers measures 'to prevent and reduce risks' as prevention. A more consistent attitude with regard to the repertory of concepts used would be welcome here too.

§3.03 PERSONAL SCOPE AND DUTIES OF THE PARTIES INVOLVED

[A] The Five Main Actors

An important part of the Belgian well-being legislation is based upon the rules determining the relations between the different well-being actors. The five main actors involved in the well-being policy are: the employer [1], the internal services for prevention and protection at work [2], the (members of the) hierarchical structure [3], the workers [4] and the Committee for Prevention and Protection at Work [5].

[1] The Employer

The employer has the obligation to pursue a well-being policy and will always be the prime person liable for the well-being in the workplace.[31] Since the employer has the authority in the workplace, s/he also remains the person ultimately liable even if s/he calls upon experts, which is in accordance with Article 5, 2 of the Framework Directive.

According to Article 5 § 1 of the Well-Being Act, employers have to take the necessary measures to promote the well-being of the workers at work. For this purpose, they have to apply twelve general principles of prevention.[32] These principles are mainly the same principles as those mentioned in Article 6 of the Framework Directive. However, here the employer's duty to provide information is defined more explicitly than in the Framework Directive. According to the Well-Being Act, the employer has

31. Article 15 Royal Decree of 27 Mar. 1998 on the policy of well-being of the workers at work, *Belgian Official Journal*, 31 Mar. 1998.
32. The Well-Being Act instructs the employer to take care of the well-being of his or her employees in less precise terms than the European Framework Directive. Article 5 of the Framework Directive states that the employer shall have a duty to ensure the health and safety of workers in every aspect related to the work which can be considered as an obligation to guarantee a certain result (L. Vogel, *Droit de la santé au travail* (Université catholique de Louvain 2008), 203), whereas Article 5 of the Well-Being Act stipulates that the employer shall take the necessary measures to promote the well-being of the workers at work, which can be considered as an obligation to perform to the best of one's abilities. However, this contradiction between the Framework Directive and the Well-Being Act was never brought before the Court of Justice.

the duty to inform the worker about the nature of his or her work, the associated residual risks and the measures aimed at preventing or limiting these hazards at the commencement of the employment and whenever necessary for protecting their well-being; s/he must give appropriate instructions to the workers and establish guidelines to guarantee compliance with these instructions; and s/he has to provide or ascertain the existence of appropriate health and safety signs at work whenever risks cannot be avoided or adequately limited by the general technical protective measures or by measures, methods or practices regarding work organization.

When defining the well-being measures the employer should not limit him or herself to technical measures. S/he also needs to consider the economic, ethical and social feasibility of the prevention measures.[33] The different interests that may exist, must always be taken into account as well. One could speak about a general duty of care which has to be interpreted in a reasonable manner.

According to Article 5 § 2 of the Well-Being Act, the employer has to determine the means and manner in which the policy regarding the well-being of the workers at work can be pursued and the competence and liability of the persons in charge of the implementation of that policy. Authority and responsibility go hand in hand; the one who has the authority to purchase is also liable for making the right purchase.

Moreover, the employer has to implement a dynamic well-being policy which will be adapted in the light of past experience and the development of the working methods or the working conditions.

Considering the fact that science and technology are evolving constantly, Article 5 of the Well-Being Act states that the general prevention principles could be defined and worked out in more detail by a Royal Decree.[34] This involves more specifically general prevention principles laid down in specific European directives such as, for example, the directives on working with carcinogenic and biological agents.

[2] The Internal Services for Prevention and Protection at Work

[a] Obligation for Each Company to Have Atleast One Prevention Service

Every employer needs to have an internal service for prevention and protection at work.[35] Therefore, every employer needs to have at least one prevention advisor.[36]

The compulsory training of the internal prevention advisor, the tasks that the company has to carry out, the tasks which may be delegated and, in some cases, have to be delegated to an external service are based on the division of the companies into four different groups. This division is linked to the kind of risks and the number of workers.

33. Explanatory Memorandum to the Bill concerning the well-being of workers during the performance of their work, *Parliamentary Documents*, Chamber of Deputies, Extraordinary Session 1995, nos. 71/1, 9.
34. Article 5 § 3 Well-Being Act.
35. Article 33 § 1 Well-Being Act.
36. This obligation is copied from Article 7 Framework Directive.

Group A of employers consists of all employers that have more than 1000 workers on their pay roll. Depending on the kinds of risks this number is reduced to 500, 200 and 50 respectively. Group B consists of all employers that have between 200 and 1000 workers on their payroll. Depending on the sector and the kinds of risks these numbers can be reduced to 20 and 50. Group C consists of employers who have less than 200 workers on their payroll and who are not classified in Group A or B. Group D consists of those employers who, irrespective of the sector, have less than 20 workers.[37] In this group, the employer can take up the function and liability[38] of prevention advisor himself.[39,40]

[b] Legal Status of the Internal Prevention Advisor

Appointment and dismissal:[41] The employer has to appoint the internal prevention advisor amongst his or her own workers; if s/he calls upon an external person,[42] this person has to be put on the company's payroll, i.e., the employer is required to conclude an employment contract (private sector) with this person, even if this is only a part-time contract. The employer also has the authority to replace the prevention advisor, to appoint another one temporarily and to release the prevention advisor from his or her duty. However, the employer is only allowed to carry out these actions with the agreement of the Committee for Prevention and Protection at Work. In case no agreement can be obtained, the competent labour inspector has to be asked for advice. After listening to the points of view of both parties the labour inspector will have to reconcile these. The employer is only allowed to take a final decision after s/he has submitted the advice of the inspector to the Committee for Prevention and Protection at Work for the private sector and the competent consultation committee for the public sector.

In case the employer wants to end the employment contract with the prevention advisor or wants to give him or her another function for reasons not related to the prevention advisor's independence or for reasons related to his or her incompetence to execute the prescribed tasks, s/he has to follow the specific procedures described in the

37. Article 3 Royal Decree of 27 Mar. 1998 concerning Internal Services for prevention and protection at work, *Belgian Official Journal*, 31 Mar. 1998 (hereinafter 'Royal Decree') on the internal services.
38. Belgian legislation uses the term function; the Framework Directive on the other hand uses the term responsibility.
39. Article 7, 7 of the Framework Directive stipulates that Member States may define, in the light of the nature of the activities and size of the undertakings, the categories of undertakings in which the employer, provided he is competent, may himself take responsibility for the measures referred to in para. 1. Belgian legislation only takes into account the criterion of the number of people on the payroll and does not specify any requirements about the activities of the companies in which the employer is allowed to take the responsibility of prevention advisor. The requirement that the employer can take up the responsibility of prevention advisor if he is competent is not mentioned in Belgian legislation.
40. Article 33 Well-Being Act.
41. Chapter 4 Royal Decree on the internal services.
42. For instance, a person who works mainly for an external prevention service.

Act of 20 December 2002 concerning the protection of prevention advisors.[43] For instance, a private sector employer who wants to dismiss a prevention advisor without the consent of the Committee has to bring this matter before the labour tribunal.[44]

The law provides for a compensation for prevention advisors who are unlawfully dismissed. This compensation equals a normal two year salary if the prevention advisor has had this position for less than fifteen years and a three year salary if the prevention advisor has held this function for over fifteen years. For those prevention advisors who combine this position with another job, the compensation is calculated proportionally.

The protection against unlawful dismissal is not valid in the case of dismissal for compelling reasons, closure of the company or a collective dismissal of part of the workers, or when the prevention advisor is still in his or her probationary period or resigns him or herself.

Independence: Prevention advisors must fulfil their assignment in total independence with regard to the employer and workers[45] and they may not experience any disadvantages as a consequence of their activities as prevention advisors.[46] At the request of the prevention advisor, the employer or the workers, the differences of opinion on this actual independence will be submitted for advice to the civil servant in charge of inspection.[47]

The prevention advisor responsible for the management of the service is directly answerable to the person in charge of the daily management of the undertaking or institution and must have direct access to the person or persons in charge of the daily management of the technical operational unit or units.[48]

Training: The basic competences which all prevention advisors must have are specified in Article 21 of the Royal Decree on the internal services. They must have sufficient knowledge of the legislation regarding the well-being of the workers at work that is applicable to the undertaking or institution where they fulfil their assignments and have the necessary technical and scientific knowledge to be able to fulfil their activities. More specifically, this knowledge concerns the techniques relating to risk analysis; the co-ordination of the prevention activities in the internal service; the co-ordination between the internal and external services and with the employers and workers of the external undertakings; the measures regarding hygiene in the workplaces; the organization of the first aid and emergency care of victims of an accident or a sudden disease and the measures to be taken in the case of serious and imminent danger; the specific assignments of the prevention advisors referred to in the Royal Decree of 3 May 1999 on the assignments and operation of the Committees for Prevention and Protection at Work and the reporting.

---

43. Law of 20 Dec. 2002 concerning the protection of prevention advisors, *Belgian Official Journal*, 20 Jan. 2003, first edition.
44. Article 7 § 1 and § 2 Law of 20 Dec. 2002 on the protection of the prevention advisors, *Belgian Official Journal*, 20 Jan. 2003.
45. Article 25 Royal Decree on the internal services.
46. Article 43 Well-Being Act.
47. Article 25 Royal Decree on the internal services.
48. Article 16 Royal Decree on the internal services.

Depending on the category of employer the prevention advisor must have had additional training. A prevention advisor working for an employer of category A for instance has to take a Level 1 training programme which includes 400 hours of courses.

Logistics and Time use:[49] After consulting with the Committee, the employer determines the composition of the internal service, the technical and scientific resources, the rooms, the administrative staff attached to the internal service and the minimum time the prevention advisors need to spend on their work in order for the internal service to fulfil its assignments completely and efficiently at all times.

[c] Tasks of the Internal Prevention Services

The job of prevention advisor is one of the best described jobs of all Belgian professions. His or her or her duties are mainly although not solely, mentioned in the Articles related to the tasks of the internal prevention departments.[50] A fully staffed prevention department has the necessary know-how as well as prevention advisors specialized in the following five well-being domains: work safety; occupational medicine; ergonomics; operational hygiene and the psychosocial aspects of work. Only when these five expertise domains are available in the company itself, the company does not have to work with external prevention advisors (i.e., contract an external prevention services).

The assignments and tasks of the internal prevention services can be subdivided in four categories: assignments related to the dynamic risk management system, assignments that must be carried out by specialists in occupational medicine,[51] a set of tasks[52] that, in order to fulfil the abovementioned assignments, constitutes a minimum requirement and a set of tasks at any rate to be carried out by the section responsible for medical surveillance.

Assignments related to the dynamic risk management system:[53] Among the most important tasks related to the dynamic risk management system are: collaboration with regard to the determination of dangers; advice on the results following from the risk assessments; proposing measures with a view to having a permanent risk analysis available; advice on and formulation of propositions for the draft, implementation and guidance of the overall prevention and yearly action plans, participating in the study of the factors influencing the cause of accidents or incidents and the study of the causes that are crucial in any accident leading to occupational disability; and participation in analysing the causes of occupational diseases. It is clear that learning from incidents and accidents is one of the cornerstones of the dynamic risk management system.

---

49. Article 17 Royal Decree on the internal services.
50. Article 5-8 Royal Decree on the internal services.
51. Prevention advisors who meet the requirements provided in Article 22, 2° of the Royal Decree of 27 Mar. 1998 on the external services for prevention and protection at work.
52. Depending on the category of employers the employer belongs to, certain tasks always have to be carried out by the internal service and cannot be delegated to the external service for prevention and protection at work.
53. Article 5 Royal Decree on the internal services.

Assignments that have to be carried out by prevention advisors specialized in occupational medicine:[54] Prevention advisors specialized in occupational medicine have the duty to examine the interaction between the person and the job s/he performs, to guarantee health surveillance of workers, in order to, amongst others, detect work-related occupational diseases as early as possible and to supervise the organization of first aid and emergency care for workers who become victims of accidents or are afflicted by disease.

Set of tasks constituting a minimum requirement to fulfil the assignments:[55] Within the context of permanent risk analysis, and the drawing up and adjustment of the overall prevention plan and the annual action plan the prevention advisors must, *inter alia*, inspect the workplace frequently and systematically and at least once a year carry out a thorough inspection of the workplaces and the workstations. Within the context of the management of the service the prevention advisors must, amongst others, make monthly or three-monthly reports, an annual report, complete the occupational accident index card and take care of the secretarial work of the Committee.

Tasks to be carried out by the section responsible for medical supervision: These tasks include making sure that workers who have an accident or are afflicted by disease receive first aid and emergency care[56] and the reporting of occupational diseases. From the preceding it is clear that although the prevention advisors inform, advice and support the other actors, they are not liable for the implementation of the policy. There is only one exception to this principle: in the case of urgency and when it is impossible to call on the management, the prevention advisors have to take the necessary measures themselves to prevent the causes of the hazard or trouble.[57]

[3] The Hierarchical Structure

Persons such as foremen, supervisors, team leaders, heads of departments, managers who have the authority to give orders and assignments are all members of the hierarchical structure. So the hierarchical structure consists of all persons who have authority over one or more persons or a department.

Notwithstanding the fact that there is no legal definition of the term hierarchical structure,[58] it is of crucial importance for the prevention policy. For this reason the legislator has assigned a number of specific tasks to the hierarchical structure which, because of its specific position, is able to play a role in the making of the well-being policy, the planning, the implementation and auditing of the prevention measures.

As mentioned before, the employer is the first person liable for the well-being of his or her workers in the workplace. However, the hierarchal structure is responsible

54. Article 6 Royal Decree on the internal services.
55. Article 7 § 1 Royal Decree on the internal services.
56. Unless other medical services established in implementation of the Law of 10 April 1971 on occupational accidents are responsible for this.
57. Article 7 Royal Decree on the internal services.
58. The term 'hierarchical structure' is mentioned in the Royal Decree on the well-being policy.

for the execution of the company's well-being policy as defined by the employer. In other words, the hierarchical structure has to take care of the daily management of the well-being of the workers. This means that the hierarchical structure is liable for the execution of its part of the well-being policy and could also be held accountable for this.

The legislator has clarified this general duty with a number of specific assignments and tasks which are listed in Article 13 of the Royal Decree of 27 March 1998 on the policy of well-being of workers at work.[59]

According to this Royal Decree, the hierarchical structure has, amongst others, the duty to exercise effective supervision of work equipment, collective and personal protective equipment and substances and preparations used, for the purpose of determining irregularities and it has also the duty to take measures to put a stop to this. The hierarchical structure must obtain timely recommendation from the services for prevention and protection at work, monitor compliance with the instructions to be given in implementation of the legislation regarding the well-being of workers at work and ensure that the workers properly understand and put into practice the information they receive in implementation of the legislation regarding the well-being of workers at work.

As a result of its supervising role, the hierarchical structure cannot only control the effectiveness of the training and information that has been given to the workers but if necessary, it can also adjust and complement this information and training.

In order to be able to play its role as it is supposed to, the hierarchical structure itself should receive the necessary information and training. That way it will be able to give advice and feedback on the well-being policy, be aware of the risks and the measures which have to be taken, be able to recognize and evaluate the risks as well as to analyse accidents and near-accidents, and finally be able to give its staff members instructions and to control compliance with and effectiveness of these instructions.

[4] The Workers

The Well-Being Act is applicable to employers and workers. According to Article 2 of the Well-Being Act the following persons are equated with workers: persons who perform work under the authority of another person, other than in terms of an employment contract;[60] persons who follow a vocational training curriculum that provides a form of work which may or may not be performed within the educational institution; persons bound by an apprenticeship contract; trainees; pupils and students who follow an educational curriculum that provides for a form of work that is performed in an educational institution.

This concept of worker is broader than in EU Law. The element of remuneration is not a necessary element. However, there always has to be a kind of authority over the

59. Article 13 Royal Decree of 27 Mar. 1998 on the policy of well-being of workers at work, *Belgian Official Journal*, 31 Mar. 1998.
60. In this way, people working in the public sector, people who carry out labour during their imprisonment and even persons working for a local employment agency, except when they work as a domestic worker, fall within the scope of the regulations on well-being.

worker. Even in the case of voluntary work one has to distinguish between the cases where there is a hierarchical relation and the cases where no such relation exists. Whenever there is a hierarchical relation the voluntary worker is considered as a worker with regard to the obligations described in the legislation on well-being.

The workers are of crucial importance for the success of all actions concerning well-being. Article 6 of the Well-Being Act lists a number of explicit obligations for the workers. It is specifically mentioned that each worker, as far as his or her conduct in the workplace is concerned, must take good care to the best of his or her ability of his or her own health and safety and that of other persons concerned in accordance with the training and the instructions given by his or her employer.

As mentioned before, the employer and the hierarchical structure are liable for the policy on well-being and the execution of this policy. As indicated in what proceeds, this does not mean that the workers have no role to play. In order to make sure that the policy and all proposed actions are really effective, the workers, based on the training they have received, must make correct use of the working systems, dangerous substances, personal and protective equipment...[61] The workers are also of key importance for the correct feedback to the employer and their prevention advisor so that the well-being policy and actions carried out can be adapted in order to make them more effective.

The user of temporary workers is liable for developing and implementing a prevention policy towards them.[62] S/he must apply the provisions of the Well-Being Act and the decrees implementing the act to temporary workers. According to Article 12*ter* of the Well-Being Act every user is obliged to refuse the services of the temporary employment agency of which s/he may know that it does not meet the obligations towards temporary workers as imposed by this Act and by the Act of 24 July 1987 concerning temporary work, temporary employment and the provision of workers to users, and the implementing decrees of these acts.

A temporary employment agency is obliged to refuse to provide temporary workers to an employer-customer of whom it may know that, as regards his or her temporary workers, this user does not comply with the obligations imposed by this Act, the Act of 24 July 1987 and the decrees implementing these Acts.[63]

[5] The Committee for Prevention and Protection at Work

Social concertation concerning well-being at work takes place at three levels.[64] At the national level, there is the High Council for Prevention and Protection. At the sectoral level, collective labour agreements often contain elements related to the prevention and protection at work. At the level of the company, the Committee for Prevention and

61. Article 6 Well-Being Act.
62. Article 19 Law of 24 Jul. 1987 concerning temporary employment, temporary agency work and making employees available to customers, *Belgian Official Journal*, 20 Aug. 1987.
63. Article 12 *quater* of the Well-Being Act.
64. M. Heselmans & J. Van Peteghem, *Meer welzijn op het werk, een eigen visie* (Garant 2007), 83–84.

Protection at Work is the most important actor, and can be considered as the fifth main actor in the Belgian system on well-being at work. The roles, responsibilities and tasks of this Committee are further explained in paragraph 4.

[B] Specific Provisions with Regard to Contractors

Before the Act of 3 June 2007 related to different labour matters[65] only 'activities carried out in the establishment of the employer by workers of outside companies or by independent persons'[66] were mentioned, there was only minor attention for chains of contractors or subcontractors. The Act of 3 June 2007 introduced in Chapter IV of the Well-Being Act special provisions regarding work performed by outside undertakings or temporary workers, the concepts of 'contractor' and 'subcontractor' and specified the concept of 'establishment', which is described as a geographically defined location that is part of a company or institute and that falls under the liability of an employer who engages workers. A contractor is defined as an employer or an independent person from outside that carries out activities in the establishment of the employer, for his or her account or with his or her permission, in fulfilment of an agreement with that employer. A subcontractor is defined as an employer or independent person from outside who carries out activities in the establishment of the employer, on the basis of an agreement with a contractor. Subcontractors also include employers or independent persons from outside that carry out activities in the employer's establishment, on the basis of an agreement with a subcontractor.

Specific tasks are listed for those employers on whose premises contractors and, where relevant, subcontractors carry out activities. Those employers must provide their contractors with the relevant information the contractors need to give to their workers and subcontractors. They must take the proper measures specific to their establishment for the organization of the introduction of the workers of the contractors and subcontractors and they have the duty to check whether the workers of the contractor have received the relevant training and the instructions concerning the matters intrinsic to the business activity.

The employers on whose premises contractors and, where relevant, subcontractors carry out activities also have a co-ordination and co-operation duty. They must co-ordinate the actions of contractors and subcontractors and ensure their mutual co-operation when carrying out the measures concerning the well-being of the workers during the performance of their job. Finally, they must make sure that the contractors, fulfil the obligations that are applicable on their premises concerning the well-being of the workers.

The contractors and, where relevant, the subcontractors must fulfil, and have the subcontractors fulfil, their obligations concerning the well-being of the workers in the performance of their job that are in force on the premises where they are carrying out

65. *Belgian Official Journal*, 23 Jul. 2007.
66. Independent persons are only indirectly and partially 'covered' insofar as they carry out activities by means of a contract with the employer and on the premises of the employer (Article 10 Well-Being Act).

their activities; they must provide information to their workers and subcontractors, inform the employer where they are carrying out their activities about the risks that are inherent to their activities and comply with their co-ordination and co-operation obligations.

The abovementioned obligations are further specified in Article 9 § 2 of the Well-Being Act. This Article stipulates that the employer shall expel any contractor about whom s/he knows or ought to know that s/he does not observe the obligations aimed at the protection of the workers. S/he must enter into an agreement with each contractor concerning, in particular, the following stipulations: the contractor undertakes to fulfil his or her obligations concerning the well-being of the workers during the performance of their job which are relevant to the establishment where s/he performs his or her activities and to ensure that these are also fulfilled by his or her subcontractors; if the contractor fails to fulfil the aforementioned obligations, or fulfils them only partially, the employer in whose establishment the activities are carried out can, in the cases stipulated in the agreement, take the necessary measures him or herself at the expense of the contractor; the contractor who, for the performance of activities in the establishment of an employer, engages subcontractor(s), commits him or herself to stipulate in his or her agreements with those subcontractor(s) the conditions mentioned. This implies that, if the subcontractor does not fulfil the obligations mentioned, or fulfils them only partially, the contractor him or herself can take, in the cases determined in the agreement, the necessary measures at the expense of the subcontractor.

If the contractor does not take these measures or if s/he fulfils his or her obligations only in part, the employer, after serving the contractor notice of default, must him or herself immediately take the necessary measures concerning the well-being of the workers during the performance of their job specific to his or her establishment. The contractors and, where relevant the subcontractors in respect of their subcontractors have the same obligations as the employer has towards his or her contractors.

§3.04 WORKERS' INVOLVEMENT IN HEALTH AND SAFETY ISSUES: THE QUEST FOR PARTICIPATION

[A] Health and Safety, an Issue of Collective Labour Law

Health and safety are issues of *collective* labour law. Although Article 6.2 (h) Framework Directive suggests that 'work' needs to be adapted to the 'individual', the management of avoiding, evaluating and combating health and safety risks requires an employer 'to give collective protective measures priority over individual protective measures'[67] Furthermore, the Framework Directive also affects *industrial relations,* insofar as Articles 10 and 11 provide for various techniques of workers' involvement. These range from mere information to consultation and even 'participation rights'. The

67. See Article 6.2. Framework Directive, as implemented by Article 5 § 1(e) of the Well-Being Act.

entitlement of these rights is ambiguous from two points of view. First, the two Articles do not grant 'rights', but prescribe obligations. Second, the beneficiaries of both Articles are 'workers and/or their representatives'. Although it is completely valid that workers should be directly informed about risks and the related protective and preventive measures, it is astonishing that the European legislator suggests an indiscriminate choice between direct and indirect procedures of workers' involvement in the managerial policy to avoid, evaluate and combat those risks. In fact, it is astonishing to see that workers are mentioned before workers' representatives, contrary to what is for example the case in Article 6 of the Collective Redundancy Directive 1998/59/EEC.

[B] The Nature of Workers' Involvement in Health and Safety Issues

The language of the Framework Directive is puzzling. Although information and consultation rights are dominant and frequent types of workers' involvement in Belgian collective labour law, the notion of 'participation' is unprecedented.

The notion of participation is confusing since it usually tends to be associated with a kind of organic participation, i.e., the ability of workers' representatives to be integrated into corporative governance.[68] Contrary to information and consultation procedures which as a general rule do not affect the managerial prerogative, here the idea seems to be that workers' representatives participate in the adoption of a decision, i.e., acquire the status of a participant in a decision, e.g., a protective or preventive measure.

However, such a prima facie interpretation is at odds with the language of the Framework Directive, which relates participation to a discussion amounting to the adoption of such measures.

Under Belgian Law, information, consultation and participation rights have primarily been granted in an *indirect way*. Article 65 of the Well-Being Act assigns workers' involvement in health and safety issues to the Committees for Prevention and Protection at Work (*Comités pour la Prévention et la Protection au travail*). The Belgian legislator has thus opted for the establishment of two distinct bodies of elected workers' representatives at the level of the undertaking, based on the object of the workers' involvement.[69] In the absence of a Committee, the local shop stewards (*délégation syndicale*) will exercise the prerogatives of the Committee (Article 52). A direct involvement of workers is only provided for insofar as no Committee or *délegation syndicale* has been established (Article 53).

Contrary to the Framework Directive, the Belgian legislator only refers to 'participation' in relation to the absence of a workers' representation. In such a scenario, the individual workers 'directly' participate in the debate on issues of health

68. For example, Article 2(k) SE Directive 2001/86.
69. This picture needs to be mitigated if no works council at plant level has been established. In such a scenario which will be very common in undertakings with an average employment of less than 100 employees, the Committee will exercise a part of the prerogatives granted to the works councils, notably those related to information and consultation in *economic* matters.

and safety. In short, participation is not construed as a mechanism of workers' involvement. It is construed with a view to appointing an agent empowered to exercise a set of prerogatives. In the Royal Decree of 3 May 1999, the word 'participation' is used in another context. It relates to the involvement of the Committee in the application of the dynamic risk management system (Article 10).

The Committee has prerogatives which transcend the traditional ambit of information and consultation. In this respect, the nature of the prerogatives of the Committee are essentially different from those of the Belgian works council (*conseil d'entreprise*) which only needs to be informed and consulted in social, economic and financial matters. The overall image of the Belgian works councils established in undertakings with a workforce of at least 100 workers, is that of a passive body. They need to be informed and consulted, but have scarcely any margin of manoeuvre to take initiatives, let alone decisions. Article 65 of the *Loi relative au bien être des travailleurs* takes a different stance. It states that the Committee has a general competence to examine and to submit proposals related to means and to contribute actively to everything that is done to improve the 'well-being' of the workers. Thus, the Committee is actually apprehended as a pro-active agent involved in the employer's policy or strategy in the field of health and safety. The Royal Decree of 3 May 1999 confirms this impression, by highlighting the right of the Committees to give advice and to submit proposals.[70] In other words, the Committee is not only a body of workers' representatives which is informed and consulted when the employer is obliged to do so. It can also take initiatives in its own right and formulate proposals (Article 2).

The fact that the Committee is involved in setting out a policy or strategy is in sharp contrast with the *object* of information and consultation in the field of social, economic and financial issues. In fact, the works council (*conseil d'entreprise*) is not entitled to be informed, let alone consulted on the business strategy of the undertaking. The overall view is that it will be consulted in the case of managerial decisions adopted *as a result* of this strategy, which may substantially affect the workers' interests. The works councils also have to be informed about the financial and economic situation of the undertaking, which inevitably will be influenced by its business strategy. The business strategy as such is a blind spot in the field of workers' involvement under Belgian Law.[71]

Belgian Law also grants strong consultation rights to the Committees. This consultation relates specifically to all the measures the employer endeavours to take with regard to health and safety.[72] The Royal Decree stresses that this consultation must take place prior to the adoption of those measures, since the Committee needs to give a preliminary advice *(un avis préalable)*. In some rare cases, the Committee has

70. Article 2 of the Royal Decree of 3 May 1999, *Belgian Official Journal*, 10 Jul. 1999.
71. A recent judgment of the ECJ mitigates this state of affairs. In *Fujitsu Siemens*, the ECJ argued that the adoption of '*strategic* decisions or of changes in activities within a group of undertakings which compel the employer to contemplate or to plan for collective redundancies gives rise to an obligation on that employer to hold consultations with workers' representatives', ECJ 10 Sep. 2009, C-44/08.
72. Article 2 of the Royal Decree of 3 May 1999, *Belgian Official Journal*, 10 Jul. 1999.

involvement rights which are akin to decisional powers.[73] Since the Committee is a mixed body, composed of the management and the elected workers' representatives, these decision rights boil down to a right of co-determination for the elected workers' representatives.

The Committee also has a role in the 'implementation' of the health and safety policy. Article 5 of the Royal Decree refers to the elaboration and the implementation of means to propagate and to the adoption of measures related to the assistance of newly recruited workers (*accueil des travailleurs*). The same competence is recognized with regard to the information and training in the field of prevention and protection.

[C] *Ratione Loci*

It is worthwhile to underline the discrepancy between the level at which the Committee is established and the level at which health and safety issues need to be examined. The latter need to be examined at the most local level. The Framework Directive refers to the workplace and even to the workstation.[74] The name of the *Comité pour la Prévention et la Protection au travail* refers to the place where the work has to be performed.[75] However, the Committee is established at the level of the undertaking (enterprise). However, this concept is defined as a technical production unit (*unité technique d'exploitation*), on the basis of economic and social criteria. In the case of a conflict, social criteria will prevail. In this respect, the notion of undertaking will come closer to that of the so-called establishment. In practice, the workplace might coincide with the establishment, but is arguable that in some cases an establishment consists of several workplaces. Though at first sight the level at which works councils and Committees have to be established are defined in an identical way, both definitions function within different statutory instruments. It is generally accepted that the interpretation of the concept of technical production unit in relation to works councils can differ from that in relation to Committees.[76]

[D] Workers' Representatives and Workers as Experts

The Framework Directive and Belgian Law tend to promote the involvement of two distinct kinds of workers in the management of health and safety risks. The first kind consists of experts, i.e., workers designated by the employer to carry out activities related to the protection and prevention of occupational risks within the meaning of Article 7 of the Framework Directive (the already mentioned internal safety advisors). These workers are qualified as competent and in possession of the necessary skills. The Framework Directive clearly expresses a preference for internal workers as opposed to

73. Article 4 of the Royal Decree of 3 May 1999, *Belgian Official Journal*, 10 Jul. 1999.
74. See Articles 6 and 10 of the Framework Directive.
75. Thus, Article 65 of the Well-Being Act which describes the competence of the Committee as related to the implementation of the employment contract.
76. See M. De Gols & S. Depuydt, *Overlegorganen in de onderneming* (M. De Gols & M. Rigaux eds., Kluwer 2009), 248.

external services[77] The second kind of workers are workers' representatives. Their legitimacy is not primarily based on their skills, but rather on their representative status. In theory, these two different modes of defining legitimacy can clash. In practice, the different prerogatives of these distinct agents could avoid such a clash. However, according to Article 11 of the Framework Directive 'workers' *or* workers' representatives with a specific responsibility for the health and safety of workers need to be consulted by the management on a series of specified issues. It is unclear to what extent the notion 'workers' with a specific responsibility also refers to these experts within the meaning of Article 7. Anyhow, Article 7, 5) also suggests that these experts need to be 'consulted'. The object of the consultation process is not specified.

Article 25 of the Royal Decree of 3 May 1999 tries to mitigate this potential conflict by prescribing that these internal experts have to participate in the meeting of the Committees without a right to vote (*voix consultative*). More importantly, Article 23 of that Royal Decree states that the internal service should be in charge of the secretariat of the Committee. Though this technique at least allows the workers-experts to take the opinion of the elected workers' representatives into consideration, it does not preclude the possibility of a clash. These independent experts concerned have to assist the employer. For this reason, they must be distinguished from the experts chosen by the workers' representatives.[78] These experts are entitled to assist the workers' representatives. In practice, the experts' assistance somewhat restores the balance within the Committee, because this way both parties in the mixed body can have recourse to an expert. At first sight, the system seems to respect the idea of balanced participation as expressed in Article 11 of the Framework Directive. However, if looked at more closely, this may raise some doubts. Although a Belgian employer cannot designate the internal experts without prior consultation[79] of the Committee, s/he maintains sufficient leeway to designate the experts even if the Committee or the social inspectorate has a different view.

For the experts of the employee representatives, Article 26 of the Royal Decree only states that the latter can only participate after the employer's agreement. This implies that the employer could refuse the presence of such experts. In other words, there is insufficient balance. Workers' representatives cannot determine who the independent internal expert will be, and moreover, they might not be able to have recourse to experts of their choice.

§3.05 CONSEQUENCES AND SANCTIONS

Article 5.1 of the Framework Directive obliges the employer to see to the health and safety of the employees as far as all work-related aspects are concerned. By applying the fourth paragraph of this provision, the Member States can restrict this responsibility with regard to 'occurrences due to unusual and unforeseeable circumstances, beyond the employer's control, or to exceptional events, the consequences of which could not

77. Article 7 of the Framework Directive, as implemented by Article 33 of the Well-Being Act.
78. See Article 26 of the Royal Decree of 3 May 1999.
79. See Article 20 of the Royal Decree of 27 Mar. 1999.

have been avoided despite the exercise of due care'. According to the Court of Justice,[80] this provision only imposes the general obligation on the employer to see to safety without requiring any form of liability, let alone a risk liability. On the contrary, the Member States maintain the freedom to determine what kind of liability rests on the employer.[81] In Belgium, all forms of liability are brought in to obtain a correct observance of the regulations on well-being.

[A] Criminal Liability

First of all violations of the well-being regulations can be *criminally* prosecuted. By choosing criminal sanctions the Belgian legislator indicates that the protection of the employees' well-being at work is a 'public interest'. At the same time this option implies a certain degree of vulnerability. Indeed, one of the characteristics of social criminal law is that criminal behaviour/negligence is not always adequately specified. But 'social' criminal law is also criminal law and therefore legislative law, which entails that on the basis of the legality principle[82] only behaviour previously declared punishable by law can be punished. This does not mean that the elements of the offence have to be described in the penalty provision but it does mean that the penalty provision itself or in the context of other provisions, should provide a precise description of the punishable behaviour. The courts will always maintain a certain degree of discretionary power. In an interesting judgment of 10 May 2006,[83] the Belgian Constitutional Court ruled that the employer's obligation – as laid down in the Belgian Well-Being Act, just like in Framework Directive – to take all necessary measures to promote the employees' well-being at work (Article 5 § 1, 1) as well as to plan and take prevention measures with a view to a systems approach is extremely vaguely worded and insufficiently specified by the other relevant provisions of the Well-Being Act, its parliamentary preparation as well as the international provisions on which it is inspired. Insofar as these breaches are made punishable on the basis of Article 81, 1° Well-Being Act, they violate the legality principle.[84] Until now the effects of this ruling are still relatively limited. The criminal courts do not consider the definition of what constitutes an offence in numerous provisions of the regulations on well-being to extend so far that the punishable nature of such conduct would be unpredictable.[85]

80. ECJ 14 Jun. 2007, C-127/05, *Commission v. United Kingdom of Great Britain and Northern Ireland* [2007] ECR I-4610 and *SEW* 2008, 151, note H. Voogsgeerd.
81. P. Mengozzi, opinion at the ECJ 14 Jun. 2007, C-127/05, *Commission v. United Kingdom of Great Britain and Northern Ireland.*
82. Articles 12 and 14 Constitution.
83. Court of Arbitration (now Constitutional Court) 10 May 2006, no. 71/2006, www.grondwettelijkhof.be.
84. For comments: F. Lagasse, Commentaire de l'arrêt rendu le 10 mai 2006 par la Cour d'arbitrage. Son implication sur la législation en matière de bien-être des travailleurs lors de l'exécution de leur travail, *Sociaalrechtelijke Kronieken/Chroniques de Droit Social* (2007): 10–13.
85. For an overview of this case law: C. Persyn & A. Van Regenmortel, Het risico op aansprakelijkheid bij inbreuken op de welzijnsreglementering "geanalyseerd", *De Preventieadviseur van morgen* (2009): 28–31.

Moreover, in addition to prosecution on the grounds of Article 128 Social Criminal Code, prosecution on the basis of Articles 418 et seq. of the Criminal Code is possible as well, more specifically because of involuntary manslaughter, blows and injuries, of which the penalty provisions are sufficiently specific, clear and predictable.

This shortcoming is remedied by the new Social Criminal Code[86] which came into force on 1 July 2011. General penalty provisions to make violations of certain types of legislation and implementing orders – often still to be adopted – punishable 'en masse' are done away with and exchanged for an exhaustive inventory of violations that could lead to punishment. Depending on their gravity, violations and sanctions are classified in a coherent four category scheme of punishments. The lightest violations (level 1) are de-penalized and can only lead to an administrative fine. Violations of moderate gravity (level 2) and serious violations (level 3) can be punished either by an administrative fine, or by a penal fine. Only for the most serious violations (level 4) imprisonment is still possible. From now on the violations regarding health and safety in the workplace are listed exhaustively in Articles 123 through 137 in the new Social Criminal Code. All these violations are considered to be 'serious' and punished by a sanction of level 3. If the violation causes damage to the employee's health or amounts to an occupational accident, this will be regarded as an aggravating circumstance. And in that case a sanction of level 4 is provided for. Given the fact that, as a rule, only violations involving injuries and deaths are prosecuted the heaviest sanction will tend to become the rule rather than the exception. As a result of the introduction of the new Social Criminal Code, Article 81, 1° Well-Being Act was withdrawn as a basis for penalization and replaced by Article 128 Social Criminal Code.

If the violation leads to deaths or injuries, the prosecution will usually take place on the basis of the penalty provision under the ordinary law provision of Article 418 Criminal Code – involuntary manslaughter, blows and injuries – as well as on the basis of Article 128 Social Criminal Code – which more specifically penalizes violations of the regulations on well-being. There are clear differences between these two penalties: First of all, with regard to the required error. For a prosecution on the basis of Article 418 Criminal Code, any lack of caution or precaution is relevant and the slightest error suffices. This error does not necessarily need to be a violation of the well-being regulations. It could as well be any other violation of the general duty of carefulness.

A second difference concerns the required damage. If the offence is described as 'involuntary manslaughter or blows and injuries', the presence of damage is inherent so that this damage can only be prosecuted on the basis of Article 418 Criminal Code. Conversely, Article 128 Social Criminal Code stipulates that the mere violation of the Well-Being Act and implementing orders are punishable without there having to be damage. So violations without injuries or deaths can only be prosecuted on this basis albeit on the condition that the material as well as the moral element[87] is proved. A

86. Act of 26 Jun. 2010 establishing the Social Criminal Code.

87. In several rulings (Cassation 12 May 1987, *Cassation Judgement* (1986–87): 1194 and *Rechtskundig Weekblad* (1987–88): 538, opinion of Adv.Gen. J. Dujardin; Cassation 13 Dec. 1994, *Rechtskundig Weekblad* (1995–96): 533; *Cassation Judgement* S.09.0053.N, 26 April 2010, not published) the Court of Cassation emphasizes that also in the event of offences whose criminalization does not necessarily need a moral element, a material and moral element is

third important difference concerns the 'punishable offenders'. For a prosecution on the basis of Article 418 Penal Code the material or impersonal imputation applies, i.e., anyone, irrespective of his or her capacity and position in the company, can be prosecuted on this basis.

For a prosecution on the basis of Article 128 Social Criminal Code, the system of legal attribution applies which means that the legislator designates, explicitly and exhaustively, the potential punishable offenders. This concerns 'the employer, his or her agents or appointees'. According to the most recent theory[88] the criminal liability of these persons is based on 'the power to take decisions'. It should be possible to call the person with the power and means to prevent violations to account for violations committed under his or her control. The fact that the employer bears the main liability with regard to well-being at the workplace explains why only s/he and his or her entourage – his or her agents or appointees – are designated as punishable offenders. The fact that the term 'agents or appointees' within the meaning of Article 128 Social Criminal Code only refers to persons who, as it were, 'propagate' the employer's authority cannot be read in the text itself but is a consequence of the settled case law of the Court of Cassation. Ever since a general ruling of 15 September 1981,[89] agents and appointees are only understood to mean 'those persons who have the necessary authority and competence to see to the compliance of the law'.

In concrete terms, this means that only persons with a certain degree of power to take decisions on the basis of Article 128 Social Criminal Code can be prosecuted. The difference between an agent and an appointee depends on the existence or non-existence of a relation of subordination. This specific restricted qualification means that neither purely operational staff nor, in principle,[90] prevention advisors are appointees. So if in Belgium these persons fail to observe their obligations this cannot result in criminal liability on the basis of Article 128 Social Criminal Code. The members of the hierarchical structure on the other hand are, as it were, predestined to be prosecuted as 'appointees'. Although the qualification as a member of the hierarchical structure in itself is, in principle, not relevant for the criminal liability, the members of the hierarchical structure certainly are, each within their competence and at their level, liable for the implementation of the well-being policy,[91] which suggests the precondition of a certain power to issue orders. However, only if this power has been confirmed in practice they can be regarded as appointees. Conversely, if it is proven that they only act under orders and only execute the decisions taken by the employer without they themselves having any power of decision in this respect, they cannot possibly be qualified as appointees.

---

always required. For that reason, the term 'material breaches', which could mistakenly give the impression that the presence of the material element would suffice, is objectionable.

88. F. Kefer, *Le droit pénal du travail* (die Keure, 1997), 237–246.
89. Cassation 15 Sep. 1981, *Rechtskundig Weekblad* (1981–82): 1123, note H. Bosly.
90. Indeed, in principle prevention advisors have a consultative capacity. Only if they have the power to decide, which is highly exceptional, they will be 'appointees' (see for more details: A. Van Regenmortel & V. Vervliet, *Actuele problemen van het socialezekerheidsrecht* (R. Janvier, A. Van Regenmortel & V. Vervliet eds., die Keure 1999), 212–220).
91. Article 3 Royal Decree Well-Being policy.

The question is whether the above system of legal attribution is corrected by means of the *criminal liability of legal persons*, the concept of *delegation* or the rules of *punishable participation*.

The answer is negative as far as the criminal liability of legal persons is concerned, which Belgium only regulated by the Act of 4 May 1999.[92] Since this Act was introduced, legal persons themselves can be prosecuted and punished directly. However, this does not refer to objective liability because the legal person is only punishable if the fact, both with regard to the material and moral element of the offence, can be attributed to it. Indeed, a legal person is only criminally liable for offences which have either an intrinsic connection with the realization of its purpose or with the observation of its interests, or which the concrete circumstances show to have been committed on its liability.[93] As far as legal persons are concerned, there needs to be an act of negligence which has a causal connection to the offence. However, under certain circumstances cumulative liability of the legal person and the natural person is still possible.[94] A requirement is that the legal person is held liable solely on account of the action of an identified natural person. Only in those situations the identified natural person can be sentenced together with the legal person, provided that s/he committed the error 'wittingly', which will be assessed on the basis of the natural person's concrete state of mind and is deemed to be the case if s/he acted consciously, voluntarily, i.e., without any coercion. If the error was not made wittingly, only the person who committed the 'gravest error' will be sentenced. Precisely the balancing of these errors is a delicate matter. According to the case law of the Constitutional Court, it is assumed that the above described criminal liability of legal persons cannot detract from the legal attribution of a natural person, if the error is attributed to this person because of his or her special competence or special qualification.[95] This is the case for violations of the regulations on well-being. So the system of legal attribution continues to be fully applicable.

The same conclusion can be drawn with regard to the concept of *delegation*. Delegation of powers and duties is possible on the condition that the delegated powers are clearly and explicitly defined. The existence and validity of delegation is not subject to any formal requirements. However, the delegated person shall have the necessary competence, authority and means to carry out his or her tasks. Through Article 5 § 2 (b) Well-Being Act, Belgian legislation explicitly obliges the employer to define the powers and liabilities of the persons in charge of implementing the well-being policy. According to a recommendation of the Belgian Federal Public Financial Service this only refers to the members of the hierarchical structure. It is generally accepted that delegation is also possible in the case of legal attribution. In concrete terms, this means that 'the employer, his or her agents or appointees' can also invoke delegation. If the delegated person belongs to the category of the legally designated punishable offenders, then the

92. Law of 4 May 1999 regarding the criminal liability of legal persons, *Belgian Official Journal*, 22 Jun. 1999.
93. Article 5, first paragraph Law of 4 May 1999 concerning the criminal liability of legal persons.
94. Article 5, second paragraph, Criminal Code.
95. For a discussion of this case law: F. Deruyck & P. Waeterinckx, *Sociaal handhavingsrecht* (G. Van Limberghen ed., Maklu 2010), 101–104.

delegated person can be punished, if necessary together with any possible misconducting delegating person. If the delegated person is not one of the punishable offenders intended by the law and there has been no error as far as the delegating person is concerned, then according to some legal doctrine this could lead to impunity.[96]

There is more discussion about the answer to the question as to whether the rules of punishable participation could not imply a correction of the system of legal attribution and whether on the basis of this concept other persons than the legally designated perpetrators, among whom the employees, could be prosecuted. The advocates of the protective nature of social law answer in the negative. The legislator expressly excluded the employees as punishable offenders so that they cannot be held punishable by means of the participating concept either. Others take the view that the specific objective of social law may not damage the intended purpose of criminal law, i.e., protecting society. Since the Well-Being Act itself does not declare the rules of punishable participation non-applicable, it is obvious that the legislator considers the protective function of social law secondary to the protective function of criminal law. Moreover, the employee is deemed to be protected sufficiently by the requirement of a moral element to the effect that the s/he has to commit the offence 'wittingly'. However, as soon as it has been proven that the employee did commit a criminal offence wittingly, s/he is said to be no longer entitled to any protection. We will have to wait and see how the courts will react in this respect. In any case, this view fits in with the tendency, also apparent in Belgium, to put more and more emphasis on the employee's own liability.

For the sake of completeness we would also like to mention that as a result of the increased attention for atypical working situations – such as working with outside companies and temporary and mobile construction sites – specific penalty provisions[97] and new potential offenders have been defined.

The decision to prosecute or not is for the Public Prosecutor to make. Only if the decision is made not to prosecute, a fine can be imposed pursuant to Article 96 § 2 Well-Being Act. This is mainly aimed at the employer.

[B] Civil Liability

The consequences, as far as civil law is concerned, differ sharply depending on whether or not the offence committed caused damage and whether or not this damage constitutes an occupational accident or disease. Indeed, in the case of damage civil liability applies. A clear distinction must be made between this civil liability and criminal liability. Nonetheless, quite often both kinds are brought up together and are usually settled in the same procedure. After all, if an accident leads to criminal

---

96. E. Roger-France, *La delegation de pouvoirs en droit penal ou comment prévenir le risqué penal dans l'entreprise*, Journal des Tribunaux (2000): 262–263; I. Verhaert & P. Waeterinckx, *Strafrechtelijke verantwoordelijkheid, een beheersbaar ondernemingsrisico? De delegatie in het strafrecht*, Rechtskundig Weekblad (2001–02): 1024; F. Kefer, *Le droit pénal du travail*, no. 266 (die Keure 1997).
97. Articles 130 and 131 Social Criminal Code (former) Articles 81, 2° through 88 Well-Being Act.

prosecution, the victim or his or her next of kin can also assert their claims at the civil law level by taking civil action.

If there is damage and this damage constitutes an occupational accident or disease one needs to take the principle of civil immunity into account. This principle is part of the historic compromise which was at the basis of the professional indemnity insurance. 'Historic' because it distanced itself from civil liability law which had traditionally governed occupational damage. 'Compromise' because both employers and employees made concessions. Henceforth, the employer was to pay compensation for any accident, regardless of who was to blame, but contrary to civil liability law, this does not involve a compensation in full but a flat-rate part of this.

Moreover, in order to maintain social peace, civil claims were excluded in the context of accidents and diseases within the firm. Except in a few special cases, the victim of an occupational accident or disease or other rightful claimants can no longer file claims against the employer, his or her agents or appointees – so mainly co-workers. This means that the victim has to be satisfied with the compensation for occupational accidents. This civil immunity also affects the insurer. Indeed, the occupational accidents insurance paying the victim is subrogated in the rights of the victim but this subrogation is limited to the rights the victim would have had under civil law for the same material damage as that covered by the occupational accidents compensation. The fact that this system introduces a different treatment of victims of professional damage on the one hand and victims of accidents/diseases under ordinary law on the other is not considered to be in contravention of the constitutional principle of equality.

In essence, civil immunity only protects the employer and his or her agents or appointees. The employer is the natural or legal person with whom the person concerned concluded an employment contract and therefore is the holder of the right of authority. Agents are persons who are tied to the victim's employer by means of a contract of agency, like for instance administrators or managers. Appointees are understood to mean all persons under the authority of the same employer as the victim. The appointees need not necessarily be tied to the employer by an employment contract. It suffices that they are under the actual authority of the employer. Civil immunity only applies to the aforementioned persons. So civil liability law remains fully applicable if 'third parties', i.e., persons extraneous to the company, violate the rules and regulations regarding well-being in such a way that this leads to accidents or diseases. The above is important with regard to precarious employment relations, such as temporary agency work. Although the temporary agency worker is tied to the temporary employment agency by an employment contract and this agency is therefore the legal employer, s/he will nonetheless work under the (actual) authority of the user.

Thus the temporary agency worker becomes an 'appointee' of the user. By acquiring this qualification s/he will be able to invoke civil immunity if an offence against the well-being regulations committed by him leads to an occupational accident or disease of the user's employees. Temporary agency workers will also be able to file a civil claim against the user and his or her appointees/agents if they falls victim to an occupational accident in the workplace. Seeing as the temporary employment agency is his or her legal employer, the user, his or her appointees/agents are third parties and therefore can never be the beneficiaries of the civil immunity guaranteed by the professional indemnity insurance. However, the employees of the user, in their capacity as employees, will be able to fall back on the - less extensive - civil immunity which is granted to them pursuant to the Employment Contract Act - see *infra*.

The Constitutional Court on the other hand considered this difference in treatment between appointees/agents of the employer and appointees/agents of an user in contravention of the principle of equality. Indeed, in both cases there is a relationship of power. The fact that the user him or herself does not enjoy immunity is not a problem. After all, civil immunity is attached to the obligation to take out an occupational accident insurance and in the case of temporary agency work only the employment agency as the legal employer has the duty to do this. Therefore, only the agency is entitled to civil immunity. So a temporary agency worker who falls victim to an accident due to a violation of the well-being regulations by the user and provided that there is a direct causal connection between this infringement and the accident, can hold the user civilly liable. Another consequence of the absence of civil immunity is that the temporary employment agency's occupational accident insurance can recover the compensation from the agency.

However, there are some exceptions with regard to civil immunity.

Civil immunity on the part of the employer and his or her agents/appointees is terminated if they caused the occupational accident *intentionally*. Whereas for intent on the part of the employer it is sufficient that s/he wanted it to happen, in the case of agents/appointees intent will only be recognized if not only the accident itself but also the harmful consequences are intentional. So a grave error is not enough. This basis for exception is hardly ever used.

A second exception concerns *traffic accidents*. Neither the employer nor his or her appointees/agents can invoke civil immunity if the occupational accident 'concerns a traffic accident involving one or more (motor)vehicles and is connected to the traffic on public roads'.

Only as regards the employer there are two additional exceptions. First of all, civil immunity does not prevent the victim of an occupational accident from claiming compensation for damage to goods like clothes and other personal possessions. Moreover, since 1999 an additional exception has been provided for in the case of *accidents due to violations of the well-being rules and regulations*. The employer loses his or her civil immunity if s/he has violated the legislative and statutory provisions regarding well-being 'in an unforgivable manner'. Besides the fact that the presence of an unforgivable shortcoming - perhaps inspired by the French concept of *'faute inexcusable'* - is required, there are also strict formal requirements. For instance, prior notice in writing by the social inspectorate containing a number of obligatory statements is required. Moreover, the employer has the possibility of 'regaining' his or her civil immunity if s/he succeeds in proving that the accident was 'also' due to the employee's non-observance of the safety instructions although the employer had notified him of these in writing in advance and the necessary safety means had been made available. The threat of loss of civil immunity is intended to urge the employer to see to prevention. Because of the all too formalistic character of this measure this provision has hardly been used so far. Therefore, the desired preventive effect of this measure has failed to occur.

The lack of success of this measure explains why in recent years the Belgian legislator has been using more and more actuarial means to strengthen the preventive function of professional indemnity insurance. Indeed, since 1 January 2009, companies with a 'disproportionately increased' risk are charged a flat-rate prevention contribution. So as to prevent this premium from being avoided by the systematic contracting out of hazardous activities to temporary agency workers, accidents involving temporary agency workers are also taken into account to calculate the frequency and gravity of the occupational accidents at the employer-customer's. The contribution is labelled as 'prevention contribution' because it has to be used for the prevention of occupational accidents at the premises of the employer concerned. To prevent this measure from promoting under-declaration, failure to report occupational accidents is now prosecuted. Similarly, as regards the occupational hazard of blue-collar workers, a no-claims bonus system was introduced on 1 January 2009. Since then, the contribution is established depending on the size and the damage statistics of the company concerned. These measures show that damage prevention is a societal duty to which the social security system has to contribute as well. However, the compensating function as the main objective of social security is not affected.

If there is damage and this damage does not amount to occupational accidents or diseases, the special immunity rules of the professional indemnity insurance do not apply. Consequently, one has to fall back on the rules of ordinary law and the deviant regulation provided for in Article 18 Employment Contract Act for the protection of the employee. This Article adopts a liability restriction, more specifically in the sense that an employee who, by neglecting the well-being regulations during the implementation of his or her employment contract causes damage to the employer or third parties, is only liable in the case of intent, gross error or frequently occurring minor errors. Vis-à-vis third parties this liability restriction cannot be deviated from. To prevent that as a result of the application of this provision the victim would lose all rights to compensation, the Court of Cassation[98] holds the principle that Article 18 Employment Contract Act leaves the application of Article 1384, 3 Civil Code intact. Under this provision, the appointing person is civilly liable for the damage caused by his or her appointees in the function for which they have been appointed. The relation between appointing person and appointee is based on a factual relationship of authority. The application of this rule has as a result that a third person who falls victim to an accident due to the slightest error or a non-frequent minor error on the part of an employee cannot address the employee causing the damage but can address the latter's employer in his or her capacity as appointing person. Still, one has to take into account that the employer/appointing person can restrict or preclude contractually his or her liability on the basis of Article 1384, 3 Civil Code. If, however, the employee commits an intentional, gross or frequently occurring minor error, the victim can address either the employer as the appointing person or the employee, or both.

98. Cassation 18 Nov. 1981, *Revue Générale des Assurances et des Responsabilités* (1982): no. 10.459.

Even if the violation of the well-being regulations does not result in damage, harmful effects cannot be avoided. The Employment Contract Act too contains safety obligations. On the one hand, there is Article 20, 2° Employment Contract Act which obliges the employer to make sure that the work is done is safe circumstances. On the basis of Article 17, 4° of that very same act the employee for his or her part is under the obligation of refraining from all things that could cause damage to his or her own safety, that of his or her co-workers, his or her employer or third parties. Employees particularly are affected by failure to comply with the aforementioned obligations. In the case of a serious violation of the well-being regulations, they run the risk of being fired, possibly even for just cause,[99] meaning without notice or compensation in lieu of notice. The internal work rules too can stipulate for disciplinary sanctions for non-compliance with obligations regarding well-being.

Special mention deserves the concept of 'the employer's civil liability for the payment of fines imposed on his or her agents or appointees for violations of the well-being regulations'.[100] This concept was introduced to correct the (former) impunity of legal persons. The financial effects of a criminal offence committed by an agent or appointee is, as it were, passed on to the legal person. Even after the introduction of the criminal liability of the legal persons this concept continued to exist. In order to prevent legal persons from getting double punishment, Article 50*bis* Criminal Code was introduced. Pursuant to this Article, no one can be held civilly liable for someone else if s/he can be held criminally liable for the same acts. The new Social Criminal Code has maintained this method.

§3.06 CONCLUSIONS

The Framework Directive did not produce a change of approach in the Belgian safety regulations. Important leading principles of the Framework Directive – such as the employer's prime liability regarding safety, a multidisciplinary approach, the impulse towards integration, the assistance of experts, employee involvement, the importance of training – had already been firmly enshrined in the Belgian safety regulations since the Prevention Decree of 1975. As regards certain points, one can safely say that the Belgian regulation is 'unique' and 'progressive'. For instance, one cannot overemphasize the importance of the 'hierarchical structure'. As early as 1975, this concept was used to stress the necessity of an integrated safety policy. This emphasis is less apparent in the Framework Directive. The concept of 'hierarchical structure' indicates that safety must be integrated into the production process and safety policy if one wants to speak of a real prevention culture. However, in spite of these good intentions one cannot but observe that we have not yet succeeded in this respect. One of the obstacles is no doubt the lack of training of the members of the hierarchical structure. The legislator is not blameless either. By failing to specify the exact efforts to be expected on the part of the employer in this regard the training may perhaps not have

99. In principle, an employee can also resign for just cause on the part of the employer but this is highly exceptional.
100. Article 91 Well-Being Act.

remained dead letter in practice, but it does seem to have been reduced to a minimum. This seriously hindered the break-through of a genuine prevention culture. Even more distressing is the situation within enterprises with less than twenty employees where the employer him or herself can act as a prevention advisor. The Framework Directive allows this possibility but it does require that for the qualification of this category of enterprises the nature of the activities and the size of the enterprise must be taken into account. Another requirement is that the employer has the necessary skills. It would seem that the Belgian situation which only uses the parameter 'size of the enterprise' – i.e., the number of employees – and does not provide for any training is not consistent with this EU Law requirement.

Also progressive is the fact that in 1996 a deliberate choice was made for the term 'well-being'. First of all, this has the advantage of underlining the increasing importance of new risks, such as psychosocial stress at work. In addition, this term also carries the significant connotation that the employee is approached as a human being and as such is also entitled to respectful treatment in an employment relation. However, it is regrettable that the recognition of the importance of psychosomatic diseases has not led to sharpened attention for the link between organization of work and well-being. Still, although both the Framework Directive and the Belgian regulation provide for 'organizational' prevention measures on top of material ones, the attention paid to the former continues to be insufficient. This is not surprising because right from the very beginning of the safety regulation development, the organization of work was considered to be the employer's prerogative. Moreover, the introduction of the professional indemnity insurance also enhanced the separation of the safety issue from the production and work organization. As a result of the (deliberate) choice for a curative and individualized social security system in the wake of Bismarck, it became clear that the subject of social policy did not refer to the prevention of health risks by interventions in the employment environment but instead to compensations schemes in purely individual cases.

The less technical nature of new risks could also increase the involvement on the part of the employees. Although, compared to the works councils the Committees were given a more pro-active role in the matter of well-being, in actual practice little use is made of this. Both the Framework Directive and the Belgian Law could stimulate this by focusing on the employees' empowerment instead of cherishing a 'culture of duties'. Indeed, the employees' experience in the field of well-being is invaluable. So it is obvious that this experience increases whenever a more enduring policy is pursued. However, this requires a greater willingness to listen on the part of the employer and the members of the hierarchical structure so that employees will feel more valued as regards their own awareness of the risks involved. A better implementation of certain rights could also contribute to this. For instance the right of the employee representatives within the Committee to demand the assistance of experts is too dependent on the employer's consent. De facto this right is even made impossible by the absence of a regulation regarding the conditions under which one is allowed to ask for these experts as well as regarding the size of the compensations to be paid by the employer.

More far-reaching than the Framework Directive is the Belgian Law's attention for the safety in the case of activities that do not exclusively take place *intra muros.*

Indeed, when workers from various undertakings use the same workplace or when outside companies are involved, more stringent safety obligations are provided for without the presence of employees being an absolute necessity. That way self-employed work gets some attention as well, albeit rather weakly. The broad range of flexible and atypical work as well as the tendency towards more self-employment is passed over. In an age where the protection of employees is cut down in favour of self-employed work, one could ask oneself if the power relation as a basis for well-being protection should not be reconsidered. Should the right to safety not be recognized as a fundamental right for all? The present situation in which many precariously employed workers do not have a collective representation and are prepared to work in unsafe circumstances for fear of losing their jobs is unacceptable.

Although the Framework Directive does not impose any sanctions, the history of the Belgian safety regulation tells us that the manner of enforcement could also influence safety policy. Due to the fact that general safety obligations are no longer subject to civil law proceedings but have been replaced by technically detailed regulations that could give rise to criminal court proceedings, the dynamics in safety considerations appear to have largely been lost. The Prevention Decree of 1975 which compelled the employer to eliminate all demonstrable risks inherent to the work was an attempt to regain these dynamics. Still, the possibility of criminal sanctions was maintained. However, the question is whether this criminal enforcement is the most adequate one. Not only do general provisions because of their vagueness appear to be at odds with the legality principle of criminal law but there is also the danger of a certain disproportion between maintaining the safety obligations of the employers on the one hand and those of the employees on the other. On the one hand, the lack of manpower of the controlling authorities has as a result that the employers are only criminally prosecuted in the event of serious damage as a result of which all other violations of the well-being regulations remain largely 'hidden' and 'unpunished'.

On the other hand, the employees' safety obligations can quite easily be used by the employer to impose far-reaching disciplinary sanctions on them. It is true that recently attempts have been made to weaken the employer's civil immunity in cases of occupational accidents due to grave shortcomings with regard to the well-being regulations, but so far these have not led to the desired outcome. So the question is whether civil law should not be given a broader role or whether alternative sanctions should not be taken into consideration? This is the more so now that the recent ruling of the EFTA Court on 10 December 2010[101] seems to indicate that the employer's prime liability requires financial punishment, either by forcing the employer to repay a part of the insurance benefit or by imposing a higher premium in the future. The fact that the employer can be punished under criminal law or administratively for violations or that the employee could enjoy benefits through a mandatory accident insurance is considered to be irrelevant. However, this ruling is hardly consistent with the previously mentioned ruling of the Court of Justice of 14 June 2007. So it will be clear that the last word on the scope of the employer's liability has not yet been spoken. Nor is this the

101. EFTA Court, 10 Dec. 2010, Case E-2/10, *Kolbeinssom v. Icelandic State*, www.curia.europa.eu.

end of the debate on the conformity or non-conformity of the Belgian regulations with the European regulations.

BIBLIOGRAPHY

De Gols, M. & S. Depuydt. *Overlegorganen in de onderneming*, edited by M. De Gols & M. Rigaux. Kluwer, 2009, 245–305.

Deruyck, F. & P. Waeterinckx. Daderschap en verantwoordelijkheid voor sociaalrechtelijke

Misdrijven. In *Sociaal handhavingsrecht*, edited by G. Van Limberghen. Maklu, 2010, 87–132.

Donceel, W. *Praktijkgids bij de Welzijnswet*. Intersentia Rechtswetenschappen 1999.

Heselmans, M. & J. Van Peteghem. *Meer welzijn op het werk, een eigen visie*. Garant 2007.

Kefer, F. *Le droit pénal du travail*. die Keure 1997.

Roger-France, E. 'La délégation de pouvoirs en droit pénal ou comment prévenir le risque pénal dans l'entreprise'. *Journal des Tribunaux*. (2000): 257–263.

Sojcher-Rousselle, M. *Droit de la sécurité et de la santé de l'homme au travail*. Bruylant 1979).

Van Regenmortel, A. & V. Vervliet. *Actuele problemen van het socialezekerheidsrecht*, edited by R. Janvier, A. Van Regenmortel & V. Vervliet. die Keure, 2003, 99–239.

Verhaert, I. & P. Waeterinckx. 'Strafrechtelijke verantwoordelijkheid, een beheersbaar ondernemingsrisico? De delegatie in het strafrecht'. *Rechtskundig Weekblad* (2001–2002): 1009–1026.

Vogel, L. *Droit de la santé au travail*. Université catholique de Louvain 2008.

CHAPTER 4

Occupational Health and Safety in France: A Good Formal Protection, but a Problematic Efficiency

Corinne Sachs-Durand

§4.01 INTRODUCTION

Historically speaking, French Law was oriented more towards the compensation of already existing damages rather than to the prevention of damages likely to occur.

The first legislation in this domain was Act of 9 April 1898. The aim of this legislation was to alleviate the burden of proof for the worker, who previously had to prove the liability of the employer in case of accident. According to the Act, the worker had only to prove that the accident had occurred during his or her work. In return, the compensation of the damage was limited to a prefixed amount (indemnity) paid by the accident insurance. This was a kind of social compromise to cope with the important occupational accident's rate in the nineteenth century, and to give a minimum relief to the worker.

Moreover, at the origin, the only risk covered by the legislation was that related to occupational accidents. Occupational diseases were not at that time recognized as such by the law.[1] Second, it was only the physical risk that was taken into account. Third, only few activities were included in the scope of the law.[2]

1. The first legal provision in this field was the Act of 25 Oct. 1919.
2. Nowadays, the situation has changed following the development of new technologies. Besides the classical sectors, like mining or building for instance, new protections against radiations, nuclear industry, biological risks, etc. appeared. Moreover, following to a chemical explosion in a factory in the centre of Toulouse, which resulted in 30 dead and 2500 wounded people, workers of the firm, inhabitants of the town or tourists (see on this H. Seillan, *Un tsunami urbain – AZF Toulouse* (Bordeaux, Préventique, 2009)). The legislator, in 2003, passed a legislation to provide a prevention of technological and natural risks (Loi du 30 juillet 2003,

During the years, the concept of health and safety has been subject to an impressive evolution, under the influence of two factors: the development of fundamental rights and the progress of European Union (EU) Law in these matters. Furthermore, the scientific and medical advances as regards the knowledge of professional harm have brought new information to be integrated into the legislation.

Concerning the influence of EU Law, one can say that the spirit of French Law has changed. The transposition[3] of the Directive 89/391/EEC (hereinafter 'Framework Directive') has led to underline the necessity to prevent any degradation of health at work, and one of the outcomes of this evolution has been to put a heavy pressure on the employer concerning health and safety issues. Case law has strongly contributed to this evolution, as we will see below. Some points can be stressed to illustrate that change: priority has been given to collective protection instead of individual one, and the security imperatives are now integrated into the conception of the machines. Besides, there is a body at the plant level, whose priority is to check the working conditions and to prevent any occurrence of occupational accidents or diseases (CHSCT – see below).

The Ministry in charge of labour issues has elaborated a national plan for health and safety at work, covering the period between 2010 and 2014, including notably the pursuit of a better knowledge of risks and a policy of prevention of risks at work.[4]

Some information can be given as regards the rate of occupational diseases and accidents at work in France: 668,847 occupational accidents (stable, in proportion of people at work), 50,688 occupational diseases (slightly increasing) and 98,420 commuting accidents (slightly increasing) were identified in 2010.[5] If we try to go a bit more into details, we can underline that the rate of occupational accidents and diseases varies a lot according to the branch in question. Accidents are – not surprisingly – far more frequent in the building sector. Other branches with a rather high rate of accidents are transport and services. Concerning commuting accidents, the most afflicted branch is that one of services, followed by transport and building. As for the occupational diseases the most affected is the branch of services, followed by the building sector and metalworking industry.[6]

relative à la prévention des risques technologiques et naturels et à la réparation des dommages – a part of this text has been integrated into the Labour Code).

3. Act no. 91-1414 of the 31 Dec. 1991 'modifiant le code du travail et le code de la santé publique en vue de favoriser la prévention des risques professionnels et portant transposition de directives européennes relatives à la santé et à la sécurité du travail'.
4. Available at http://www.travail-emploi-sante.gouv.fr/actualite-presse,42/dossiers-de-presse,46/plan-sante-au-travail-2010-2014,11031.html. As the Government changed in France in May 2012, this plan could be modified.
5. Tableaux de synthèse des statistiques nationales de la sinistralité 2010 de la branche AT-MP du régime général (Statistics of the social security institutions), issued in (May 2011; see 2011-085-CNAMTS-DRP -MSTAT-YL Tableaux synthèse sinistralité 2010.doc).
6. Tableaux de synthèse des statistiques nationales de la sinistralité 2010 de la branche AT-MP du régime général (Statistics of the social security institutions), issued in (May 2011; see 2011-085-CNAMTS-DRP -MSTAT-YL Tableaux synthèse sinistralité 2010.doc).

§4.02 BASIC CONCEPTS

Within the field of health and safety, some basic concepts have to be pointed out as they have been subject to a specific legislation or to some important academic discussions. These are: the concept of health and safety [A], moral and sexual harassment [B], stress [C], working time [D], and, more recently, *pénibilité du travail*, which could be translated as 'hardness at work' [E].

[A] The Concept of Health and Safety

The concept of health at work itself has changed. We have seen above that originally only the occupational accidents (in their physical aspects) were protected by the law. Since 1919, professional diseases have been progressively taken into account, and in 2002[7] health, which was first considered only in its physical aspect, has been regarded also in its mental meaning. So, presently, the physical and mental health is protected in French Labour and Social Security Law.[8] One can mention, as an example, the provisions about bullying.[9] The emergence of this issue does not mean, of course, that bullying did not exist before. But it came to the collective consciousness after the publication of two books,[10] which were the sign of a social reflection leading finally to legislation.

[B] Moral and Sexual Harassment

Among the different kinds of psychical risk, the most common at the moment are moral and sexual harassment. There is a specific legislation about these two issues.

The definition of moral harassment (*harcèlement moral*) is rather wide in France. Article L 1152-1 of the Labour Code (hereinafter 'LC') provides that: 'Workers shall not bear repeated behaviours aiming at, or resulting in, a deterioration of their working conditions capable to alter their rights and dignity, their physical or psychical health or to jeopardise their professional future'. One can understand, reading this wide definition, that the place left to the judicial interpretation is quite important. Consequently, there have been a lot of cases concerning moral harassment.[11]

---

7. Loi de modernisation sociale, 17 janvier 2002; on this point see *inter alia*, P. Adam, 'La prise en compte des risques psychosociaux par le droit du travail français', *Dr. Ouvr.* (2008), 313.
8. See for instance Article L 4121-1 LC.
9. Article L 1152-1 et seq. LC for bullying and L 1153-1 et seq. LC for sexual harassment.
10. M.F. Hirigoyen, *Le Harcèlement moral: la violence perverse au quotidien* (Paris: Syros, 1998). C. Dejours, *Souffrance en France. La banalisation de l'injustice sociale* (Paris: Seuil, 1998, Collection L'Histoire immédiate).
11. It is of course impossible here to sum up all this case law and all the discussions. For a synthesis, see G. Auzero, J. Pélissier, E. Dockes, *Précis de droit du travail* (Paris: Dalloz, 2012), nn. 828 et seq., 851 et seq. See also P. Adam, 'La "figure" juridique du harcèlement moral managérial', SSL (2011), n° 1482, 4; P. Adam, v° 'Harcèlement moral'(Paris: Rép. Trav. Dalloz), l. See also some recent cases as Cass. Soc. 6 Apr. 2011, n. 10-30284, 16 May 2012 (no. 10-10623 and 10-15238) in Dalloz actualités (6 June 2012).

It is difficult sometimes for the workers to distinguish moral harassment from the stress resulting from the work organization. But the success of this provision shows quite clearly that a lot of workers in France feel like facing moral harassment practices. The recent increase of the number of suicides at work is attesting the reality of the problem.[12]

There is a penal sanction for moral harassment, provided in Article 222-33-2 of the Penal Code,[13] (two years of imprisonment and EUR 30,000 fine).

Concerning now sexual harassment, there has been a decision of the Constitutional Court which has led to a major change in France: the Court has declared unconstitutional the criminal offence of sexual harassment, in its previous wording, as being too wide and not precise enough in a penal law context.[14]

This decision has been a shock in France, especially concerning the cases pending before courts.

A new text has been adopted to solve the problem,[15] taking into account the decision of the Constitutional Court. A precise definition of sexual harassment is then provided by the law, in the Article 222-33 of the Penal Code. This definition has also been integrated to the Labour law Code, in the Article L 1153-1.

Sexual harassment is defined as follows, in the Labour Code:[16]

> No employee will suffer the facts of:
> 1° either sexual harassment constituted by repeated words or behaviour of a sexual nature, which undermine his dignity because of their degrading or humiliating character, or creating an intimidating, hostile or offensive situation against him;
> 2° or related to sexual harassment, consisting of all forms of severe pressure, even if not repeatedly exercised, aiming actually or apparently at obtaining an act of sexual nature, in favour of the perpetrator or a third party.

This definition distinguishes two types of sexual harassment: on the one hand, there is an environmental harassment, identified by sexual behaviour of any kind, imposed to a non consenting person. One can assume that the condition of repetition requires that the facts were committed at least twice.[17]

12. See for instance L. Lerouge, ' Suicide du salarié et faute inexcusable de l'employeur: quelles évolutions juridiques?', *Revue de droit sanitaire et social* (2012): 373.
13. Slightly modified by the law of 6 Aug. 2012, see below.
14. Conseil Constitutionnel, QPC (Question prioritaire de constitutionnalité) 2012-240 of the 4 May 2012, removing Article 222-33 of the Penal Code. See M. Bombled, 'Inconstitutionnalité de l'article 222-33 du Code pénal réprimant le harcèlement sexuel': *Dalloz actualités* (10 May 2012).
15. Loi n° 2012-954 du 6 août 2012 relative au harcèlement sexuel (*JO* du 7 août). See on this point M. Miné, 'Loi n° 2012-954 du 6 août 2012 relative au harcèlement sexuel (JO du 7 août)', to be published in *RDT*, Jan. 2013; L. Lerouge, 'Harcèlement: nouvelles dispositions issues de la loi du 6 août 2012', *Dr. Soc.* (2012): p. 944; C. Leborgne-Ingelaere, 'Le harcèlement sexuel dans le Code du travail depuis la loi du 6 août 2012: entre avancées et imperfections', *JCPS*, no. 40 (2012): p. 12. See also Circulaire du ministre de la justice du 7 août 2012, Crim. n° 2012-15.
16. We mention here the Labour Code, as in the penal Code, if the definition is quite similar, the wording is slightly different, as people in general are concerned, and not only the workers.
17. On this point, see M. Miné, 'Loi n° 2012-954 du 6 août 2012 relative au harcèlement sexuel (JO du 7 août)', to be published in *RDT*, Jan. 2013 .

However, a single act, committed to obtain a sexual act is also considered as sexual harassment, under certain conditions.

A penal sanction is provided in the Article 222-33 of the Penal Code: this text provides two years in prison and a penalty of EUR 30,000. These sanctions are increased if the offence is committed against a vulnerable person, sensible to the pressure of authority, or due to age, illness, etc.

Concerning the outcomes of the pending actions before courts when part of the first law was cancelled, the actions were not all stopped. Some of them could be carried out on the basis of other offences, and the law of August 2012 has provided the possibility to obtain some compensation for the prejudices linked to sexual harassment.[18]

[C] Stress

A lot of studies have been made and some collective agreements concluded about problems generated by stress at work,[19] even if there is no mention of stress situations in the Labour Code (the difficulty is that stress can be linked to different factors: harassment, bullying, work organization, work conditions, fear to be dismissed for economical reasons, etc ...). Some rulings of the *Cour de cassation* mention the stress, generally linked to health problems,[20] or sometimes to work organization.[21] Usually, the reference to stress is regarded as a psychological outcome of moral harassment, resulting in health problems.

[D] Working Time

The link between working time and health and safety has become progressively more visible, be it in the legislation or in case law. For instance, in the general chapter introducing health and safety rules the obligation for the employer to list into an individual fact sheet the specific risks the worker is exposed to is provided. Among these risks, working time - in particular specific working rhythm likely to produce some harm to the worker, is mentioned.[22] Among the problematic working rhythm, work at night play an important role.

Concerning this issue, some physical or psychological difficulties have been identified for night workers, and a specific set of rules applies to this working

18. See Article 12 of the new law. On this point, see M. Miné, 'Loi n° 2012-954 du 6 août 2012 relative au harcèlement sexuel (JO du 7 août)', to be published in *RDT*, Jan. 2013 .
19. Notably after the agreement concluded by the social partners at EU level, see C. Sachs-Durand, 'La transposition dans les Etats membres de l'accord conclu par les partenaires sociaux au sein de européenne, sur le stress au travail', *RDT* (2009): 243.
20. Cass. Soc. 30 March 2011, Légifrance, n° 09-71542; Cass. Soc. 21 September 2011, Légifrance n° 10-13880.
21. See for an example, Cass. Soc. 2 Mar. 2011, no. 09-11545 (France Telecom) and Cass. Soc. 26 January 2012, Légifrance n° 10-12.183, where the Court of Cassation mentions the stress, linked to some changes in the work organization.
22. Article L 4121-3-1 LC. This text has been introduced in the legislation by the law no. 2010-1330, of the 9 Nov. 2010. On this point, see V. Pontif, 'Les rythmes de travail', *RDT* (2012): 208.

situation.[23] France has been condemned by the European Court of Justice (ECJ) for discrimination between men and women, as it prohibited work at night for women (with some exceptions).[24] Nevertheless, the Government has decided not to bring the legislation concerning women in line with the provisions concerning men, but to create new rules for night workers in general. There is a definition of work at night, which is a work done between 9 pm and 6 am (with some derogations).[25] Work at night can be used only if a collective agreement has been concluded allowing it.[26] However, some derogations are possible.[27] It is considered as an exceptional situation[28] and night workers are submitted to a special medical supervision.[29] Work at night is also considered as a factor of hardness at work.

When the worker has health problems linked to his or her work at night, the employer has to propose a new position to him or her during the day. The worker cannot be dismissed only because s/he can no longer work at night, unless any reclassification is impossible or refused by the worker.[30]

Another issue related to working time is worth to be mentioned. In June 2011, there has been an important ruling of the *Cour de cassation* concerning a specific system of working time called *forfait cadre*. It is a specific contract, for assistant managers and managers[31] to have a kind of package payment: the working time is no longer counted (within some quite high limits); an extended working time is considered as normal for a manager, and included in his or her high remuneration. This working time organization must be provided by a collective agreement, be it a branch agreement or a plant agreement. So, unless the manager works clearly over the maximum limit of time (218[32] or 235[33] days a year, maximum working time allowed in the collective agreement) s/he cannot claim for the payment of overtime. One of the problems is the difficulty to prove the extension of working time, due to the kind of work performed.

This system has been condemned by the European Committee of Social Rights[34] but till now it has not been repealed. The interesting point is that the *Cour de cassation* has decided that if the employer does not respect a national agreement about working time for managers,[35] then the manager is allowed to claim for the payment of overtime.

23. Article L 3122-29 et seq. LC.
24. ECJ 13 Mar. 1997, RJS 4/97.493.
25. Article L 3122-29 LC
26. Article L 3122-33 LC.
27. Article L 3122-36 LC.
28. Article L 3122-32 LC.
29. Article L 3122-42 LC, R 3122-18 et seq. LC.
30. Article L 3122-45 LC. This provision has been created by an Act of 9 Nov. 2010, no. 2010-1330.
31. Article L 3121-42 on a yearly basis for managers, and L 3121-38, on a weekly or monthly basis for non managers.
32. Article L 3121-44 LC.
33. Article L 3121-45 LC under certain circumstances. On this question, see J. Pélissier, G. Auzero, E. Dockes, *Droit du travail, Précis* (Paris: Dalloz, 2012), 792, n° 772 et seq.
34. Decision of the 11 Dec. 2001 (claim no. 9/2000), 7 Dec. 2004 (claim no. 22/2003), 22 Jun. 2010 (claim no. 55/2009). See *inter alia* M. Miné, 'Le droit du temps de travail à la lumière de la Charte sociale européenne', *SSL*, no. 1475 (2011): 7; S. Laulom, 'L'organisation du temps de travail sous l'influence des droits sociaux européens', *RDT* (2011): 298.
35. Agreement of the 28 Jul. 1998, on the working time organization in the branch of metallurgy.

Moreover, the *Cour de cassation* stresses that the respect of this agreement aims at protecting the health and safety of the workers.[36] In the case at stake, the Court of Appeal should have checked the number of overtime hours and condemned the employer to pay. This statement is an upheaval for a lot of enterprises, and may lead to some important modifications in the management of the high rank staff. For our subject, the important point is that a link has been established by the *Cour de cassation* between working time and health and safety at work.

[E] Hardness (*Pénibilité*)

A new issue has appeared in the last years, due to the modifications in pensionable age, as an outcome of demographic problems:[37] hardness at work (*pénibilité*).[38]

The obligation to adapt certain tasks to the elderly workers has resulted, at first, in a call by the Government to the social partners to negotiate about hardness at work. The negotiation failed and the Government decided to legislate on the subject, and an act was passed in 2010.[39] According to this legislation, the employer must prevent hardness at work.[40]

The law gives a definition of 'hardness'. It supposes the exposure to one or several factors of risk (determined by decree),[41] linked to some strong physical constraints, an aggressive physical environment, or to certain rhythms of work likely to produce durable, identified, and irreversible effects for the health of the worker.[42] The employer must then adopt a sheet for each worker exposed to such working conditions. This sheet must also mention the prevention measures taken by the employer to suppress or reduce the risk. This sheet shall be provided to the occupational health doctor; when the worker leaves the firm, this sheet is given to him or her.

Besides, Social partners shall meet every three years to negotiate on hardness[43] and this issue has been added to the preventive tasks of the CHSCT (*Comité d'Hygiène,*

36. Cass. Soc. 29 Jun. 2011 no. 09-71107. For an analysis and a debate on this ruling, see B. Van Craeynest and P. Masson, 'Comment règlementer le travail des cadres?', *RDT* (2011): 474, and M.F. Mazars, S. Laulom and C. Dejours, 'Pot pourri autour de l'arrêt du 29 juin 2011, Santé et validité des forfaits en jours', *ibid.*, 481.
37. It has to be pointed out, however, that the age of retirement is about to be modified in some aspects by the new Government. M. Touraine, in charge of this issue has presented on the 6 Jun. a new project about age of retirement, specifically for the workers who started to work early in their life.
38. S. Fantoni-Quinton, J. Quandalle-Bernard 'La pénibilité au travail: un concept à géométrie (très) variable', *RDSS* (2012): 163. F. Petit, 'La pénibilité au travail, un nouveau risque professionnel ?', *Dr. Soc.* (2011): 262.
39. See *inter alia* A. Jolivet et V. Pueyo, 'Pénibilité au travail, de quoi parle-t-on?', RDT (2010): 686. B. Platel, 'Réformes des retraites et conséquences pour l'entreprise', *JCP S* (2010): 1523; B. Lardy-Pélissier, 'La pénibilité au delà de l'immédiat et du quantifiable', *RDT* (2011): 160.
40. Article L 4121-3-1 LC.
41. Decree no. 2011-354, 30 Mar. 2011; among other risks, we can mention: manual handling of loads, night work, work in a hyperbaric environment, working with certain chemicals).
42. Article L 4121-3-1 LC, in effect at the 1 January 2012.
43. Article L 2241-4 LC.

de Sécurité et des Conditions de Travail)[44] and to the mission of the medical care service.[45] Two decrees of 2011 have given more precision to the legislation.[46]

§4.03 PERSONAL SCOPE OF APPLICATION AND DUTIES OF THE PARTIES

[A] Personal Scope of Application

Obviously, the first person to be concerned is the employer [1]. Second, of course, the worker, with differences due to his or her specific situations [2], and third some other people present in the plan [3].

[1] The Employer

The first responsible of health and safety is the employer who owns the plant. Yet, also subcontractors and enterprises of services which operate outside their premises are subject to specific rules [a].

Besides, the temporal liability of an employer is limited when the worker develops an occupational disease after having been employed in several firms [b] and the spatial scope of application of the legislation may change the liability of the employer depending on in which part of the enterprise the risk occurs [b].

Last but not least, when the enterprise is big enough to deploy its activity in several places, there can be a delegation of power to a manager. This problem will be addressed by dealing with the worker's situation.

[a] The Subcontractors and Enterprises of Services

Concerning subcontractors and service providers, health and safety rules oblige the employer 'user' and the employer subcontractor or supplier of services to co-operate to set up a system of information and organization for health and safety at the workplace.[47] We cannot detail here all the provisions, but just give two significant indications: first of all the user is in charge of the general co-ordination of the prevention measures taken by him or herself or by the enterprise posting workers.[48] Such a co-ordination aims at preventing the risks induced by an interference between several enterprises operating in the same workplace.[49] This is an overall principle;

44. Article L 4612-2 LC.
45. Article L 4622-2 LC (Act of the 20 Jul. 2011)
46. Decrees no. 2011-824, JO 9 Jul. 2011 and no. 2011-823, JO 9 Jul. To have a general view of the present legislation, see the following summary: 'Prévention de la pénibilité: mise en oeuvre des obligations des enterprises et pénalités', *JCP S*, no. 29 (2011): 303; see also, 'Prévention de la pénibilité au travail: précisions administratives', *JCP S*, no. 45 (2011): 1517, about an administrative guideline.
47. Article R 4511-1 et seq. LC.
48. Article R 4511-5 LC.
49. Article R 4511-7 LC.

more precisely, we can mention the text providing a general duty of inspection of the workplace by the user and by the enterprise posting workers:[50] the law provides precisely the aspects to be checked and the decisions to be taken, such as the areas of danger, the ways where the circulation is possible and so on.[51]

These rules do not apply to the building sector, which has its own detailed regulation.[52] The building sector as well as the nuclear sector,[53] for instance, are looked at as particularly dangerous, deserving for that reason specific rules. The provisions for the building sector are very detailed, each problem linked to building activity being subject to several rules before the starting of the work, during and after the work, and for various types of activities like underground work,[54] electricity or excavation.[55] Some prescriptions are covering the activity of self-employed in this sector.[56]

[b] The Temporal and Spatial Personal Scope of the Legislation

As a rule, for occupational accidents, the present employer is liable. But s/he is not liable for occupational accidents or diseases that have occurred before, as the worker was employed by another company, unless the relapse of the consequences of an accident is due to the working conditions by the present employer.[57]

In case of occupational diseases, it is usually the last employer who is liable, unless s/he can prove that the disease results from the working conditions in a company where the worker was previously employed.[58] However, in case of exposition to the same professional risk in different companies, it becomes impossible to determine the respective liabilities. In that situation, the expenses due for the occupational disease are borne by a special found managed by the accident insurance and financed out of employers' contribution.

The employer is liable of course for the workplace, but also for the extensions of the plant, such as the canteen - if it is run by the employer - or the car park. If an accident occurs in such areas, the employer is liable unless the worker performs an activity not linked at all to his or her work.[59]

If an accident occurs during the way to work or back from work (commuting accident), the employer is no longer liable, but the worker can get compensation from the social security system, under certain conditions (normal way from home to work

50. Article R 4512-2 LC.
51. Article R 4512-3 LC.
52. Article L 4531-1 et seq., R 4532-1 et seq.
53. Article R 4521-1 LC, R 4451-36 LC et seq.
54. Article R 4534-40 LC for instance.
55. Article R 4534-24 LC for instance
56. Article R 4535-1 et seq. LC.
57. See D. Asquinazi-Bailleux, 'Détermination de l'entreprise dans laquelle la maladie a été contractée', comm. under Cass. Civ. 21 Oct. 2010, no. 09-67494, *JCP S*, no. 51 (2010): 1559.
58. Cass. Civ. 21 October 2010, no. 09-67494 above mentioned.
59. For all these issues of social security, see J.J. Dupeyroux, M. Borgetto, R. Lafore, R. Ruellan, *Droit de la sécurité sociale* (Paris: Dalloz Sirey, coll. Précis, 2011); JP. Chauchard, *Droit de la sécurité sociale* (Paris: LGDJ, collection manuel, 2010).

and vice versa, or way to the restaurant at lunch time etc.[60]). In this case, there is no presumption of occupational accident. The worker has to prove that s/he was on his or her way to work or home, or whatever the admitted venue.

Another situation is the accident occurring when the worker has been sent for a specific mission to another place: in that case, the accident is qualified as occupational accident, which means that the worker can be indemnified as if the accident had occurred at the workplace. The reasoning is the same if the accident occurs during the way to the mission or when having some daily activities, and even sometimes, when the worker has taken some shortcuts[61] during his or her mission.

A last situation must be addressed here: suicide (which has been an important problem in France in the last years) can be considered as an occupational accident if one essential reason of it is linked to work.[62] If the suicide is committed at the enterprise, the presumption that it is an occupational accident will operate. In that case, the employer who wants to escape any liability, has to prove that the reason of the suicide is completely other from the professional situation of the worker.[63] However, if the suicide is committed at home, the presumption does not operate and the link with the job must then be proved by the survivors.[64]

[2] The Worker

The worker is protected by the law. Some specific rules are provided for young workers[65] and for women.[66]

However, it is no longer only the worker of a specified employer, as contractor, who is submitted to the health and safety rules: employers, when they are operating in the same place, can be held liable for health and safety, and their responsibilities are covering their own workers, as well as other workers participating in the working process.

The employer is liable for the safety of his or her own workers, regardless the typology of their employment contract (fixed term, temporary, apprenticeship etc.).

In certain branches, like building sector, self-employed workers are protected too.[67]

---

60. A decision of the Cour de cassation admits that an accident during daily current activities may be taken into account as commuting accidents: see Cass. Soc. 19 Jul. 2001 no. 99-21536 Bull. 2001 V no. 285 p. 228.
61. Cass. Soc. 3 Jun. 2010, no. 09-65552.
62. G. Pignarre, 'Souffrance au travail et réparation des risques professionnels', *RDT* (2010): 305. L. Lerouge, 'Suicide du salarié et faute inexcusable de l'employeur: quelles évolutions juridiques?', *RDSS* (2012): 373.
63. Cass. Soc. 16 Oct. 1997, no. 96-11214.
64. Cass. Soc. 22 Feb. 2007, no.05-613771.
65. Articles L 3161-1 et seq. for instance dedicated – *inter alia* – to working time or different activities.
66. See L 1225-1 et seq. LC for protection of pregnant women, for instance.
67. Article L 4532-2 LC for prevention, and L 4535-1 LC during the working period, for instance.

A specific provision takes into account trainees, in some circumstances.[68]

The employer is liable - to a certain extend - for posted workers (of course in case of temporary work[69] but also for subcontractor's employees in the workplace). Yet, the employer of the posted workers (provider) remains the main liable person as regards certain aspects of health and safety.[70]

If an accident occurs to a temporary worker, the costs of the accident are normally divided between the user and the provider.[71]

[a] Executive Managers with a Delegation of Power

According to Article L 4741-7 LC, the employer is liable in case of condemnation of his or her managers and other workers.

Yet, in big firms, the employer may delegate powers to one or more of his or her managers: in this situation, the manager becomes liable, at least concerning criminal liability for the infringement of health and safety rules,[72] however civil liability remains on the employer except in case of heavy fault of the delegate.[73]

However, the liability of the manager - in health and safety or in other fields - can be recognized only if the delegation is clearly defined and the manager has the necessary knowledge and experience to be reasonably given this kind of delegation. This is not the case if the delegation has been given to a worker of twenty-one, employed within the company since only one year.[74]

Very often, as a consequence of the infringement of their duties, manager will be a dismissed for heavy misconduct.[75]

[3] Anybody Who Finds Him or Herself at the Workplace

Sometimes the liability of the employer exists towards people completely outside of the working process, i.e., customers or visitors.[76]

There are very few provisions in the Labour Code, mentioning these people. In the general prescriptions on health and safety, for instance, people likely to be present in the plant have to be taken into account when defining safety rules.[77]

68. Article L 4154-3 LC, providing the responsibility of an inexcusable fault, for the employer, for fixed term contracts, temporary workers, and workers in training period if they did not have the relevant information and training for the job.
69. Article L 1251-21 LC.
70. Article L 1251-22 LC for temporary workers, for instance. The temporary work agency is responsible, for instance for the obligations concerning the occupational medicine.
71. Article L 241-5 SSC.
72. See for instance Cass. Crim. 4 Jan. 2011, no. 10-82622.
73. Cass. Crim. 28 Mar. 2006 no. 05-82975 Bull. Crim. 2006 no. 91 p. 349.
74. Cass. Crim. 8 Dec. 2009, no. 09-82183 Bull. Crim. 2009, no. 210.
75. See for instance, Cass. Soc. 22 Mar. 2011, Légifrance n° 09-42317.
76. See C. Sachs-Durand, 'Evaluation des effectifs et droit à la santé', *RDT* (2006): 106.
77. Article R 4227-3 LC. In nuclear plants for instance, a system of safety in case of fire is provided for all the people occupied in the plant (L 4525-1).

[B] Duties of the Parties

Employer's duties [1] are, of course, much more important than worker's duties [2].

[1] Duties of the Employer

The first duty of the employer is to prevent the workers from any professional risk in the firm: this obligation is provided in Articles L 4121-1 et seq. LC. The general provision on this field is provided in Article L 4121-3: the employer must assess the general risks due to the working process, the machines, substances including chemical ones, used. Following this assessment, s/he has to organize the best level of protection for workers, including managers.[78] Some specific information and training must also be provided. If any change occurs in the work organization, the employer must take measures to adapt the working process.[79] Article L 4121-2 is one of the most important in this field: it is deriving from the Framework Directive of 1989, which has been implemented in France in 1991.[80] One of the duties of the employer, according to this text is to avoid the risks and to assess the risks that cannot be avoided in the enterprise.[81] It also requires employers to give priority to collective protection over individual protection[82] and to make a prevention plan integrating all the risks possible in the plant.[83] The employer has also the obligation to join an occupational health service, whose missions are defined in Article L 4622-3 LC: the occupational health doctor must prevent any alteration of the workers health.

The impact of these texts, and mainly the result of the implemented Framework Directive on health and safety, has been reinforced by the case law. According to the judges, the main duty for the employer is to keep his or her workers safe according to the 'safety obligation'[84] (*obligation de sécurité*) (2a1), established by the *Cour de cassation* interpreting Article L 4121-1 LC.

Stemming from the transposition of the Framework Directive of 1989,[85] Article L 4121-1 LC provides that 'the employer shall take the necessary decisions so that the security and the protection of the physical and mental health of the workers is provided.'

In a very famous decision of the 28 February 2002,[86] in the well-known affair of asbestos, the *Cour de cassation* by interpreting Article L 4121-1 LC, has laid down the

78. Article L 4121-3 LC.
79. Article L 4121-1 LC
80. Loi n° 91-1414 du 31 décembre 1991.
81. Article L 4121-2 1° and 2° LC.
82. Article L 4121-2 8° LC
83. Article L 4121-7° LC
84. P.Y. Verkindt, 'L'obligation de sécurité de résultat...encore et toujours !!' *Cahiers de droit de l'entreprise,* no. 1 (2011): file 5. M. Blatman, 'L'obligation de sécurité', *Dr. Soc.* (2011): 743.
85. Directive 89/391/EEC of 12 Jun. 1989 on the introduction of measures to encourage improvements in health and safety of workers at work.
86. Cass. Soc. 28 Feb. 2002, no. 99-17201, Bull. 2002 V no. 81 p. 74, A. Lyon-Caen, 'Une révolution dans le droit des accidents du travail', *Dr. Soc.* (2002): 446.

above mentioned general obligation for the employer, which has known, since then, relevant developments.[87]

In this ruling, the *Cour de cassation* has stated that 'due to the contract of employment, the employer has an "obligation to get a security result" (*obligation de sécurité de résultat*) concerning his worker (...)'. This means that the employer must succeed in protecting his or her workers against professional diseases[88] as well as occupational accidents.[89] Since then, this obligation has been confirmed many times and case law on the subject has known a wide extension in France: it has been used, for instance, in case of violation of the smoking ban at the workplace (in favour of a barman exposed to the smoke of cigarette[90]). Moreover, employer's liability may exist even if no infringement of rules or fault occurs. For example, if the worker is victim of bullying at work, the employer is condemned, even if s/he had taken measures to put an end to the bullying.[91] Besides, under certain circumstances, a type of organization which is likely to be dangerous for a worker - even if no specific rule has been violated - may be suspended by the judge.[92]

This is an audacious interpretation, very favourable to the worker. No doubt that the *Cour de cassation* is willing to protect worker's health, being aware of the situation in France, where a lot of problems raised in health and safety domain are not seriously identified and thus correctly handled.[93]

Concerning prevention, it must be said that there are national institutions dealing with prevention at work: for instance, we can mention the *Institut National de la Recherche et de Sécurité* (INRS) dealing with prevention of risks at work, or the *Caisse Nationale d'Assurance Maladie* (CNAM, in its specific branch dealing with occupational accidents and diseases.[94]

[2] Duties of the Worker

Usually, the worker is not liable for health and safety as the liability lies only on the employer. But there have been some discussions about a possible liability of workers,

87. See for instance S. Brimo, 'L'indemnisation des victimes de l'amiante', *Dr. Ad.*, no. 12 (2011): comm. 102.' Nouveau dispositif réglementaire de protection des travailleurs de l'amiante', *JCP S*, no. 20 (2012): 242.
88. Cass. Soc. 28 Feb. 2002, above mentioned.
89. Cass. Soc. 11 Apr. 2002 no. 00-16535, Bull. 2002 V no. 127 p. 133.
90. Cass. Soc. 6 October 2010, Légifrance n° 09-65103.
91. Cass. Soc. 3 Feb. 2010, no. 08-40144, M. Vericel, 'L'obligation patronale de sécurité de résultat: un regime renforcé', *RDT* (2010): 303.
92. Cass. Soc. 5 Mar. 2008 no. 06-45888, Bull. 2008, V, no. 46. In this case, the organization of the work had been modified in a plant classified 'Seveso' and the security rules had been alleviated, leaving the worker in charge of safety problems alone in the site at certain periods.
93. This is due to different factors (the worker does not want to be fired as s/he declares the problem, the enterprise does not wish to have a high rate of accidents or diseases, for various reasons, etc.
94. We can also mention the Anses, Agence nationale de sécurité sanitaire de l'alimentation, de l'environnement et du travail, created in 2010, which has a general role in the society but deals also with professional risks.

due to the wording of a provision included in the Framework Directive.[95] Article L 4122-1 LC provides that the worker shall take care of his or her safety and of that one of other persons who may be affected by his or her activity. The issue at stake here is to know whether the worker might be held liable for the infringement of safety regulations.[96] The judges, in a few cases have recognized the liability of the worker on the ground of Article L 4122-1 LC (be it a simple fault or a heavy misconduct, like to refuse to wear the security helmet or whatever security equipment it may be).[97] But mainly, in these situations, the only outcome would be a dismissal for fault or heavy misconduct,[98] not a condemnation for civil liability. If the act is also qualified as a criminal offence, a penal sanction will be inflicted. But in no ways this relieves the employer from his or her general liability.

Another point must be stressed. The worker is allowed to stop his or her work when s/he thinks that it is dangerous. This is called *droit de retrait* (right to leave the premises).[99] According to this, the worker has to alert the employer if s/he thinks that s/he is exposed to a heavy and imminent danger for his or her life or health. Under such a condition, s/he is allowed to stop working.[100] Thus, the employer must pay the worker during this time and cannot sanction or dismiss him or her for that reason. A point to underline is that the fact of not using this provision is not a fault[101] for the worker and does not discharge the employer of his or her liability.

Yet, if the worker uses this possibility in an inappropriate way, s/he might be sanctioned by the employer.[102] Case law in this matter is rather abundant: the issue at stake here is to determine whether the worker had a serious reason to think that his or her life or health was in heavy and imminent danger o not. According to case law, this situation must be appreciated with reference to the person of the worker, his or her age, professional skills, etc. As for the danger, it can be internal (for instance a medical problem of the worker, like an allergy)[103] or also – as it is most of the time – external thus including various situations such as the use of toxic paint[104] or an absolutely excessive working time.[105] If the provision is not applicable to the situation the worker may be sanctioned or even dismissed.[106]

95. Article 13 of the Framework Directive on health and safety, 89/391/CEE of the 12 Jun. 1989.
96. On this issue, see Y. Saint-Jours, 'L'obligation de sécurité du salarié: prévention et responsabilité', *Dr. Ouvr.* (2007): 560.
97. Cass. Soc. 28 May 2008, no. 06-40629; Cass. Soc. 21 Sep. 2011, no. 10-19260.
98. Cass. Soc. 23 Mar. 2005, RJS 2005, 454 no. 641. See also Cass. Soc. 21 Sep. 2011, above mentioned.
99. Article L 4131-1 LC.
100. See for instance, Cass. Soc. 2 Mar. 2010, no. 08-45086.
101. Cass. Soc. 9 Dec. 2003 'obs. Ph. Waquet', *Dr. Soc.* (2004): 313. Cass. Soc. 18 December 2007, Légifrance n° 06-43801.
102. Cass. Soc. 19 December 2007, Légifrance n° 06-43143.
103. Cass. Soc. 20 Mar. 1996, 'obs. J. Savatier' *Dr. Soc.* (1996): 684.
104. Cass. Soc. 22 September 2011, Légifrance n° 10-10928.
105. Cass. Soc. 2 March 2010 Légifrance n° 08-45086.
106. Cass. Soc. 23 June 2010 Légifrance n° 08-45308.

Anyhow, this possibility must be combined with the specific safety obligation that relies on the worker. Using his or her *droit de retrait*, the worker must not put at risk other workers or other people present in the plant.[107]

Moreover, according to Article L 4131-1 LC, the worker has to inform the employer of any situation of danger for him or herself, as well as any bad working protection system s/he has noticed. It is unclear whether the worker could be held liable for not having informed the employer of such a situation.

Anyhow, when the alert has been given and an accident occurs, the employer is held liable for an inexcusable fault.[108]

§4.04 WORKERS REPRESENTATIVES

Workers representatives play an important role in the field of health and safety, especially the *Comité d'Hygiène, de Sécurité et des Conditions de travail* (hereinafter 'CHSCT') which is dedicated to this specific mission (3a). Other institutions might also have some role to play, but in a minor way (3b).

[A] The CHSCT

The CHSCT is set up in companies employing at least fifty workers.[109] Under this threshold, the *délégués du personnel*[110] (another worker's representatives body) are in charge of its duties.

The members of the CHSCT are chosen by the members of the works council and the *délégués du personnel*,[111] according to their skills on health and safety issues.[112] Any competent worker of the plant may be appointed. The employer chairs this body, as well as the works council.

First of all, the CHSCT is informed and consulted in case of any change having some influence on the organization of work and the working conditions. The CHSCT must analyse the modifications of working conditions to assess whether they are likely to result in medical problems for the workers,[113] as, for instance, an important modification in working time (especially when the work is done at night or on three shifts).

The consultation takes place approximately according to the same model used for the works council: the CHSCT must be informed in a satisfactory way, and is consulted.[114] However, the employer is not obliged to follow its advice. In any case, the CHSCT has the right to visit regularly the premises in order to check the working

107. Article L 4132-1 LC.
108. Article L 4131-4 LC.
109. Article L 4611-1 LC.
110. Articles L 2311-1 et seq.
111. Article L 4613-1 LC.
112. Articles L 4613-1 et seq.
113. Articles L 4612-1 et seq.
114. Article L 4612-8 LC.

conditions[115] and to make inquiries. If the CHSCT identifies any risky situation for the workers, it has to give an alert to the employer.[116] Then a procedure has to be started by the employer in order to check whether the danger is real. The CHSCT has no possibility to stop the machines, even if is some danger exist. In such a case, it is up to the worker to make use of his or her *droit de retrait*.

The CHSCT may have recourse to experts in the field of health and safety.[117]

This recourse takes place in different situations, i.e., when an important modification of the work organization might have some consequences on the health and safety of the workers or when a specific risk has been identified in the plant.[118] The expert is paid by the employer: normally the agreement of the latter is required. But if the employer disagrees, the CHSCT may sue him or her before court, according to a summary procedure. It is up to the judge to decide whether the recourse to the expert is justified, but in a normal situation (without any abuse in the choice of the expert), the judge shall not question the choice of the CHSCT.[119]

As we said before, CHSCT has no power to stop the work if any danger is observed;[120] there was a few years ago a debate on this point, but in the end this power was not really wanted even by workers' representatives, as they feared to be considered liable in case of an occurring damage.[121]

[B] Other Institutions

As we have seen before, the *délégués du personnel* may operate in the field of health and safety [1]. Trade unions may also play some role, but in a different way [2].

[1] The délégués du personnel

Délégués du personnel elected on trade unions lists, are the only form of workers' representation present in undertakings between the thresholds of eleven and fifty workers. So its mission is very wide, as expressed in the Labour Code.[122] The *délégués du personnel* must promote the respect of the Labour legislation, concerning *inter alia*, health and safety. They have the possibility to inform the labour inspectorate of any claim or remarks in the field of their competence.[123] The *délégués du personnel* may

115. Article L 4612-4 LC.
116. Article L 4131-2 et L 4132-2 LC.
117. Articles L 4614-12 LC et seq.
118. As an example, we can mention Cass. Soc. 26 January 2012, Légifrance no. 10-12183, which is in between the two possible situations: in this firm, a new organization of work was resulting in a strong stress for the workers. In that case, the Court has considered that the recourse to an expert was justified.
119. Cass. Soc. 26 Jun. 2001, Bull. Civ. V no. 231, *RJS* (2001): 782.
120. But, as we shall see later on, the judge has this power, in case of danger.
121. For more information, see below.
122. Article L 2313-1 LC.
123. Article L 2313-1 al.2 LC.

submit claims and observations also on behalf of workers operating within the plant without being employed by the company.[124]

It must be said that they have no information and consultation power, except on some limited issues, and *a fortiori* no codetermination power. Anyhow, when they replace the CHSCT, when the latter has not been appointed in an enterprise over fifty workers then they have an information and consultation power on the topics of health and safety.[125]

Besides, a specific duty of the *délégués du personnel* must be mentioned. If they note an infringement of the rules concerning the personal rights of a worker (including the health and safety), they must immediately inform the employer. A procedure is then started with the employer to remedy to this situation. If no agreement can be found on the importance of the problem, or the relevant solution, the worker, or the delegate, if the worker agrees, may go to court.[126]

[2] Trade Unions

In France, trade unions may settle in the plant.[127] Their main role is to present claims to the employer about working conditions, wages etc. and to bargain on them. The health and safety domain is included. However, it must be said that in this field agreements cannot derogate but only improve the legal protection.

Nevertheless, in the field of working time for instance, some derogatory agreements, which are allowed under certain conditions, may have some consequences on health and safety.

As an example of agreements including provisions on health and safety, we can mention the agreement concluded in December 2011 in the building sector, on hardness.[128]

§4.05 CONSEQUENCES ON THE PARTIES AND REMEDIES IN CASE OF VIOLATION OF HEALTH AND SAFETY PROVISIONS

[A] Entitlement and Enforcement vis-à-vis Compliance with Health and Safety Provisions

The first actor whose liability is in question is of course the employer [1] and there are some institutions whose mission is to point out the infringements of the employer [2].

124. Article L 2313-3 LC.
125. Article L 2313-16 LC.
126. In that case, it will be the *Conseil des prud'hommes*, taking a decision in a specific form (Article L 2313-2 *in fine*).
127. Article L 2142-1 LC.
128. See a summary of this agreement in the web site Dialogue social, Article of the 27 Feb. 2012, by H. El Fattah, University of Strasbourg. http://www.dialogue-social.fr.

[1] The Employer

The approach to liability in this domain has also changed. Limited at first, the liability of the employer has increased, as a result of the EU influence on French labour law and of the interpretation of the *Cour de cassation*.

Moreover, criminal liability of the employer has been developed, as well as a shared liability in some cases. Besides, corporate persons can now also be held liable under criminal law.

[a] Liability of the Employer as an Individual

In France, like elsewhere in Europe, different types of liabilities exist. As said before, the employer is liable for occupational accidents or diseases occurring at work, in two ways: the obligation to get a result in safety field (*obligation de résultat*) which has been previously considered as a contractual liability, but is now more and more regarded as a tort liability, and a criminal liability.

For a very long time, the civil liability of the employer has been limited by the system of indemnification.[129] The political compromise at the origin of this system was a kind of objective liability, by which the employee had not to prove the fault of the employer. This insurance system was at first built up for occupational accidents. Concerning occupational diseases, this was implemented by a system of tables which has been subject to a significant evolution.[130]

This organization was quite satisfactory for the workers and the employers for a certain period of time. But in comparison with the evolution of liability in other fields of law, the social security system looked old fashioned and unsatisfactory for the workers. That is one of the reasons of the development of the concept of inexcusable fault of the employer, used to increase the indemnification of the worker, being the recourse to civil liability precluded (although with some exceptions).[131] Thus the possibilities to compensate the damages have progressively increased.[132]

The notion of inexcusable fault is a key notion in this field. This notion is used in social security law[133] and has an influence on the compensation provided to the worker after an occupational accident or disease, whoever the author may be (the employer or

129. Here there was a presumption qualifying the accident of work accident at that moment.
130. The number of occupational diseases recognized by the law has increased a lot since 1919.: Presently, there are around one hundred, and there is, since 1993, a possibility to obtain some compensation for occupational illnesses that is not listed in the tables.
131. On this question, see G. Viney, 'L'évolution du droit de l'indemnisation des victimes d'accidents du travail et de maladies professionnelle', *Dr. Soc.* (2011): 964.
132. The last step in such a direction has been made by the Constitutional Court that has recently decided that a claimant could obtain a full compensation of the damages linked to an occupational accident or disease – in case of an inexcusable fault of the employer – beyond the indemnification provided by the social security law. See Cons. Const., no. 2010-8, QPC, 18 Jun. 2010, point 16.
133. L 452-1 Social Security Code (SSC). The Labour Code refers also to this notion (for instance in Article L 4131-4 LC).

the worker).[134] There is no definition of this fault in the law. The judge has given a first definition in 1941[135] which was substantially revised in 2002. This definition varies depending upon the author of the fault.

If the author is the employer, there is an inexcusable fault 'when he was or should have been aware of the danger existing for the worker and has not taken the relevant measures to protect him'.[136] Besides, the judge states that it is no longer necessary that the fault of the employer have been determining as regards the damage: it is enough that his or her fault has been necessary to cause the damage.[137] In addition, the employer may be held liable, even if s/he has taken some measures to stop the danger.[138]

We can give some illustrations of this concept of inexcusable fault. For instance, if a tool has been left after some work done in a supermarket, and the worker is harmed by this tool, there is an inexcusable fault, as the employer should not have opened the supermarket without taking all the tools away.[139] In another case, the *Cour de cassation* stated that if the day is cold and snowy and if there are icy patches in the close access to the firm, the employer is liable of an accident even if s/he had taken measures to protect the workers (like salt, sand etc.). The judge considers that the employer has not taken the right measures, if icy patches were still there.[140]

If the judge recognizes the inexcusable fault, then the worker has the right to a supplementary compensation after an accident or a professional disease and not only to the fixed allowance provided in general cases. The social security tribunal allows, in that case, some complements to the indemnity. But we must keep in mind that even this complemented indemnification does not reach the amount of full compensation which had resulted from the application of general principles on civil liability. As seen before, this situation is likely to change.[141]

The social security institutions, which pay the rent or the indemnity to the worker, claim afterwards a reimbursement by the employer, for the additional payment.

If the worker is the author of an inexcusable fault, this may reduce the social security benefit.[142] The *Cour de cassation* defines the inexcusable fault of the worker as

134. J.-L. Gillet, D. Loriferne, X. Pretot. *Assurance et protection sociale* (Paris, Dalloz, 2011).
135. Cass. Ch. Réunies, 15 July 1941, n° 00-26.836.
136. Cass. Soc. 28 Feb. 2002, no. 99-17201 Bull. 2002 V no. 81 p. 74 ; see also Cass. Soc. 11 Apr. 2002, no. 00-16535, Bull. 2002 V no. 127 p. 133 see for a recent example, Cass. Soc. 10 May 2012, D. 2012, p. 1342.
137. Cass. Ass. Plén. 24 June 2005, n° 2005 n° 03-30038, Bull. 2005 A. P. n° 7 p. 16.
138. Cass. Soc. 3 Feb. 2010, no. 08-40144, Bull. 2010, V, no. 30.
139. Cass. Civ. 1 June 2011, Légifrance, n° 10-19147.
140. Cass. Civ. 1 June 2011 Légifrance n°10-20029 .
141. See the decision of the Constitutional Court, 18 Jun. 2010, at fn. See also Cass. Civ. 30 Jun. 2011 no. 10-19475P which adopts the same position. Some modification of the law have been proposed, but at the moment they have not yet been discussed in the Parliament. On this problem, see G. Viney, 'L'évolution du droit de l'indemnisation des victimes d'accidents du travail et de maladies professionnelles', *Dr. Soc.*, no. 702 (2011): 964.
142. Article L 453-1 of the Social Security Code. See Cass. Civ. 27 Jan. 2004, no. 02-30693, Bull. 2004 II no. 25 p. 20.

'a deliberate fault, of an exceptional seriousness, creating without any relevant reason a danger, from which the worker should have been aware of'.[143]

However, the interpretation of the inexcusable fault of the employee is quite restrictive.

On the contrary, the *Cour de cassation* gives now a wide interpretation of the inexcusable fault of the employer which was first restrictively defined. By doing so, the Court tries to make the employer more conscious and concerned with the problem, and tries also to give a better protection to the worker.

[b] Liability of the Company: Introduction of Criminal Liability

As far as civil liability is concerned rules are quite classical.

On the contrary, concerning criminal liability, there has been an evolution. Criminal liability has been introduced for companies in 1992, to a certain extent,[144] and then generalized in 2005[145] to all offences (except specific ones listed within the law);[146] CRIMINAL liability may be shared between a worker of the company (often under the condition that s/he has a delegation of power) and the company itself. We have an example of such a shared liability in a decision of the *Cour de cassation*.[147] In the specific case, a worker walked on a path where a crane was in operation and died from an accident due to the violation of safety rules. The manager and the company have been both condemned.[148]

In case of delegation of power, the person who delegates is not liable, but the company still is.

[2] Compliance and Control

First of all, there are some institutions whose traditional mission is to control the respect of the law, among others, in the field of health and safety. The Labour inspection is one of them [a], the Health medical service is another one [b], and the social security institutions have also a role to play [c]. Besides, we must mention a new actor in the plant, who is in charge of prevention, and health and safety: it is a worker, specifically appointed for this mission [d].

143. Cass. Civ. 17 January 2007, Légifrance n° 05-17701.
144. A. Coeuret, 'La responsabilité en droit pénal du travail: continuité et rupture', *RSC* (1992): 475. E. Fortis, A. Coeuret, *Droit pénal du travail* (Paris: Lexis Nexis, 2012).
145. Article 121-2 of the Penal Code, see also A. Vallat et A. Coeuret, 'Droit pénal et accident du travail après la loi Perben II', *SSL*, no. 1206 (2005): p. 7.
146. Law 9 Mar. 2004, Articles 223-1 NCP (new Penal Code) for general infringements and Article 221-6 NCP, as well as for specific violations of the law.
147. Cass. Crim. 15 January 2008, Légifrance n° 06-87805.
148. Cass. Crim. 24 May 2011, Légifrance n° 10-82285. F. Duquesne, 'Imputation de la faute du délégataire commun à la personne morale: vers la clarification? – Cour de cassation, chambre criminelle, 23 nov. 2010', *Dr. Soc.*, no. 698 (2011): 361.

[a] Labour Inspection

Labour inspectors and labour controllers (mainly operating in small enterprises) are civil servants depending upon the *Direction régionale des entreprises, de la concurrence, de la consommation, du travail et de l'emploi* [DIRECCTE].[149]

The role of labour inspectors is to check the application of labour law, in all its aspects, in the enterprises, included of course, rules regarding health and safety.

They have the power to enter the premises and verify the working conditions, the duty to advise the employer and the workers if they notice any problem. They can discuss with the workers and workers representatives, and ask the communication of any document of the firm, in the scope of their competence.

Inspectors have the possibility to take some samples of the product used in the enterprise or ask for a specialist to check the installations in the plant.[150]

They also have a coercive power vis-à-vis the employers: the power to make some injunctions, and in exceptional cases, to stop the work.[151] But this last possibility exists only in the building sector and in case of use of certain chemical products.[152] In other cases, they have to refer to the judge, who has the power to stop the activity of the company.[153] Labour inspectors may observe some violations of the rules constituting penal offences. In such cases, they advise the enterprise to suppress the irregularities within an allotted time,[154] and if there is no result, transmit an official report to the judge.

The big problem in France is that very often these claims have no effect as the judge usually decides not to bring any action against the employer for different reasons (overwork, protection of enterprises, especially small ones, etc.).

Employers must keep the documents related to the actions and demands of the labour inspector for five years,[155] and they have the obligation to transmit these documents to the CHSCT and other institutions of social security or to some specific institutions, like in the building sector the Professional Organism of prevention (*Organisme Professionnel de Prévention du Bâtiment et des Travaux Publics* – OPPBTP).

[b] The Labour Health Service

Other important actors are the Labour Health Services, among which are the occupational health doctors. Their status has been modified very recently.[156] The role of the

149. Regional direction of enterprises, competition, consumption, work and employment.
150. Article L 4722-1 *and seq*. LC.
151. It is the case in the building sector, Article L 4731-1 LC.
152. Article L 4731-1 LC for building industries and L 4731-8 for chemicals.
153. Article L 4732-1 LC.
154. Article L 4721-1 et seq. LC. The director of the DIRECCTE is also allowed to do so.
155. Article D 4711-3 LC.
156. Loi n° 2011-867 du 20 juillet 2011 relative à l'organisation de la médecine du travail, Article L 4622-1 LC et seq. See also the two decrees 2012-135 and 2012-137 of the 30 Jan. 2012, in force on the 1 Jul. 2012. On this reform, see M. Delaunay et G. Lefrand, 'Rapport d'information sur la mise en oeuvre de la loi n° 2011-867 du 20 juillet 2011 relative à l'organisation de la médecine du travail' (Paris, Assemblée Nationale, rapport n° 4462, 7 mars 2012). See also M.

occupational health doctor is only preventive. S/he has no right to medical prescription, but s/he must give advices to the employer, and take decisions about the fitness or unfitness of workers.

The recent reform is due to the lack of occupational health doctors, in relation with several factors: the wages, the difficulties of their mission, the fact that they cannot really treat the workers, which can be frustrating in a way, their difficult position, between the pressure of the employer and the interests of the worker. The reform did not change deeply the role of the occupational health doctor, but suppressed the quotas of workers to be checked in a certain period of time, as well as some specific health controls for a certain type of workers[157] and provides some other measures which do not seem always very favourable for the workers.

[c] The Caisse Nationale d'Assurance Maladie (CNAM)

The *Caisse Nationale d'Assurance maladie* (CNAM), the national institution in charge of general health care has to manage also the professional risks, but these risks are taken in charge by a specific branch (that of *Accidents du travail et maladies professionnelles – ATMP*). This institution must at the same time pay the indemnities to the workers, maintain the financial balance among the various branches[158] and develop a policy of prevention as regards risks at work.

[d] A Worker (or Another Competent Person) Specifically Appointed

This new actor has been introduced by an Act of the 20 July 2011,[159] amending the Labour Code.[160] According to the new text, the employer shall appoint one or several workers competent on the field of health and safety, in order to develop an activity of protection and prevention from occupational risks.

These workers may receive, on their demand, a training adequate to their mission.

If there are no competent workers in the company, the employer may ask, after a consultation of the CHSCT, to a person belonging to the Health service covering the enterprise, or to other institutions operating in the field, to play this role. Apparently, the employer is relatively free to decide whether or not there is someone competent in the company.[161]

Caron, P.Y. Verkindt, 'La réforme de la médecine du travail n'est plus (tout à fait) un serpent de mer', *JCP S*, no. 39 (2011): 1421.

157. '*Arrêté*' of the 2 May 2012, repealing 12 '*arrêtés*' providing a specific health control for certain workers.
158. In this respect, the *ATMP* contributes every year to the general health fund in order to compensate what has been spent for the undeclared occupational accidents or diseases, taken in charge by the *CNAM*.
159. Nr 2011-867 du 20 juillet 2011 – Article 1 (V).
160. Article L 4644-1 LC.
161. On this new institution, see the Article on the health service reform.

This provision raises some problems. First of all, nothing is said about the protection of the workers appointed. Second, no threshold is provided, meaning that this person should be designated in any enterprise. Last but not least, there could be a risk of competition with the CHSCT, when present in the company.

[B] Sanctions and Remedies

Sanctions are linked to the liability system. As mentioned before, in the field of health and safety the liability might be civil or criminal. As regards civil liability, the employer is liable for him or herself of course, but also in case of condemnations of his or her managers.[162] In case of inexcusable fault of the employer, normally the rate of social security contribution due for occupational accidents or diseases is increased, and the employer shall compensate the additional prejudice to the victim or to his or her survivors.

Moreover, there are some practical sanctions, such as stopping the work (see above), or suspending a new organization of the work.[163]

Criminal offences as regards health and safety are gathered in a specific chapter of the Labour Code.[164] The employer can be condemned[165] as well as a manager with a delegation of power.[166] The sanctions are a fine, or, in case of second offence, prison.

Some additional obligations or sanctions may be decided to put the plant in conformity with health and safety rules,[167] such as informing workers and customers that the company has been condemned, by posting the text of the decision at the gates of the company. The prohibition to perform certain functions within the company or within controlled companies for a maximum of five years may also be decided.[168]

However, the condemnation to prison is very rare in practice, and, should it be, it would probably be with a suspended sentence.

The third type of sanction is the increase of the contribution due by the employer, i.e., the amount of money s/he has to pay in consideration of the rate and gravity of the occupational accidents and diseases in the firm, or in the branch. The system is rather complicated and has been recently modified. This legislative modification is effective since the beginning of 2012.[169] It modifies the rules without changing completely the general framework of the system.

The reform refers to three main points. First of all, the threshold relevant to determine whether an enterprise is submitted to an individual, mixed or collective tariff has been modified. The collective tariff is calculated on an average basis, the mixed is

162. Article L 4741-7 LC.
163. Cass. Soc. 8 mars 2008, above mentioned.
164. Articles L 4741-1 et seq. LC.
165. B. Lapérou Scheneider, 'La responsabilité pénale de l'employeur personne physique et la présomption de faute', *Dr. Soc.* (2012): 273.
166. Article L 4741-1 LC. A. Coeuret, 'La délégation de pouvoirs et les responsabilités pénales dans l'entreprise', *RJS*, no. 1 (2012): Chron. 3.
167. Article L 4741-4 LC.
168. Article L 4741-5 LC.
169. Decree of the 5 Jul. 2010, JO 7 Jul., p. 12320.

partly collective, partly individual and the individual is calculated on the costs of occupational accidents and disease of the enterprise itself in correspondence with the salaries of the workers in this enterprise. The threshold for small companies in order to apply the collective rate, has been increased from nine workers to nineteen workers. The threshold of application of the individual rate (for big enterprises) has been lowered to 150 workers, instead of 200. In between, the enterprises from 20 to 149 workers are submitted to the mixed rate. Besides, it is now possible for companies articulated in several establishments, to choose a unique rate of payment instead of a specific one for every single part of the undertaking.

However, the main, rather controversial, modification refers to the new mode of imputation of the payment for occupational accidents and diseases. There will be no longer an assessment on the real costs of each accident but the evaluation will be on the basis of an average cost, calculated with reference to the branch. The essential criticism to this provision is that a real prevention strategy will no more be incentivized, as the payment will no longer be related to the situation within the single company.[170]

We can also mention a specific duty of the employer, which is not really a sanction – or it has not been looked at like that – because it is not always the consequence of a fault of the employer, i.e., the mandatory reinstatement of the worker who has been victim of an occupational accident or disease linked to his or her work. If the worker is found unfit to resume his or her previous job, the employer must find a new position as comparable as possible to the old one, but consistent with his or her new capabilities.[171] The worker has the possibility to refuse the proposal,[172] and then the employer must find a new convenient position. When the reinstatement is impossible within the company, the employer is allowed to dismiss the worker.[173] Yet, in case of such a dismissal, the indemnification is higher than that provided for dismissals on other grounds.[174]

§4.06 CONCLUSIONS

Drawing some conclusions, it is possible to identify two specificities of the French system.

First of all, it is inspired to a 'written law conception', which means that all the rules are laid down in the law in a precise and detailed way. Each branch with specificities concerning health and safety has its own rules (chemicals, nuclear industry and, above all, the building sector which probably has the most important set of rules, as it is also the branch where there is the maximum rate of occupational accidents).

170. On this reform, see P. Bodin, D. Nauleau, 'La réforme des règles de tarification: quelles conséquences pour l'entreprise?' *JCP S*, no. 47(2010): 1494.
171. Article L 1226-10 LC for occupational accidents or diseases, and L 1226-2 for non occupational accidents or diseases.
172. See for instance Cass. Soc. 30 Nov. 2010, *Dr. Soc.* (2010): 215, obs. F. Favennec-Héry.
173. See for instance Cass. Soc. 30 Nov. 2010, aforesaid.
174. On this issue, see the Articles L 1226-10 and *seq.* of the LC.

Legal provisions are going into details like the rate of dust in the air, the light in the work place or the allowed quantity of some dangerous chemicals which may be used by workers, etc. Yet, the plethoric number of texts is not always a guarantee of efficiency.

However, thanks to EU Law, some fundamental principles too have been integrated within French legislation. The role of case law is essential in that field and it can be said that the judge has had a very important influence in the enforcement of a due protection of people at work.[175]

The second characteristic of French Law on health and safety is the widespread use of penal sanctions. The rate of infringements of the law, the rate of occupational accidents and the very frequent violation of the rules has led to take this path to fight professional risks. This tendency is not so controversial in France. Indeed it is generally admitted – by the workers and trade unions at least – as a necessary mean to fight what is sometimes called the 'delinquency of employers'. It seems to work like a symbol of justice for the workers, and a psychological threat for the employers.

Despite this favourable context, one must be conscious that a huge part of occupational accidents and diseases are not identified, and if they are identified, they might not always be compensated. Two explanations can be given to this situation. First of all, a percentage of occupational accidents is not even declared as such for different reasons.[176] The rate of undeclared situations seems to be much higher for occupational diseases. Second, compensation in case of occupational diseases is not so easy to obtain, due – *inter alia* – to the table system.

Moreover, the labour inspectors are not many – even after the reform of 2008, not enough in fact to visit regularly the enterprises; besides, we have seen that their action is not always resulting in effective outcomes because of the reluctance of the judge to sanction the employer. The economic imperatives tend to prevail on the social ones, like elsewhere in labour law.

BIBLIOGRAPHY

Adam, P. 'La prise en compte des risques psychosociaux par le droit du travail français', *Dr. Ouv.* (2008): 313.

Adam, P. 'La "figure" juridique du harcèlement moral managérial'. *SSL* (2011): n° 1482, 4.

Adam, P. v° *'Harcèlement moral'*. Paris: Rép. Trav. Dalloz, *l.*

Allain, E. 'Harcèlement sexuel: les six (premières) propositions de loi'. *Dalloz actualités*, 4 juin 2012.

175. Ph. Coursier, 'Enjeux et mutations autour des risques professionnels', *JCP S*, no. 30 (2010): 1320.
176. But this percentage, by definition, is difficult to know. See, on this question Rapport de la Commission instituée par l'Article L. 176-2 u code de la sécurité sociale, Juin 2011, p. 54 et seq., available on the web, http://www.annuaire-secu.com/pdf/rapport-commission-diricq2011.pdf

Asquinazi-Bailleux, D. 'Détermination de l' entreprise dans laquelle la maladie a été contractée'. comm. under Cass. Civ. 21 October 2010, n° 09-67494, *JCP S* n° 51, (2010): 1559.

Auzero, G., J. Pélissier & E. Dockes. *Précis de droit du travail.* Paris: Dalloz, 2012.

Blatman, M. 'L'obligation de sécurité'. *Dr. Soc.* (2011): 743.

Bodin, P. & D. Nauleau. 'La réforme des règles de tarification: quelles conséquences pour l'entreprise?'. *JCP S* n° 47 (2010): 1494.

Bombled, M. 'Inconstitutionnalité de l'article 222-33 du Code pénal réprimant le harcèlement sexuel'. *Dalloz actualités,* 10 May 2012.

Brimo, S. 'L'indemnisation des victimes de l'amiante'. *Dr. Ad.* n° 12, (2011): comm. 102. 'Nouveau dispositif réglementaire de protection des travailleurs de l'amiante'. *JCP S* n° 20, (2012).

Caron, M. & P.Y. Verkindt. *'La réforme de la médecine du travail n'est plus (tout à fait) un serpent de mer'*. *JCP S* n° 39 (2011): 1421.

Chauchard, J.P. *Droit de la sécurité sociale.* Paris: LGDJ, collection manuel, 2010.

CNAMTS, 2011-085-CNAMTS-DRP -MSTAT-YL Tableaux synthèse sinistralité 2010.doc.

Coeuret, A. 'La responsabilité en droit pénal du travail: continuité et rupture'. *RSC* (1992): 475.

Coeuret, A. 'La délégation de pouvoirs et les responsabilités pénales dans l'entreprise'. *RJS* n°1 (2012): Chron. 3.

Coursier, Ph. 'Enjeux et mutations autour des risques professionnels'. *JCP S,* n° 30, (2010): 1320.

Dejours, C. *Souffrance en France. La banalisation de l'injustice sociale.* Paris: Seuil, 1998, Collection L'Histoire immédiate.

Delaunay, M. & G. Lefrand. 'Rapport d'information sur la mise en œuvre de la loi n° 2011-867 du 20 juillet 2011 relative à l'organisation de la médecine du travail'. Paris: Assemblée Nationale, rapport n° 4462, 7 mars 2012.

Dupeyroux, J.J., et al. *Droit de la sécurité sociale.* Paris: Dalloz Sirey, coll. Précis, 2011.

Duquesne, F. 'Imputation de la faute du délégataire commun à la personne morale: vers la clarification? – Cour de cassation, chambre criminelle, 23 November. 2010'. *Dr. Soc.* n° 698 (2011): 361.

El Fattah, H. comm. of the agreement concluded in December 2011 in the building sector, on hardness (Strasbourg 2012, 27 February) University of Strasbourg. http://www.dialogue-social.fr.

Fantoni-Quinton, S & J. Quandalle-Bernard. 'La pénibilité au travail: un concept à géométrie (très) variable'. *RDSS* (2012): 163.

Favennec-Héry, F. Obs. sous Cass. Soc. 30 November 2010, *Dr. Soc.*(2010): 215.

Fortis, E. 'Harcèlement moral en droit pénal et en droit du travail, unité ou dualité?'. *SSL* (2011): n° 1482, 4.

Fortis, E. & A. Coeuret. *Droit pénal du travail.* Paris: Lexis Nexis, 2012.

Gillet, J.L., D. Loriferne & X. Pretot. *Assurance et protection sociale.* Paris, Dalloz, 2011.

Hirigoyen, M.F. *Le Harcèlement moral: la violence perverse au quotidien.* Paris: Syros, 1998.

Jolivet, A. & V. Pueyo. 'Pénibilité au travail, de quoi parle-t-on?'. RDT (2010): 686.

Kaltenbach, Ph. 'Proposition de loi tendant à qualifier le délit de harcèlement sexuel'. Paris: Sénat, (11 may 2012): doc. n° 536.

Lapérou Scheneider, B. 'La responsabilité pénale de l'employeur personne physique et la présomption de faute'. *Dr. Soc.*(2012): 273.

Lardy-Pélissier, B. 'La pénibilité au delà de l'immédiat et du quantifiable'. *RDT* (2011): 160.

Laulom, S. 'L'organisation du temps de travail sous l'influence des droits sociaux européens'. *RDT* (2011):298.

Leborgne-Ingelaere, C. note sous, Cass. Soc. 24 September 2008, Nr 06-46517, *JCP S* (2008):1537.

Lerouge, L. 'Suicide du salarié et faute inexcusable de l'employeur: quelles évolutions juridiques'?. *RDSS*, (2012): 373.

Gillet, J.L., D. Loriferne & X. Pretot. *Assurance et protection sociale*. Paris: Dalloz. 2011.

Lyon-Caen, A. 'Une révolution dans le droit des accidents du travail'. *Dr. Soc.* (2002): 446.

Mazars, M.F., S. Laulom & Dejours C. 'Pot pourri autour de l'arrêt du 29 juin 2011, Santé et validité des forfaits en jours'. *ibid.*, 481.

Miné, M. 'Le droit du temps de travail à la lumière de la Charte sociale européenne'. *SSL*, n° *1475*, (2011): 7.

Petit, F. 'La pénibilité au travail, un nouveau risque professionnel ?'. *Dr. Soc.* (2011): 262.

Pignarre, G. 'Souffrance au travail et réparation des risques professionnels'. *RDT* (2010): 305.

Platel, B. 'Réformes des retraites et conséquences pour l'entreprise'. *JCP S* (2010):1523.

Pontif, V. 'Les rythmes de travail'. *RDT* (2012): 208.

Radé, C. 'Abrogation du délit de harcèlement sexuel': quelles conséquences en droit du travail'. *D.* (2012): 1392.

Rapport de la Commission instituée par l'article L. 176-2 du code de la sécurité sociale, Juin 2011, 54 et seq., available on the web, http://www.annuaire-secu.com/pdf/rapport-commission-diricq2011.pdf.

Sachs-Durand, C. 'La transposition dans les Etats membres de l'accord conclu par les partenaires sociaux au sein de européenne, sur le stress au travail'. *RDT* (2009): 243.

Sachs-Durand, C. 'Evaluation des effectifs et droit à la santé'. *RDT* (2006): 106.

Saint-Jours, Y. 'L'obligation de sécurité du salarié: prévention et responsabilité'. *Dr. Ouvr.* (2007): 560.

Seillan, H. *Un tsunami urbain – AZF Toulouse*. Bordeaux, Préventique, 2009.

Vallat, A. & A. Coeuret. 'Droit pénal et accident du travail après la loi Perben II', *SSL* (2005): n° 1206, 7.

Van Craeynest, B. & P. Masson. 'Comment règlementer le travail des cadres?'. *RDT* (2011): 474.

Vericel, M. 'L'obligation patronale de sécurité de résultat: un régime renforcé' Obs. sous Cass. Soc. 3 February 2010, n° 08-40144, *RDT* (2010): 303.

Verkindt, P.Y. 'L'obligation de sécurité de résultat...encore et toujours !!' *Cahiers de droit de l'entreprise* n° 1, (2011): file 5.

Viney, G. 'L'évolution du droit de l'indemnisation des victimes d'accidents du travail et de maladies professionnelle'. *Dr. Soc.* (2011): 964.
Waquet, Ph. Obs. sous Cass. Soc. 9 December 2003. *Dr. Soc.* (2004): 313.

Chapter 5

Occupational Health and Safety in Germany: A Dual System in Change

Olaf Deinert

§5.01 INTRODUCTION

The German system of safeguarding and promoting occupational health and safety is a dual system including tasks for public authorities as well as for the autonomous bodies of the occupational accident insurance (*Berufsgenossenschaften*). The latter has a longstanding tradition, dating back to the end of the nineteenth century. In that period, the German government started to deal with social security of the working class in order to weaken the Socialist opposition by solving the problems of their potential supporters, the workers. In that context, the Occupational Accident Insurance (hereinafter 'the Insurance') was created. It took the place of the former *Reichshaftpflichtgesetz* which had established a strict liability in tort for factory owners. Departing from this idea of liability, the Insurance provides that the employer has to pay contributions being therefore not liable for damages caused by occupational accidents to his or her employees.

Apart from the Insurance, according to the Civil Code of 1900, the employer is obliged to take care of health and safety of the employee. This general obligation gets more concrete requirements by the public law obligations of the employer which are transposed into contractual duties (see below). These public law obligations have an increasing outcome, especially since the Directive 89/391/EEC (hereinafter 'the Framework Directive') and the following specific directives on health and safety have been transposed into German law.

§5.02 BASIC CONCEPTS

[A] The System of Health and Safety Rules

The Insurance system is compulsory for any employer. S/he has no choice to conclude an alternative contract with a private insurance company. This is not in contradiction with European Union (EU) Competition Law as the European Court of Justice (ECJ) pointed out that the German occupational accident insurance bodies are no enterprises in the sense of competition law due to their solidarity element.[1] The compulsory system complies with the freedom of service, too, because of the aim of holding its financial balance.

Nevertheless, the replacement of liability does not dispend the employer from duties for safeguarding the workers. The employer is liable for health and safety from the point of view of contract law as well as from the point of view of public law. On the other hand, it is a substantial interest of the Insurance to prevent occupational accidents. Therefore, the *Sozialgesetzbuch VII* – SGB VII (Social Code Book VII) which deals with the Insurance, stipulates in § 14 the priority of prevention. For this purpose the insurance bodies have the power to adopt *Unfallverhütungsvorschriften* – UVV (accident prevention rules) according to § 15 SGB VII. These are binding to the employers and employees. Complementarily, the insurance bodies have a monitoring and advisory duty. With concern to industrial diseases they have furthermore the duty to use any possible mean to cope with dangers of appearance or re-appearance or worsening of an illness. If this is not possible they have to try to convince the employee to stop the unhealthy work, according to § 3 section 1 of the *Berufskrankheitenverordnung* – BKV (Ordinance on Occupational Diseases). This task of the Insurance is complemented by the health promotion duty of the health insurance (*Krankenkasse*) in the plant as described in § 20a SGB V. *Krankenkassen* shall co-operate with the Insurance and support its prevention activities.

The described role of the Insurance in the field of occupational health and safety represents the autonomous route in the before mentioned dual system. It is autonomous because of the structure of the Insurance which works under self-administration by bodies with an equal number of members elected by the employers and by the employees.

The other route in the dual system is the public authoritative one. The *Arbeitsschutzgesetz* – ArbSchG (Act on Health and Safety at Work) stipulates the main obligations of employers and employees in the field of occupational health and safety. It has been adopted by the Federal parliament in 1996. It is the basic codification of technical labour protection and has been amended several times.

The ArbSchG creates only a general framework of health protection rules. In line with the concept of the Framework Directive, the ArbSchG includes in §§ 18 and 19 competences for regulations in specific fields of health and safety. This is an easy way to create new standards, especially if necessary because of the implementation of

1. Case C-350/07, *Kattner Stahlbau GmbH v. Maschinenbau- und Metall-Berufsgenossenschaft* [2009] ECR I-1513; see as well BSG (11 Nov. 2003), NZS 2004, 490.

European directives.[2] This has happened several times, e.g., in the cases of the *Lastenhandhabungsverordnung* (regulation on carrying heavy loads) and the *Bildschirmarbeitsverordnung* (regulation on work at display screens). Rules on safety at construction sites are laid down in the *Baustellenverordnung* (construction sites regulation).

EU Law does not adopt a dual system like the described German one. It only indicates aims to be reached. In order to fulfil its EU obligations, the German legislator has introduced more and more public law regulations on health and safety. This leads to the result that the room for autonomous rules is becoming smaller and smaller. This has on the other hand the result that public law rules become less concrete and more goal-oriented. Therefore, we can draw a somewhat curious picture of 'de-nationalization of law by public law regulation' (*Entstaatlichung des Rechts durch Verstaatlichung*).[3]

The autonomous route has the advantage that the branch organized Insurance, is managed by bilateral bodies which include a lot of expertise from the workplace level. On the contrary, the up-growing public authoritative model stands far away from the workplace. In order to benefit from the knowledge from the side of the insurance bodies the legislator has established a national common strategy of health and safety in §§ 20a, 20b ArbSchG and in § 20 SGB VII.[4]

The autonomous route is also realized through collective bargaining. The freedom of bargaining collectively and concluding collective agreements includes the power to regulate health and safety matters. These agreements have a binding effect on employers and employees if both are members of the contracting parties. In practice, agreements are normally used to upgrade the protection of employees while the technical details for specific work are regulated by UVV.

The described system prevails in the private as well as in the public sector. Nevertheless, some peculiarities which cannot be described in this chapter exist in the public sector. Irrespective of this, each branch has its own special rules with concern to the concrete obligations of the parties. For example, the construction industry has also its own rules, namely the construction sites regulation and the UVV for construction sites (see below). In general, these rules also apply irrespective of a special protection requirement of the employee as a member of a weaker group like foreigners or disable workers. Nevertheless, special acts on protection of young people and children, of pregnant women, of disable workers etc. exist, which partly include special protection for health and safety.

---

2. Michael Kittner, *Arbeits- und Sozialordnung*, 37th ed. (Frankfurt/Main: Bund Verlag, 2012), 216 et seq.
3. Olaf Deinert & Michael Kittner, 'Die arbeitsrechtliche Bilanz der Großen Koalition 2005-2009', *Recht der Arbeit* (2009): 271.
4. The strategy aims to co-ordinate the action of the different actors (the Insurance, Regional and Federal labour inspectorates) in the field auf health and safety. In order to achieve this, actors meet in a national conference on labour protection. They have to fix the goals, evaluate the achievements and identify the fields in which they should act in the first place. Furthermore, they have to coordinate their activities, namely they have to tune their regulations.

To summarize, the German approach should be characterized as follows:

- dual system in change due to the implementation of EU Law;
- preventive approach;
- re-integration instead of compensation (see below).

[B] The Concepts of Health and Safety

Due to the dual system, it is not easy to define the concepts of health and safety.

The Insurance follows the insurance rules. This means that the concept of risks which shall be covered is precisely defined with reference to occupational accidents. This is, according to § 8 SGB VII, an event with impact on the body from outside in a limited time period and which causes death or a damage to health.

In contradiction to this rather wide definition of safety, the protection of health follows a very narrow field of application. According to § 9 SGB VII, an occupational disease has for precondition that someone gets an illness which is listed in the occupational diseases list provided by the Addendum to the BKV. Only illnesses which can be caused to specific groups of workers more than to the rest of the people can be included in this list. If a specific illness is not mentioned in the list, the worker will not get any benefit even if it is proved that the disease has been caused by his or her work. For example, if someone gets ill due to harassment at the workplace s/he will get no compensation from the Insurance because the physical and psychical consequences of the harassment are not specified as industrial diseases in the list. There is only one exception for the case that according to new medical knowledge a disease meets the criteria to be included into the industrial diseases list. But this knowledge must have been acquired after the drafting of the list. Due to the list principle regarding occupational diseases, the concept of health in the Insurance system is therefore very narrow.

However, the concept of health and safety in the public law system of the ArbSchG is a very broad one. According to § 4 no. 1 ArbSchG, the employer has to organize the work in a way which leads to a minimum of danger for health and safety. The optimum is the aim. This is also expressed in § 2 section 1 ArbSchG, according to which labour protection shall include – besides protection from accidents – measures to protect health, including work arrangements with respect to the worker as human being. Moreover, one should stress that the list principle has indeed for result that a risk, respectively a damage, which is not covered by the list will not lead to compensation of the Insurance. However, since the employer is obliged to protect the employee against the risk, s/he may be liable according to civil law. For example, if an employer or one his or her staff member causes damage to the employee's health or dignity by bullying, s/he remains liable for material and immaterial damages,[5] irrespective of the fact that this damage is not covered by the insurance. This example gives one further indication: The decent work approach is not only an aspect of health

5. BAG (16 May 2007), NZA 2007, 1154.

and safety. The employer has also to respect the dignity and personality of the employee as a contractual duty.

As described above, this general and indefinite duty of minimizing danger is more and more specified by regulations and UVV. Nevertheless, dangers are very different at different work places. Therefore, the legislator uses some kind of referral to normative standards. They put obligations to the employers by referring to the state of the technology and of occupational science. These technical clauses have ambivalent effects.[6] On the one hand, it is not easy to precise the specific obligation of the employer. On the other, this concept of referral allows the legislator to keep step with the technical and scientific evolution. The Federal Constitutional Court has pointed out that this is often the only pragmatic way for the legislator in the field of technical security.[7] Therefore, this method of legislation seems to be in line with constitutional standards.[8]

To summarize, the public law rules on health and safety create a dynamic high level standard of health and safety, oriented at the possibilities and knowledge of today. The total amount of occupational accidents has decreased since the 1960s up to the beginning of this millennium from 3 million to round about 1 million.[9] The number of death by work accident was 765 cases in 2008.[10] In the same year, nearly 64,000 cases of suspicion of industrial diseases have been notified to the insurance bodies, 13,500 industrial diseases have been certified.[11] Totally 2,430 people died due to an industrial disease.[12] The evolution of occupational accidents and diseases as well as the unemployment rate is to be seen in the following table (Table 5.1 Evolution of occupational accidents and diseases and of the unemployment rate). It follows from the data that the decrease of occupational accidents and diseases has nothing to do with the employment number or unemployment rates which increased. The increase of diseases in 1995 is to be seen in the light of the re-union of the country in October 1990, the number increased especially in the years 1992–1994. (Figure 5.1 Works Accidents; Figure 5.2 Fatal Accidents; Figure 5.3 Industrial diseases; Figure 5.4 Unemployment rates).

6. Wolfhard Kohte, '§ 290 Durchführung des Arbeitsschutz- und Kündigungsrechts', in *Münchener Handbuch für Arbeitsrecht Band 2*, 3rd ed. by Reinhard Richardi et al. (Munich: C.H.Beck, 2009), margin no. 16.
7. BVerfG (8 Oct. 1978), NJW, 1979, 359.
8. Olaf Deinert, 'Arbeitsunfallschutz und Ausgleich von Personenschäden', *Arbeitsrecht im Betrieb* (1998): 89.
9. BMAS in co-operation with BAuA (eds), *'Sicherheit und Gesundheit bei der Arbeit 2008'* (Dortmund, Berlin & Dresden: BAuA, 2010), 150.
10. BMAS in co-operation with BAuA (eds), *'Sicherheit und Gesundheit bei der Arbeit 2008'* (Dortmund, Berlin & Dresden: BAuA, 2010), 77.
11. BMAS in co-operation with BAuA (eds), *'Sicherheit und Gesundheit bei der Arbeit 2008'* (Dortmund, Berlin & Dresden: BAuA, 2010), 89.
12. BMAS in co-operation with BAuA (eds), *'Sicherheit und Gesundheit bei der Arbeit 2008'* (Dortmund, Berlin & Dresden: BAuA, 2010), 95.

Table 5.1 Evolution of Occupational Accidents and Diseases and of the Unemployment Rate.

Year	*Registered Occupational Accidents*	*Fatal Accidents*	*Occupational Diseases*	*Unemployed*	*Unemployment Rate*
1960	2,711,078	4,893		270,678	1.3%
1965	2,655,363	4,784	registrations	147,352	0.7%
1970	2,391,757	4,262	since 1978	148,846	0.7%
1975	1,760,713	3,137		1,074,217	4.7%
1980	1,917,211	2,597	13,092	888,900	3.8%
1985	1,536,090	1,795	7,886	2,304,014	9.3%
1990	1,672,480	1,558	10,384	1,883,147	7.2%
1995	1,813,982	1,596	24,298	3,611,921	10.4%
2000	1,513,723	1,153	18,689	3,889,695	10.7%
2005	1,029,520	863	16,519	4,860,909	13.0%
2009	974,642	622	16,657	3,423,283	9.1%
Source	Federal Government	Federal Government	Federal Government	Federal Employment Agency	

The evolution in graphical pictures:

Figure 5.1 Works Accidents Since 1960

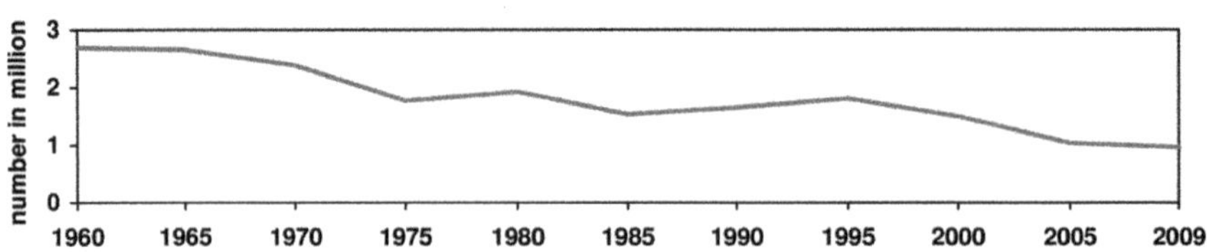

Figure 5.2 Fatal Accidents Since 1960

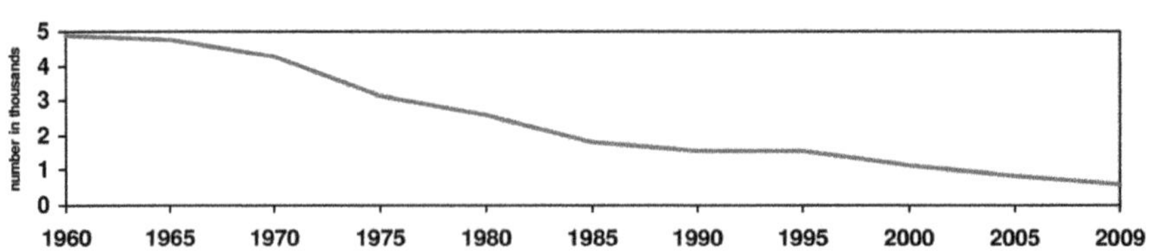

Figure 5.3 Industrial Diseases Since 1960

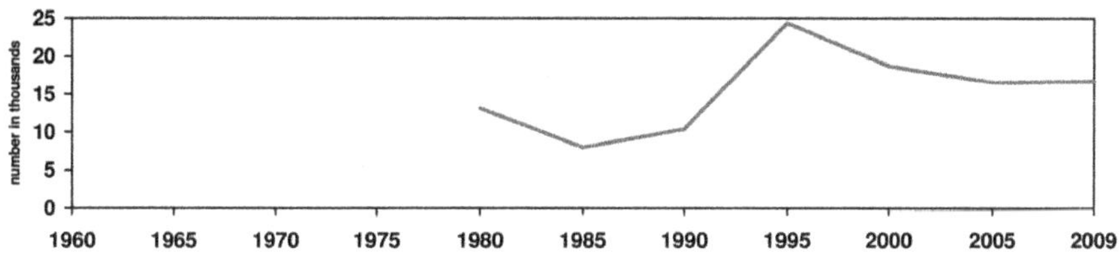

Figure 5.4 Unemployment Rates Since 1960

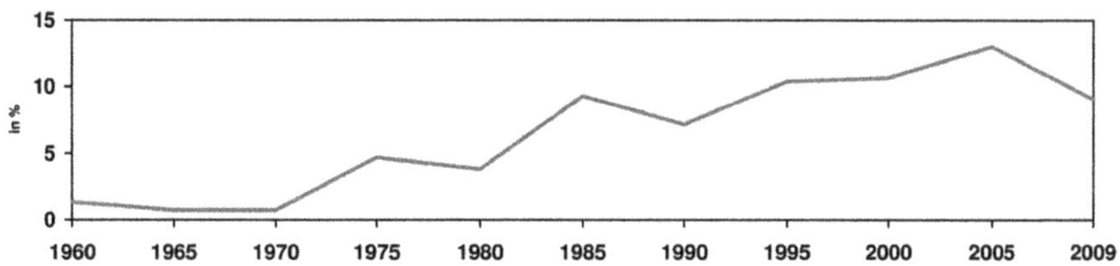

Both the autonomous route via Insurance and the authoritative route, create public law obligations on the employer and on the employees. But they do not regulate the reciprocal obligations of the individual contract parties. This field of labour law is fragmentarily regulated in the *Bürgerliches Gesetzbuch* - BGB (Civil Code). According to § 618 BGB, the employer has to organize the work environment and the work process in a way that the employee is protected against health and safety hazards as far as possible with respect to the nature of the work to be carried out. It could be very difficult to define to what extent the nature of work could lower the protection standards. But this problem is solved by a juridical device. The standard of health and safety the employer has to comply with face to his or her employee is always at least the same s/he has to comply with following the public law. This is the result of the 'transformation doctrine' (*Transformationslehre*) according to which any public law obligation in the field of health and safety has to be regarded as contractual obligation towards the employee, provided that this obligation can constitute the content of a clause in the individual labour contract.[13] Although in given cases the civil law obligation to protect health and safety may exceed the public law standards, this will not be easily proved by the employee since the employer is supposed to have adopted every necessary measure if s/he complies with the public law duties. (Table 5.2 The System of Health and Safety Law).

13. BAG (12 Aug. 2008), DB 2008, 2030.

Table 5.2 The System of Health and Safety Law

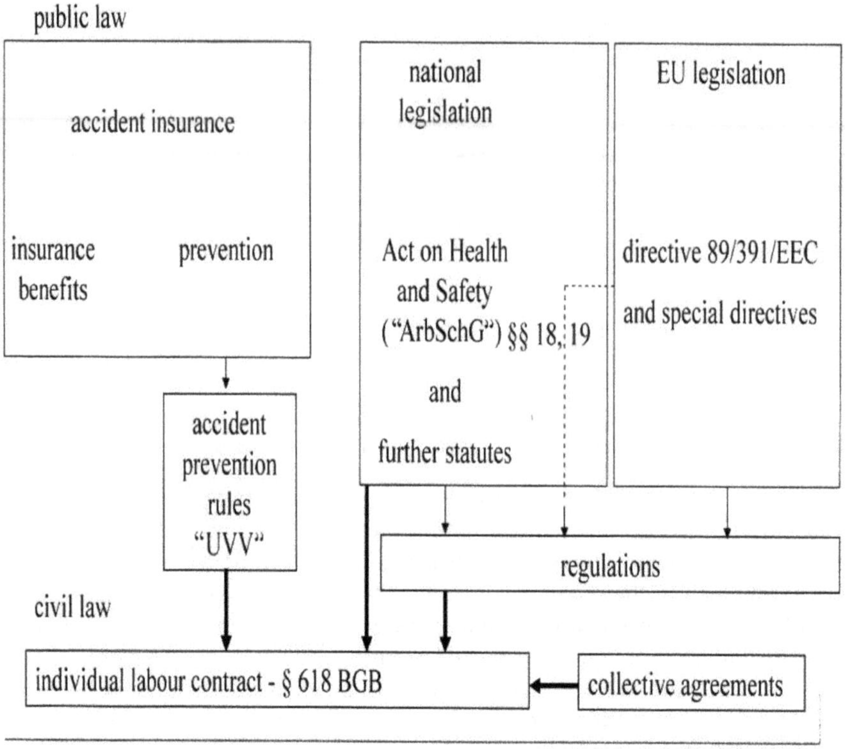

The contractual obligations of the employee in this field are not regulated at all. According to § 241 section 2 BGB, the employee has to take care of the rights and interests of the other part and according to § 242 duties shall be carried out in good faith with respect to common usage. To specify these standards, reference to public law obligations of the employee is necessary as well.

§5.03 PERSONAL SCOPE OF APPLICATION AND DUTIES OF THE PARTIES

[A] Scope of Public Law Labour Protection

The ArbSchG is applicable according to § 2 section 3 to employers who engage *Beschäftigte*. The notion of *Beschäftigt* in this context is much wider than that of *Arbeitnehmer*. The latter is a worker who has to carry out his or her work under a civil law contract and under the instructions of the employer. On the contrary, *Beschäftigt* in the sense of the ArbSchG includes, according to § 2 section 2, besides white collar and blue collar workers and apprentices, soldiers, civil servants and judges. Furthermore the ArbSchG is applicable to employee-like persons. These are not bound to employers' instructions but are economically dependent from the employer and need to a certain extent the same protection of employees. Furthermore, some legal provisions expand the scope of health and safety law. The most important is § 11 section 6 *Arbeitnehmerüberlassungsgesetz* – AÜG (Act on Agency Work) which extends the health and safety duties in the case of agency work to the user company in whose plant the work is carried out. The same applies to so-called 1-€-jobs according to § 16d *Sozialgesetzbuch II* – SGB II (Social Code Book II). These are work opportunities for the unemployed in order to promote their chances in the labour market. They do not lead to an employment contract according to the before mentioned act, but the labour protection law shall be applied.

[B] The Scope of the Insurance

Any employee in the sense of the social security provisions (§ 8 *Sozialgesetzbuch IV* – SGB IV [Social Code Book IV]) is by law covered (§ 2 section 1 no. 1 and 2 SGB VII) by the Insurance irrespective if the employer has declared him or her to the insurance body and even irrespective of any contributions paid for him or her. Those employees are designated as *Beschäftigte* too, although the precise meaning is not exactly the same as under the ArbSchG. Nevertheless dependent workers under a civil law contract and apprentices are covered by the Insurance as well as agency workers. Employers may be included into the system by ordinance of the insurance body or join it voluntarily. Soldiers, public servants and judges are insured under a special public scheme.

According to § 2 section 2 SGB VII the Insurance covers also persons who act 'like' an employee (*Wie-Beschäftigte*). They are not employed by the employer. Nevertheless, they act for the interest of the enterprise in a way in which an employee could have done as well. This includes e.g., people who help for a co-operative or altruistic reason. If a pedestrian helps a lorry driver when connecting the trailer, s/he is entitled to benefits provided by the construction industry insurance if s/he gets injured.

[C] Duties of the Employer

With respect to the above mentioned people, the employer has to take any necessary measure for secure working conditions (§ 3 section 1 ArbSchG). S/he has to concretize these measures for the individual employee. For this purpose the employer has to respect the following basic principles (§ 4 ArbSchG):

- Dangers for health and safety should be avoided as far as possible.
- If a danger cannot be avoided it should be minimized as much as possible.
- The adopted measures must take into account technical, medical and hygienic standards.
- Dangers are to be combated at their roots.
- Individual protection measures should be *ultima ratio* in comparison with any other measure.

For the purpose of finding out the necessary protection measures, the employer has to arrange a risk assessment (§ 5 ArbSchG). This obligation is often ignored in practice. The reason may be an information deficit. But in many cases employers are not aware of the practical necessity or even try to avoid costs. As the risk assessment is the precondition for the effectiveness of health and safety measures, it is very important that the Federal Labour Court has decided that any individual employee is entitled to claim for the arrangement of a risk assessment for his or her work place.[14]

Besides the risk assessment and the necessary health and safety measures, the employer must define the liabilities at workplace level.

The employer has to ensure that, while working, the employee is able to respect the rules and measures of health and safety (§ 7 ArbSchG). Furthermore s/he has to instruct the employees on occupational health and safety. This must take place during the working time. If the workers wish so, the employer has to offer occupational health care examinations unless a health damage is excluded.

As described, the ArbSchG is the 'Fundamental Act on Safety at Work'[15] but the employer has the obligation to transpose this fundamental obligation at workplace level. Special obligations derive from particular regulations:

- Regulation on occupational health care (*Arbeitsmedizinvorsorgeverordnung*).
- Regulation on requirements of the place of work (*Arbeitsstättenverordnung*).
- Act regulating job safety (*Arbeitssicherheitsgesetz*).
- Working time Act (*Arbeitszeitgesetz*).
- Construction sites regulation (*Baustellenverordnung*).
- Regulation on screen handling (*Bildschirmverordnung*).
- Regulation on dangerous materials (*Gefahrstoffverordnung*).
- Regulation on biological materials (*Biostoffverordnung*).
- Noise and vibrations protection regulation *Lärm- und Virbrations-Arbeitsschutzverordnung*).
- Regulation on load handling (*Lastenhandhabungsverordnung*).
- Regulation on handling personal security equipment (*PSA-Benutzung sverordnung*).

Sometimes these regulations are not very precise as well. For example, the construction sites regulation (*Baustellenverordnung*) contains in § 5 a clause referring to the general guidelines of § 4 ArbSchG. The regulations' rules on employers' duties contain very vague formulas: The employer must take the necessary measures with reference to:

- maintenance of the equipment;
- arrangements for storage and disposal of work materials, in particular dangerous substances;
- adjustment of working time taking into account the peculiarities of the situation at the construction site;
- cooperation of employers and enterprises without employees;

14. BAG (12 Aug. 2008), DB 2008, 2030.
15. Markus Vogl, 'Das neue Arbeitsschutzgesetz', *Neue Juristische Wochenschrift* (1996): 2754.

- inter-action between construction work and other work at the constructions site or nearby.

A special rule on co-ordination at construction sites is laid down in § 3 of the regulation. According to that a co-ordinator has to be appointed. S/he must coordinate the work with concern to security, create a plan for health and safety and supervise different employers' compliance with their duties.[16]

Very detailed obligations are laid down in a huge amount of UVV which deal partly with special kinds of work and partly with general or special work conditions.[17] For example, the UVV 'GUV-V C 22' dealing with construction work has in its § 12 precise provisions in case a protection against fall is needed or in § 8 section 2 a duty to create a special work place for work on an area disposed more than 45°.

In order to allow an appropriate application of labour protection in compliance with the peculiarities at plant level, the *Arbeitssicherheitsgesetz* – ASiG (Act on Security at Work) stipulates that the employer must appoint occupational health doctors (*Betriebsärzte*) and health and safety officers (*Fachkräften für Arbeitssicherheit*) (§§ 2-4; 5-7). In alternative, according to § 19 ASiG, the employer may entrust external medical and safety services (*Überbetriebliche Dienste*).

A contract law provision with special concern to occupational health and safety is laid down in § 84 section 2 *Sozialgesetzbuch IX* – SGB IX (Social Code Book IX). It states the employer's duty to organize a workplace re-integration management (*betriebliches Eingliederungsmanagement*) in the event of inability to work due to illness.[18] This provision is built on an older provision for prevention of dismissals of disabled persons laid down in § 84 section 1 SGB IX. The logic of that rule was to create a procedure to cope with problems in the labour relation of disabled persons to avoid dismissal. Following the logic that it is much easier to search for solutions for keeping the position of a disabled person then searching for a new job in case s/he get dismissed, the legislator has now established a general procedure of workplace re-integration management in the case of inability to work due to illness: If a worker is unable to work for more than six weeks during one year the employer has to clear the possibilities to get over the inability unless the worker refuses. Such a clearing procedure includes the worker and the employer, the workers representatives and, in the case of disabled persons, their representatives. If necessary the occupational health doctor shall be involved. The re-integration management shall focus on any measure could get over the inability including benefits and facilities. The Local common service centres of the social security bodies (*Gemeinsame Servicestelle der Sozialversicherungsträger*) must be involved if benefits or facilities are to be considered.

Re-integration management is intended to avoid dismissals on the ground of inability to work due to illness. Therefore, the question is which are the consequences if the employer fails to follow the procedure. The Federal Labour Court[19] has decided

16. For the question of liability for damages in the case of cooperation see 5.2.3.
17. Overview and download http://regelwerk.unfallkassen.de/regelwerk/index.jsp.
18. Wolfhard Kohte, 'Betriebliches Eingliederungsmanagement und Bestandsschutz', *Der Betrieb* (2008): 582 et seq.
19. BAG (10 Dec. 2009), DB 2010, 621.

that employer who fails will be penalized with respect to the burden of proof. Normally, a worker who cannot fulfil his or her duties anymore may be dismissed by the employer unless there is another position which fits to his or her residual capacities. This has to be proved by the worker. In case the employer has ignored the workplace re-integration management procedure, s/he may not claim that the employee is no more able to perform his or her tasks. Instead the employer must prove that there is no appropriate position left for the employee in the company as a whole.

However, the Court has decided that if the employer has arranged a re-integration management that meets the legal criteria and if s/he has taken measures for getting over the inability, the worker may not claim for other measures under the *ultima ratio* principle as provided by dismissal law unless these measures have show to be feasible after the conclusion of the re-integration process. This decision is highly disputable because the Court creates legal rules turning the re-integration management into the opposite of its legal aim: Following the interpretation of the Federal Labour Court, the re-integration management becomes an instrument for facilitating dismissals for health reasons.[20]

The employer may delegate obligations imposed by the law, but s/he remains liable according to civil law (§ 278 BGB). S/he has also to assure compliance with public law obligations in his or her enterprise; if s/he fails and a violation of the duties occurs, s/he will be liable according to § 130 of the Administrative Offences Act (*Ordnungswidrigkeitengesetz*). If the employer is a corporate body it cannot fulfil employer's obligations on its own, it will act via its organs and therefore is liable for any misconduct of the organs (§ 31 BGB).

[D] Duties of the Employee

Employees have two kind of duties. On the one hand, they have duties with concern to their own health and safety. On the other hand, they have duties faced to their colleagues, including the employer him or herself.

In the first place, employees have to take care for their own health and safety (§ 15 ArbSchG). In this respect, they have to follow the instructions of the employer. They have to use the work materials and equipment in accordance with their goals and functions (§ 15 section 2 ArbSchG).

Employees have the duty to take care for the health and safety of those who could be affected by their behaviour (§ 15 section 1 subsection 2 ArbSchG). Besides other employees and the employer, this should include third parties like customers or people walking by.

Employees are obliged to inform immediately the employer and superior staff about dangers and damages at protection systems (§ 16 section 1 ArbSchG).

20. Olaf Deinert, 'Kündigungsprävention und betriebliches Eingliederungsmanagement', *Neue Zeitschrift für Arbeitsrecht* (2010): 974.

§5.04 WORKERS REPRESENTATIVES

According to § 80 no. 1 of the *Betriebsverfassungsgesetz* – BetrVG (Works Constitution Act), the works council (*Betriebsrat*) has the general task to supervise the compliance with statutes, regulations, UVV and collective agreements set in favour of the workers in plants employing more than five workers. This includes surely any statute, regulation and agreement on occupational health and safety. Therefore, the works council is a very important player in the field of supervision. If a works council has not been established the single worker is entitled to claims.

However, the works council is not the only actor. As described above, the employer has to appoint occupational health doctors and health and safety officers. On the other hand, they have not a dominant position face to the works council: occupational health doctors and health and safety officers can only be appointed after approval of the works council (§ 9 section 3 ASiG). During their mandate they have to work together with the works council (§ 9 section 1 ASiG). Furthermore they must inform the works council in any important matter in the field of health and safety (§ 9 section 2 ASiG).

In companies with more than twenty employees, according to § 22 SGB VII, the employer must appoint a safety advisor (*Sicherheitsbeauftragte*). This is a commissioner who shall support the employer in matters of prevention from occupational accidents and disease, supervise the existence and correct usage of security equipment and indicate dangers.

In plants with more than twenty employees, a Health and safety committee (*Arbeitsschutzausschuß*) has to be established according to § 11 ASiG. It is composed by the employer or his or her representative, two members of the works council, the occupational health doctor, the health and safety officers and the safety advisor. The Health and safety committee has the task to discuss health and safety issues and shall meet four times the year.

Besides the supervision and advisory function, the works council has co-determination rights. Further to the above mentioned involvement in the appointment of the occupational health doctor, the health and safety officer and the safety advisor, § 87 section 1 no. 7 BetrVG regulates a co-determination right on rules of prevention of occupational accidents and diseases and on health protection in the framework of legal rules and UVV. The co-determination right exists unless a statutory rule or a collective agreement is in place. If employer and works council do not agree on a solution, an arbitration committee (*Einigungsstelle*) will take the decision. The right to co-determination is given only in domains in which the employer has a discretionary power (no co-determination without determination) and it does not go beyond the compulsory level of protection.[21] However, in any case of adoption of health and safety measures, the employer must involve the works council. This concerns in particular security instructions.[22]

21. BAG (8 Jun. 2004), NZA 2004, 1175.
22. BAG (16 Jun. 1998), NZA 1999, 49.

If the employer wants to modify or change the workplace, the work process or the work environment, the works council has no possibility to oppose since it as no real co-determination rights in this field. However, the works council and the employer must bargain an agreement of balance of interest (*Interessenausgleich*) in specific situations regulated in §§ 111, 112, 112a BetrVG. This means that the employer may freely decide to alter or even withdraw the intended measures. In these cases, however, the works council has a co-determination right on measures for economic compensation aimed at coping with the disadvantages deriving to the workers from the employer's decision (the so called *Sozialplan*). Additionally, if those modifications or changes are apparently in contradiction to acquired ergonomic knowledge and if they have a specific negative impact on the workers, the works council has the right to claim for prevention, lowering or compensation. If works council and employer do not agree on such a measure the above mentioned arbitration committee will take a decision.

It should be stressed that workers involvement and even co-determination rights have the only function to promote compliance and to improve the health and safety standard but that these rights are nevertheless workers protection rights which therefore do not limit the employer's liability according to the ArbSchG and § 618 BGB.

§5.05 CONSEQUENCES ON THE PARTIES AND REMEDIES IN CASE OF THE VIOLATION OF HEALTH AND SAFETY PROVISIONS

[A] Entitlements and Enforcement vis-à-vis Compliance with Health and Safety Provisions

The employee has an entitlement to claim against the employer for measures to guarantee occupational health and safety. But if the employer has a discretionary power, the worker can only claim for appropriate measures in general[23] and not for a specific measure because the decision which measure shall be adopted is up to the employer and – not to forget – to the works council under its co-determination right (see above).

A very important instrument to find out which measures are necessary and if there is an employer's discretionary power, is the entitlement of the employee to ask for the risk assessment. In order to avoid danger for his or her health, the employee is also entitled to occupational health care examinations.

According to contract law (§ 273 BGB), if someone does not fulfil his or her obligations, the other part may refuse to fulfil its own obligations. Therefore, the employee may refuse work if the employer fails to comply with health and safety obligations.[24] This has for consequence that the employee may claim for wages although s/he did not work because it was the employer's fault that has made the fulfilment of the work obligation impossible (§ 326 section 2 BGB).

23. Olaf Deinert, 'Arbeitsunfallschutz und Ausgleich von Personenschäden', *Arbeitsrecht im Betrieb* (1998): 91.
24. BAG (19 Feb. 1997), BB 1997, 1364.

If there is a direct and considerable danger, according to § 9 section 3 ArbSchG, the employee has the right to leave the workplace and to look for security. It is understood that s/he may claim for wages in this case too (§ 615 section 3 BGB).

If the employee claims for secure working conditions without success, s/he may bring an action to the labour court. However, the employee may be reluctant to sue the employer during the labour relationship. Therefore, s/he may be interested in someone else acting on his or her behalf. Since no right is recognized to the trade unions to act on behalf of their members, the employee could lodge a complaint via the public authority, hence the Federal Constitutional Court has decided that the right to complain is protected by the Constitution.[25] Nevertheless, the Federal Labour Court is of the opinion that, according to the weighing of the interests principle (*Interessenausgleich-prinzip*), such a complaint may constitute an infringement of contractual obligations.[26] This may be the case if the employee did not ask the employer for interventions in advance. However, in grievous cases, namely if the employer has committed a crime, the employee may address directly the public authorities.[27] In any case, the Federal Labour Court shall revise its negative attitude towards 'whistleblowing' taking into account the decision of the European Court of Human Rights in the *Heinisch* case.[28]

Public authorities which enforce labour protection are the Regional labour ministries and the *Gewerbeaufsichtsämter* (public authorities for labour protection). They may force the employer to fulfil his or her public law obligations in the field of health and safety. Furthermore they can impose fines. Many regulations and acts state a competence for imposing fines in the event of violation of labour protection rules. For example, according to § 25 ArbSchG a fine up to EUR 25,000 may be imposed. This shows that, although there is no tradition for enforcement by criminal and disciplinary law in Germany, the field of health and safety is an exception where the state intervenes via those instruments. If the violation takes place for several times or in case of intent, criminal sanctions are foreseen, too, e.g., in § 26 ArbSchG. If an injury occurs, the general criminal law will be additionally applicable, e.g., § 223 *Strafgesetzbuch – StGB* (Criminal Code) which deals with bodily harm.

These remarks indicate that the German legislator does not believe in effective enforcement by the employee on his or her own. Instead, the enforcement is in the hands of stronger actors like the public authorities for labour protection. In this framework, the works council should be mentioned as well. Furthermore, the Insurance bodies are obliged to employ inspectors in order to comply with their supervisory task (§ 18 SGB VII). These inspectors – although employed by the self-administrated Insurance bodies – have public law powers in order to avoid accidents and dangers for health (§ 19 SGB VII).

The employer is entitled to claim for compliance of the employee with his or her obligations in the field of health and safety. This refers first to the obligation of the

25. BVerfG (2 Jul. 2001), NZA 2001, 888.
26. BAG (2 Jul. 2003), NZA 2004, 427.
27. BAG (7 Dec. 2006), DB 2007, 808.
28. ECHR (3 Feb. 2011) Case no. 28274/98.

employee to follow employer's instructions and to avoid dangers for third parties like the employer him or herself, other employees or costumers, people walking by etc.

If the employee does not comply with these obligations, the employer may give him or her a notice of warning or in grieve cases or in cases of continuance after notice of warning even dismiss the employee.

[B] Compensation and Reparation in the Event of Damage

[1] Insurance Benefits

The Insurance provides benefits in cash and kind in cases of occupational accidents and diseases. As described above, an occupational accident is an event with impact on the body from outside in a limited time period and which causes death or damage to health. An occupational accident can only occur during work.[29] If it takes longer, the event could only cause an occupational disease.

Several preconditions have to be fulfilled for an accident to be qualified as occupational by the Insurance. First of all, the behaviour of the worker must fall in a working activity.[30] This precondition tends to limit the risk which shall be covered by the Insurance. If someone acts at his or her own interest, e.g., a construction worker who has an accident when borrowing some equipment for working at home, s/he will not fall under the insurance protection. The worker will be protected if s/he acts while following the instructions of the employer or even if s/he acts on his or her own intention but at the interest of the employer or even if s/he only thinks that his or her work is at the employer's interest.

The worker will get benefits, regardless if the accident is typical of the work s/he has carried out. If a worker e.g., is a victim of a criminal offence while working, s/he will be covered by the insurance protection as well.[31]

Accidents during lunch time or drinking are at private interest are therefore not covered by the Insurance.[32] Whereas, accidents on the way to or from the workplace are covered by the Insurance according to a special clause contained in § 8 section 2 SGB VII.

The working activity must lead to the accident. In this field, the problem arises that the insurance system follows an 'all-or-nothing criterion'. If there is an occupational accident covered by the Insurance the worker will get full compensation irrespective if there have been other sources for the accident. And the system has no rule for the case of contributory negligence. Therefore, the Federal Social Court has created a doctrine which looks at the essential source.[33] If the work has been the essential source of the accident the worker will get compensation regardless of other sources. If a worker e.g., gets a heart infarction at work but the origin of this infarction

29. BSG (26 Sep. 1961), BSGE 15, 112.
30. BSG (10 Oct. 2002), NZS 2003, 268.
31. BSG (20 Oct. 1962), BSGE 18, 106.
32. BSG (10 Oct. 2002), NZS 2003, 268.
33. BSG (12 Apr. 2005), NZS 2006, 154.

lies only in his or her body and has nothing to do with the work, s/he or his or her survivors are not entitled to benefits.

Even very stupid behaviours during the work may not lead to a loss of benefits. For example if someone looks for a faster way up and uses the belt conveyer and this causes an accident, s/he is covered by the Insurance. An exception is made if s/he works, at the same time, on behalf of the interest of the employer and on behalf of his or her own interests, e.g., a co-driver takes a sun-bath on the cargo area of the lorry: s/he accompanies the driver and s/he relaxes. In those cases, the self-made danger is an important circumstance when deciding whether the work at the employer's interest was the essential source of the accident.[34]

If someone works while being drunk, it is to be decided whether the intoxication was the essential source of the accident. Otherwise s/he will be entitled to benefits.[35] If someone is so drunk as to be incapable of action, no work at the employer's interest is carried out and the employee will not be covered by the Insurance protection.[36]

Furthermore, the accident must have been the essential source of the health damage in the before described sense. This includes psychological reactions. If someone gets ill because of the wish to get a pension following to an accident (*Rentenneurose*), s/he will not be entitled to benefits because it is the wish of the victim, and not the accident, that have caused the psychological damage.[37] It is questionable if the accident has been the essential source of the damage if someone already ill died. The Federal Social Court has decided that the accident has to be seen as the essential source of the damage if the employee would have lived without the accident for more than one year.[38]

An occupational disease will only be covered by the Insurance if the worker has got ill while working at the employers' interest. For example, if somebody cleans the clothes of a worker which are contaminated by asbestos s/he will get no benefit in the event of asbestosis.

As described above, the work at the employers' interest must lead to an occupational disease which is enlisted within the BKV (see above).

Moreover, the work shall represent the essential source of the disease, regardless whether the illness has appeared after the termination of the employment relationship.[39] The impact must have taken place at work. If someone e.g., gets asbestosis by inhaling asbestos at home and not at work there will not be an occupational disease (although the list of diseases contains asbestosis). This is not easy to prove. For this reason, the Federal Social Court has decided that it suffices that the employee proves that it is acceptably presumable that the work has been the essential source of the illness.[40]

34. BSG (2 Nov. 1988), BSGE 64, 159.
35. BSG (30 Mar. 1982), BSGE 53, 215.
36. Jochem Schmitt, *SGB VII Gesetzliche Unfallversicherung*, 4th ed. (Munich: C.H.Beck, 2009), § 8 margin no. 111–112.
37. BSG (18 Dec. 1962), BSGE 18, 173.
38. BSG (30 Jun. 1960), BSGE 12, 247.
39. BSG (30 Jun. 1993), BSGE 73, 1.
40. BSG (2 Apr. 2009), NZS 2010, 288.

The occupational disease must be the essential source of the damage in the above described sense.[41]

If an occupational accident or disease appears under the described criteria, the employee is entitled to therapy. This is no entitlement to compensation for the costs of therapy. Instead the Insurance bodies have to guarantee it. The employee can go to the doctor or hospital and the latter will be paid by the insurance.

The employee is furthermore entitled to adjuvant and medicaments and s/he will get benefits for rehabilitation. This includes medical rehabilitation like psychotherapy (§§ 27 section 1 no. 7 SGB VII, 26 section 2 no. 5 SGB IX) and wage compensation, if necessary. Furthermore professional rehabilitation like further education (§ 35 SGB VII) is foreseen, including wage compensation if necessary. And even social rehabilitation benefits are possible, e.g., benefits for a car (§ 40 SGB VII).

In the event of need of care, a compensation or realization of care is foreseen (§ 44 SGB VII).

If the employee becomes partially or totally incapable to work, s/he will get a pension up to 2/3 of the former wages (§ 56 SGB VII).

In the case of death, the partner and children will get pensions up to 30% of the former wages for each relative and 80% for all surviving relatives (§ 65-70 SGB VII). Additionally they will get a funeral allowance (§ 64 SGB VII).

[2] Wage Continuation

As described above, an employee may have an accident or get sick and s/he will not get any insurance benefits. This may have several reasons, e.g., work at his or her own interest, absence of work as essential source, perhaps in the case of self-made danger, illness which is not listed as occupational disease etc. In these cases, the employee will get therapy by the health insurance (*Krankenkasse*). If s/he is unable to work because of the accident or the disease, s/he will get six weeks wages by his or her employer according to § 3 *Entgeltfortzahlungsgesetz* - EFZG (Act on Wage Continuation). If someone else is liable, the employer is entitled to regress (assignment by law of the entitlement of the employee, § 6 section 1 EFZG).

If the illness is self-procured, the employer may refuse wage continuation. Namely the violation of rules on health and safety by the employee may be such a case.[42] But a self-made sickness is only given in the case of gross fault.[43] A violation of UVV is in no case such a gross fault. This depends on the specific standards.

[3] Civil Law Entitlements

If the employee causes damage to the employer, s/he will be liable under contract law (§ 280 BGB) as well as under tort law (§ 823 BGB) in the case of fault. But this civil law

41. BSG (2 Apr. 2009), NZS 2010, 288.
42. BAG (25 Jun. 1964), BB 1964, 1044.
43. BAG (25 Jun. 1964), BB 1964, 1044.

system of liability does not take into account the specific circumstances of carrying out employed work. Those are characterized by the fact that the employer decides on the arrangements of the work process and of the work environment. Therefore, the employee may not take any measures to prevent damage. S/he has to accept the work organization as it is. For this reason, case law follows a doctrine of limited liability of the employee because the employer has to take the operational risks of the enterprise.[44] In former days, this was only applicable to work which was carried out under circumstances with increased danger or risk of damage. Nowadays this rule applies to any work at the employers' interest in the above described sense.[45] Following this doctrine, the employee is liable in the case of intention as well as normally in the case of gross fault while in the case of slight negligence the employee will not be liable at all. In case of normal negligence, employee and employer have to share the damage. Therefore, a weighing of interests is necessary, taking into account any circumstances of the individual case, in particular the operational risk of the employer and the fault of the employee.

If the employee causes damage to a colleague or to a third party, his or her liability will not be limited.[46] But it is often by accident that the employee hurts the employer or someone else. And it is the employer who organizes the work process and work environment. And if the employer would have acted s/he would have been liable. For these reasons, the employee is entitled to compensation from his or her employer for compensation paid to the third party.[47] This compensation depends on the dimension of liability towards the employer if the latter would have been the one who was hurt.[48]

In contrast, the employer is liable for any damage s/he causes to the employee under contract law and tort law.

The described doctrine is to a certain extent overruled by §§ 104-110 SGB VII:

> If the employer causes a occupational accidents and diseases to the employee s/he will be liable only if s/he has acted with intention (which is given even in the case that the employer instructs the employee to risky work knowing the risks and being able to better protection[49]) or if the accident occurs on the way from or to work (§ 104 SGB VII). The reason for this exception is, in the first place, that the employer already pays contributions to the Insurance (§ 150 section 1 SGB VII).[50] The 'compensation' by the insurance corresponds approximately to the civil law compensation. If the employer had to pay compensation – even in the case of assignment by law to the Insurance body – at the end of the day s/he would have paid twice.

44. BAG (25 Sep. 1957), AP Nr. 4 zu §§ 898, 899 RVO.
45. BAG (27 Sep. 1994), AP Nr. 103 zu § 611 BGB *Haftung des* Arbeitnehmers.
46. BGH (10 Sep. 1989), BGHZ 108, 305.
47. BAG (25 Sep. 1957), AP Nr. 4 zu §§ 898, 899 RVO.
48. The employer would have been liable for the damage in the case that s/he would have acted. So one has to compare the (real) situation in which the employee has caused the damage and the (hypothetic) situation in which the employer caused the damage.
49. BAG (28 Apr. 2011), AP Nr. 6 zu § 104 SGB VII.
50. The employer has to pay the contributions alone. For the complex system of contributions in detail see Wolfgang Spellbrink, 'Das Beitragsrecht der Gesetzlichen Unfallversicherung', *Soziales Recht* (2012): 17.

> If the employee causes damage to a colleague a paradox would appear. The employee must pay compensation to the colleague. But in the event of slight or normal negligence s/he would be entitled to compensation against his or her employer insofar. This would lead to the before mentioned double payment of the employer: contributions and compensation. The Federal Labour Court had decided in former days that this result was not intended by law and that therefore colleagues are not entitled to compensation.[51] They are a *Gefahrengemeinschaft* (community of dangers) which means that any worker has to take the risk of being injured for the purpose of not being liable for damage of his or her (risky) work.[52] The legislator followed this reasoning and create a legal rule which – today in § 105 SGB VII – states an exclusion of liability in the event auf causing damage to someone who is covered by the Insurance.

The same rule applies in the case when the employee causes damage to the employer. This is unproblematic in the case of an employer who is covered by the Insurance too. If the employer is not covered, s/he will nevertheless get compensation by the insurance body according to § 105 section 2 SGB VII.

In general, liability is excluded between workers and employers in the plant. This can promote a harmonious climate at workplace since there will be no disturbance for the reason of claims for damages. The courts[53] and some scholars[54] identify this as the legislator's aim of the described rules.

The above mentioned exclusions concern damages of workers or employers of the same plant. But the legislator has amended the rule provided by § 106 section 3 SGB VII. If employers or employees who are covered by the Insurance work at a 'common workplace' they will not be liable for the damages, too. It has been highly discussed how to interpret the notion of 'common workplace'. A very wide interpretation would focus on the aspect that two workers are at the same location, irrespective what they are doing. A very narrow interpretation would ask for an aim common to the workers. The Federal Court of Justice (*Bundesgerichtshof*) found an intermediary solution, based on the concept of *Gefahrengemeinschaft*.[55] Two workers are meant to carry out their work at a 'common workplace' if there is a community in danger. If the possibility that one hurts the other is given in both directions a 'common workplace' is given. If e.g., a roofer falls down from a scaffold s/he will be entitled to compensation towards the

51. BAG (25 Sep. 1957), AP Nr. 4 zu §§ 898, 899 RVO.
52. Work is always accompanied by risks of being injured and risks of injuring someone. Therefore someone will not get compensation from the producer of the damage and will not be liable vice versa. On the other hand, both will get insurance benefits.
53. BVerfG (7 Nov. 1972), BVerfGE 34, 118; BGH (25 Oct. 1952), BGHZ 3, 298; BAG (25 Sep. 1957), AP Nr. 4 zu §§ 898, 899 RVO.
54. Wolfgang Gitter, *Schadensausgleich im Arbeitsunfallrecht - Die soziale Unfallversicherung als Teil des allgemeinen Schadensrechts* (Tübingen: J.C.B. Mohr, 1969), 239–240; Christian Rolfs, *Die Haftung unter Arbeitskollegen und verwandte Tatbestände* (Karlsruhe: Verlag Versicherungswirtschaft, 1995), 50 et seq.; Frank Maschmann, 'Haftung und Haftungsbeschränkung bei Arbeitsunfällen nach neuem Unfallversicherungsrecht (§§ 104 ff. SGB VII)', *Die Sozialgerichtsbarkeit* (1998): 59; against: Olaf Deinert, *Privatrechtsgestaltung durch Sozialrecht, Begrenzungen des Akzeptanz- und Vermögenswertprinzips durch sozialrechtliche Regelungen*, (Baden-Baden: Nomos, 2007), 267.
55. BGH (16 Dec. 2003), NZA 2004, 983.

scaffolder in the case of fault. In such a case, there is no 'common workplace' because the scaffolder may cause damage to the roofer but not vice versa.

To the extent that liability of workers or employers is excluded, there is no entitlement to compensation, irrespective if the Insurance covers the whole damage or not. The latter may be the case as far as immaterial damage not covered by the Insurance are concerned. According to the Federal Constitutional Court (*Bundesverfassungsgericht*) this is justified by the reasons on which liability is excluded in general.[56] However, the insurance 'compensation' may go beyond the real damage. This may happen in particular when an employee has a loss of work capacity and therefore s/he is entitled to a pension although s/he has suffered no loss of income in reality.

A special problem arises in the case of compensation between joint debtors. If an employee causes damage to a third party, the employer is liable according to § 831 BGB as principal and the employee him or herself is liable according to tort law (§ 823 BGB). Both are joint and several debtors. If the employee worked at a 'common workplace' in the sense of § 106 section 3 SGB VII with the colleague, his or her liability is excluded. So the injured colleague could only claim for compensation against the employer. Following the legal rule for joint and several debtors in tort law, the employer would get in the internal relationship with the employee full reimbursement because the employee is the one who acted directly (see § 840 section 2 BGB). Again the problem of double claim arises. The Federal Court of Justice decided that the injured worker should loose his or her entitlement vis-à-vis the employer, too.[57] This would not harm too much because the injured worker gets compensation by the Insurance.

To summarize: The liability of employees for occupational accidents and diseases of their colleagues and of their employer is excluded as well as the liability of the employer for damages against his or her employees. Liability may remain in cases of health damage which are not related to a work accident (e.g., while the worker was acting at his or her own interest) or to an occupational disease (e.g., an illness which is not listed into the BKV) and in cases of other damages (e.g., destroyed car or damaged clothes). In these cases, the liability of the employee face to his or her employer may be limited or excluded according to civil law doctrine. If the employee has caused damage to a colleague or a third party, s/he remains liable but may get full or partial compensation for his or her payments.

This shows that the liability may depend upon the question whether the damage is the result of an insured event. If the insurance body rejects the claim and a labour or civil court decides as a preliminary question that there has been an accident in the sense of the accidents insurance law, the victim would get neither compensation nor the insurance benefit. In order to avoid such undesirable result, the civil and labour courts are bound to the decisions of the insurance bodies according to § 108 SGB VII. In the case given, they have to suspend the proceeding as long as the insurance body decides whether or not to grant the benefit to the claimant.

56. BVerfG (7 Nov. 1972), BVerfGE 34, 118; BVerfG (8 Feb. 1995), NJW 1995, 1607.
57. BGH (11 Nov. 2003), NJW 2004, 951.

[4] Regress by the Accident Insurance

If the employee is injured by someone who is neither the employer nor a colleague of the and s/he is entitled to insurance benefits, § 116 *Sozialgesetzbuch X* – SGB X (Social Code Book X) is applicable which regulates assignment by law (regress). The Insurance is the successor in the position of the employee face to the causer.

If the causer is a colleague or the employer this assignment by law would be senseless because liability is excluded. Therefore, a specific rule for damages caused by colleagues and employers is laid down in § 110 SGB VII. Those causers are obliged to pay the expenses of the Insurance in cases of intention or gross fault. But the Insurance may quitclaim at its discretion according to § 110 section 2 SGB VII.

The regress is limited to the fictive obligation of the causer. This limitation is normally of no significance because the liability of the employee is usually not limited in cases of gross fault. But the Federal Labour Court makes an exception in cases of a serious disproportion between the damage risk and the wages, generally speaking if the damage amount to more than three month earning.[58]

Such a limitation of the civil law liability has indirect effects if the Insurance claims for regress in cases of damages caused to the employer. If the employee caused damages to a colleague, there would be no limitation to regress because the liability towards the colleague is not limited (see above). But if the Insurance could claim for regress, the employee could claim for compensation towards employer. Again the problem of double claim appears because the employer would have been obliged to compensation payments although s/he has paid contributions to the Insurance. This problem has not yet been solved by courts. The best solution seems to be a use of the possibility to quitclaim according to § 110 section 2 SGB VII.[59]

§5.06 CONCLUSIONS

The German system is to be characterized as a dual channel system of public law rules on health and safety and autonomously produced rules. An increasing state's intervention, in particular due to EU Law, leads nowadays to a reduction of autonomous rules.

Although the German labour law system stands in a contract law tradition, legal rules on health and safety are of public interest and of public law character. The link to contract law is established via transposition of public law obligations into contractual obligations which produces a double character (public and private law) of health and safety provisions.

58. BAG (15 Nov. 2001), AP Nr. 121 zu § 611 BGB *Haftung des Arbeitnehmers*.
59. Wolfgang Gitter, *Schadensausgleich im Arbeitsunfallrecht - Die soziale Unfallversicherung als Teil des allgemeinen Schadensrechts* (Tübingen: J.C.B. Mohr, 1969), 257; Olaf Deinert, 'Arbeitsunfallschutz und Ausgleich von Personenschäden', *Arbeitsrecht im Betrieb* (1998): 99; against: Horst Kater & Konrad Leube, *Gesetzliche Unfallversicherung, SGB VII* (Munich: C.H.Beck, 1997), § 110 margin no. 20.

A new focus - partly induced by EU Law - takes more care of the situation of the individual worker. Examples for this evolution of the health and safety concept are the risk assessment and the re-integration management.

Health and safety is a topic of private contract law as well as of social security law. The Insurance covers and connects both fields. It is, on the one hand, an insurance for damages of the employee. On the other hand, it works as a liability insurance for the labour contract parties and leads to essential modifications in contract law.

BIBLIOGRAPHY

BMAS in co-operation with BAuA (eds). *Sicherheit und Gesundheit bei der Arbeit 2008.* Dortmund, Berlin & Dresden: BAuA, 2010.

Deinert, Olaf. 'Arbeitsunfallschutz und Ausgleich von Personenschäden'. *Arbeitsrecht im Betrieb* (1998): 85–99.

Deinert, Olaf. 'Kündigungsprävention und betriebliches Eingliederungsmanagement'. *Neue Zeitschrift für Arbeitsrecht* (2010): 969–975.

Deinert, Olaf. *Privatrechtsgestaltung durch Sozialrecht, Begrenzungen des Akzeptanz- und Vermögenswertprinzips durch sozialrechtliche Regelungen.* Baden-Baden: Nomos, 2007.

Deinert, Olaf & Michael Kittner. 'Die arbeitsrechtliche Bilanz der Großen Koalition 2005-2009'. *Recht der Arbeit* (2009): 265–279.

Gitter, Wolfgang. *Schadensausgleich im Arbeitsunfallrecht, Die soziale Unfallversicherung als Teil des allgemeinen Schadensrechts.* Tübingen: J.C.B. Mohr, 1969.

Kater, Horst & Konrad Leube. *Gesetzliche Unfallversicherung, SGB VII.* Munich: C.H.Beck, 1997.

Kittner, Michael. *Arbeits- und Sozialordnung.* 37th edn. Frankfurt/Main: Bund-Verlag, 2010.

Kohte, Wolfhard. 'Betriebliches Eingliederungsmanagement und Bestandsschutz'. *Der Betrieb* (2008): 582–587.

Kohte, Wolfhard. '§ 290 Durchführung des Arbeitsschutz- und Unfallverhütungsrechts'. In *Münchener Handbuch für Arbeitsrecht Band 2*, 3rd edn, edited by Reinhard Richardi et al. Munich: C.H.Beck, 2009.

Maschmann, Frank. 'Haftung und Haftungsbeschränkung bei Arbeitsunfällen nach neuem Unfallversicherungsrecht (§§ 104 ff. SGB VII)'. *Die Sozialgerichtsbarkeit* (1998): 54–63.

Rolfs, Christian. *Die Haftung unter Arbeitskollegen und verwandte Tatbestände.* Karlsruhe: Verlag Versicherungswirtschaft, 1995.

Schmitt, Jochem. *SGB VII Gesetzliche Unfallversicherung.* 4th edn. Munich: C.H.Beck, 2009.

Spellbrink, Wolfgang. 'Das Beitragsrecht der Gesetzlichen Unfallversicherung'. *Soziales Recht* (2012): 17.

Vogl, Markus. 'Das neue Arbeitsschutzgesetz'. *Neue Juristische Wochenschrift* (1996): 2753–2757.

CHAPTER 6

Occupational Health and Safety in Hungary: Changes in the Name of Competitiveness?

Erika Kovács

§6.01 INTRODUCTION, HISTORICAL AND LEGAL BACKGROUND

In 2010, a new conservative government came into power in Hungary,[1] which started to reorganize nearly the whole legal system. It elaborated a new Constitution, new Labour Code, new regulation for persons with reduced capability to work and reorganized the authorities responsible for occupational health and safety. This strong desire for change brought some modification in the system of health and safety. Time will tell whether the amendments have been successful.

The Act XXII of 1992 on the Labour Code (in the following: old Labour Code) has been in force for nearly twenty years and was amended more than fifty times. On 1 July 2012, *the new Labour Code* (in the following: new Labour Code) has entered into force and has changed the regulation of employer's liability for occupational accidents and diseases. Both the old and the new Labour Code include regulations on nearly all individual and collective labour issues. They regulate all rights and obligations of the employment relationship (even atypical employment) and all collective labour issues (trade unions, collective agreement, works council, collective dispute resolution), apart from the strike.[2] Main aim of the new Labour Code was to make the regulation more flexible and employer-friendly in order to enhance the competitiveness of the Hungarian economy. It emphasizes the significance of individual and collective

1. The conservative government of Viktor Orbán has won more than two-thirds of the seats in the Parliament.
2. Separate acts regulate the status of some professional groups, e.g., civil servants, public sector employees, armed employees, judges, etc.

agreements and gives both the partners of the employment contract and of the collective agreement broader rights to self-regulate employment relationship. This principle goes in line with less protection of workers.

In Hungary, *no significant change* or even a shift of approach can be noticed *in the distribution of responsibilities* in the field of occupational health and safety. The state and the employer are the main actors and the employer has the major liability to guarantee safe and healthy working conditions. The law provides the opportunity for workers' representatives and trade unions to participate in the protection of workers. However, trade unions and workers' representatives are in general very weak and this reflects their absolutely minor role in the field of health and safety, as well. Self-regulation of the employers is important, but it is significantly limited by the legal framework provided.

As the annual report of the National Authority for Labour points out, the number of occupational accidents and particularly the number of fatal accidents has been dramatically dropped directly after the fall of the communism. This happened mainly due to the decline of the dangerous heavy industry and appearance of unemployment. However, since 1998 the number of accidents per 1,000 people has not significantly changed.[3] In a survey made be the authority in 1995, the result came out that employers did not register at least one-fourth of the accidents. This statement is likely to be true for today as well. Main reason for this is the high level of black economy. As far as the fatal accidents are concerned, the absolute number is slightly decreasing every year, but counting per 100,000 workers it is nearly stable since 1991.[4] The conclusion is that *there has not been real development* in occupational health and safety protection in the last fifteen to twenty years.

[A] History of Health and Safety Legislation in a Nutshell

[1] The Initial Development

The development of health and safety protection until the World War II was similar to the evolution in the Western European countries, even with some delay.[5] *Workers in the factories enjoyed extended protection*, while those in the agriculture have been completely left alone by the state. None of the several acts regarding labour law, health and safety and insurance applied to workers in the agriculture. Consequently, they lived and worked under very poor and dangerous conditions. The protection of this group was introduced only in the socialist system.

3. The number of occupational accidents was about 80.000 in 1989; 45.230 in 1992; 28.688 in 1998; 23.971 in 2005; 18.454 in 2009 and finally 17.295 in 2011. See the homepage of the European Agency for Safety and Health at Work: http://osha.europa.eu/fop/hungary/hu/statistics/index.html, accessed on 15 Jun. 2012.
4. The number of fatal occupational accidents was the following: in 2005: 125; in 2006: 123; in 2007: 118; in 2008: 116; in 2009: 99; in 2011: 80.
5. See in more detail: M. Sarolta, 'A munkavédelem és a munkaegészségügy intézményrendszerének kialakulása és fejlődése Magyarországon a II. világháborúig, különös tekintettel a munkáltató felelősségére', *Publ. Univ. Miskolc. Sect. jur. et pol. Tomus*, no. 28 (2010): 415–436.

First acts on occupational health and safety have been established in the nineteenth century. The XVII Act of 1840 on the relationship of factories restricted children's working time.[6] The first Act on industry in 1872 contained more detailed rules on the restriction of child work.[7] The second Act on industry - Act XVII of 1884 - has extended the protection of young workers, prescribed a compulsory annual inspection in every factory and promoted the creation of workers' social funds, which were financed by the workers and employers and intended to support the workers.

The compulsory health insurance of industry workers was established in 1891 and the compulsory accident insurance in 1907.[8] The Act XXI of 1927 on the compulsory insurance regarding accident and disease contained a modern, broad definition of occupational accident and ordered compensation also for people suffering from certain occupational diseases.

The Act XXVIII of 1893 on the protection of industry and factory workers against accidents and on the labour inspectors[9] is considered being the first act on occupational health and safety and it was applied for more than fifty years. The act obliged the employer to undertake the necessary protective measures to prevent accidents at the workplace and established a system of inspectors.

In the first decades of the twentieth century, further protective regulations were enacted, particularly regarding certain groups. The Act V of 1928 provided protection for children, young workers, pregnant and breastfeeding women. The mentioned women enjoyed protection against dismissal. The Act XXI of 1937 regulated that the daily working time may not exceed eight hours and the weekly working time forty-eight hours. It set six days of paid annual leave and the possibility to set the official minimum wage by the minister.

[2] In the Socialist System

In the socialist system, the approach to occupational health and safety regulation has been dramatically changed. In 1950, *trade unions took over the duty to organise and*

6. Children under the age of 12 were allowed to be employed only for professions, which are not dangerous for their health and development. Children under the age of 16 may not work more than nine hours a day and this shall include one hour rest period. In: Article 6 of Act XVII of 1840.
7. It stated in general that the employer is obliged to create and maintain all circumstances which serve to guarantee the life and health of the workers. Employers were controlled and fine could be imposed if they did not comply with this rule. Article 69 of Act VIII of 1872 on industry.
8. The first forms of insurance were developed in the mining industry in the thirteenth to fourteen centuries. Miners created their own fund to support each other in case of accident or disease with medical treatment, medicines, financial support and provided regular benefits for the disabled or the orphans and widows. In the fifteenth century, the mining funds even employed an own doctor. See: G. Galgóczy, 'A foglalkozás-egészségügy központi intézménye', *Hippocrates*, no. 1 (1999): 26. The Act of 1891 applied only to workers in the industry, but not to workers in the agriculture, public employees and employees of state factories.
9. 1893. évi XXVIII. törvénycikk az ipari és gyári alkalmazottaknak baleset elleni védelméről és az iparfelügyelőkről.

control occupational health and safety.[10] This was based on the principle that primarily workers should take care of the adherence of these regulations. The Labour Code of 1967 confirmed the main responsibility of trade unions for occupational health and safety. The Central Council of Trade Unions was the main organ responsible for the central organization and the inspection.[11] This included the participation in the legislation, research, training, international relations and also supervision of the compliance with the law, as quasi authority. The Central Council of Trade Unions delivered acts in 1965 and 1967, which contained the general rules for occupational health and safety and based on them several occupational rules for certain professions have been enacted by the ministers.

The 47/1979 Decree of the Council of Ministers[12] changed the system and whereas trade unions have been left responsible for the control and supervision, *the guidance of occupational health and safety became the state's responsibility*. In 1984, the Central Occupational Safety Authority was founded. It has been reorganized and renamed several times, but is even now responsible for the supervision in this field. The occupational health has been supervised by a separate body until 2007.[13]

[3] After the Change of Regime

In 1993, *a new comprehensive act on occupational health and safety* has been introduced (Act XCIII of 1993 on Labour Safety, in the following: Labour Safety Act or LSA). The authorial supervision in this field has been regulated in 1996.[14]

In 2004, Hungary joined the European Union and in the accession process it *implemented the whole range of European legislation* in this field. The regulations and directives of the European Union (EU) initiated several modifications in the law. For example, the structure of workers' representation has been changed in 2004 due to the implementation of the EU Law. In the process of the implementation of the *acquis communitaire*, new terms have been introduced, for example 'sensitive group', which includes especially young or elder workers, protected women and people with reduced capability to work. Parallel to this, the legislation gradually included new risks which have arisen from the development of technology. For example working with display screen has been first mentioned in 1996 in a decree as a risk for the health and has been regulated in 1999 in a separate ministerial decree.[15] In this way, the Council Directive

10. Based on the 240/1950. (IX. 23.) MT rendelet a balesetelhárítás szabályozása tárgyában (240/1950. (IX. 23.) Decree of the Council of Ministers on the prevention of accidents).
11. Agriculture was an exception, as until 1969 it was controlled by the local governments. Mining industry has had its separate authority.
12. 47/1979. (XI. 30.) MT rendelet a munkavédelemről (47/1979. (XI. 30.) Decree of the Council of Ministers on occupational health and safety).
13. Állami Népegészségügyi és Tisztiorvosi Szolgálat (National Public Health and Medical Officer Service).
14. 1996. évi LXXV. törvény a munkaügyi ellenőrzésről (Act LXXV of 1996 on labour inspection).
15. 50/1999. (XI. 3.) EüM rendelet a képernyő előtti munkavégzés minimális egészségügyi és biztonsági követelményeiről (50/1999. (XI. 3.) Decree of the Ministry for Health on the minimum health and safety requirements on work with display screen equipment).

90/270/EEC of 29 May 1990 on the minimum safety and health requirements for work with display screen equipment has been transposed in the Hungarian law.

As mentioned above, in 2011 several parts of the legal system have been amended and this affected also occupational health and safety regulation.

[B] The Hungarian Legal Framework on Occupational Health and Safety

[1] The Legal Framework

The earlier Constitution - being in force until 2011 - mentioned occupational safety, as one of the means to achieve the highest possible level of physical and mental health which is the fundamental right of persons living in the territory of the Hungarian Republic.[16] The new Constitution - which entered into force in 2012 - declares quite the same. Pursuant to Article XX, everybody has the right to physical and mental health and Hungary promotes the enforcement of this right *inter alia* by organizing occupational health and safety.

Main regulation of occupational health and safety in Hungary is the *Labour Safety Act* (hereafter 'LSA'), which contains the basic general provisions in this field. Labour Safety Act regulates both health and safety. LSA contains the basic principles, the state's duties, the requirements of the healthy and safe performing of work, the duties of the employer and employee, the registration and examination of occupational accidents, the rights of the workers' representatives and the common rules of the supervisory authorities in the field of health and safety. Detailed provisions regarding the performance of work, work procedures, workplaces, technological processes, work equipment and personal protective equipment are contained in other legal regulations, ministerial decrees and in standards.[17] Due to its high importance we should emphasize the 5/1993 (XII.26.) Decree of the Ministry of Labour on the enforcement of the Labour Safety Act.[18]

The Labour Code regulates the employer's liability for the employees' damages related to the employment relationship. The Act LXXX of 1997 on social security provides definitions of occupational accident and disease and includes the social security rules for these cases. The act on health care determines the objectives and the tasks of the occupational health care and the function of the supervisory body in this field.[19]

---

16. Article 70/D. (1) and (2) of the Act XX of 1949 on the Constitution of the Hungarian Republic. The Constitution was completely revised in 1989.
17. National standards concerned with occupational safety are considered as national regulations. Therefore where the employer's solution differs from the Hungarian national standard, the employer shall prove – in the event of a dispute – that the solution s/he used offers the same or better protection in terms of occupational safety as the requirement of the standard.
18. 5/1993. (XII. 26.) MüM rendelet a munkavédelemről szóló 1993. évi XCIII. törvény egyes rendelkezéseinek végrehajtásáról (5/1993. (XII. 26.) Decree of the Ministry for Labour on the enforcement of certain provisions of the Act XCIII of 1992 on labour safety).
19. Articles 53-55. and 85-85. 1997. évi CLIV. törvény az egészségügyről (Act CLIV. of 1997 on health care).

It is worth mentioning the 219/2011 (X. 20.) governmental decree on protection against serious accidents caused by dangerous materials.[20] Aim of the decree was to strengthen the regulation after the serious accident on 4 October 2010, when the dam at a reservoir of toxic red sludge broke and the sludge poured out killing ten people, injuring 150 and causing an ecological catastrophe.

The ministers have enacted several decrees concerning their field of activities. The 22/2010. (V.7.) Decree of the Ministry for Health regulates the minimum health and safety requirements regarding artificial optical radiation affecting workers.[21] This was enacted to implement the Directive 2006/25/EC. The 16/2006. (III.23.) Decree of the Ministry for Health on the protection of workers exposed to risks related asbestos implemented the EU provisions on asbestos.[22]

[2] The Role of Self-regulation

LSA supports employers' self-regulation by declaring that employers shall define the manner of implementation of health and safety requirements within the framework of legal regulations and applicable standards.[23] Employers shall establish the procedure of protection against the sources of hazard. They have freedom in enforcement, i.e., they can decide how to achieve the objectives set by the legislator. Consequently, the employer shall elaborate his or her own regulation on health and safety. This can be a comprehensive regulation on occupational health and safety or a rule only for a special field (e.g., providing personal protecting equipment) or an instruction plan (e.g., rescue plan) or an individual instruction (e.g., providing the operating instructions for a machine with the instruction to follow them).[24] The employer's comprehensive statute can regulate the tasks and competences of employees, workers' representatives and employers. It can describe the procedures and define the responsible persons, behavioural rules, the operation of the workers' representatives etc.[25] The employer's statute is a legal source of occupational health and safety. The consent of the Labour

20. 219/2011. (X. 20.) Korm. rendelet a veszélyes anyagokkal kapcsolatos súlyos balesetek elleni védekezésről. Other important sources of law are the following: 354/2009. (XII.30) Korm. rendelet a munkabiztonsági szakértői tevékenységről (354/2009. (XII. 30.) Decree of the Government on the activities of the occupational safety experts); 193/2009. (IX. 15.) Korm. rendelet az építésügyi hatósági eljárásokról és az építésügyi hatósági ellenőrzésről (193/2009. (IX.15.) Decree of the Government on the procedures and supervision of the construction authority); 191/2009. (IX.15.) Korm. rendelet az építőipari kivitelezési tevékenységről (191/2009. (IX. 15.) Decree of the Government on construction activities).
21. 22/2010. (V. 7.) EüM rendelet a munkavállalókat érő mesterséges optikai sugárzás expozícióra vonatkozó minimális egészségi és biztonsági követelményekről (22/2010. (V. 7.) Decree of the Ministry of Health on the minimum health and safety requirements regarding the optical radiation affecting employees.
22. 12/2006. (III. 23.) EüM rendelet az azbeszttel kapcsolatos kockázatoknak kitett munkavállalók védelméről (12/2006. (III. 23.) Decree of the Ministry of Health on the protection of employees exposed to risks of asbestos).
23. Article 12 LSA.
24. L. Varga, (ed.), *A munkavédelmi törvény magyarázata* (Interpretation of the Labour Safety Act), (Budapest: Complex Kiadó, 2005), 48.
25. Varga, *supra* n. 24, at 49.

safety representative (committee) is required for the enactment of the statute.[26] Furthermore, the employer is obliged at least fifteen days prior to his decision to ask the works council for its opinion on the plan of the health and safety statute.[27] S/he is however not obliged to take its opinion into consideration by decision-making.

§6.02 BASIC CONCEPTS

[A] Basic Definitions

LSA contains the definition of the most important terms, like occupational health and safety (Article 1), employer, employee, workers' representative, accident, occupational disease, risk, psychosocial risk, prevention, industrial accident, dangerous, hazardous material (all in Article 87.). The major objective of the LSA is to protect the health and ability to work of persons in organized employment and humanize working conditions, thereby preventing industrial accidents and occupational diseases. The term 'humanization of working conditions' indicates that health and safety has to be interpreted in a broad way, namely including dignity, prohibition of any form of discrimination at the workplace, special protection of young workers and women etc.[28] However, the act does not reflect this broad approach, as it does not regulate the mentioned issues, but concentrates rather on the traditional fields of occupational health and safety.

The new Act CXCI of 2011 on the benefits of persons with reduced capability to work provides *a modern definition of health.* To determine the health situation of a person we have to take into consideration the impairments due to a disease, accident or born damage, which causes a lasting or permanent modification of the physical, mental and social well-being of a person.[29]

The Act LXXXIII of 1997 on the benefits of the compulsory health insurance regulates the benefits which the employee who suffered an occupational accident or disease is entitled to. The health care includes health service and two kinds of benefits depending on the length of incapability to work. The act provides a broad definition of occupational accident, but a very narrow definition of occupational disease.

Occupational accident is an accident, which affects the employee while carrying out work in the framework of his or her profession, or in connection with it, or on his or her way to the workplace or back home.[30] A special feature of the definition is that it excludes certain cases from the scope of the act, when the employee's behaviour has significantly contributed to the accident. The accident is not considered as occupational accident, if it happened only due to the drunkenness of the person, or during activities, which do not belong to the work obligations performed without permission.

26. Article 72(4) LSA.
27. Article 264(2)e) new Labour Code.
28. See Varga, *supra* n. 24, at 27.
29. Article 1(2)1. 2011. évi CXCI. törvény a megváltozott munkaképességű személyek ellátásairól és egyes törvények módosításáról (Act CXCI. of 2011. on the benefits of persons with reduced capability to work).
30. Article 52.

It is not considered as an occupational accident, if it occurs during the use of a machine without permission or during affray at the workplace. The same applies for the accidents occurring on the way to or from the workplace, if the employee does not take the shortest way or interrupt the way without any good cause.

Occupational disease is determined in a very narrow way in the act on the benefits of the compulsory health insurance. This reflects the earlier approach to occupational diseases, which has its roots in the socialist time. A disease is considered being an occupational disease, when it occurs due to the special dangerous character of the profession of the insured person. A governmental decree contained the list of occupations, which have been seen as especially dangerous. The regulation of occupational diseases became controversial very late first, namely in 2006. The Constitutional Court ruled that the government did not have competence to make such a list and suspended it.[31] The list contained thirty-five diseases caused by the work, e.g., poisoning from several chemicals, various kinds of cancer due to dangerous working conditions, infection at the workplace etc. The list has been inherited from the communist system. The Act II of 1975 on social insurance elaborated this list of the occupational diseases and it was maintained for three decades.[32] Consequently, until 2007 only diseases specified in this list were qualified as occupational diseases.

Following the decision of the Constitutional Court a modern definition of 'occupational disease' has been introduced into the LSA and the decree, which regulates the registration and supervision of occupational diseases.[33] 'Occupational diseases' is a rapid or enduring damage of health, emerging in the process of performing work or after it, and is either a) related to the profession, or to the performance of work and was caused by a physical, chemical, biological, psychosocial or ergonomic factor of the working process or performing work or b) is the consequence of the too low or too high workload demanded to the worker.[34]

Although the old Labour Code did not use the term 'occupational disease', the settled case law was of the opinion that the employer is liable for all the damages the worker suffered in connection with the employment relationship. Published judgments of the Supreme Court on occupational diseases due to psychosocial circumstances rarely exist. Cases usually deal with physical factors causing occupational disease.[35] The new Labour Code maintained the principle of the employer's main liability for occupational diseases, which will be described below.

31. 21/2006. (V. 31.) AB határozat (21/2006. (V. 31.) decision of the Hungarian Constitutional Court).
32. II. Annex 1975. évi II. törvény a társadalombiztosításról (Act II of 1975 on the social insurance).
33. 27/1996. (VIII. 28.) NM rendelet a foglalkozási betegségek és fokozott expozíciós esetek bejelentéséről és kivizsgálásáról (27/1996. (VIII. 28.) Decree of the Ministry of Welfare on the registration and examination of the occupational diseases and increased exposed cases).
34. Article 87. 1/D. LSA
35. For example, a nurse was infected with Hepatitis B during her work and the Court ruled the employer's liability for this. See: EBH 2010. 2158. Legf. Bír. Mfv. I. 11.061/2008. Published in *Legfelsőbb Bíróság Határozatainak Hivatalos Gyűjteménye,* 1 (2010). In another case a dental technician suffered in silicosis diseases: BH 2009. 88. Legf. Bír. Mfv. I. 10.670/2007.

[B] The Risks Covered

The term 'risk' has been introduced in the LSA first in 1998, in the process of implementation of the Directive 89/391/EEC (hereafter the Framework Directive). It is similar to other expressions used earlier, like danger, source of danger etc. The meaning of the term has been modified several times. Recently, pursuant to Article 87 1/F. LSA the term 'risk' shall mean the combined effect of the probability and gravity of injury or harm in an emergency situation.

LSA mainly concentrates on the traditional fields of occupational health and safety. New risks, especially psychosocial risks have appeared quite lately in the legislation. First a ministerial decree of 1981 mentioned that in certain workplaces long working time can cause 'increased psychical demand', which can lead to the damage of health.[36]

The 33/1998 (VI. 24.) Decree of the Ministry for Social Affairs introduced a definition of the *psychosocial risk factors*. These are such durable social risk situations (e.g., social isolation, conflict with the boss or colleagues), or forced events and experiences of changing the lifestyle (e.g., migration, loss of something/somebody important), which can cause - in the existence of certain risk disposition - mental or psychosomatic diseases, accidents and difficulties to integrate into the society. The fifth annex of the decree provides a list of the activities having increased mental strain. Such circumstances are for example: bearing heavy responsibility for people or asset; the necessity to make decisions in new situations, based on missing or contradictory information, complicated rules; creative intellectual activity in the environment of disturbing inducements or based on complicated instructions; routine intellectual work under time pressure; precise, quickly changing use of working equipment or technological processes under time pressure; working at workplaces with screen; dealing with people on conflict situations; dealing with disabled persons; working in hospitals and mental hospitals etc. The sixth annex specifies workers who are delivered to the psychosocial risk factors, like workers being far from their family, migrant workers, homeless workers, workers coming from children's home three years after starting to work etc. These rules have not been changed since 1998.

The term 'psychosocial risk' has been introduced in the LSA first on 1 January 2008. This late date indicates that for a long time the significance of psychosocial risks in occurring occupational accidents or occupational diseases was not accepted. Pursuant to Article 87 1/H. LSA 'psychosocial risk' shall mean the circumstances (conflicts, organization of work, work schedule, uncertainty of employment etc.) which influence the reaction of workers to these effects, or in consequence of which stress or occupational accidents may occur, and psychosomatic symptoms (relating to or involving both the mind and body) may develop.

36. 6/1981. (VII. 23.) EüM-MüM együttes rendelet egyes egészségre ártalmas munkakörülmények között foglalkoztatott dolgozók napi munkaidejéről (6/1981. (VII. 23.) Common Decree of the Ministries for Health and Labour on the daily working time of employees working under dangerous working conditions).

Unfortunately, there is no official data, to which extent does psychosocial risk play a role in the causes of occupational accidents or occupational diseases. Furthermore, psychosocial risk is barely present in the case law of the courts as the ground for occupational disease or accident.

The National Authority for Labour carried out a survey in 2010 on occupational stress and psychosocial risk factors.[37] Majority of the respondents agreed that stress is caused mainly by the insecurity of the job (78.1%), followed by too much or little workload (45.7%). 78.7% of the participants answered that stress influences the increased absence from workplace, 77.8% was of the opinion that stress brings about accidents and 67.6% said that it influences the chronicle diseases.

Misrecognition of *musculoskeletal disorders* as occupational disease is still typical due to two circumstances. First, the exhaustive list of occupational diseases mentioned above specified only one form of musculoskeletal disease, which harms the knee. Other diseases of this kind have not been recognized as occupational diseases until 2007. The other reason of small numbers of recognized musculoskeletal diseases is the fact that the direct connection between the work and disease is not always obvious or difficult to prove. In case of a musculoskeletal disease, it is often hard to assess whether it is connected to and caused by the work, or it has a natural reason. This is particularly the case when it takes several decades, until the complaints reach the seriousness that the worker becomes (partly) incapable to work. Courts always rely on the medical experts to assess whether the disease is connected to the work. When the musculoskeletal disease was partly caused or worsened by the work, the court orders the allocation of the compensation and thus the employer has to pay only for the part of the damage, which was caused by the employment.[38]

[C] Risk Assessment

Since 1998, the employers' major duty is to *conduct a specific risk assessment*. The rules have been elaborated in more detail and modified in several times in the process of implementation of the EU Law. Employers shall evaluate the qualitative and quantitative aspects of risks jeopardizing the health and safety of employees, with particular regard to the applied work equipment, the hazardous materials and preparations, the potential strain on the employees and the arrangement of the workplace. In the evaluation, the employer shall identify the potential risks, the persons jeopardized by risks, and the extent of endangerment based on the characteristics of the danger. Employers shall carry out a risk assessment before starting the operation. Risk assessment shall be carried out in every third year and when necessary, i.e., when any

37. It was not a representative survey, carried out by 361 workers in micro, small and medium enterprises. Országos Munkavédelmi és Munkaügyi Főfelügyelőség, Összefoglaló „a munkahelyi stressz és a pszichoszociális kockázati tényezőkkel kapcsolatos munkáltatói és munkavállalói ismeretek, attitűdök országos vizsgálatához' végzett tájékozódásról, 2009. http://www.ommf.gov.hu/index.php?akt_menu=311, accessed on 15 Jun. 2012.
38. See e.g., BH 2007. 240. Legf. Bír. Mfv. I. 10.500/2006. The worker concerned performed heavy physical work in the agriculture. He was declared incapable to work, but he had suffered from musculoskeletal complaints since the age of 15.

substantial change in the risks occurs.[39] Employers must record the assessment of risks in an appropriate way including all important data. Risk assessment is controlled by the labour inspectors and thus the compliance with this rule is quite high. Over 90 % of the employers have provided risk assessment, which is a very high proportion in comparison to the Member States of the EU.[40]

Every employer is obliged to provide a basic Occupational Health Service in order to fulfil his or her duties under the LSA. This service can be operated directly by the employer or by a service provider, with whom the employer concludes a contract.[41] Without prejudice to the liability of employers, the Occupational Health Service shall participate in the development of healthy working conditions and in the prevention of damage to health. The compulsory system of having an occupational health doctor in every company has been established in the communist system and has been preserved.

[D] Basic Principles of Occupational Health and Safety

[1] State as the Main Regulator

The LSA states the main principles in the field of health and safety. As first, it declares that it is the task of the state to define the basic health and safety requirements, as well as the institutions for the control and supervision of such, and enact a national programme for the protection of occupational health and safety.[42] This obligation includes that the state shall promote and supervise the enforcement of the regulations on occupational safety and health by way of the supervisory bodies established for this purpose.[43]

[2] General, Objective and Primary Liability of the Employer

The LSA clarifies[44] that the implementation of health and safety requirements shall be the duty and obligation of employers. The employer has general, objective and primary liability. It is a general liability, because it includes all regulations; objective, as it cannot be devolved to someone else; and primary, because in general it prevails on the employee's liability.[45]

Both employers and employees have independent liabilities. Even if the employees have obligations in the field of health and safety, their liabilities shall have no effect on the employer's obligation and thus the employee's breaches of obligation cannot release the employer from his or her liability.

39. Article 54 LSA.
40. See: European Agency for Safety and Health at Work, 'European Survey of Enterprises on New and Emerging Risks, Managing Safety and Health at Work': 26. https://osha.europa.eu, accessed on 15 Jun. 2012.
41. Article 58 LSA.
42. Article 2(1) LSA.
43. Article 3 LSA.
44. Article 2(1) LSA.
45. Varga, *supra* n. 24, at 33.

In order to protect the employees, the LSA lays down that the costs and other burdens (e.g., wage, working time etc.) associated with this obligation must not be devolved upon the employees, i.e., occupational health and safety is only the cost and burden of the employer.[46]

Pursuant to Article 18(2) LSA, employers may not provide pecuniary or other form of compensation to employees in lieu of meeting the requirements of occupational safety and health standards. Employers should not be able to evade the law by paying for the employee.

Furthermore, it is the employers' liability to provide the opportunity to all employees to become familiar with the relevant regulations pertaining to health and safety in the language they understand. If the employee cannot understand the Hungarian language the employer is obliged to translate the regulation concerned.

[3] Principle of Co-operation

Major principle of the regulation of occupational health and safety is the principle of co-operation.[47] Every stakeholder shall co-operate in the course of exercising their rights and fulfilling their obligations. Employer is obliged to co-operate with other employers, his or her employees, the workers' representatives and state authorities. The authorities shall co-operate with other authorities; the workers' representatives with the employer and employees. Finally, employees must co-operate with each other and shall perform their work without endangering their own health and the health of others.[48] Co-operation covers particularly the obligation to provide information to the other parties in due time to be able to exercise rights and fulfil obligations; to abstain from hindering the other actors in the exercise of their rights; to promote the fulfilment of the obligation of others and to take into consideration the interests of the others.[49] Employees have several obligations in the frame of co-operation, like participation in the medical examination prescribed to him or her and in the occupational training, report any potentially dangerous irregular situation to the employer etc.[50]

[4] Right to the Reconciliation of Interests

The representation of workers' interests is one of the basic principles in this field. LSA provides for the reconciliation of interests related to occupational safety, defining the rights and obligations of the representatives of occupational safety.

46. Article 2(1) LSA.
47. Article 6 LSA.
48. Article 60(1) LSA.
49. Varga, *supra* n. 24, at 34.
50. Article 60. LSA.

[5] Data Protection

LSA regulates the protection and disclosure of data related to occupational health and safety. Data shall be protected in the exercise of rights and fulfilment of obligations in the field of occupational health and safety and during both the internal (made by the employer) and external (made by an authority) processes. The protection of data (personal, special and public data, classified information, industrial and business secrets) shall be guaranteed during the course of all proceedings related to occupational safety.[51] LSA refers to the data protection regulation. In this field, the Act CXII of 2011 on the right of informational self-determination and freedom of information is the most important legal source.[52] Also, the Labour Code contains certain special rules on data protection (Article 8-10). The person managing data – who is usually the employer or the authority – is responsible for the protection of data. The data may be used for statistical purposes, and may be disclosed for the purposes of statistical use in a manner which precludes identification.

§6.03 PERSONAL SCOPE OF APPLICATION AND DUTIES OF THE PARTIES

[A] The Personal Scope of the LSA

The personal scope of the LSA contains employees, employers, workers' representatives and certain state authorities. Employee is any person who performs work within the framework of 'organized employment'. The term *employer* shall cover user undertakings hiring workers, temporary employers of posted persons, employers providing hands-on vocational training and any private entrepreneur (also those operating as sole proprietor, who does not employ others and performs his or her work only by him or herself).[53] However, these categories shall be considered as employers only regarding the provisions on the protection of persons in the proximity of the area where the work is performed. As regards non-resident employers without Hungarian tax number, employer means the person or body, who (which) exercises effective control, or bears the ultimate liability for the workplace, or in the absence of these, the person on whose property the work is performed.

[B] All Forms of 'Organized Employment'

The Labour Safety Act shall apply to all forms of 'organized employment', irrespective of the organizational and ownership form involved. *Organized employment* means every form of performing work under the direction of the employer. Subordination is

51. Article 7 LSA.
52. Relevant is further the Act XLVII of 1997 on the management and protection of data on health and related personal data (1997. évi XLVII. törvény az egészségügyi és a hozzájuk kapcsolódó személyes adatok kezeléséről és védelméről).
53. Article 87(8) LSA.

the determining feature of 'organized employment'. 'Organized employment' includes all kinds of employment in the private and public sector, the relationship of trainees, students' legal relationship in vocational schools while carrying out the requirements of their professional training, work in the prison, charity work and voluntary work organized by the employer.[54]

Some forms of performing work do not belong to 'organized employment'. Work performed without subordination is out of the scope of the LSA (e.g., contract of service, contract for work). Similarly, a natural member of a company, who carries out activities in favour of its own company, is not under the scope of the act.

Self-employed are in general not covered by the Act, but they are also obliged to fulfil certain requirements of the Act. The place of work of a self-employed (also if operating within the framework of a sole proprietorship), who performs all work by him or herself and who does not employ any other persons can be controlled by the authorities in order to protect persons being in the proximity of the workplace.[55]

[C] Exceptions in the Public Sector

The Labour Safety Act does not provide for a general derogation for the employment in the public sector. Nevertheless, *in exceptionally justified cases* different requirements and procedural rules may be established for labour activities carried out within the framework of a legal relationship for work or service or as a public sector employee in the police, fire and rescue service, civil national security services, the disaster relief forces, National Tax and Customs Office, the Hungarian Army and the Army Security Service. Special rules can be applied under extraordinary working conditions (e.g., rescue or state of emergency). The ministers responsible for the field concerned have taken advantage of these rules and enacted several decrees determining the terms 'exceptionally justified case' and 'extraordinary working conditions' and setting special provisions.[56]

[D] Persons Being in the Proximity of the Workplace

LSA stipulates that regulations pertaining to health and safety shall be defined in such a way that they provide adequate protection not only to employees, but also to other persons in the proximity of the area where the work is performed, and to persons using the services.[57] The Labour Safety Act *extends its personal scope* by a territorial rule

54. Article 87(9) LSA. It does not include any work performed in the household of natural persons as employers of domestic personnel within the framework of the so-called simplified employment.
55. Article 87(5) LSA and Article 9(2) LSA.
56. See for example: 70/2011. (XII. 30.) BM rendelet a belügyminiszter irányítása alá tartozó rendvédelmi szervek munkavédelmi feladatai, valamint foglalkozás-egészségügyi tevékenysége ellátásának szabályairól (70/2011. (XII. 30.) Decree of the Home Secretary on the rules of occupational health and safety tasks and activities for the police, fire and rescue services and disaster relief forces).
57. Article 4 LSA.

saying that certain rules of it shall apply to persons being in the area where the work is performed in order to protect them from risks arising from it.[58] The law expressly mentions passers-by, visitors and persons using services. Furthermore neighbours can be affected by this rule. The workplace affects a broader geographic area, in which the technology, the material or by-product predictably might cause danger of accident or disease.[59]

Only certain provisions of the Act shall apply for this group of persons explicitly listed in the Act. For the protection of employees and other persons in the area, safety and health protection alarm signals shall be installed at such places, where it is justified by the nature of the hazard.[60] Noise emission and vibration, dust and chemicals, as well as radiation and lower or higher atmospheric pressure may not result in the injury of the employees or other persons in the work area.[61] Work procedures, technological processes, work equipment and materials shall be selected so as to prevent any imminent hazard to the health and safety of employees and to other persons in the area of the performance of the work. In respect of work procedures where persons staying in the scope of the work performed may be exposed to the effects of sources of hazard, effective protection shall be provided by the application of closed technological process, or if such processes cannot be provided, by the application, or joint application if necessary, of safety equipment, personal protective equipment and organizational measures.[62]

[E] Special Groups

The Hungarian law contains special provisions for certain groups of employees who shall be particularly protected. Two groups of employees have to be emphasized, namely *young employees* and pregnant women, women, who have recently given birth and those, who are breastfeeding (in the following: *protected women*). Regulations on the increased protection of these groups during work are scattered in the LSA, the Labour Code and the 33/1998. (VI. 24.) Decree of the Ministry for Social Affairs.[63] The last one was intended to implement the Directive 92/85/EEC and the Directive 94/33/EC.

LSA gives a definition of a 'sensitive group'.[64] *Sensitive group* shall cover workers with special physical or mental characteristics or condition, who are thereby increasingly exposed to risks related to work, and who themselves represent extra risk in their

58. Article 9(2) LSA.
59. See similar: Varga, *supra* n. 24, at 43.
60. Article 26/A. LSA. At workplaces where the employee or other persons in the work area are at risk from falling objects, protection shall be provided by fencing, covering or any other appropriate way. Article 28(1) LSA.
61. Article 32. LSA.
62. Article 44. LSA.
63. 33/1998. (VI. 24.) NM rendelet a munkaköri, szakmai, illetve személyi higiénés alkalmasság orvosi vizsgálatáról és véleményezéséről (33/1998. (VI. 24.) Decree of the Ministry for Social Affairs on the medical examination and evaluation of the professional and personal hygienic aptitude for the position).
64. Article 87(8/A.) LSA.

employment (e.g., young people, protected women, elderly people, workers with any degree of incapacity). Besides this definition the LSA does not contain any special rule for this group of workers, only refers to the special regulation, according to which these people shall be particularly protected.

LSA includes special provisions for *tele-workers*.[65] These rules have been introduced in 2004 in order to implement the European framework agreement on tele-work of 2002. Special rules are necessary, as the worker usually works at home or at a place defined by him or herself, using his or her own working tools. These circumstances require special regulation, as the employer has little control over the actual workplace. LSA regulates, how the employer and the authorities can control the working conditions.

Pursuant to Article 1 of the 33/1998 (VI.24.) Decree of the Ministry for Social Affairs young people, protected women and elderly people belong to the sensitive group. Persons who have reached their retirement age are considered as elderly persons. Young and elderly workers have to take part annually in occupational medical examination. The decree specified several circumstances and risks, which are too dangerous for young or elderly workers and protected women, including increased mental or physical workload, working with chemical materials, suffering some kind of pollution etc.[66]

The old Labour Code contained *special rules for young workers* and the new Labour Code adopted basically the same regulation. Employees under the age of eighteen years shall be considered as young workers. The main rule is that only persons being at least sixteen years old can enter into an employment relationship. People under this age can conclude an employment contract in two cases. One is for special professions, such as modelling, sport or advertising activities. In this case, the young worker needs the permission of the competent authority responsible for children. Another opportunity is that any person of at least fifteen years being in full-time school education may enter into an employment relationship during school holidays. In both cases, the permission of the parents or other legal guardians is necessary.[67]

For young workers different working time provisions apply. They shall have longer breaks and daily rest periods as normal workers. Their working time must not exceed eight hours daily or forty hours weekly. The reference period shall not be longer than one week. For the purpose of working time limits, the length of work performed for more employers shall be added. Young persons cannot be used for night work, overtime or for stand-by or on-call duty. The new Labour Code basically applies the same rules.[68] The old and the new Labour Codes state that when a person under the age

65. Chapter VII/A. Differing occupational safety rules for telework, Article 86/A. LSA.
66. 8. Annex of the 33/1998. (VI. 24.) NM rendelet a munkaköri, szakmai, illetve személyi higiénés alkalmasság orvosi vizsgálatáról és véleményezéséről (33/1998. (VI. 24.) Decree of the Ministry for Social Affairs on the medical examination and evaluation of the professional and personal hygienic aptitude for the position).
67. Article 72 old Labour Code. Hungary has adopted the Convention No. 78. and 138. of the International Labour Organisation dealing with minimum age and child labour.
68. Article 114 new Labour Code.

of eighteen years performs work not in an employment relationship, some protective provisions of the Labour Code still have to be applied for him or her.

Both the old and new Labour Codes contain *special provisions for protected women*. These rules are scattered in the code and are intended to implement the Directive 92/85/EEC. The new Labour Code slightly reduced the rights of this group, but even so they enjoy broad protection.

Article 60 of the new Labour Code formulates a compulsory modification of the employment contract. The pregnant employee shall be offered a job fitting to her state of health from the time her pregnancy is diagnosed until her child reaches one year of age. The pregnant employee shall be released from her work duty if no position appropriate for her medical condition is available.[69]

Different working time rules have to be applied to the pregnant women, the women and the single parents until his or her child reaches the age of three. For them an irregular work schedule may be used only with the employee's consent; weekly rest days may not be allocated irregularly; overtime work, night work or stand-by duty cannot be ordered.

An employee may not be transferred to work at another location without his or her consent from the time pregnancy is diagnosed until her child reaches three years of age and in case of a single parent until the child reaches sixteen years of age.

The new Labour Code contains some protection against dismissal of the protected women. The employer may not terminate the employment relationship by notice during pregnancy, maternity leave, a leave of absence taken without pay for caring for a child, and in the case of women, while receiving treatment related to a human reproduction procedure[70] for up to six months from the beginning of such treatment. Protection during pregnancy and human reproduction procedure shall apply only if the worker has informed the employer thereof before the notice was given.[71]

A mother and a single father shall be protected against dismissal until the child reaches the age of three, even if the employee is not taking up maternity leave or leave of absence without pay for the purpose of caring for the child. The employment relationship of such workers may be terminated on the grounds of the workers' behaviour in relation to the employment relationship only for extraordinary reasons without notice.[72] The employment relationship of such workers may be terminated in connection with workers' ability or for reasons in connection with the employer's

69. The worker shall be given the basic wage normally paid for the position offered, which may not be less than her basic wage fixed in the employment contract. The basic wage shall be payable for the duration of leave of absence, except if the job offered is refused without good reason. According to the old Labour Code the woman could not suffer any financial loss, as she was entitled anyway for her earlier average wage. This was changed by the new Labour Code. Furthermore, the employer needed the consent of the woman to modify the contract concerning her position and recently this is not necessary any more.
70. This protection shall be available to the mother, if unpaid leave has been taken by both parents.
71. Article 65 new Labour Code.
72. An employer or employee may terminate an employment relationship without notice if the other party: (a) wilfully or by gross negligence commits a grave violation of any substantive obligations arising from the employment relationship; or (b) otherwise engages in conduct that would render the employment relationship impossible.

operations, if the employer has no vacant position available at the workplace suitable for the worker affected in terms of skills, education and/or experience required for his or her previous job, or if the worker refuses the offer made for his or her employment in that job.[73]

Mothers shall be entitled to twenty-four weeks of maternity leave.[74] The duration of maternity leave, except where entitlement is specifically connected to work, shall be recognized as time spent at work. Employees shall be entitled to unpaid leave at the time requested by the worker for the purpose of taking care of his or her child, until the child reaches the age of three. In addition, employees shall be entitled to unpaid leave for providing care for a child in person until the child reaches the age of ten, during the period of receiving child-care allowance.[75]

§6.04 WORKERS' REPRESENTATIVES

[A] Organization of the Workers' Representation Related to Health and Safety

The role of workers' representatives and the law regarding them has been significantly changed in the last decades, also due to the implementation of the EU Law. In the first version of LSA, the election of workers' representatives was only a possibility of the workers. Recently, it became the obligation of the employers, who can be fined, if they do not initiate the election. Besides, a new procedure, namely the consultation was introduced in the process of implementation of the EU Law.

Workers' representation at the workplace in the field of occupational health and safety *has three forms,* namely: (1) consultation with workers, (2) interest representation (by the Workers' Representative or Health and Safety Committee), and (3) a body equally represented by the workers and employers (called: Common Representative Body). Consultation is the most common and thus most important kind of workers' representation operating in the majority of the companies. Interest representation by the Workers' Representative or a Health and Safety Committee works only in a few workplaces and it is estimated that only one-sixth of the employees are represented in this way.[76] Finally, a Common Representative Body exists only in companies, where there are some kind of interest representation and at least fifty employees.

73. According to the old Labour Code, the employer could not terminate the employment relationship by ordinary dismissal of these workers with any reason. Thus, the new law loosened the protection.
74. Maternity leave shall also be provided to a woman, who has been given custody of a child for the purpose of adoption.
75. Articles 128-130 new Labour Code.
76. Varga, *supra* n. 24, at 445.

[B] Consultation with Workers

Consultation with workers, as a general, independent, substantive right of the workers was introduced into the LSA in the process of implementation of the *acquis communautaire* in 2004. Prior to 2004 information of the workers and consultation with them have been guaranteed only in a few special cases. The modification of the LSA in 2004 declared consultation as an autonomous right and as the first, general level of workers' representation.

Pursuant to Article 70, employers shall consult workers and/or their representatives and allow them to take part in discussions in advance and in good time on all questions relating to the employers actions for the sake of safety and health at work. 'In good time' means that the workers, their representatives and the employer have to provide the information to each other in a time, that allows the other party to act efficiently with reference to the issue at stake.[77] The time necessary for consultation is not regulated by law.

About the *composition of the consultative body* the law is reticent. It states that in the consultation the balanced participation of the stakeholders and the right of workers and/or their representatives to make proposals shall be ensured. The law regulates neither the meaning of 'balanced participation' in detail nor the number of participants. Article 70(4) LSA states that the employer shall be represented in the consultation by a person vested with authority in questions of occupational safety. By this rule shall be guaranteed that the employer is represented by a person having appropriate competence in this field. The right to consultation is exercised by the Workers' Representative, the Health and Safety Committee or in lack of these directly by the workers. Due to this rule, even if there is no representative body, the workers are entitled to the whole range of consultation rights. This rule causes great concerns in practice. It is difficult to clarify how this right shall be ensured for all workers, since for organizational reasons it is usually impossible to invite all of them for a consultation at the same time. As the employer is liable for the exercise of the right to consultation, s/he has to control, whether the result of the consultation has reached every worker.[78]

Subjects of the consultation are the conditions of performance of healthy and safe work. In all circumstances regarding occupational health and safety workers can be consulted, but it is useful to specify certain issues in an agreement with the employer (e.g., guarantee protection for the individual, the procedure to deal with complaints, comprehensive assessment of the situation of occupational health and safety, consultation on risk assessment, etc.). The law specifies some obligations of the employer on which consultation is compulsory. Workers or their representatives shall be consulted on the designation, employment and activities of workers who carry out the duties related to occupational safety[79] and on the planning and organization of the training

77. Varga, *supra* n. 24, at 451.
78. Varga, *supra* n. 24, at 450–451.
79. Workers should be designated, who deal with health at work, safety at work, and who are required to implement the evacuation of workers, for fire-fighting and disaster rescue operations in the event of emergency (s 54/A. LSA). Besides, employers shall designate or employ persons with specialized occupational safety qualifications (s 57 LSA).

concerned with occupational safety.[80] LSA also provides special regulations on which information should be provided.[81] Subject of the consultation is, in which form, when and by which representative of the worker shall provide the information.

LSA set some rules on *the method of consultation.*[82] The exercise of this right requires the employer's activity, i.e., s/he has to provide information on the most important contents (danger, risks, protective measures, liability). On the worker's request, the employer has to provide information regularly or in more detail. The law specifies some issue, where the employer is obliged to initiate consultation, like in case of introduction of a new technology, or if workers use display screen equipment.

[C] Workers' Representatives and Health and Safety Committee

[1] Workers' Representatives

Pursuant to Article 70/A, LSA workers shall have the right to elect a representative or representatives to represent their rights and interests related to health and safety (hereinafter referred to as Workers' Representative). Every worker being in organized employment shall have this right, with the exception of persons working in penal institutions. It is not clear in the Hungarian law, whether the temporary agency worker has the right to participate in the election and be elected as Workers' Representative. As generally the user undertaking is liable for the health and safety of the agency workers, in my opinion such workers should have the right to elect and be elected, if they fulfil the general requirements. Concerning temporary agency workers, the user should have the obligation to initiate and organize the election of the Workers' Representative.[83]

LSA gives a definition of 'Workers' Representative'. It shall mean a person elected by the workers to represent their rights and interests in dealings with the employer concerning health and safety.[84]

Election of Workers' Representative is compulsory in two cases. First, s/he shall be elected in all the enterprises with at least fifty workers. The provision of the conditions for the election and the organization of it, is the responsibility of the employer. The election of a Workers' Representative is however the substantive right of every worker, which is not depending on any minimum number of workers. Therefore, in case of an employer with less than fifty employees, the employer is obliged to organize the

80. Article 55 LSA.
81. For example, in case of hazardous work procedures and technological processes, employees shall be informed on the manner of protection against such and the occupational safety and health conditions (s. 42a LSA). All employees involved shall be instructed regarding the rescue plan pertaining to their workplace (s. 45(2) LSA). Employers shall inform workers or their representatives on the conclusions and findings of the hazard evaluation, the recording and reporting procedures of occupational accidents and occupational diseases, and the information concerning occupational safety received from the Health and Safety Authority, particularly on the findings of their inspections (s. 59 (2) LSA).
82. Consultation can occur not only by discussion, but also by performance or illustration.
83. Varga, *supra* n. 24, at 458–459.
84. Article 87/A. (6/A) LSA.

election, when it is initiated by the local branch of the trade union or the works council, or failing this, by the majority of workers.

The question arises, whether Workers' Representative shall be elected at an independent establishment or division of a company. LSA regulates that they may be elected if the manager of the independent establishment or division is vested with some or all of the rights concerning occupational health and safety.[85] It is not an obligation, but a possibility of the workers.

Workers' Representatives are elected by equal, secret and direct ballot for a term of five years. As regards the election, mandate, dismissal and recall of Workers Representatives, and the area of competences, the provisions of the Labour Code pertaining to members of works councils and to employee delegates shall be duly applied.[86]

The LSA does not determine the number of Workers' Representatives. Earlier this number was bound to the number of the works council's members, but this rule was cancelled. However, this regulation should be taken into consideration also in the future, as the only guideline.

Although Workers' Representatives enjoy a broad list of rights, in practice they hardly play any role. In many enterprises, their role is limited to the participation of the examination of occupational accidents and signing the report of the accident.[87]

[2] The Employer's Obligation

The employer has far-reaching obligations concerning the election. S/he has to initiate the election, contact the workers and inform them about the procedure. The employer or the workers can initiate the establishment of an independent election committee, which is in charge of organizing the election. The Supreme Court has confirmed[88] that the employer is obliged to organize the election and this obligation is not depending upon the initiative of the worker. If the health and safety inspector controls the employer, a fine can be imposed in lack of the Workers' Representative. The employer cannot successfully argue that it is not his or her task to inform the workers on their right to elect a representative. Even if the workers and trade unions have not initiated the election, the employer is in default, if s/he does not take care of the organization of the election. The employer has to organize the election and provide the appropriate conditions e.g., room, time, communication equipment, provision of the list of eligible

85. Article 70/A.(1) LSA
86. Article 70/A.(3) LSA
87. Országos Munkavédelmi és Munkaügyi Főfelügyelőség, 'Összefoglaló „a mikro-, kis- és közép vállalkozások munkavédelmi helyzetképének (profiljának) felmérése' tárgyban végzett tájékozódásról', 2009. http://www.ommf.gov.hu/index.php?akt_menu=311
88. EBH 2008. 1812. Legf. Bír. Mfv. II. 10.411/2007. Published in Legfelsőbb Bíróság Határozatainak Hivatalos Gyűjteménye, 1 (2008).

and electable workers.[89] However, the employer is not responsible for failing the election, if it occurs due to the inactivity of the workers.[90]

[3] The Workers' Representative's Rights

Workers' Representatives have a wide range of rights. They are entitled to *check the enforcement of occupational safety and health requirements* at workplaces. Article 72 LSA emphasizes certain fields, where they shall control, e.g., the safety of workplaces, work equipment and personal safety equipment; the implementation of measures taken in order to protect health and to prevent occupational accident and diseases; and finally the preparation and preparedness of employees for occupational health and safety. Workers' Representatives can participate in the employer's decisions making process in this field. This right does not include the right to co-decision, as the objection is not compulsory for the employer. It means rather the right to participate in the preparation of decision-making. Workers' Representatives can request information from employers regarding any and all issues pertaining to health and safety as well as express their opinion, and request the employer to take the necessary measures. Employers shall take measures or respond within eight days to the initiatives of Workers' Representatives. If they disagree with the initiative, employers shall explain their position in writing, with the exception of cases requiring immediate action. Workers' Representatives can take part in the investigation of occupational accidents, and in the exploration of the circumstances of occupational diseases. In justified cases, the Workers' Representatives can make a petition to the competent occupational health and safety administration and express their opinion to the person conducting the inspection.

In one important issue, the Workers' Representatives have the right to co-decision. If an employer defines the manner of enforcement of the health and safety requirements in an health and safety statute, the consent of the workers' representatives (or the Health and Safety Committee) is required. Workers' Representatives (or the Health and Safety Committee) are entitled to propose the employers to prepare an occupational safety programme.

Having in mind the tasks of the Workers' Representatives, the division of responsibility between the stakeholders will be clear. The employer has primary, main and objective liability. Workers' Representatives participate in the enforcement of health and safety requirements, but do not overtake the liability for it. Workers' Representatives suggest, but do not give instructions; they participate in the inspection, but do not lead it; they check, but do not overtake the task of control; in the risk assessment they call attention on the risk, but do not assess it.[91]

89. EBH 2008. 1812. Legf. Bír. Mfv. II. 10.411/2007. Published in Legfelsőbb Bíróság Határozatainak Hivatalos Gyűjteménye, no. 1 (2008). If the employer does not fulfil his obligations, the worker can turn to the authority and initiate the procedure.
90. Varga, *supra* n. 24, at 456.
91. Varga, *supra* n. 24, at 466.

[4] Promotion and Protection of Workers' Representatives

LSA contains special protective rules for the Workers' Representatives. Employers shall ensure the appropriate conditions in order to guarantee that they can exercise their prerogatives. They should particularly guarantee leave of absence with average pay, as necessary for carrying out their duties, meaning at least 10% of the monthly working time; the necessary equipment for their work, such as operating, technical and material conditions; and finally the opportunity to attend, within one year of being elected, a training course of at least sixteen hours during an election cycle, and the opportunity of annually taking eight hours of training subsequently. The employer has to bear the costs for these conditions. These are the compulsory minimum, but parties can agree on a higher degree of protection. It is useful to regulate the details in an agreement between the employer and the Workers' Representative (or the Health and Safety Committee).

The activities of the Workers' Representatives can necessarily provoke conflicts between them and the employer, as those can cause additional financial burden for the employer. Workers' Representatives can carry out their activities only, if they are protected in their status. The protection of Workers' Representatives is guaranteed in many ways.

LSA regulates a special protection for Workers' Representatives adopting the rules of the Labour Code. Workers' Representatives enjoy a special protection against ordinary dismissal, namely the provisions pertaining to elected trade union officials shall be duly applied to them. In order to ordinary dismiss a Workers' Representative, the employer shall require the consent of the Health and Safety Committee or in the absence thereof, the body of employees who elected the person.[92] The same rule applies, if the employer wish to employ the Workers' Representative temporarily under different conditions, than those one set in the employment contract, e.g., the employer wants to post him or her in another town for some weeks. The requirement of the consent can causes difficulties in practice in big companies, as it is extremely difficult to get the consent of the workers. In companies with many workers, it is not possible to ask every worker for his or her consent. The only legal solution would be to hold a secret election to get the consent of the workers, but it is difficult to imagine that the employer would get the consent of the workers in this way. The Workers' Representative enjoys this protection during his or her mandate and six months after the termination of it, provided s/he has been in this post for at least one year.[93]

Pursuant to the LXXV Act of 1996 on Labour Inspection, this authority had the competence to control the employers concerning the compliance with the protective rules of Workers' Representatives. This competence of the authority was cancelled with effect of 1 January 2012. Prior to that time the inspector had the right - after the

92. See also the judgment of the Supreme Court: BH 2000. 463. Legf. Bír. Mfv. II. 10.990/1998. sz.
93. Workers' Representatives enjoy broad protection against discrimination guaranteed by Article 76 LSA and a separate act on discrimination (2003. évi. CXXV. törvény az egyenlő bánásmódról és az esélyegyenlőség előmozdításáról, Act CXXV of 2003 on equal treatment and promotion of equal opportunities).

notice – to control, whether the employer complies with the rules and in extreme cases to impose fine. The cancellation of this right of the inspectors is a clear sign for the loosening up of the protection of the Workers' Representatives.

[5] Health and Safety Committee

If there are at least three Workers' Representatives by the employer, they may establish a local Health and Safety Committee. The establishment of a committee is voluntary and can occur only with the consent of all Workers' Representatives. If a committee is established, the prerogatives of Workers' Representatives shall be exercised by it in favour of all employees.[94] The employer, or its duly authorized representative, shall attend the committee sessions if so requested by the Committee. As regards the election, mandate, dismissal and recall of Workers' Representatives and their area of competence, the provisions of the Labour Code concerning the members of works' councils shall be duly applied.

[6] Common Representative Body

If there are more Workers' Representatives in the independent establishments or divisions of the employer, a Common Representative Body shall be set up, with an equal number of delegates representing the employer and the workers.[95] The employer shall take the initial step for the foundation of the Common Representative Body and shall provide for the conditions of the election and the operation of it. The employer shall delegate an executive employee with decision-making authority to the Common Representative Body and a person appointed to carry out some or all of the employers' obligations concerning occupational safety. The experts providing occupational safety related services to the employer on a regular basis shall be invited to the Common Representative Body's sessions in an advisory capacity.

The number of regular and alternate members, expiration of term, conditions for dismissal, the order of chairmanship and the rules of operations shall be decided by discussion between the workers' representatives and the employer.

The Common Representative Body has the right to analyse the working conditions and the working environment for the health and safety of workers on a regular basis, but at least once a year, as well as the related measures that may be necessary; to discuss the programme of occupational health and safety and shall monitor its implementation; to present its opinion on the drafts of internal regulations concerning safety at work.

94. Article 70/A.(4) LSA.
95. Article 70/B. LSA. The term of all members of the representative body (regular and alternate) shall be five years.

[7] National Health and Safety Committee

National reconciliation of interests related to occupational health and safety shall be performed by the National Health and Safety Committee consisting of the interest representation organs of employees and employers, as well as the representatives of the government. This Committee operates since 1990. Until 2011 it was a specialized committee of the National Council of Interest Reconciliation, which was the tripartite negotiating body at national level composed of the representatives of the trade unions, employers' organizations and government. After the reorganization of the mentioned council, the National Health and Safety Committee became an independent organ.

The National Health and Safety Committee has several prerogatives. It express its prior opinion on bills of legislation and drafts of other provisions and measures; it takes part in drawing up the national programme for occupational safety and the annual schedule and timetable for the execution of the programme; it adopts recommendations concerning occupational safety requirements exceeding those contained in occupational safety regulations; it provides information to the general public concerning its work; it supports the occupational safety information network by providing comments and information from its own database.

§6.05 COMPLIANCE AND CONTROL

[A] National Authority for Labour

The structure of the authorities responsible for the enforcement and supervision of health and safety regulations has been changed recently. Major tendency is the *centralisation of the authorities*. In 2011, with effect from 1 January 2012 the main authority has been restructured and renamed - called earlier National Supervisory Authority for Occupational health and safety (*Országos Munkavédelmi és Munkaügyi Főfelügyelőség*). The new name is National Authority for Labour (*Nemzeti Munkaügyi Hivatal*, hereafter NAL). The NAL has integrated three bodies working earlier in separate organizations, namely the Authority for Employment, the National Supervisory Authority for Occupational health and safety as well as the National Institute of Vocational Training and Adult Education. Within the single authority two separate directories have been established, namely the Directorate of Occupational health and safety and Labour Issues as well as the National Employment Service. The first one is the major body responsible for occupational health and safety in every branch and for every profession.[96] Only the mining sector has a separate authority maintaining its competence.

96. The operation and structure of the new NAL is regulated by the governmental Decree No. 323/2011. (XII. 28.) Kormányrendelet a Nemzeti Munkaügyi Hivatalról és a szakmai irányítása alá tartozó szakigazgatási szervek feladat és hatásköréről (323/2011. (XII. 28.) governmental Decree on the tasks and competences of the National Authority for Labour and the bodies being under its occupational guidance).

NAL is the major central body being responsible for occupational health and safety in Hungary and the national leader of the supervisory activities in this field. The NAL supports the work of the regional bodies by guidelines and recommendations and operates the uniform electronic system with registered data. It organizes training, education and examination of the supervisors. At national level, NAL takes part in the elaboration of new acts and the national policy, the annual assessment of the situation of national occupational health and safety, the preparation of the annual report and the information of people. Furthermore, it takes part in the activities of European and international organizations and co-operates with the organizations of other states.[97]

Parallel to the restructuring at national level, also *the organization at regional level* has been changed. The Regional supervisory bodies of occupational health and safety have lost their autonomy and recently constitute a part of the Regional governmental agencies next to several other authorities. The regional supervisory bodies for occupational health and safety (*munkavédelmi felügyelőség*) carry out the general tasks of the authority. They are the authorities at first instance and perform *inter alia* the supervisory activities.[98]

[B] Other Authorities in the Field of Occupational Health and Safety

Since 2007 the National Supervisory Authority for Occupational health and safety has overtaken the responsibility also for the supervision of occupational health.[99] Until this time there was a separate institution in this field, namely the *National Public Health and Medical Officer Service* (*Állami Népegészségügyi és Tisztiorvosi Szolgálat*).[100] It is composed of the Office of the Chief Medical Officer, its regional offices and the national institutes under its direction. The Office of the Chief Medical Officer participates in elaborating the national public health programmes and manages, organizes, co-ordinates the execution of the regional and local tasks of the programmes.[101] The regional governmental offices fulfil several tasks in the field of public health (environmental health, hygiene of settlements, food and nutritional health, radiohygiene and radiobiology, chemical safety), epidemiology, health development (health protection, education, promotion), management and co-ordination of public health and pharmaceutical administration. Until 2007 it had the task to elaborate the personal and material conditions of performing work which are not dangerous for the health (healthy working conditions); to develop by research the limits which are permissible

97. Articles 13-14. governmental Decree No. 323/2011. (XII. 28.).
98. Articles 17-19. governmental Decree No. 323/2011. (XII. 28.).
99. 362/2006. (XII.28.) Kormányrendelet az Állami Népegészségügyi és Tisztiorvosi Szolgálatról és a gyógyszerészeti államigazgatási szerv kijelöléséről (Governmental Decree No. 362/2006. (XII.28.) on National Public Health and Medical Officer Service and determination of the administrative body of pharmaceutics).
100. 323/2010. (XII. 27.) Korm. rendelet az Állami Népegészségügyi és Tisztiorvosi Szolgálatról, a népegészségügyi szakigazgatási feladatok ellátásáról, valamint a gyógyszerészeti államigazgatási szerv kijelöléséről (Governmental Decree No. 323/2010. (XII. 27.) on National Public Health and Medical Officer Service, performing the public health activities and determination of the administrative body of pharmaceutics).
101. Articles 6-7. governmental Decree No. 323/2010. (XII. 27.).

from the health point of view, the so-called limit values (e.g., for vibration and noise), the methods to reduce the risks for health and the provisions concerning occupational diseases, poisoning; to control and supervise the enforcement of the rules in the field of health at work.

There is a separate authority for mining, namely the Hungarian Mining and Geology Office and its six regional offices.[102] These are responsible for the protection of life and health of the workers in mining and they perform supervisory tasks in this field.[103]

The government enacted a decree in 2011, in which it determined the supervisory bodies responsible for occupational health and safety for certain professions in the public sector.[104]

The decree specifies the police, bodies of the penal system, the Hungarian Army, National Security Office, the National Rescue Service and the municipal fire service and determines the first and second instances of health and safety issues concerning public employees of these bodies.

[C] Common Rules for the Authorities

Chapter VII, Articles 81-86. LSA set common rules for the authorities performing supervisory activities in this field. In the following, only the most significant rules will be described.

The occupational health and safety authorities carry out their supervisory activities based on the supervisory guidelines published by the head of the authority. The guide contains the primary fields of supervision in the year concerned, the concrete activities, the objectives to be achieved and the main activities, branches or occupations, which shall be controlled.

The *supervisory activities include* the following: (a) the monitoring of compliance with the duties and obligations of employers and employees in connection with occupational safety and health; (b) enforcement of the requirements related to the establishment of workplaces, operation of work equipment, technological processes and materials applied, and to personal protective equipment; (c) investigation, reporting and registration of occupational accidents, occupational diseases, and cases of extreme exposure, as well as measures taken with a view to prevention.[105]

The authority *can impose fines* against the employers failing to fulfil the requirements of occupational health and safety, and thereby seriously endangering workers'

---

102. 1993. évi XLVIII törvény a bányászatról (Act XLVIII. of 1993 on mining); 267/2006. (XII. 20.) Korm. rendelet a Magyar Bányászati és Földtani Hivatalról (Governmental Decree No. 267/2006. (XII.20.) on the Hungarian Mining and Geology Office); 320/2011. (XII. 27.) Korm. rendelet a Magyar Bányászati és Földtani Hivatallal összefüggő egyes kormányrendeletek módosításáról (Governmental Decree No. 320/2011. (XII.27.) on the modification of certain governmental decrees relating the Hungarian Mining and Geology Office).
103. Article 43(2) 1993. évi XLVIII törvény a bányászatról (Act XLVIII. of 1993 on mining).
104. 373/2011. (XII. 31.) Korm. rendelet a munkavédelmi hatósági feladatokat ellátó egyes szervek kijelöléséről (373/2011. (XII. 31.) Governmental Decree on the determination of bodies carrying out administrative activities of occupational health and safety).
105. Article 81(4) LSA.

life or health.[106] The law provides for a list of cases which particularly causes serious endangerment of the life or health of workers, e.g., failure to carry out the periodic safety inspection; failure to make the risk assessment in connection with employers included in the highest hazard category and in case of workers potentially at risk from certain sources of danger; the necessary protective equipment and personal safety equipment are inoperable or absent; failure to perform the prescribed assessment of capacities and biological monitoring of workers working at dangerous work places, operating dangerous work equipment or working in dangerous technological processes, including the jobs where workers can be exposed to the effects of certain sources of danger defined in specific other legislation and workers in sensitive risk groups; employment in an environment where workers are exposed to the effects of factors in excess of the prescribed limit; failure to conduct the assessments prescribed in specific other legislation where workers might be exposed to carcinogens etc. Labour safety fines between 50,000 and 10,000,000 HUF (around EUR 200–40,000) may be imposed. The authority shall impose occupational safety fines on each business location of an employer if according to the findings of the inspection the endangerment is imminent in more business locations of the employer in violation of the same legal regulation.[107]

Until 2012, LSA provided for that the fines imposed shall be paid to a *special account of the treasury*. Half of the fines from the penalty payments was allocated by public tender exclusively for the objectives of occupational health and safety, while the remaining portion was used for the operation of the central occupational safety public information system. This rule has been cancelled and thus recently the fines can be used by the state for any possible objectives.

The occupational safety and health authority shall maintain *an official register* with a view to be able to verify employers' compliance with occupational safety regulations in connection with proceedings before other bodies conducted in accordance with specific other legislation (e.g., providing budgetary subsidies for the employment of workers with some degree of incapacity).[108] The authority publishes on its official website the name, registered office and tax number of employers sanctioned

106. The law provides for short term for the procedure, which is in general forty-five days from the time of the opening of the proceedings in connection with accidents at work, occupational diseases and cases of harmful exposure etc.

107. The amount of the fine shall be imposed based on the extent of endangerment, number of endangered persons, length of endangerment and its periodic feature, number of the breached rules, possible consequences of the endangerment, extent of the injury and harm, number of employed people and net revenues of the employer as well as the personnel and material circumstances related to the negligence. Article 82 LSA. More detailed rules contains the 273/2011. (XII. 20.) Korm. rendelet a munkavédelmi bírság mértékére és kiszabására vonatkozó részletes szabályokról (governmental Decree No. 273/2011. (XII. 20.) on the detailed rules of the amount and imposition of the fines).

108. The official register shall contain the particulars of employers sanctioned and upon whom the competent authority has imposed occupational safety fines by final and executable decision for any breach of occupational safety regulations. The authority shall erase data from the register after two years from the operative date of the resolution. Article 83/B (1) LSA.

for breach of occupational safety regulations, including the description of the infringement, the amount of the fine imposed, the date of the resolution on the infringement and the date when it becomes operative.[109]

Inspectors of the occupational health and safety authority *have substantial prerogatives*. Inspectors are entitled to hold inspections at any workplaces; to investigate occupational accidents, without prejudice to the employers liability; to instruct employers to satisfy the requirements of occupational health and safety; to compel employers to eliminate any deficiencies which are determined, within the deadline specified; to prohibit employees from working under conditions which are in grave violation of the regulations pertaining to occupational health and safety; to order the suspension of hazardous activities, and/or the operation and use of divisions, sections and work equipment, and the use of any dangerous substances; to declare an accident as an occupational accident; to request information from the persons present, to the extent required for the inspection; to request police assistance in the event of any attempt made to prevent the inspection; to order the inspection of occupational hygiene.[110] Furthermore, based on the facts stated, inspectors have competence to declare an employment-related legal relationship between the employer and a person working at the workplace as 'organized employment'.

§6.06 REMEDIES

[A] Criminal Remedies

The Hungarian Criminal Code (IV. Act of 1978.) regulates an offence concerning occupational health and safety, called *occupational endangerment*. Pursuant to Article 171, any person who *breaches the rules of his or her profession* and thus directly jeopardizes the life, bodily integrity or health of another person or persons or causes bodily harm, commits an offence and shall be punished by imprisonment for up to one year. In case the offender invokes the direct danger deliberately, s/he commits a crime and shall be punished by longer imprisonment. Similarly, the imprisonment shall be longer, when the crime results in permanent physical disability or a grave injury to health, if it causes death or a (fatal) mass catastrophe. The rule has been nearly unchanged since 1978, only the kinds of punishment were modified. Until 2009 the offenders could be punished by fine or by compulsion to work. Since 2009, however, the only possible punishment is imprisonment. In fact, offenders are often placed on probation.

The term 'rules of his or her profession' covers also occupational safety rules and includes all written and unwritten regulations, guidelines of the profession concerned, which the person practising it have to take into consideration. Every regulation on health and safety is considered constituting occupational rules, i.e., not only the Act on health and safety, but also the ministerial decrees and standards. According to the

109. Article 83/B. (5) LSA.
110. Article 84 LSA.

settled case law, if performing work is considered as 'organized' work, the employer is liable for the compliance with the occupational rules. Consequently, if the worker suffers an accident due to the breach of the occupational law, the question arises whether the employer committed this offence.

As this rule is in force for several decades, great case law has been developed by the Supreme Court. In case of voluntary work for friends or relatives, often arises the question, *who is the employer* liable for the adherence of occupational health and safety rules. According to the opinion of the Supreme Court in such cases 'employer' is the person, who not only asks the people for carrying out work in his or her interest, but also gives instructions for these people.[111] In one case, a person without any occupational knowledge asked people to help him to demolish his house without having any expert and did not take care of the information of these workers on occupational safety rules. Due to the unprofessional performance of work one person suffered fatal accident. The Supreme Court decided that the person instructing the demolition was the employer and he breached the occupational rules. In another case, the Supreme Court came however to the opposite conclusion, when the 'employer', whose interest the worker has performed work, charged an expert having special knowledge with leading of the work. As in this case the employer chose an appropriate expert and engaged him to instruct the employees, he fulfilled his obligation as he could rely on the knowledge of the expert. Therefore in this case, the employer was not liable for committing the crime, when the accident occurred. However, according to the Supreme Court also the engaged expert has not committed the offence of occupational endangerment, as he was not the employer, but only the instructor of the work.[112]

In order to ascertain the employer's liability, it is necessary to examine, *whether the employer actually gives instructions* to perform the work. The fact that the person is the legal employer is not enough. It is necessary to be the actual instructor of the constructions work, where the accident happened. The employer commits the occupational endangerment crime only, if s/he – as the actual leader of the construction work – is aware of the fact that the workers perform their work without complying with the occupational safety rules and a worker has a serious accident.[113]

[B] Co-ordination of Workers of More Employers and their Liability

It often occurs that workers of more employers work at one workplace. In such cases, it is very important to co-ordinate workers and to determine the employer who is liable

111. BH 2005. 236. Legf. Bír. Bfv. I. 725/2004. sz.
112. BH 2005. 236. Legf. Bír. Bfv. I. 725/2004. sz.
113. In the case concerned, a limited liability company had two managers. The actual work, during which the accident happened, was carried out under the instructions of the first manager. He noticed that the workers perform the work without any protecting equipment and without complying with the safety rules and he did not raise objection against this procedure and did not stop the work. As the first manager was liable for managing the work, he was liable to guarantee the safe and healthy working conditions. As he violated his obligations, he committed the offence. BH 2001. 7. Szabolcs-Szatmár-Bereg Megyei Bíróság 3. Bf. 1103/1999. sz. See also: BH 2005. 309. Legf. Bír. Bfv. II. 210/2004. sz.

for the co-ordination and adherence of occupational health and safety rules. Already the decree in 1979 contained the rule that at such workplaces the work has to be co-ordinated in order to guarantee health and safety and the obligations of the parties have to be stated in an agreement.[114] The LSA took over this provision and connected the parties' liability to it, which has been modified later.

Main principle is that at workplaces where *the employees of different employers are working simultaneously*, work shall be co-ordinated as to prevent any imminent hazard to the employees and to other personnel in the area. The original rule of 1993 laid down that for the enforcement of this principle the employer is liable, who is designated in the contract by the parties; in lack of such clause, the general (main) contractor; and finally in lack of the latter, the person, on whose property the work is performed. In 2004, the rule was completed by the declaration that within the framework of co-ordination the employees affected, workers' representatives and other personnel in the area concerned shall be informed of any risks regarding occupational safety and health and of the preventive measures introduced. Adoption of this rule served the implementation of the Framework Directive.

Since 2012 the regulation of liability has been changed. The new rule states that for the enforcement of the co-operation *primarily the general contractor* pursuant to the Civil Code is liable; in lack of this any other person or body, who is actually in charge of work or has the main liability for the workplace; and finally in lack of the latter that person, in whose interest the work is performed.

In the past, the regulation caused great problems in cases where *several subcontractors* have worked with mixed group of workers at the same workplace performing certain parts of the construction work. In these cases, the liability of the subcontractors was often only partially regulated in the agreements between the general contractor and the subcontractor and was difficult to clarify.[115] The Supreme Court called attention to the fact that according to the earlier provision the single employers of the workers - the subcontractors - separately had no liability to the enforcement of the occupational health and safety provisions during the common performance of work.[116]

The new rule serves the workers' interests by providing clear liability. Basically the general contractor has to take care of the co-ordination of work and the enforcement of the occupational health and safety provisions. The earlier regulation provided the possibility to shift the liability to one or more subcontractors, who possibly were not in the actual situation to co-ordinate the work. The recent rule restricts the possibility to evade the law.

The law clarifies that the term 'general contractor' shall be interpreted based on the Civil Code. The only problem is that the Civil Code does not use the term 'general conductor', only the expressions 'conductor' and 'subconductor'.[117] Consequently, it

114. Article 30(5) of the 47/1979. (XI.30) MT rendelet a munkavédelemről (47/1979. (XI.30) Decree of the Council of Ministers on labour safety).
115. See e.g., the following case: BH 2007. 73. Legf. Bír. Bfv. II. 276/2005. sz.
116. BH 2007. 73. Legf. Bír. Bfv. II. 276/2005. sz.
117. Pursuant to Article 391 Civil Code the conductor is entitled to engage subcontractors. The conductor is liable for the legally engaged subcontractor in the same way, as if he would have performed the work.

can causes serious trouble to identify the general contractor, if not only one company conclude an agreement with the client.

There are several judgments dealing with accidents occurred in the construction sector, when it was controversial, which conductor shall be liable for the damages.[118] The Supreme Court stated that the rule on the liability of the general conductor did not mean that every single employer would not be liable for the enforcement of the occupational health and safety provisions concerning his or her own workers. *The employer's primary liability* to guarantee safe working conditions *is not affected* by the general contractor's main liability to co-ordinate workers. Every employer has to guarantee the safety of his or her workers and make sure that the circumstances guarantee these, even if the general contractor has the main liability.[119] Also, the judgment EBH 2004. 1151[120] clarifies that the rule on the co-operation of employers does not exempt the employer from the obligation to provide appropriate protective equipment for the workers. From the approach of occupational health and safety and especially from the point of view of the labour inspections it is decisive, whether each employer fulfilled his or her obligations arising from the LSA and other provisions and took care of the guaranteeing of a safe workplace. This obligation is not affected by the obligation to co-operate.

[C] Labour Law (Private Law) Remedies

[1] The Old Labour Code

Both the old and the new Labour Codes contain separate chapters on the liability of employers. The old Labour Code regulated a very strict system with extremely narrow possibilities of the employer to relieve from the liability. *Employers have had full and objective liability for all damages of the workers* caused in connection with the employment relationship.[121]

In case of an occupational accident or disease, the social insurance system provides health care, accident benefit and long-term accident benefit to the worker. The employer has to reimburse the health insurance system for the amount of the care and benefits, if the employer or his workers breached the rules of health and safety and thus have been liable for the cause of the accident (or occupational disease). If the employer and his workers observed the provisions of occupational health and safety,

118. See for example: EBH 2010. 2261 Legf. Bír. Mfv. II. 10.394/2010. Published in Legfelsőbb Bíróság Határozatainak Hivatalos Gyűjteménye, no. 2 (2010) and EBH 2004. 1151. Legf. Bír. Mfv. I. 10.659/2003. sz. Published in Legfelsőbb Bíróság Határozatainak Hivatalos Gyűjteménye, no. 2 (2004).
119. EBH 2009. 1996. Legf. Bír. Mfv. II. 10.206/2008. Published in Legfelsőbb Bíróság Határozatainak Hivatalos Gyűjteménye, no. 1 (2009).
120. EBH 2004. 1151. Legf. Bír. Mfv. I. 10.659/2003. sz. Published in Legfelsőbb Bíróság Határozatainak Hivatalos Gyűjteménye, no. 2 (2004).
121. 'In connection with the employment' means that the employer has liability for damages occurring during activities preparing for or ending of the actual work, conducts necessary for performing work (e.g., eating, washing etc.) and activities outside the workplace (e.g., during posting). See: Opinion No. 29. of the Labour Committee of the Supreme Court.

he is exempted from the reimbursement of the costs of the health insurance. However, the social insurance system does not pay compensation, so the employer is still obliged to pay full compensation for the worker. Therefore, it can for example occur that both the social insurance and the employer have to pay some kind of long-term benefit for the worker. In general, I have to emphasize that the employer has a wide liability which is caused mainly by the fact that in Hungary there is no compulsory liability health insurance for occupational accidents, like in Germany or Austria. This is a major difference compared to the most Western European states. That means that the compulsory health insurance of workers does not cover liability insurance and does not exempt the employers from civil liability in case of occupational accident or disease. Employers, of course have the possibility to cover voluntary liability insurance in order to exempt form the compensation. This is however only a possibility.

This strict system of the employer's liability has its roots obviously in the socialist time. The liability rules of the social security insurance system are principally inherited from the socialist system. The rules of the Labour Code of 1967 on liability have been overtaken to a large extent into the Labour Code of 1992. The employer was and still is liable for all material and immaterial damages caused for the employee, including any physical and mental problems arising due to an occupational accident or disease. It is objective liability regardless of the employer's culpability, i.e., the employer had to compensate for the full damage independently of the fact, whether s/he caused the damage deliberately, negligently or without any negligence. The employee is obliged to prove the damage, its amount and the connection between the damage and the employment relationship.

Pursuant to the old Labour Code, *the employer could relieve from his or her liability* only in two narrow cases, namely if s/he was able to prove that the damage was caused by an unavoidable event outside his or her field of operations or solely by the unavoidable conduct of the aggrieved party.[122] The term 'employer's field of operation' was interpreted in a very broad way by the court.[123] The Opinion No. 29 of the Labour Committee of the Supreme Court[124] stated that a reason causing damages shall only be considered *falling outside 'the employer's field of operation'*, if it is completely independent of the employer's activity, i.e., there is no connection between the cause of the damage and the employer's activity. The employer's field of operation was not limited to the employer's premises especially since in many cases the worker performs his or her work usually outside the employer's premises. The employer was also objectively liable, if the cause was outside the employer's field of operation, but it could have been objectively avoided by the employer. The opinion of the Supreme Court mentions the following interesting example: A thunderbolt is obviously out of the employer's field of operation, it can however be expected from the employer to

122. Article 174 old Labour Code.
123. Article 174(5) old Labour Code provides the following definition: 'The employer's field of operations shall particularly include the causes arising from the conduct related to the activities pursued by the employer in the course of his duties and from the characteristics, properties, movement and operation of the materials, equipment and energy involved.'
124. Legfelsőbb Bíróság Munkaügyi Kollégiumának 29. számú állásfoglalása (Opinion No. 29. of the Labour Committee of the Supreme Court).

install appropriate equipment to avoid it. According to the Supreme Court an influence or circumstance, which causes the damage is unavoidable, if at the recent level of technology and engineering in the given time it is not possible to prevent it. Consequently, the employer was nearly always liable for any damages caused for the worker in connection with the employment relationship.[125]

In the second case of the exemption of the employer, *exclusively the conduct of the worker* caused the damage and his or her conduct was unavoidable. To estimate whether the exclusive cause of the damage was the worker's conduct, also the adverse effects of the workplace and the special circumstances of performing work had to be taken into account. Even if the accident was caused by the worker's inattention, clumsiness, the breach of technological rules or feeling of faintness, it does not mean that this was the only reason of the damage. All of the mentioned facts can be in connection with the characteristics of work and workplace. If so, the damage was not exclusively caused by the worker's behaviour. Attention of the worker can be influenced by lots of circumstances. In many cases, the employee is affected by a quick impact, which is the reason of distraction causing the accident. For the exemption of the employer the worker's exclusive behaviour should be unavoidable by the employer.

The employer's full and objective liability does not mean that s/he cannot shift the amount of compensation partly to the person caused the damage. Thus, the Labour Code provided the possibility to allocate the compensation for damages in case of the worker's negligence. The employer shall not compensate for the part of the damage, which was caused by the worker's contributory negligence.

[2] The New Labour Code

The new Labour Code restructured and slightly limited the employer's liability. *The main rule remained unchanged*, i.e., the employer has full and objective liability without the estimation of his or her culpability for any damages caused to its employee in connection with his or her employment. However, the new Labour Code redefines the grounds of exemption. The term used earlier 'field of operation' was replaced by the expression *field of control*. The aim of this change was to reduce the employer's liability elaborated in an extremely broad way by the case law of the labour courts. The legislator intended to exclude the employer's liability in cases, when the employer could not have had even an indirect effect on the cause of the damage.

The employer cannot exempt from his or her liability, if the damage occurs within the field of his or her control. Pursuant to the official explanation of the new Labour Code, the employer's 'field of control' includes all those circumstances (devices, energy, workers), on the operation of which the employer has influence.

125. Until 2009, there was a derogation from this rule for private individual employers employing a maximum of ten full-time employees who were liable only for damages caused to the employee by negligence or deliberately. The decision No. 41/2009. (III.27.) of the Constitutional Court ruled that from the workers' point of view it was unconstitutional to make a difference between employers' liability and thus between the compensation of workers.

The employer has two opportunities to exempt from liability, if the damage occurs beyond its field of control. In the first case, the damage was caused by an external circumstance beyond its control which the employer did not have to reckon with and s/he was not expected to avoid the emergence of the circumstance giving rise thereto or to eliminate the same. In the second case, the damage was caused exclusively by the aggrieved party and his or her behaviour was not evitable.[126]

As earlier, the employer shall compensate the employee for the totality of his or her damage; however, two new rules have been introduced to reduce the amount of compensation. The novelty is that no compensation shall be provided for the damages, in connection with which the employer proves that its emergence was not foreseeable at the time it was caused.[127] The *foreseeability rule* was overtaken from the regulation of the Civil Code into the new Labour Code. No compensation shall be provided for the part of the damage which was caused through the employee's culpable conduct or which arose from the fact that the employee failed to meet his or her obligation of damage alleviation. Important new rule is that the court may release the employer from the compensation for damages under special circumstances. The court shall consider particularly the financial situation of both parties, the gravity of the infringement and the consequences of paying compensation.[128] This rule enables the labour courts to reduce the amount of compensation *on ground of equity*. According to the official explanation of the new Labour Code, this rule can be applied in case of small employers (particularly natural person employer), where a high amount of compensation would endanger their existence as an employer or - in case of natural persons - even their subsistence.

The other rules on compensation for damages have not been significantly modified in the new Labour Code. The employer has to reimburse all the lost income of the worker received in or outside of the employment relationship. Not only the lost wages and the value of any regular benefits shall be taken into consideration to which the employee was entitled over and above his or her wages on the basis of his or her employment. The employer has to reimburse the worker for any other regular income outside the employment lost by virtue of the injury. Compensation shall also be provided for any loss or damage that the employee averts through his or her extraordinary effort and performance in spite of his or her significant disability arising from the injury. No compensation shall be provided for the value of the benefits which, by virtue of their purpose, are only due in case of the performance of work, or for amounts received as a cost allowance.[129] Compensation may also be awarded in form of monthly benefit. A regular benefit shall be awarded if it serves to support or to supplement the maintenance of the worker or his or her relative entitled to maintenance vis-à-vis the worker.[130]

126. Article 166(2) new Labour Code.
127. Article 167(1) new Labour Code.
128. Article 167(3) new Labour Code.
129. Article 169. Labour Code. The Labour Code provides for detailed rules on the calculation of the amount of compensation for damages in Articles 170-175. new Labour Code.
130. Article 171-173(1) new Labour Code.

Employer has to compensate all damages of the workers arising from an occupational accident or disease. However, only proven damages can be compensated. Non pecuniary damages can only be compensated, if these are proven usually by a medical opinion. Mental problems due to an accident or disease are only compensated, if they are proved by an expert. In Hungary, lump sum compensation for immaterial damages does not exist.

The new Labour Code regulates the *compensation for damages of the relatives* of the aggrieved party similarly to the old Labour Code. Employer shall also provide compensation for any loss sustained by the employee's relatives in connection with the loss. If the employee dies due to the occupational accident or disease, beyond the compensation for loss, a dependant relative may also demand compensation for maintenance of an amount that, also with regard to his or her actual wages and income or the wages or income expected to be attained, ensures that all his or her needs are satisfied to the standards enjoyed before the injury occurred.[131]

An important innovation of the new Labour Code is that it makes the Civil Code applicable as a background regulation. In cases where the Labour Code does not contain any rules, compensation for damages shall be governed by the rules of civil law relating to compensation for damages.[132] Another important rule is that unlike the general rule, a collective agreement may only depart from the provisions of the Labour Code on the employer's liability in the employee's favour.

§6.07 REINTEGRATION EFFORTS REQUIRED FROM THE EMPLOYER AND EMPLOYEE

[A] Workers Having Reduced Capability to Work

In the occupational health and safety policy, it should be a major objective to reintegrate those people who suffer from the consequences of occupational accident or disease and are therefore only partially able to work.

The system of regulation of persons having reduced capability to work has been completely changed recently with *the introduction of a new act*, namely the Act CXCI of 2011 on the maintenance of persons with reduced capability to work and the modification of certain acts.[133] Main aims of the new act are twofold, namely to create a unified system for persons with reduced capability to work and to motivate people to return to the labour market instead of receiving disability benefit. The act emphasizes the significance of individual recovery and introduces a new, comprehensive notion of 'health', which considers not only physical, but also mental and social circumstances.

131. Article 171 new Labour Code.
132. Article 177(1) new Labour Code.
133. 2011. évi CXCI. törvény a megváltozott munkaképességű személyek ellátásairól és egyes törvények módosításáról (Act CXCI of 2011 on the maintenance of persons with reduced capability to work and the modification of certain acts). Beside the regulation of persons with reduced capability to work, this act modifies more than 40 acts dealing with other issues. In Hungary such acts are called 'salad act', as they are adopted at the end of year and cover huge pile of provisions dealing with different issues.

Stressing the rehabilitation of such persons with the aim to return into the labour market is a new approach in the Hungarian law, which was mainly started with the Act LXXXIV of 1997 on the rehabilitation benefit.[134] Prior to 1997 this category of benefits did not exist. Even the names of the earlier forms of assistance for people having reduced capability to work did not suggest the motivation or aim to recover (disability pension, accidental disability pension, accidental benefit). The first change came by the mentioned act and it was significantly enhanced by the new act of 2011.

Two definitions of person with reduced capacity to work exist. Pursuant to the Act IV of 1991 on promotion of employment and unemployment benefit[135] the expression 'a person with reduced capacity to work' covers persons being physically or mentally disabled, and those whose chances to get a new job or keep the old job after the medical rehabilitation have been reduced due to physical or mental damages. Act LXXXI of 1997 on social insurance contains also a definition, which refers to the new act.[136] Persons with reduced capacity to work are those, whose health condition is maximal 50% pursuant to the new CXCI Act of 2011. Consequently, the new regulation includes only those persons, who lost at least 50% of their health condition meaning a disadvantageous modification of the physical, mental or social well-being of a person.

[B] Reintegration Efforts Required from the Employer and Employee

The new Act CXCI of 2011 changed the whole system of maintenance of persons with reduced capability to work with effect from 2012. It cancelled the earlier categories of disability pension and accidental disability pension and introduced two new forms of benefits depending on the medical assessment, making a difference on the basis, whether the person concerned can be rehabilitated and re-employed or not. In the first case, the person will receive a *rehabilitation benefit*, in the second case a *disability benefit*. Persons getting rehabilitation benefit have to co-operate with the rehabilitation authority. Co-operation includes especially active search for a job, compulsory participation in trainings and acceptance of a job offer for a suitable workplace.

The last obligation covers also the so-called public employment. *'Public employment' in this special sense* means the employment of unemployed persons especially by the state and the municipalities, but also by registered associations and churches for activities in the interest of public, like providing assistance for old people, reconstruction of buildings, cleaning of the public areas or hiking path.[137] However, this kind of 'public employment' has also been used for example in building motorways (!), which indicates that in some cases it is an extremely cheap alternative workforce. The

134. 2007. évi LXXXIV. törvény a rehabilitációs járadékról (Act LXXXIV of 1997 on the rehabilitation benefit).
135. Article 58 m) 1991. évi IV. törvény a foglalkoztatás elősegítéséről és a munkanélküliek ellátásáról (Act IV of 1991 on promotion of employment and unemployment benefit).
136. Article 4(1) k) 1997. évi LXXXI. törvény a társadalombiztosítási nyugellátásról (Act LXXXI of 1997 on social insurance).
137. 2011. évi CVI. törvény a közfoglalkoztatásról és a közfoglalkoztatáshoz kapcsolódó, valamint egyéb törvények módosításáról (Act CVI of 2011 on 'public employment' and the modification of other acts).

declared aim of this form of employment is to lead people being in active age and capable of work back to the labour market and thus reduce long-term unemployment.

The way of implementing this aim has been criticized regarding the low wages. Persons being in 'public employment' receive lower wages (71,800,-HUF/month gross, which is about EUR 243), than the official minimum wage is in Hungary (in 2012 93,000,-HUF/month gross, which is about EUR 315).[138] The government particularly encourages the 'public employment' in part-time - for four hours a day -, where the person receives 34,546,-HUF (about EUR 117). Another criticism is that the new act allows to employ the persons of 'public employment' very far from their home. When the length of the return journey between the home and the workplace of the worker exceeds six hours, the employer has to provide free accommodation and food. By this rule, the government wanted to support great projects.

Rehabilitation benefit can be granted for a maximum of three years. The amount of it shall be between 30% and 50% of the minimum wage. It shall be terminated, if the person receives any regular income, does not co-operate with the authority, does not accept the offer for employment etc. In the last mentioned case, the person receiving the benefit has to pay back the benefits received in the last twelve months.[139] Disability benefit can be higher, and it shall be between 30% and 150% of the minimum wage.

In my opinion, *the major concern of the structural change* is that the majority of people receiving now disability benefit received earlier disability pension. This was part of the pension system and constituted a special form of pension. The amount of disability pension was counted based on the length of service and the amount of wages earned by the person concerned during his or her active life. Recently, disability pension was restructured into rehabilitation benefit and disability benefit. The major concern is that the amount of the disability benefit became completely independent of the earlier income. Consequently, as disability benefit does not constitute a form of pension any more, there is no vested right on it and the government can cancel it at any time.

Nevertheless, fact is that the *employment rate in Hungary* was *the lowest* of the European Union being 55.4% (60.4% of men and 50.6% of women) in 2010 compared with the EU average of 64%.[140] Thus, it is a justified objective to lead people being able to work back to the labour market. It is often argued that one significant reason for the extremely low employment rate is the fact that many people have applied for disability

138. See: 170/2011. (VIII. 24.) Korm. rendelet a közfoglalkoztatási bér és a közfoglalkoztatási garantált bér megállapításáról (170/2011. (VIII.24) governmental Decree on the determination of the wage and guaranteed wage in public employment) and 298/2011. (XII. 22.) Korm. rendelet a kötelező legkisebb munkabér (minimálbér) és a garantált bérminimum megállapításáról (298/2011. (XII.22.) governmental Decree on the determination of the minimum wage and the guaranteed minimum wage).
139. Article 9-10. 2011. évi CXCI. törvény a megváltozott munkaképességű személyek ellátásairól és egyes törvények módosításáról (Act CXCI of 2011 on the benefits of persons with reduced capability of work and on the modification of certain acts).
140. See: the statistics of the Hungarian Central Statistical Office under: http://www.ksh.hu/docs/hun/xstadat/xstadat_hosszu/mpal9807_01_02a.html or the statistics of the Eurostat under: http://epp.eurostat.ec.europa.eu/tgm/table.do?tab=table&language=en&pcode=tsiem010&tableSelection=1&footnotes=yes&labeling=labels&plugin=1, accessed on 15 Jun. 2012.

pension in order to avoid unemployment. However, in 2006 the proportion of people getting disability pension was 16.4% of the people getting some kind of pension. This is not extremely high percentage, as in six Member States there are higher proportions.[141] According to another survey, 7.24% of the people being in active age received some kind of disability pension or benefit in 2009.[142]

[C] Incentives for the Employers

The government intends to motivate employers to employ more persons with reduced capability. Various incentives have been elaborated. One of them is the financial support of the employers who employ such persons through *direct wage subsidies*. The concrete subsidies are changing from time to time.[143]

Since 1996, the employer is obliged to pay a special tax, the so-called *rehabilitation tax*, if s/he does not employ enough persons with reduced capability to work. Employers having more than twenty-five employees have to pay this tax, if they do not employ at least 5% of their workforce such employees. Small enterprises are excluded from this rule. The amount of the tax is counted based on the number of missing employed persons multiplied by 964,500,-HUF/per year/person (about EUR 3270).[144]

[D] Protection against Dismissal

According to the old Labour Code, *ill workers enjoyed far-reaching protection against ordinary dismissal*. Employer might not terminate an employment relationship by ordinary dismissal during the period of incapacity to work due to illness, not to exceed one year following expiration of the sick leave period. Furthermore, the employee was protected for the entire duration of incapacity, if it was the result of an occupational accident or disease.

As the ordinary dismissal was prohibited, the only possibility of employers to terminate the employment relationship of these workers was to give notice of termination with immediate effect, which was called extraordinary dismissal. It could be used only in special cases. The employer was allowed to use extraordinary dismissal in

141. Data in 2009. See: Országos Nyugdíjbiztosítási Főigazgatóság, Közgazdasági Főosztály, *Nyugdíjas ellátottak az EUROSTAT adatbázisa alapján*, ed. M. Hablicsekné Richter (Budapest 2009), 25.
142. See: Országos Nyugdíjbiztosítási Főigazgatóság, Közgazdasági Elemzések Főosztálya, *A 2009-es új rokkantsági nyugdíjasok és rehabilitációs járadékosok összetételének vizsgálata*, eds. J. Hollósné Marosi & Gy. Császár (Budapest 2011) 5. table (5. táblázat) www.onyf.hu, accessed on 15 Jun. 2012.
143. The procedure of the subsidies is regulated in the following decree: 15/2005. (IX. 2.) FMM rendelet a megváltozott munkaképességű személyek foglalkoztatásához nyújtható költségvetési támogatás megállapításának részletes szabályairól (Decree No. 15/2005. (IX.2.) of the Ministry for Employment and Labour on the detailed rules of the subsidiaries provided for the employment of persons with reduced capability to work).
144. Article 23 2011. évi CXCI. törvény a megváltozott munkaképességű személyek ellátásairól és egyes törvények módosításáról (Act CXCI of 2011 on the benefits of persons with reduced capability of work and on the modification of certain acts).

the event that the employee deliberately or by gross negligence seriously violated any substantive obligations arising from the employment relationship, or otherwise engaged in a conduct rendering further existence of the employment relationship impossible. Consequently, protection against dismissal of ill workers has been nearly absolute. The labour courts have taken very seriously these provisions. In practice, workers often abused the mentioned rule. When they have taken notice of the employer's intention to terminate their employment relationship, they went to the doctor, who signed them off work. In this way, they could evade or at least postpone the dismissal.

The *new Labour Code* wanted to set an end to this practice and *modified the rule.* Recently, the employer can hand over the notice on dismissal at any time, but the notice period will start only after the expiry of the sick leave. Consequently, if the worker is incapable of work, the employer can use ordinary dismissal, but the employment relationship remains to exist during the sick leave, by no later than one year following expiration of the sick leave period.[145] Important is that in the calculation of the length of the notice period, the time in which the worker has not received wages consecutively at least thirty days long, e.g., during sick leave, is not taken into account.

The second rule relating to the protection of ill workers in the old Labour Code provided a *ban on ordinary dismissal during* the whole period of *receiving rehabilitation benefit* and being incapable of work.[146] After the expiry of this protection, the law provided further restrictions on dismissal. Pursuant to Article 89 (7)-(8) old Labour Code, an employer was only in particularly justified cases allowed to terminate the employment relationship by ordinary dismissal of a person, who was drawing rehabilitation benefits, but was not incapable to work. The expression 'particularly justified cases' was interpreted meaning that the employer may terminate the employment relationship by ordinary dismissal if the worker can be no longer employed in his or her original position and no other job considered appropriate for his or her medical condition is available, or if the worker refuses to agree with the amendment of his or her contract of employment to accept a job in such other position.[147]

As far as the workers receiving rehabilitation benefit are concerned, the new Labour Code only slightly modified the rule. The employer can terminate their employment relationship by ordinary dismissal on the ground of their incapacity to work under certain conditions. This is only allowed, if the worker can no longer be employed in his or her original position, and the employer cannot offer him or her another position, which is suitable for his or her medical condition, or the worker rejects the position offered by the employer without having a solid ground.[148]

---

145. Article 68(2) new Labour Code.
146. Article 90(a) and (g) old Labour Code.
147. Article 91 old Labour Code. The employer cannot amend the contract of employment unilaterally without the consent of the worker. In case the worker did not agree with the modification of his or her contract to be employed in a new position, the employer can dismiss him/her, but cannot instruct her/him to work in another position. See also the settled case law of the Supreme Court, e.g., EBH 1999. 142. Legf. Bír. Mfv. I. 10. 594/1998/3. Published in 'A Legfelsőbb Bíróság határozatainak hivatalos gyűjteménye', no. 2 (1999). BH 2005. 118. Legf. Bír. Mfv. I.10.078/2004. BH 2008. 131. Legf. Bír. Mfv. I. 10.176/2007.
148. Article 66(7) new Labour Code.

[E] Other Rules in the Labour Code Concerning Workers with Reduced Capability

The old Labour Code contained another rule to protect and support workers with any degree of incapacity. It stated that the employer shall continue to employ workers, whose capacity to work has been impaired in the course of employment, in a position suitable their condition.[149] This rule included two obligations at the same time. First, the employer was not allowed to terminate the employment relationship of a worker suffered an accident or disease during the employment relationship and second, the employer had to modify the worker's position or working conditions according to his or her needs. The new Labour Code does not contain the same rule and thus the employer does not have the obligation to modify the worker's position.

When creating a workplace where disabled or physically handicapped persons are employed, the physical environment shall be designed to accommodate such disabilities.[150] Workers with any degree of incapacity are included in the sensitive group, as they are particularly exposed to risk and they constitute a risk themselves. Employees having suffered a damage to health to an extent of at least 50% as diagnosed by the body of rehabilitation experts shall be entitled to five working days of extra holiday a year.[151]

The employer is not obliged to elaborate a plan for the rehabilitation of workers with health impairment and persons with reduced ability to work. However, if s/he decides to do so, s/he has to consult works council fifteen days prior to adopting a decision.[152]

§6.08 CONCLUDING REMARKS

The law on occupational health and safety has been continuously changing in the last decades. Implementation of the legislation of the European Union has played a major role in this process. Parallel to the development of the technology, new risks has emerged, which can affect both the physical and mental health of workers. The fact is thought-provoking that the relative number of occupational accidents to thousand workers has not changed for several years. New term has been introduced in the regulation, namely the psychosocial risk. There is rarely any survey, to which extent psychosocial risk plays a role in the cause of an occupational accident and particularly of an occupational disease. It barely occurs that the authority recognizes a mental disease as occupational disease. The same is true for musculoskeletal diseases.

The law on workers' representation concerning occupational health and safety has been significantly changed in the last decades, mainly due to the implementation

149. Article 85(3) old Labour Code.
150. Article 19(4) LSA.
151. Article 120 new Labour Code.
152. Article 264 new Labour Code.

of the EU Law. Workers' representatives enjoy several rights and protection. Nevertheless, in spite of the detailed and comprehensive regulation, workers' representatives play a minor role in practice.

The new Labour Code slightly reduced the employer's liability for damages suffered by the workers and their relatives due to an occupational accident or disease. However, even so the employer has full and objective liability for all damages occurring in connection with the employment relationship.

In 2012, a new legal system has been introduced concerning persons with reduced capability to work. Clear objective was to motivate people to return to the labour market. Incentives are various, e.g., the very low amount of rehabilitation benefit, support to find a new job, 'forced' employment in the framework of 'public employment', etc. Employers are motivated by wage subsidies and a special kind of tax. This movement includes the withdrawal of disability pensions of several thousand people, which affects the subsistence of these people. As it is a brand new programme, the effects of it are still not measurable.

The authorities responsible for the supervision of occupational health and safety rules have been centralized lately. With some minor exceptions – in mining and in certain public sectors – one single authority, namely the National Authority for Labour is responsible for both the safety and health at work having broad supervisory tasks. As the changes introduced in 2011 and 2012 are brand new, the experiences after some years will show whether the reform attempts have been successful.

BIBLIOGRAPHY

European Agency for Safety and Health at Work. 'European Survey of Enterprises on New and Emerging Risks, Managing Safety and Health at Work'. www.osha.europe.eu

Galgóczy, G. 'A foglalkozás-egészségügy központi intézménye'. *Hippocrates* 1 (1999): 26–28.

Sarolta, M. 'A munkavédelem és a munkaegészségügy intézményrendszerének kialakulása és fejlődése Magyarországon a II. világháborúig, különös tekintettel a munkáltató felelősségére'. *Publ. Univ. Miskolc. Sect. jur. et pol. Tomus* 28 (2010): 415–436.

Országos Munkavédelmi és Munkaügyi Főfelügyelőség. 'Összefoglaló 'a munkahelyi stressz és a pszichoszociális kockázati tényezőkkel kapcsolatos munkáltatói és munkavállalói ismeretek, attitűdök országos vizsgálatához' végzett tájékozódásról, 2009. http://www.ommf.gov.hu/index.php?akt_menu=311, 15 June 2012.

Országos Munkavédelmi és Munkaügyi Főfelügyelőség. 'Összefoglaló 'a mikro-, kis- és közép vállalkozások munkavédelmi helyzetképének (profiljának) felmérése' tárgyban végzett tájékozódásról', 2009. http://www.ommf.gov.hu/index.php?akt_menu=311, 15 June 2012.

Országos Nyugdíjbiztosítási Főigazgatóság. 'Közgazdasági Főosztály'. In *Nyugdíjas ellátottak az EUROSTAT adatbázisa alapján,* edited by M. Hablicsekné Richter. Budapest, 2009.

Országos Nyugdíjbiztosítási Főigazgatóság. 'Közgazdasági Elemzések Főosztálya'. In *A 2009-es új rokkantsági nyugdíjasok és rehabilitációs járadékosok összetételének vizsgálata*, edited by J. Hollósné Marosi & Gy. Császár. Budapest, 2011, 5. table (5. táblázat), www.onyf.hu, 15 June 2012.

Varga, L. (ed.). *A munkavédelmi törvény magyarázata* (*Interpretation of the Labour Safety Act*). Budapest: Complex Kiadó, 2005.

CHAPTER 7

Italy: From Occupational Health and Safety to Well-being at Work

Edoardo Ales, Luca Miranda and Alessia Giurini

§7.01 INTRODUCTION

In the Italian legal order, the regulation of occupational health and safety is characterized by an imposing body of legislation. Initially the means of realizing certain constitutional precepts,[1] this then formed the fulcrum for implementing the Community preventive standards outlined by Directive 89/391/EEC (hereinafter 'Framework Directive') and its so-called daughter Directives.[2]

Recently, the adoption of Legislative Decree no. 81/2008 has drastically reorganized the national law by pursuing the *reductio ad unum* of the general and sectoral legislation on the subject.[3]

1. See G. Natullo, *La tutela dell'ambiente di lavoro* (Turin: Utet, 1995); L. Galantino (ed.), *La sicurezza del lavoro* (Milan: Giuffrè, 1996); L. Montuschi (ed.), *Ambiente, salute e sicurezza,* (Turin: Giappichelli, 1998); M. Rusciano & G. Natullo (eds.), *Ambiente e sicurezza del lavoro,* (Turin: Utet, 2007); P. Albi, *Adempimento dell'obbligo di sicurezza e tutela della persona. Art. 2087 c.c.*, Vol. of *Il Codice Civile. Commentario. Fondato da P. Schlesinger,* ed. F.D. Busnelli (Milan: Giuffré, 2008).
2. M. Biagi (ed.), *Tutela dell'ambiente di lavoro e direttive CEE* (Rimini: Maggioli, 1991).
3. P. Pascucci, 'Dopo la legge n. 123 del 2007. Titolo I del d.lgs. 9 aprile 2008, n. 81 in materia di tutela della salute e della sicurezza nei luoghi di lavoro', http://csdle.lex.unict.it/Archive/WP/WP%20CSDLE%20M%20DAntona/WP%20CSDLE%20M%20DAntona-IT/20111025-121117_pascucci_n73-2008itpdf.pdf>, vol. 73; M. Tiraboschi (ed.), *Il testo unico della salute e sicurezza nei luoghi di lavoro. Commentario al decreto legislativo 9 aprile 2008, n. 81* (Milan: Giuffrè, 2008); L. Zoppoli, P. Pascucci & G. Natullo (eds.), *Le nuove regole per la salute e la sicurezza dei lavoratori, Commentario al D.lgs. 9 Aprile 2008, no. 81* (Milan: Ipsoa-Wolter Kluwer 2008); L. Montuschi (ed.), *La nuova sicurezza sul lavoro. D.lgs. 9 aprile 2008, n. 81 e successive modifiche* (Bologna: Zanichelli, 2011); M. Persiani (ed.), *Il nuovo diritto della sicurezza sul lavoro* (Turin: Utet, 2012).

In the light of such facts, it appears appropriate briefly to reconstruct the subject-matter's regulation, before considering the cardinal principles and important novelties that Legislative Decree no. 81/2008 contains.

A first area of analytical reconstruction cannot fail to take account of the fundamental principles contained in Articles 32(1), 35(1) and 41(2) of the Constitution, which are closely interconnected.

First of all, health is the object of specific provision in Article 32(1) of the Constitution.[4] The emphasis here is on a consideration of the protection of citizens' health (and, consequently, of working conditions themselves) as a fundamental social right. It therefore follows that the right to a safe workplace and to safe working conditions may be included, at a national level, within the wide-ranging catalogue of fundamental personal and workers' rights.

Article 32(1) of the Constitution has another, equally important feature. Its effect is simultaneously 'individual' (creating as it does a fundamental personal right) and 'collective' (referring as it does to the collective interest in health protection). It specifically places health at the centre even of relationships between private parties[5] and permits the natural tensions between the merely individual/contractual dimension of the work relationship and the affirmation of collective rights to be overcome.[6] This is by virtue of the intrinsically collective value of the right to health protection.

Article 35(1) of the Constitution is closely connected to this fundamental provision. Although included amongst the provisions regarding economic relations, it asserts the need to protect work in all its forms and practices.

As a consequence, the combined effect of Articles 35(1) and 32(1) requires the protection of health not only in situations of employed work but also in those of self-employed work or, in any event, work carried out in the context of another party's production organization.[7] Such fact is closely linked to one of the most important innovations contained in the recent preventive legislation, namely, the distinctive concepts of worker and employer for the purposes of safety. These concepts (to which we will return) impose a duty to protect safety at work also on figures that in the past were uncovered and under-protected.

Another aspect of primary significance may be found in the close interrelation between Articles 32(1) and 41(2)[8] of the Constitution, if one only considers how the

---

4. 'The Republic shall protect health as a fundamental right of the individual and as a collective interest, and guarantees free medical care to the poor.'
5. See, by way of example, *Corte Costituzionale,* Judgement no. 88/1979, *Giustizia Civile,* no. 3 (1979): 121, according to which health, as a good, 'is protected by Article 32 of the Constitution, not only as a collective interest but also, and above all, as a fundamental individual right, so that it takes the form of a primary and absolute right that is also fully operative in relations between private parties. It is certainly to be included amongst the subjective positions that are directly protected by the Constitution'.
6. See G. Balandi, 'Individuale e collettivo nella tutela della salute nei luoghi di lavoro: l'art. 9 dello Statuto', *Lavoro e Diritto,* no. 1 (1990): 222 et seq.; C. Zoli, 'Sicurezza del lavoro: contrattazione e partecipazione', *Rivista Giuridica del Lavoro,* no. 3 (2000): 614 et seq.
7. See *Corte di Cassazione* (*Sezione Lavoro*), Judgement no. 45/2009.
8. The first two sections of Article 41 specify, 'Private-sector economic initiative shall be free. It cannot be conducted in conflict with social usefulness or in such a manner that it could damage safety, liberty or human dignity.'

identification of workers' safety as a good of constitutional importance requires entrepreneurs to put the safety of the person supplying such service before their own profit or economic saving.

It follows that health, being both a fundamental worker's right and a collective interest, cannot be considered merely desirable or just one aspect of production management. Its protection constitutes, rather, a specific *sine qua non* for the pursuit of an activity.[9]

At the level of ordinary legislation, Article 2087 of the Civil Code constitutes the cornerstone of the national preventive structure. Although it precedes the constitutional provisions in temporal terms, Article 2087 has, through its constitutionally-oriented interpretation by legal academics and in the case law, become the alpha and omega of the whole system protecting health and safety at work.

By providing that 'in the exercise of their business activity, entrepreneurs shall adopt the measures that, according to the particular nature of the work, experience and technology, are necessary to protect the physical, mental and moral integrity and general dignity of their employees', Article 2087 identifies the employer's contractual duty to ensure safety as a general principle (or, more correctly, duty) from which all the specific duties of prevention derive.[10] We will subsequently consider this point in detail.

Alongside such fundamental provisions, the national legislation preceding transposition of the Community provisions was highly fragmented. Such fact is evidenced by Presidential Decrees nos. 547/1955 and 303/1956, which identified specific, detailed employers' duties closely correlated with the 'business' sector of production and the nature of activity pursued.

Together with a close application and evolutive interpretation of Article 2087 of the Civil Code, the legislation of the 1950s permitted a particularly stringent system for protecting workers' health to develop and, thanks to the *Corte di Cassazione*, in particular, one capable of constituting a solid point of reference for preventive safety at work.

Nevertheless, it proved impossible to adapt the said preventive system of the 1950s to the different approach towards protection of occupational health and safety contained in the Framework Directive. The Italian legislator was therefore obliged to meet the Directive's requirements by issuing further, new legislation on the subject. This took the form of Legislative Decree no. 626/1994, which implemented the Framework Directive and a part of its so-called daughter directives.

As a result, the national preventive system was characterized for more than a decade by a highly fragmented regulation that risked seriously compromising both certainty as to the law and full regulatory adaptation to important new categories of risk.

Such problems, combined with persistently worrying statistics both for accidents and deaths at work and for occupational diseases, pushed the government to exercise

9. L. Montuschi, *Diritto alla salute e organizzazione del lavoro* (Milan: Giuffrè, 1980).
10. On this point, see *Corte Costituzionale*, Judgement no. 384/2005 and Judgement no. 50/2005, published in the Official Journal on 19 October 2005 and 2 February 2005, respectively.

the power delegated under Act no. 123/2007 and pass Legislative Decree no. 81/2008. This was a particularly innovative regulatory text and one capable of meeting the need to reorder and rationalize the preventive legislation at both a general and a sectoral level.

§7.02 BASIC CONCEPTS

[A] Risk Prevention and Protection from Risk

Italian health and safety law is characterized by an imposing number of definitions that expound the preventive legislation's fundamental concepts. The definitions of 'risk prevention', 'protection from risk' and 'health' are particularly important for our purposes.

In truth, the national legislation takes the definition of 'prevention' contained in the Framework Directive, understood as 'all the steps or measures taken or planned at all stages of work in the undertaking to prevent or reduce occupational risks'. It extends its meaning, however, adapting it to the interpretation of Article 2087 of the Civil Code evolving in the case law over the years.

Thus, the legislator defines 'prevention' as 'all the steps or measures that are necessary, including according to the particular nature of the work, experience and technology, to prevent or reduce occupational risks whilst fully respecting the health of the population and the integrity of the environment' (Article 2(1)(n) of Legislative Decree no. 81/2008).

Precisely the reference to the particular nature of the work, experience and technology allows the problem of the lack of an express definition of 'protection' to be overcome. In this way, an approach to the protection of occupational health and safety that is capable of adapting to different new possible sources of risk is imposed.

This evolutive approach to the concept of 'prevention' is closely linked to the extension of the legislation's scope of application to all undertakings and all types of risk, as well as to employers' general duties. It permits the presumption of an employer's safety duty that borders on the Community concept of precaution (see Article 191(2) TFEU[11]). Such a duty will exist in the presence of three key elements: identification of all the potential risks; a full and rigorous scientific assessment carried out on the basis of all existing data and, lastly, the lack of a scientific certainty that permits the presence of identified risks to be reasonably excluded. All these elements are present in the national system managing health and safety at work.

11. 'Union policy on the environment shall aim at a high level of protection taking into account the diversity of situations in the various regions of the Union. It shall be based on the precautionary principle and on the principles that preventive action should be taken, that environmental damage should as a priority be rectified at source and that the polluter should pay.'

[B] The Concepts of 'Health', 'Danger' and 'Risk'

The national legislator then enunciates other important concepts. The first of these is the concept of 'health', taken from the definition given by the World Health Organization (WHO) in the preamble to its Constitution. Pursuant to Article 2(1)(o) of Legislative Decree no. 81/2008, 'health' is 'a state of complete physical, mental and social well-being and not merely the absence of disease and infirmity'.

The equating of health with the mere absence of disease and infirmity is therefore rejected and the concept of 'well-being', in the triple sense of physical, mental and social, is adopted as the point of reference. Thus, the tie with the notion of 'well-being at work' outlined by the European Commission in its last two Strategies on Health and Safety at Work (2002–2006, 2007–2012)[12] is recovered.

In this way, what some legal academics define as 'an "integral" notion of health that includes both physical and mental well-being and therefore combats the stress, monotony and repetitiveness of work'[13] is achieved.

Thanks to these spurs deriving mainly from international and Community contexts, the need for a global approach to well-being at work is beginning to emerge. It is no longer sufficient to intervene with the sole aim of reducing the number of accidents or the incidence of occupational disease. Prevention of the so-called social risks, namely, stress, depression, anxiety, harassment in the workplace, mobbing and discrimination, is also fundamental.

To the abovementioned definitions should be added the further notions of 'risk' and 'danger'. From these it is possible to derive a dynamic, wide-reaching and evolutive protection of the working environment that is capable, in theory, of constantly adapting the protection of workers to the particular nature of the workplace.

In this perspective, 'danger' is the capacity that any given entity has to cause a worker harm, whereas 'risk' pinpoints the probability that an event that is harmful to a worker may occur. To such end, Article 2(s) of Legislative Decree no. 81/2008 defines 'risk' as the probability of reaching the 'potential harm' level in working conditions or of being exposed to a given factor or agent or to a combination of the two. Risk arises when there is, contemporaneously, a danger and a worker exposed to it.[14]

Analogously, Article 2(r) of Legislative Decree no. 81/2008 defines 'danger' as the property or quality of a given factor that is intrinsically capable of causing harm. The term therefore refers to a given entity's harmful potential, understood in the sense of a work process, a working tool, work equipment or a chemical, physical or biological agent.

It is, therefore, essential to distinguish between the concepts of 'danger' and 'risk'. These are fundamentally different, insofar as danger encompasses the certainty that the adverse event will occur, whereas risk only implies the probability, with the

12. Cf. COM (2002) 118 (not published in the Official Journal) and COM (2007) 62 – final.
13. B. Caruso, 'L'Europa, il diritto alla salute e l'ambiente di lavoro', in *Ambiente, salute e sicurezza*, ed. L. Montuschi (Turin: Giappichelli, 1998), 4 et seq.
14. M. Masi, 'Il rischio: le ragioni di una disciplina in materia di sicurezza', in *Il testo unico della salute e sicurezza nei luoghi di lavoro*, ed. M. Tiraboschi (Milan:Giuffrè, 2008), 377 et seq.

consequence that it will not be possible to eliminate risk as long as a source of danger exists.

[C] Risk Deriving from Interference and Its Relevance in the Building Sector

One of the most important innovations in the national legislator's recent modifications to the system governing occupational health and safety concerns risk deriving from interference i.e., those risks present when the activities of other undertakings or self-employed workers become part of an employer's organized production structure 'through outsourcing "within" the production process'.[15]

Legislative Decree no. 81/2008 therefore also includes 'risk deriving from interference' amongst the basic concepts in occupational health and safety. This particular concept gives rise to a series of stringent duties for the so-called commissioning entity, the most important of which is the duty to assess the situation and draw up a consolidated risk-assessment document.

The legislator's intervention is thus a response to the need to tackle a phenomenon constituting a considerable threat to the safety of workers operating in complex business structures. This is the contemporaneous presence of the specific risks attaching to individual undertakings' activities and risks deriving both from the concomitant presence of employees from different undertakings and from reciprocal interference with the various activities carried out in the same theatre of work.[16]

The above assumes a particular importance with regard to the duties imposed on undertakings operating in the building sector. Risk deriving from interference constitutes one of the greatest sources of danger for these undertakings, by virtue of the potentially multiple forms of work that may be performed on worksites where there is contracting or sub-contracting and that involve a main undertaking, sub-contractors, self-employed workers and suppliers of materials.

The risk deriving from interference with the commissioning entity's workplace (and, possibly, workers) caused by the activity of workers from 'outside' undertakings, therefore assumes a central importance in the building sector and one that the legislator has deemed necessary to refer to specifically. Legislative Decree 81/2008 explicitly creates the duty to draw up a specific document on the risk from interference and requires such document to be given to all contractors entrusted with works under procurement contracts and works contracts.

15. As the *Corte di Cassazione* (*Sezione lavoro*) expressed it in its Judgement no. 45/2009.
16. P. Pascucci, 'Dopo la legge n. 123 del 2007. Titolo I del d.lgs. 9 aprile 2008, n. 81 in materia di tutela della salute e della sicurezza nei luoghi di lavoro', http://csdle.lex.unict.it/Archive/WP/WP%20CSDLE%20M%20DAntona/WP%20CSDLE%20M%20DAntona-IT/20111025-121117_pascucci_n73-2008itpdf.pdf >, vol. 73, 112; L. Angelini, 'La tutela della salute e sicurezza negli appalti prima e dopo il d. lgs. n. 81/2008', < www.uniurb.it/olympus >, 12 June 2008.

§7.03 PERSONAL SCOPE OF APPLICATION

[A] The National Legislation's Application to All Sectors of Production and All Types of Risk

An important feature of the current national legislation is its application to all sectors of activity, whether private or public, and to all types of risk.[17]

Legislative Decree no. 81/2008 thus does not draw distinctions based on the nature (i.e., entrepreneurial or non-entrepreneurial) or the type of activity pursued or, indeed, the dimensions of a business.

It is precisely this extension of the law's scope of application to all private and public sectors and all types of risk that constitutes one of the most important innovations of the new national preventive legislation. The other is its emphasis on the employer's primary duty to assess risks in the workplace.

The novelties introduced by Legislative Decree no. 81/2008 have a twofold effect. On the one hand, they remedy the highly fragmented nature of the previous legislation with its different rules for different sectors of production. On the other, they tend to guarantee that the law assumes a global approach to the problem of protecting safety at work. This is done by establishing an action programme that is truly centred on a culture of accident prevention and on a constant, planned protection of occupational health and safety, rather than intervening solely to establish rights and duties for employers and workers.

One of the most important innovations introduced by Legislative Decree no. 81/2008 is the lack of importance attached to the size of a business. This aspect is closely linked to the legislation's applicability to all types of activity and all risks, as well as the innovative all-inclusive concept of 'worker'.[18]

Indeed, even in the only case where the size of a business becomes important, the innovative concept of 'worker' has a decisive influence on the application of the related law. This occurs by virtue of sections 3, 4 and 7 of Article 47 of Legislative Decree no. 81/2008. In specifying the size limit of fifteen workers (which constitutes the dividing line between the election or designation of representatives from among a business's trade union representatives on the one hand and their direct election from among the workers, on the other), these no longer use the concept of 'employee' but, rather, that of 'worker'.

Thus an all-inclusive concept of 'worker' is also embraced for the purposes of applying the provisions on the participation and consultation of workers' representatives. This has the effect that, at least at the formal level and in the absence of any other

17. See P. Pascucci, 'Dopo la legge n. 123 del 2007. Titolo I del d.lgs. 9 aprile 2008, n. 81 in materia di tutela della salute e della sicurezza nei luoghi di lavoro', http://csdle.lex.unict.it/Archive/WP/WP%20CSDLE%20M%20DAntona/WP%20CSDLE%20M%20DAntona-IT/20111025-121117
_pascucci_n73-2008itpdf.pdf>, vol. 73, 30 et seq.

18. P. Pascucci, 'Il campo di applicazione soggettivo delle nuove regole sulla salute e sicurezza nei luoghi di lavoro', *Diritti lavori mercati*, no. 2 (2008): 297 et seq.

legislative specification, representation includes parties not typically covered at the collective/trade union level.

[B] Specific Legislation Regarding Pregnant Women

The general applicability of Legislative Decree no. 81/2008 does not exclude further legislation governing specific categories of worker. One such category is that of pregnant workers and workers who have recently given birth or are breast feeding, in relation to whom specific regulations apply. These are currently contained in Chapter II of Legislative Decree no. 151/2001, which implemented Community Directive 92/85/EC.[19]

Law of this kind is based on the need to protect the health of female workers during the moment of greatest vulnerability i.e., during pregnancy and maternity. Indeed, such conditions necessitate specific and more detailed legislation than that envisaged either for female workers in general or for male workers.[20]

More specifically and in compliance with Directive 92/85, Legislative Decree no. 151/2001 requires employers to carry out a specific risk-assessment for pregnant workers and for workers who have recently given birth or are breast feeding. It further expressly identifies the measures that employers must adopt in order to avoid the related risks.

In the first place, employers must temporarily alter such workers' working conditions or working time (Article 12(1)). Should that prove impossible for organizational reasons or reasons linked to production, they must assign them different work duties.

An analysis of the Italian legislation clearly reveals the complementary relationship between Legislative Decree no. 151/2001 and Legislative Decree no. 81/2008.[21] The close continuity between them is evidenced by a series of references that Legislative Decree no. 151/2001 makes to Legislative Decree no. 626/94 (subsequently absorbed by Legislative Decree no. 81/2008), primarily regarding risk-assessment and workers' right to information. Without forgetting that Article 15 of Legislative Decree no. 81/2008 expressly provides that, 'Save as otherwise provided for by the present Chapter, the provisions contained in Legislative Decree no. 626 dated 19 September 1994, as subsequently amended, as well as every other provision regarding safety and health in the workplace, shall remain in force.'

This provision not only expresses the complementary relationship between Legislative Decree no. 151/2001 and Legislative Decree no. 81/2008 but also indicates the special nature of the law contained in the former vis à vis the general health and safety legislation.

19. See F. Adinolfi & R. Bortone, 'Commento alla direttiva europea sulla tutela della maternità', *Giornale di diritto del lavoro e di relazioni industriali*, no. 1 (1994): 373 et seq.; M. Lai, 'Tutela della maternità nella legislazione europea', *Diritto e pratica del lavoro*, no. 2 (1993): 77 et seq.
20. F. Borgogelli, *Il lavoro femminile tra legge e contrattazione* (Milan: Franco Angeli, 1987), 189 et seq.
21. L. Calafà, 'Le lavoratrici madri', in *Ambiente e sicurezza del lavoro*, eds. M. Rusciano & G. Natullo (Turin: Utet, 2007), 394.

[C] The Distinctive Concepts of 'Worker', 'Employer' and 'Commissioning Entity'

Another important feature of the national legislation is the use it makes of wholly distinctive concepts of 'worker', 'employer' and 'commissioning entity'. Such use ensures that the prevention duties and the protection afforded parties extend beyond the confines of the contractual relationship.

To such end, the concept of 'worker' contained in Article 2(1)(a) has a preliminary importance. For the intents and purposes of Legislative Decree no. 81/2008, 'worker' is to be understood as the 'person who, independently of the type of contract, carries out a remunerated or unremunerated working activity within a public or private employer's organization, even for the sole purpose of learning a trade, art or profession, excluding domestic workers and family members'.

Thus an all-inclusive concept of 'worker' is embraced.[22] This not only covers employed workers (regardless of the various types of contract) but also workers that, albeit not employed by the employer, are nevertheless subject to his or her managerial authority (agency workers). It also covers those workers who have drawn up a contract for services with an employer, provided (obviously) that the service that is the object of such contract potentially (insofar as its provision is within the employer's organization) exposes them to health and safety risks deriving from the activity pursued by the employer.[23]

As regards self-employed people, however, the said extended application is important in relation to certain specific duties. These include the duty to procure individual preventive and protective equipment personally and use it within the business organization, over and above the general duty to equip themselves with personal identity cards.

Analogously and following the same line of interpretation, the legislator has also embraced a broad concept of 'employer'. Focussing on the function of managing safety at work, this figure is defined as the person who has full and effective power to organize the business.

To such end, Article 2(b) of Legislative Decree no. 81/2008 identifies a private sector 'employer' as the party in charge of the working relationship with the worker or, in any event, the party that, according to the type and set up of organization in which the worker provides his or her services, has responsibility for the organization itself or for the relevant production unit, insofar as s/he exercises the decision-making and spending power. In this respect, a recent judgment of the Criminal Division of the *Corte di Cassazione*[24] has upheld the macro-concept of 'employer' and included within the category whomsoever has, in all reality, the de facto role of employer. For safety

22. D. Venturi, 'Lavoratore: definizione e obblighi', in *Il testo unico della salute e sicurezza nei luoghi di lavoro*, ed. M. Tiraboschi (Milan: Giuffrè, 2008), 247 et seq.
23. P. Pascucci, 'Dopo la legge n. 123 del 2007. Titolo I del d.lgs. 9 aprile 2008, n. 81 in materia di tutela della salute e della sicurezza nei luoghi di lavoro', http://csdle.lex.unict.it/Archive/WP/WP%20CSDLE%20M%20DAntona/WP%20CSDLE%20M%20DAntona-IT/20111025-121117_pascucci_n73-2008itpdf.pdf>, vol. 73, 33 et seq.
24. See *Corte di Cassazione (Penale, IV Sezione)*, Judgement no. 36878/2009.

purposes, therefore, the position of 'employer' vis à vis workers reaches well beyond the existence of a formal employment relationship, whatever the nature of the contractual relationship may be. It coincides with the full and effective exercise of an employer's powers.

Within public authorities, however, employer's duties regarding workplace safety fall on the public servant with managerial authority or on an officer without managerial rank, but solely in those cases where the latter is in charge of an autonomous office, has been appointed by the individual authority's governing body and enjoys autonomous decision-making and spending powers.

Lastly, the same need for certainty as to the figure responsible for work organization is met by the distinctive concept of commissioning entity. This extends preventive duties considerably in favour of self-employed workers, particularly people involved in co-ordinated project work and of agency workers.

To such end, Article 26 of Legislative Decree no. 81/2008 generally identifies the 'commissioning entity' as the person who entrusts works to a contracting undertaking or to self-employed workers or who uses agency workers within his or her own business.[25]

For the purposes of extending these important responsibilities, the determining factor is therefore the fact that a person carries out a working activity on the 'commissioning entity's' premises. In such cases, as the Criminal Division of the *Corte di Cassazione* has confirmed,[26] it is the commissioning entity who must organize appropriate measures for guaranteeing the safety of workers not directly employed by him or her but operating within his or her business. This is when such measures are generic and not tied to the worker's specific activity.

A similar extension of the safety duty to those responsible for the business structure in which work is performed relates specifically to the building sector. Here, pursuant to Article 89(1)(b), the commissioning entity is identified as the party on whose behalf the whole work is being realized. This independently of possible sub-divisions of the work involved. In the case of public procurement, however, the commissioning entity is identified as the party vested with decision-making and spending power for the contract.

In such a context, the commissioning entity must therefore be identified as a physical person who will have duties carrying a criminal liability. In the case of public or private juristic persons, this person corresponds to the person authorized to sign procurement contracts for the execution of works.

25. F. Bacchini, 'Committenti e appaltatori', in *Il testo unico della salute e sicurezza nei luoghi di lavoro*, ed. M. Tiraboschi (Milan: Giuffrè, 2008),183 et seq.
26. So held the *Corte di Cassazione* (*Penale, IV Sezione*) in its Judgement no. 12348/2008.

§7.04 DUTIES OF THE PARTIES INVOLVED

[A] The Employer's Main Duty: Assess Risks to Workers' Safety and Health

Turning to consider the employer's safety duties more closely, ever since Legislative Decree no. 626/1994 (which specifically transposed the provisions of Article 6(3)(a) of the Framework Directive), the legislator has identified[27] the employer's principal, non-delegable duty to evaluate risks in the workplace as the cornerstone of a business's preventive system.[28]

Indeed, it is primarily through an emphasis on both the preventive assessment of all risks and the adoption of all the necessary collective protective measures (and, subordinately, the individual ones), that the concept of safety at work assumes a revolutionary dimension in comparison to the law previously in force at a Community and national level.

This insofar as the Community legislator clearly intended to abandon the idea that had characterized both the preceding legislative framework and subsequent Community legislation.[29] That is to say, the idea of protecting workplace safety by setting technical work standards.

A purely industrial and highly technical concept of safety at work has therefore been replaced by an a-technical acceptation of the safety of working conditions. In this way, it becomes possible to pass from the industrial perspective of workplace hygiene to one that seeks to guarantee occupational health and safety as a fundamental good.[30]

To such end, Article 4(1) of Legislative Decree no. 626/94 provided for a generalized duty on employers to assess all the risks potentially present in workplaces and not only those remaining after preventive and protective measures had been adopted. Thus not only the risks 'caused by work' but also those arising 'during work' counted as, indeed, they do now.

In the current legislative context, one of the strong points of the reform realized by Legislative Decree no. 81/2008 is the new formulation of the risk-assessment provisions.[31] In establishing the law on risk-assessment in Article 28, the legislator not

27. Not without difficulty, given that, in this respect the European Court of Justice, in its judgement in Case C-49/00, held the Italian State to be in breach of Art. 6 (3)(a) insofar as the initial formulation of Legislative Decree no. 626/1994 did not extend the risk-assessment duty to all the potential sources of harm to the health of workers present in the workplace.
28. See G. Natullo, *La tutela dell'ambiente di lavoro* (Turin: Utet, 1995); L. Galantino (ed.), *La sicurezza del lavoro* (Milan: Giuffrè, 1996); L. Montuschi (ed.), *Ambiente, salute e sicurezza,* (Turin: Giappichelli, 1998); M. Rusciano & G. Natullo (eds.), *Ambiente e sicurezza del lavoro,* (Turin: Utet, 2007); C. Smuraglia, 'Sicurezza del lavoro e obblighi comunitari. I ritardi dell'Italia nell'adempimento e le vie per uscirne', *Rivista italiana di diritto del lavoro,* no. 1 (2002): 221 et seq.
29. Which may be epitomized, by way of an example, in the provisions of Directive EEC/80/1107.
30. See J. Aparicio Tovar, 'Sicurezza sul lavoro', in *Dizionario di diritto del lavoro comunitario,* ed. A. Baylos Grau, B. Caruso, M. D'Antona & S. Sciarra (Bologna: Monduzzi, 1996), 568.
31. G. M. Monda, 'La valutazione dei rischi per la sicurezza e la salute dei lavoratori', in *Le nuove regole per la salute e la sicurezza dei lavoratori,* ed. L. Zoppoli, P. Pascucci & G. Natullo (Milan: Kluwer - Ipsoa, 2008), 336 et seq.

only expressly refers to technical risks deriving from production processes but also widens the employer's duty by referring to all the risks to workers' health and safety. These include, by way of example, those linked to work-related stress (in the light of the 'Framework Agreement on work-related stress', signed by the European social partners in October 2004), those regarding pregnant workers (in accordance with the provisions of Legislative Decree no. 151/2001) and those linked to gender differences, age and provenance from other countries.[32]

The provision has created not a few problems of interpretation. For the most part, these concern the assessment of risks associated with work-related stress and are caused by the lack of specific guidelines on how to carry out this type of assessment. Legislative Decree no. 106/2009 (correcting Legislative Decree no. 81/2008) sought to respond to these problems by attributing to a Permanent Advisory Committee the task of providing guidelines on methods of assessment. Nevertheless, serious difficulties in application persist, even following adoption of the guidelines dated 17 November 2010.

Article 28 of Legislative Decree no. 81/2008 is further distinguished by the greater detail in which the risk-assessment document's obligatory provisions must be expressed, as well as the fundamental requirements that the document be in written form and of certain date. In this way, it renders the employer's related duty far more meaningful. Indeed, the specificity of the law responds to a clearly perceived need to clarify the safety measures that the employer must adopt in order to eliminate the widespread practice (particularly amongst small and medium-sized enterprises) of carrying out a purely formal risk-assessment and considering the drafting of the risk-assessment document merely a bureaucratic duty.

Last but not least, it is appropriate to note how Article 28 also substantially affects the duty to revise and update the risk-assessment document.[33] To such end, the provision does not limit itself (as the previous law did) to providing that the assessment must be revised after changes to a business's production processes or work organization that are relevant to workers' health and safety. Clearly echoing the provisions of Article 2087 of the Civil Code, it also requires the revision to be carried out in relation to the extent to which technology, prevention and protection have evolved, as well as after significant accidents or when the results of health surveillance show the need. Furthermore, it is also necessary to update the preventive measures. Thus Article 28 makes the wide-ranging and evolutive nature of the employer's duty explicit, in line with Directive.

32. P. Pascucci, 'Dopo la legge n. 123 del 2007. Titolo I del d.lgs. 9 aprile 2008, n. 81 in materia di tutela della salute e della sicurezza nei luoghi di lavoro', http://csdle.lex.unict.it/Archive/WP/WP%20CSDLE%20M%20DAntona/WP%20CSDLE%20M%20DAntona-IT/20111025-121117_pascucci_n73-2008itpdf.pdf>, vol. 73, 122 et seq.

33. P. Pascucci, 'Dopo la legge n. 123 del 2007. Titolo I del d.lgs. 9 aprile 2008, n. 81 in materia di tutela della salute e della sicurezza nei luoghi di lavoro', http://csdle.lex.unict.it/Archive/WP/WP%20CSDLE%20M%20DAntona/WP%20CSDLE%20M%20DAntona-IT/20111025-121117_pascucci_n73-2008itpdf.pdf>, vol. 73, 122.

[B] The Entrepreneur and Management: The Delegation of Duties

The employer is the person with the general duty to ensure the safety of all his or her workers. As already pointed out, in the Italian legal order, this principle has its origins in Article 2087 of the Civil Code, by virtue of which an entrepreneur has a general duty to protect his or her workers' mental and physical integrity. Such rule can be read in two ways.[34] On the one hand, Article 2087 is the 'alpha' of all the legislation, since it has 'the function of permanently adapting the legal order to the underlying socio-economic reality'.[35] On the other, it may also be considered the legislative system's 'omega' provision, since it is the vehicle for filling possible lacunae and adapting the legislation to concrete, changing situations.[36]

This general principle creates a double duty for the employer: on the one hand, to adopt general measures protecting the mental and physical integrity of workers and, on the other, to manage risk prevention in the workplace.

Alongside such fundamental general provision, both the general and the specific employer's duties created by Legislative Decree no. 81/2008 are to be noted for their significantly analytical nature. In particular, Article 17 picks out the two fundamental duties that an employer cannot delegate to anyone else, namely, risk-assessment and appointment of the person responsible for the preventive and protective service. Article 18 then specifies in detail those general duties that an employer may share with his or her *alter ego* in the management of workplace safety, the manager.

In comparison with Legislative Decree no. 626/1994 (which previously regulated such profile), Article 18 of Legislative Decree 81/2008 eliminates references to the supervisor, whilst the law governing this figure's specific duties and tasks is set out separately in Article 19. This is justified by the wide range of provisions contained in Article 18, since these do not fit the figure of the supervisor as s/he is identified by Article 2(e) of Legislative Decree no. 81/2008. In this way, what has been defined as a contribution to simplification through the identification 'of the respective tasks and related responsibilities'[37] is guaranteed.

Pursuing this objective, the two provisions define the powers and responsibilities of the manager and the supervisor. The manager is the person who, within the limits of the hierarchical and functional powers vested in the mandate received, implements the employer's health and safety directions and monitors working activities. The supervisor, however, supervises working activities and guarantees implementation of

34. L. Montuschi, 'La sicurezza nei luoghi di lavoro ovvero l'arte del possibile', *Lavoro e diritto*, no. 2 (1995): 413 et seq.
35. As the *Corte di Cassazione* put it in its Judgement no. 5048/1988.
36. P. Pascucci, 'Dopo la legge n. 123 del 2007. Titolo I del d.lgs. 9 aprile 2008, n. 81 in materia di tutela della salute e della sicurezza nei luoghi di lavoro', http://csdle.lex.unict.it/Archive/WP/WP%20CSDLE%20M%20DAntona/WP%20CSDLE%20M%20DAntona-IT/20111025-121117_pascucci_n73-2008itpdf.pdf>, vol. 73, 103 et seq.
37. P. Pascucci, 'Dopo la legge n. 123 del 2007. Titolo I del d.lgs. 9 aprile 2008, n. 81 in materia di tutela della salute e della sicurezza nei luoghi di lavoro', http://csdle.lex.unict.it/Archive/WP/WP%20CSDLE%20M%20DAntona/WP%20CSDLE%20M%20DAntona-IT/20111025-121117_pascucci_n73-2008itpdf.pdf>, vol. 73,108.

the directions received, checking that they are carried out correctly.[38] It should further be noted that, prior to Legislative Decree no. 81/2008, no express statutory definition of supervisor existed. Such fact led to uncertainty about how the figure was to be identified and, consequently, about potential liability should the related legislation be breached.

The objective of actually, concretely dividing the tasks and responsibilities between the employer and his or her manager and supervisor, however, is achieved principally through the formal act of appointing such figures i.e., the so-called delegation of functions.

By its very nature, delegation can create problems. It can be difficult to identify the delegated person correctly and, more importantly, to divide responsibility between the same and the delegating party. Precisely for this reason, Articles 16 and 17 of Legislative Decree no. 81/2008 define the limits of delegation clearly.[39]

In the first place, as we have seen, Article 17 identifies the two activities that the entrepreneur is under a mandatory duty to carry out personally: risk-assessment and appointment of the person responsible for the preventive and protective service.

In the second place, Article 16 establishes the specific formal and substantive pre-requisites for valid delegation. In particular, the law provides that delegation must be effected through a clearly dated, written instrument that has been expressly accepted by the delegated person in writing. Likewise, it must establish specific subjective requirements: the delegated person must possess professional qualifications that are appropriate to exercise of the delegated authority. Furthermore, for the purposes of avoiding the creation of a so-called empty delegation, it must ensure that the delegated person receives the powers needed to exercise the delegated authority, including appropriate spending powers.

In closing, the law provides that delegation does not, of itself, exclude the employer's duty to ensure that the transferred duties are carried out correctly.

[C] The Role of Internalized and Externalized Preventive and Protective Services

As we have seen, the Italian health and safety system involves a variety of figures in the protection of workers' mental and physical integrity. Of those involved, the Preventive and Protective Service enjoys a role of primary importance.[40]

38. M. R. Gentile, 'I dirigenti e i preposti', in *Il testo unico della salute e sicurezza nei luoghi di lavoro*, ed. M. Tiraboschi (Milan: Giuffrè, 2008), 208 et seq.
39. P. Pascucci, 'Dopo la legge n. 123 del 2007. Titolo I del d.lgs. 9 aprile 2008, n. 81 in materia di tutela della salute e della sicurezza nei luoghi di lavoro', http://csdle.lex.unict.it/Archive/WP/WP%20CSDLE%20M%20DAntona/WP%20CSDLE%20M%20DAntona-IT/20111025-121117_pascucci_n73-2008itpdf.pdf>, vol. 73,103 et seq.; A. Russo, 'La delega di funzioni e gli obblighi del datore di lavoro non delegabili', in *Il testo unico della salute e sicurezza nei luoghi di lavoro*, ed. M. Tiraboschi (Milan: Giuffrè, 2008), 217 et seq.
40. R. Bortone, 'Il servizio di prevenzione', in *Ambiente, salute e sicurezza*, ed. L. Montuschi (Turin: Giappichelli, 1998), 135 et seq.

The Service comprises a *Responsabile* (Person Responsible) and one or more *Addetti* (Officers). The professional pre-requisites for these figures are expressly indicated in the highly innovative Article 32 of Legislative Decree no. 81/2008. Both figures may be enlisted from either inside or outside the undertaking. More specifically, employers have a duty to enlist external persons when it is not possible to identify persons possessing the legally required qualifications within their own business. Furthermore, and irrespectively of such a situation, Article 31(6) of Legislative Decree no. 81/2008 identifies the circumstances in which the Preventive and Protective Service must be internal,[41] relating them to the hazardousness of the business's activity.[42]

Of the tasks assigned to the Service, that of identifying and assessing both the risk factors and the occupational health and safety measures assumes a particular importance. To such end, *Responsabile* and *Addetti* collaborate closely with the employer in the drafting of the risk-assessment document. Furthermore, it is the Preventive and Protective Service that formulates the safety procedures, proposes the workers' training and information programmes and provides workers with information about risks and the undertaking's safety system.

An undertaking's Preventive and Protective Service therefore ensures a series of transversal activities geared to co-ordinating the safety measures adopted by the employer with the real needs for prevention and protection in the workplace. It likewise participates in consultations on health and safety protection and prepares the documentation needed for continuous training on the undertaking's safety situations.

[D] The Training Duty

The principle of the division of competences in work-related health and safety management that informs the Framework Directive has, after some initial ambiguities in Legislative Decree no. 626/1994, been fully applied by the current law in Legislative Decree no. 81/2008.

This provides that workers' involvement is to be achieved not only by informing and consulting the workers' representatives and the workers themselves but also through the employer's fundamental duty to train his or her own employees.[43]

41. P. Pascucci, 'Dopo la legge n. 123 del 2007. Titolo I del d.lgs. 9 aprile 2008, n. 81 in materia di tutela della salute e della sicurezza nei luoghi di lavoro', http://csdle.lex.unict.it/Archive/WP/WP%20CSDLE%20M%20DAntona/WP%20CSDLE%20M%20DAntona-IT/20111025-121117_pascucci_n73-2008itpdf.pdf>, vol. 73, 128 et seq.
42. For example: (a) in industrial businesses covered by Article 2 of Legislative Decree no. 334/1999 that have a duty to issue notifications or reports pursuant to Articles 6 and 8 of the same decree; (b) in thermoelectric power plants; (c) in plants and installations covered by Articles 7, 28 and 33 of Legislative Decree no. 230/1995; (d) in businesses involved in the manufacture and separate storage of explosives, gunpowder and munitions; (e) in industrial businesses with over 200 workers; (f) in mining industries with over fifty workers and (g) in public and private health-care structures with more than fifty workers.
43. O. Di Monte, 'L'informazione e la formazione dei lavoratori', in *La sicurezza sul lavoro*, ed. M. Ricci (Bari: Cacucci, 1999), 169 et seq.

Indeed, the prevention of accidents at work requires knowledge and awareness of the risks. To such end, the duty to train workers is amply covered by Articles 36 and 37 of Legislative Decree no. 81/2008, which include a reference to practical training geared to teaching the worker 'the correct use of apparatus, machines, plant, substances and equipment, including personal protective equipment, as well as the working procedures' (Article 37(4) and (5)).

The new national regulatory framework is also to be noted for how it defines the concept of 'training'. Indeed, under Article 2(1)(aa) of Legislative Decree no. 81/2008, the latter is understood as the:

> instructive process through which the knowledge and procedures that aid both the acquisition of competences for safely carrying out their respective tasks within the undertaking and the identification, reduction and management of risks are to be transferred to workers and the other parties in the undertaking's prevention and protection system.

In the said context, training therefore constitutes one of the most important general measures for protecting workers.[44] More specifically, the employer must guarantee not only that the training occurs but also that it is realized using the means that are most appropriate for getting through to the trainees. Indeed, entrepreneurs are under a general duty to ensure results i.e., to check that the trainees have assimilated the concepts.[45] To such end, Article 37 of Legislative Decree no. 81/2008 requires the employer to ensure that each worker receives a sufficient and appropriate training in health and safety, including with regard to his or her linguistic abilities.

In this light, the provision in Article 37(13) of Legislative Decree no. 81/2008 assumes a particular importance. Referring to immigrant workers, it states 'where training concerns immigrant workers, it shall take place after prior verification of the understanding and knowledge of the vehicular language used in the training programme'.[46]

In addition to requiring the employer to make sure that each worker receives sufficient and appropriate health and safety training, Article 37 of Legislative Decree no. 81/2008 makes the content of the training explicit. In particular, training must refer to: (a) the concepts of risk, harm, prevention, protection and organization of the undertaking's prevention, as well as rights and duties attaching to the various figures within the undertaking and the bodies responsible for supervision, inspection and assistance; (b) the risks attaching to individual tasks and the possible forms of harm,

---

44. The case law is in agreement, holding that the employer may discharge his or her duties to provide information and training through competent third parties and/or through figures that will typically contribute to protecting safety at work. In this sense, see *Corte di Cassazione (Penale, III sezione)*, Judgement no. 41609/2006, in *Igiene e sicurezza del lavoro*, no. 2 (2007): 104 et seq.
45. M. Lai, *La sicurezza del lavoro tra legge e contrattazione collettiva* (Turin: Giappichelli, 2002), 201 et seq.; L. Miranda, *La dimensión colectiva de la tutela del entorno de trabajo en Europa: perfiles comparativos entre Italia, Francia y Eslovenia* (Albacete: Bomarzo, 2012), 165 et seq.
46. A similar provision is contained in Article 36(4), with reference to information, according to which, 'Where the provision of information concerns immigrant workers, it shall take place after prior verification of the understanding and knowledge of the language used in the information process.'

as well as the attendant preventive and protective measures and procedures that are characteristic of the sector or sub-sector to which the business belongs.[47] Furthermore, training must be provided during working hours and may not be at the workers' expense.

The legislator further specifies when the employer must discharge his or her training duty. The relevant moments are: upon recruitment, in the event of a transfer or a change of job and in the event of the introduction of new work equipment, any new technology or new hazardous substances or preparations. Thus the dynamic nature of the training duty may be inferred. It is not discharged in any one given moment but 'shadows' the changes characterizing not just individuals' work but the whole production process. Indeed, pursuant to Article 37(6) of Legislative Decree no. 81/2008, training must be re-proposed as risks evolve or as new risks arise.[48]

It is likewise appropriate to observe that if the main beneficiary of the information and training duty is doubtless the worker, the same reasoning must be applied to all parties involved in protecting health and safety in the workplace. Thus the workers' safety representatives, the managers, the persons in charge and the persons responsible for the Preventive and Protective Service must all receive detailed information and training that is specific to their roles.

The training of workers' safety representatives is accorded particular importance by Article 37(10) of Legislative Decree no. 81/2008. Such training is seen as the means of ensuring the necessary competences for exercising the powers to check health and safety that are proper to the role. Representatives must have such techno-cultural qualities and general training as allow a real exchange of information and fully conscious, preventive and timely consultation regarding the management of work safety. More generally, training must make the conscious participation of workers' specialist representatives possible. It therefore assumes a function that is closely linked to the establishment of important rights concerning information and consultation and constituting the 'balanced participation' referred to in Framework Directive.

As regards the other parties who are to receive training, Legislative Decree no. 81/2008 provides for special training for the workers given responsibility for fire prevention, fire-fighting, evacuation from the workplace in the event of serious and imminent danger, rescue work, first aid and the management of emergencies.

Furthermore, one of the novelties of Legislative Decree no. 81/2008 is its provision of an appropriate, specific training for supervisors (Article 37(7)). Furthermore, pursuant to Article 15, such figures have the right to receive the information and training that are appropriate to one of the primary roles in health and safety protection.[49]

---

47. P. Pascucci, 'Dopo la legge n. 123 del 2007. Titolo I del d.lgs. 9 aprile 2008, n. 81 in materia di tutela della salute e della sicurezza nei luoghi di lavoro', http://csdle.lex.unict.it/Archive/WP/WP%20CSDLE%20M%20DAntona/WP%20CSDLE%20M%20DAntona-IT/20111025-121117_pascucci_n73-2008itpdf.pdf>, vol. 73, 132.
48. L. Carollo, 'Informazione e formazione dei lavoratori', in *Il testo unico della salute e sicurezza nei luoghi di lavoro*, ed. M. Tiraboschi (Milan: Giuffrè, 2008), 493 et seq.
49. M. R. Gentile, 'I dirigenti e i preposti', in *Il testo unico della salute e sicurezza nei luoghi di lavoro*, ed. M. Tiraboschi (Milan: Giuffrè, 2008), 215.

Indeed, the numerous functions attributed to the health and safety supervisor demand that such figure be given an adequate training. In particular, the training must allow the supervisor to operate independently when applying the preventive and protective measures that have been prepared by the employer and manager.

Analogously with the supervisor, the manager has the right to receive training that is appropriate to his or her role in the undertaking and in relation to application of the accident-prevention rules.

Lastly, the persons responsible for the Preventive and Protective Service also fall within the employer's training duty.[50] It should be noted that Legislative Decree no. 81/2008 does not contain any particular novelties in this area, if compared to the previous legislation. Thus, in addition to the general training courses, these persons must attend specific courses on the prevention of and protection from risks (including ergonomic risks), the organization and management of technical-administrative activities, business communication techniques and industrial relations. The real novelty lies in the duty on persons responsible for the Preventive and Protective Service to attend training courses on the risks deriving from work-related stress: pursuant to Article 28 of Legislative Decree no. 81/2008, employers have to take account of these risks when drawing up the risk-assessment document.

[E] Duties on Contractors and Sub-contractors, with Specific Reference to the Building Sector

It has already been observed how, by virtue of the broad concept of commissioning entity as defined by Article 26 of Legislative Decree no. 81/2008, those who entrust works to contracting undertakings or to self-employed workers or who use agency workers within their undertakings are considered by the current national legislation to fall within the definition. These very different categories are therefore united by the respective duties accorded them by the health and safety law.

Such a decision is based on the undisputed fact that workers' safety can be even more precarious in cases where, through the award of contracts, the activities of other undertakings or self-employed workers become an inextricable part of an employer's production organization. Indeed, in this way, to the normal risks of the workplace are added new ones deriving from reciprocal interference with the various activities.[51]

Precisely by virtue of this fact, commissioning entities have duties to supervise, co-ordinate and co-operate with the parties to whom they entrust works. In particular and as already observed, the commissioning entity's fundamental duty is to draw up a

50. P. De Vita, 'Formazione dei lavoratori e dei responsabili/addetti', in *Il testo unico della salute e sicurezza nei luoghi di lavoro*, ed. M. Tiraboschi (Milan: Giuffrè, 2008), 495 et seq.
51. See P. Pascucci, 'Dopo la legge n. 123 del 2007. Titolo I del d.lgs. 9 aprile 2008, n. 81 in materia di tutela della salute e della sicurezza nei luoghi di lavoro', http://csdle.lex.unict.it/Archive/WP/WP%20CSDLE%20M%20DAntona/WP%20CSDLE%20M%20DAntona-IT/20111025-121117
_pascucci_n73-2008itpdf.pdf>, vol. 73, 112 et seq.

consolidated risk-assessment document, which must be attached to procurement contracts or contracts for personal professional services.

It is clear that such a provision acquires particular importance in certain working contexts where the risk deriving from interference not only is more probable but will also have a greater impact on workers' health and safety. This is particularly the case in the building sector, which is characterized by the considerable recourse it has to procurement contracts and contracts for personal professional services.

In the light of the above, it may be helpful to analyse one of the commissioning entity's fundamental duties, namely, the duty to appoint safety co-ordinators. Article 90(3) of Legislative Decree no. 81/2008 establishes, in relation to temporary and mobile worksites, the cases in which the commissioning entity is under a duty to designate co-ordinators during planning and during execution of the works. Under the previous legislative framework, appointment of such figures was required solely in cases where more than one undertaking was present on-site (not necessarily contemporaneously) or in cases where the magnitude of the project was equal to or greater than 200 man-days, or in the case of projects below that threshold but presenting particular risks for workers' health and safety, of the types indicated in Annexe II to Legislative Decree no. 494/1996.

In contrast, the new provision establishes the commissioning entity's duty to appoint safety co-ordinators in the case where the (not necessarily contemporaneous) presence of more than one undertaking is envisaged. The further condition linked to the magnitude of the project or the more hazardous nature of the activity is therefore no longer relevant.

Section 11 of the same Article 90 provided for a specific derogation from such provision in the case of those private works that do not require a building permit but have to present a simple 'declaration of an activity's commencement' (the so-called DIA – *Dichiarazione di Inizio Attività*). In its judgment dated 7 October 2010 in case no. C-224/09, the Court of Justice of the European Union found against the Republic of Italy on this point, for having exempted commissioning entities from the duty to appoint safety co-ordinators for the planning stage, in breach of Article 3(1) of Directive 92/57/EEC.

The commissioning entity's duties are also important as far as checking is concerned. In addition to checking the technical/professional suitability of both the undertakings and the self-employed workers entrusted with the works, it must also require the undertakings carrying out the works to supply a declaration regarding the average annual number of staff, classified according to qualifications. Such declaration must be accompanied by the details contained in the declarations made by workers to the National Social Security Institute (INPS – *Istituto nazionale della previdenza nazionale*), the National Occupational Accident Insurance Institute (INAIL – *Istituto nazionale per l'assicurazione contro gli infortuni sul lavoro*) and the *casse edili* (bilateral bodies established by collective agreements), as well as a declaration regarding the collective agreement stipulated by the comparatively most representative Trade Unions, applied to the employed workers. Before the works commence, the commissioning entity is likewise required to send the names of the undertakings executing the works and the documentation acquired from them to the competent public authority. In the absence of the said documentation, the undertaking's activity is suspended.

[F] Workers' Duties

As stated earlier, the Italian legal order has also embraced a broad concept of 'worker'. So broad that it includes employees, self-employed and those involved in co-ordinated project work. This all-inclusive concept results in an extension of the related health and safety duties to all workers, regardless of the type of contract they have.

In connection with these duties, it should first be observed that the Italian framework, in line with the Community legislation, is based on the two fundamental principles of employer's liability and workers' involvement.

In the light of the latter, the idea of a worker who not only has health and safety rights but also specific duties is enunciated, with the aim of spreading that 'safety culture' that is the bedrock of the current national legislation.

Thus workers have become an active part of the prevention system, since they are under a duty to take care of their own health and safety, as well as that of all the people who may be affected by their actions or omissions at work. Indeed, people have come to talk of 'the worker's safety duty towards him or herself'. This would have its origins in Article 32 of the Constitution, in its assertion of the individual's right to have his or her health protected.[52]

It should nevertheless be observed that workers' involvement does not in any way limit the principle of employers' liability as expressly established both by Legislative Decree no. 626/1994 and by Legislative Decree no. 81/2008.[53] In truth, the legislator's intention was not to extend the employer's duties to workers but, rather, to increase the latter's awareness of their own role within the prevention system, thereby avoiding those negligent attitudes that workers all too often have, particularly regarding the use of personal protective equipment.[54] Workers therefore have the duty, amongst others, to respect the safety provisions established by employers or commissioning entities.

§7.05 COMPLIANCE AND CONTROL

One of the most controversial aspects of the law governing health and safety in the Italian legal order is the role it accords the inspection and monitoring bodies.[55]

52. S. P. Emiliani, 'Il dovere di sicurezza del lavoratore verso se stesso alla luce della normativa comunitaria', *Argomenti di diritto del lavoro*, no. 1 (2009): 104 et seq.
53. See M. Corrias, *Sicurezza e obblighi del lavoratore* (Turin: Giappichelli, 2008), 71.
54. A. Pinna, 'Obblighi, compiti e diritti dei lavoratori', in *La tutela della salute nei luoghi di lavoro,* ed. G. Loy (Padua: Cedam, 1996), 32 et seq.
55. P. Rausei, 'Le contravvenzioni e le sanzioni amministrative', in *Il testo unico della salute e sicurezza nei luoghi di lavoro,* ed. M. Tiraboschi (Milan: Giuffrè, 2008), 729 et seq.

At present, pursuant to Article 13 of Legislative Decree no. 81/2008, observance of the work safety rules is monitored by the Local Health Authorities (*Aziende sanitarie locali* – hereinafter also 'ASL'), each one exercising jurisdiction in its own territory; by the national Fire Brigade, in relation to its own specific jurisdiction; by the Ministry of Economic Development, in relation to the mining sector (until the latter jurisdiction is effectively transferred, pursuant to Legislative Decree no. 300, dated 30 July 1999, as subsequently amended) and by the Regions and autonomous provinces of Trento and Bolzano, as regards the secondary mining industries and both mineral and thermal waters.

In workplaces managed by the Armed Forces, the Police Forces and the Fire Brigade, monitoring of the application of the occupational health and safety legislation is carried out exclusively by health services and technical services set up within the said organizations i.e., by structures within the same authority.

It should be added at this point that, as regards matters either directly or indirectly linked to the protection of workplace safety (such as the subject of working time, for example), such bodies are flanked by inspectors at the Ministry of Labour and Social Policies (Labour inspectors) who have both a general jurisdiction and the fundamental task of co-ordinating jurisdictions and/or functions when these overlap with those of other inspectorates.

Labour inspectors monitor the application of the health and safety provisions in the following areas of activity: construction and civil engineering; works using caissons in compressed air conditions and underwater works, and those other work activities carrying particularly high risks that have been identified by decrees of the President of the Council of Ministers.

It is therefore evident that the institutional system for the national, labour law monitoring of occupational health and safety involves different parties with basically different inspection duties that sometimes overlap.

Against this particularly complex, articulated backdrop, the powers attributed to individual administrative inspection and monitoring bodies for the purposes of facilitating undertakings' compliance with legal obligations stand out in sharp relief.

In an occupational health and safety context, such function is carried out through three fundamentally important legal tools that turn to account the role of the administrative inspection bodies that is geared to fostering the re-establishment of lawful situations, rather than exclusively punishing infringements.

Such tools are the *interpello* (formal query), governed, generally, by Article 9 of Legislative Decree no. 124/2004, as amended by Decree Law no. 262/2006 and, as regards occupational health and safety, by Article 12 of Legislative Decree no. 81/2008; the *prescrizione* (prescribing notice), which is governed by Article 20 of Legislative Decree no. 758/1994 – to which Article 301 of Legislative Decree no. 81/2008 specifically refers with regard to the protection of workplace safety – and which provides for the concessional extinguishment of administrative offences following regularization (governed, in its turn, by Article 301*bis* of Legislative Decree no. 81/2008, as introduced by Article 143(1) of Legislative Decree no. 106, dated 3 August 2009) and, last but not least, the power to issue provisions (governed by Article 302*bis* of Legislative Decree no. 81/2008, as introduced by Article 144(2) of Legislative Decree no. 106, dated 3 August 2009).

In pride of place stands the *interpello* regarding matters of occupational health and safety, which has only been fully realized during the last few months. Constituting one of the main novelties introduced by Legislative Decree no. 81/2008, this permits the associative bodies of the local authorities, the trade unions and the most representative employers' organizations at a national level, as well as the professions' national councils, to send queries of a general nature regarding application of the occupational health and safety legislation to a special '*Interpello* Committee' established at the Ministry of Labour by way of a Ministerial Decree dated 28 September 2011 (issued pursuant to Article 12(2) of Legislative Decree no. 81/2008).

When it receives questions of a general nature regarding workplace safety, the '*Interpello* Committee' answers them, issuing directions that are binding on the majority of undertakings operating in the national territory. Such directions also constitute interpretative and executive guidelines for the competent monitoring bodies when carrying out their activities.

In this respect, the tool of the 'safety' *interpello* differs from the same-named 'ordinary' *interpello* (governed, regarding other work and social security issues generally, by Article 9 of Legislative Decree no. 124/2004, as amended by Decree Law no. 262/2006) whereby an undertaking's compliance with the reply from the Inspectorate's Head Office to an 'ordinary' *interpello* precludes the applicability of the prescribed civil, criminal and administrative penalties.

The difference is fully justified by the constitutionally protected nature of the matters that are the object of the 'safety *interpello*', as well as by the impossibility of defining a priori the adequacy of the workplace safety measures adopted. This partly in relation to the position of guarantor that the employer has towards his or her own employees, which, by virtue of the constantly dynamic conditions in the workplace and as constantly stated by the case law, persists even when the monitoring inspections have been passed.

Thus *interpelli* regarding workplace safety have a far more limited effect than their more general counterparts. Such fact does not diminish their usefulness, however. This is especially so when they raise issues of a formal nature, as in the cases regarding the reporting of occupational diseases (*Interpello* no. 5/2009, dated 06 February 2009) and delivery of the risk evaluation document (*Interpello* no. 58/2008, dated 19 December 2008).

In a perspective similar to that of the *interpello* (since it, too, is chiefly directed at guaranteeing the re-establishment of legality within businesses), the tool of the *prescrizione* (prescribing notice) has been grafted onto the national preventive system. It is governed by Articles 20 et seq. of Legislative Decree no. 758/1994, as specifically referred to – regarding the protection of workplace safety – by Article 310 of Legislative Decree no. 81/2008 (amended by Article 142(1) of Legislative Decree no. 106, dated 3 August 2009, confirming the tool's value insofar as it relates to workplace safety).

In cases where a breach of workplace safety provisions has been ascertained by the competent supervisory inspection bodies, a *prescrizione* allows the enterprise a concessional extinguishment of those offences that are punished with alternative penalties of imprisonment or a fine, or with just a fine, once it has re-established a lawful situation by complying with the directions issued by the administrative monitoring bodies.

More specifically, pursuant to Article 20 of Legislative Decree no. 758/1994, the administrative monitoring body with specific competence that ascertains a breach of the law when carrying out its judicial police functions, is called to communicate a crime report to the Public Prosecutions Office, pursuant to Article 347 of the Code of Criminal Procedure, and subsequently send the enterprise a detailed 'performance notice' prescribing the performance of specific actions, whilst fixing a deadline for its implementation. This for the purposes of reconciling the need to re-establish lawful business safety conditions, on the one hand, with the need to impose a penalty through the payment (on concessional terms) of the established fine, on the other.

When compliance with the duties is particularly complex or objectively difficult, the timeframe granted to the enterprise may, pursuant to Article 20(1) of Legislative Decree no. 758/1994, be extended at the offender's request. That said, the time extension cannot in any circumstances exceed six months, except where specific circumstances not imputable to the offender cause an objective delay in the regularization. In such a case, at the offender's request, an extension may be permitted just once, for a period of no longer than another six months. It is granted by way of an administrative measure that includes a statement of reasons, which measure is immediately communicated to the Public Prosecutions Office.

Pursuant to Article 21 of Legislative Decree no. 758/1994, the monitoring bodies are then required to verify whether the breach has been eliminated according to the specifications and within the timeframe indicated in the 'prescribing notice'. This must be done within sixty days of expiry of the deadline established by the *prescrizione*.

When the assessment is positive, the offender will be permitted to pay the related fine, reduced to an amount equal to one-quarter of the maximum statutory fine for the infringement committed, within thirty days. At the same time and within 120 days of expiry of the timeframe fixed in the *prescrizione*, the monitoring body communicates both the fact of compliance with the latter and the payment of the fine to the Public Prosecutions Office. Pursuant to Article 24(1) of Legislative Decree no. 758/1994, these together result in the infringement's extinguishment and the consequential dismissal of the criminal proceedings at the public prosecutor's request.

Late compliance will allow relief from the penalties applicable to the offender, even if it is achieved in a manner other than that indicated by the monitoring body, provided that it is adequate (according to the meaning attributed to the term through reference to the criterion set out in Article 20(1) of Legislative Decree no. 758/1994 i.e., the elimination of the harmful or dangerous consequences ascertained). This by virtue of the provisions contained in Article 162*bis* of the Criminal Code, concerning the payment of fines. In such a case, the sum to be paid is reduced to one-quarter of the maximum fine established for the offence committed.

Should the administrative monitoring body ascertain a persisting failure to comply with the *prescrizione*, however, then pursuant to Article 21(3) of Legislative Decree no. 758/1994, it communicates such fact both to the Public Prosecutions Office and to the offender within ninety days of expiry of the timeframe in the 'prescribing notice'. As a consequence, the legally established penalties will be applied to the offender in their entirety.

The issue of whether the provisions governing powers to issue notices (identified by Article 9 of Presidential Decree no. 520, dated 19 March 1955, and Article 13 of Legislative Decree no. 124/2004) could be extended – in favour of the inspectors working at the Provincial Labour Offices – to those breaches of prevention law punished as administrative offences, proved far more problematic (partly as a result of the enactment of Legislative Decree no. 81/2008).

Indeed, notices constitute one of the essential tools for guaranteeing the right balance between the need to re-establish lawful situations and the punitive function of the penalties system. This they achieve by offering the extinguishment of the administrative offence through the payment of a sum equal to the minimum penalty, after regularization of the ascertained breach within the timeframe indicated in the minutes of the first inspection.

That the tools may be extended to the prevention rules falling within the jurisdiction of the Ministry of Labour's 'administrative monitoring' bodies was subsequently made clear by Circular no. 9, dated 23 March 2006. This provided that, 'there being no limitation whatsoever in this respect, the power to issue notices applies to all matters falling within the jurisdiction of the labour inspectors and, therefore, also to matters – such as workplace safety, in particular – where there exist residual State assessment powers'.

Despite the power of the Provincial Labour Offices' inspection bodies to issue notices regarding preventive matters, no similar power had been legally established in favour of the ASL's inspection bodies, even after Legislative Decree no. 81/2008 had been enacted.

It was only with the insertion of Article 301*bis* into the text of Legislative Decree no. 81/2008 (effected by Article 143(1) of Legislative Decree no. 106, dated 3 August 2009) that the problem of the notice tool's applicability to preventive matters was solved.

Indeed, by inserting an instrument totally analogous with the abovementioned notice, Article 301*bis* of Legislative Decree no. 81/2008 introduces the tool of the so-called concessionary extinguishment of administrative offences following regularization, by virtue of which, 'in all cases of non-compliance with the [preventive – Ed.] duties punished with an administrative fine', the offender is allowed to pay a reduced amount equal to the minimum provided for by law, if s/he regularizes the position within the deadline set in the minutes of the first inspection access.

In fact, following the rule's insertion into Legislative Decree no. 81/2008, one can witness the institution of a special 'preventive notice' that is applicable solely and exclusively to breaches regarding workplace safety that are punished with administrative fines.

On the one hand, the rule solves the problem of the ASL's lack of powers in this area by extending the tool's application to all the monitoring bodies with competence for preventive matters. On the other and in a purely simplifying manner, it introduces a new instrument that, whilst following the implementation specifications contained in

Article 13 of Legislative Decree no. 124/2004, potentially guarantees the special nature of the workplace safety provisions. This is consistent with the aim of focussing on the effectiveness of a punitive reaction centred not so much on fining breaches as on re-establishing lawful conditions (the aim that had already inspired Parent Act no. 123/2007).

Lastly, as far as inspectors' compliance tools are concerned, brief mention must be made of another tool grafted onto the system through Article 302*bis* of Legislative Decree no. 81/2008 (via Article 144(2) of Legislative Decree no. 106, dated 3 August 2009). This is the provision-issuing power attributed to the workplace safety inspection bodies regarding the application of technical rules and good practices where the latter are voluntarily adopted by employers.

The legislation accords the supervisory inspection bodies the power to issue executive orders ensuring the application of technical rules and good practices in cases where the latter have been voluntarily adopted by employers and are expressly referred to by the same during an inspection, should the monitoring bodies find that such rules and good practices have not been put into practice correctly. Failure to comply with an order results in application of the legally established penalties, in those cases where a breach has clearly occurred.

Owing to the voluntary nature of technical rules and good practices, the law also envisages a mechanism for appealing against executive orders within thirty days of their issue. Appeals are made to the relevant authority with immediate hierarchical superiority, which authority decides the appeal within fifteen days.

§7.06 WORKERS REPRESENTATIVES

[A] The Role of Workers' Representatives in the Management of Occupational Health and Safety

The Italian legal order has always accorded workers' general representatives a primary role in the management of undertakings' preventive systems. They flank and sometimes overlap with the representatives accorded specific responsibility by the health and safety legislation.

In this context, it should be noted that the national model of representation is characterized by a single channel system through which the collective protection of workers is exclusively entrusted to the trade unions, to which both participation and negotiating powers are devolved.[56]

The powers of collective bargaining with undertakings are therefore attributed to the trade unions' representative bodies within such undertakings. In the context of safety at work, however, these powers are subject to obvious limitations created by the very real risk that bargaining about preventive measures could lead to what is referred

56. See, L. Miranda, *La dimensión colectiva de la tutela del entorno de trabajo en Europa: perfiles comparativos entre Italia, Francia y Eslovenia* (Albacete: Bomarzo, 2012), 104 et seq.

to as the 'monetization of safety'[57] i.e., that the parties may agree to a qualitative reduction of workplace safety standards in exchange for contractual concessions.[58]

For this reason, occupational health and safety in general and work hygiene and safety conditions, in particular, are non-negotiable issues in the collective bargaining context.[59] Such fact does not mean, however, that collective bargaining never concerns matters closely related to occupational health and safety. It plays a primary part in managing working conditions whenever there is a need to regulate specific issues linked to work organization.[60]

As a consequence, collective bargaining at every level may only implement the safety standards established by law. Although it has a primary role, influencing occupational health and safety both directly and indirectly every time it concerns itself with work organization (particularly the management of working time), it has different aims from the participatory one provided for by the Framework Directive. This insofar as it presupposes a natural conflict of the parties' interests, which finds its settlement in the agreement that is negotiated.

It is onto this complex structure that the first collectively representative figure was grafted in the 1970s. Defined by Article 9 of Law no. 300/70 (the so-called Workers' Statute), this figure has duties to protect working conditions.[61] More specifically, under the rubric 'protection of health and physical integrity' and disregarding how the representative bodies it refers to are legally established, Article 9 provides that:

> Workers, through their representatives, have the right both to check that the rules for preventing accidents and occupational diseases are applied and to promote research into and the development and implementation of all the appropriate measures for protecting their health and physical integrity.

The 'total functional overlapping' of 'their representatives' with the workers' safety representatives created by the recent legislation results in the 'Article 9 representatives'

57. See, G. Balandi, 'Individuale e collettivo nella tutela della salute nei luoghi di lavoro: l'art. 9 dello Statuto', *Lavoro e Diritto,* no. 1 (1990): 222 et seq.; C. Zoli, 'Sicurezza del lavoro: contrattazione e partecipazione', *Rivista Giuridica del Lavoro,* no. 3 (2000): 624 et seq.; A. Lo Faro, 'Azione collettiva e tutela dell'ambiente di lavoro in Europa', *Giornale di diritto del lavoro e di relazioni industriali,* no. 1 (1991): 162 et seq.; M. Ricci, 'Sicurezza sul lavoro: controllo e partecipazione sindacale tra iure condito e de iure condendo', *Il lavoro nella giurisprudenza,* no. 2 (2008): 113 et seq.
58. L. Montuschi, 'Verso il testo unico sulla sicurezza del lavoro', *Giornale di diritto del lavoro e di relazioni industriali,* no. 2 (2007): 799 et seq.
59. G. Balandi, 'Individuale e collettivo nella tutela della salute nei luoghi di lavoro: l'art. 9 dello Statuto', *Lavoro e Diritto,* no. 1 (1990): 222.
60. See, L. Miranda, 'Organizzazione del lavoro e sicurezza negli accordi del Gruppo Fiat-Chrysler. Verso un comune modello di rappresentanza paritetica per la sicurezza?', *Diritti lavori mercati,* no. 1 (2012): 93 - 113.
61. G. Ghezzi, 'Art. 9 (Tutela della salute e dell'integrità fisica)', in *Statuto dei diritti dei lavoratori. Art. 1-13,* ed. U. Romagnoli et al. (Bologna: Zanichelli, 1979), 2nd ed.; U. Prosperetti, 'Commentary on the Judgement issued by the Milanese Pretura, 28 July 1971', *Massimario di giurisprudenza del lavoro,* no. 1 (1972): 20; E. Ales, 'L'articolo 9 statuto dei lavoratori alla luce della legislazione più recente in materia di salute e sicurezza: Partecipazione o controllo?', *Rivista italiana di diritto del lavoro,* no. 1 (2011): 57 et seq.; P. Pascucci, 'Salute e sicurezza: dalle rappresentanze dell'art. 9 ai rappresentanti del d.lgs. 81/08', *Diritti lavori mercati,* no. 4 (2010): 670 et seq.

being subjectively identified with the more recent ones. If, on the one hand, such fact 'legitimises the suspicion that Article 9 of the Workers' Statute no longer has any reason to continue existing',[62] on the other, it reinforces the exercise of the rights recognized by Article 9 – through the development of instrumental employers' duties such as those relating to information, training, consultation and participation for workers and their representatives.[63]

In the existing framework, therefore, the rights recognized by Article 9 of Act no. 300/1970 integrate and complete the collective workplace safety toolkit attributed to the technical-specialist representatives identified by Legislative Decree no. 81/2008.

Deferring an analysis of the specialist representatives' powers to the following Article, it is appropriate here and now briefly to mention how the representatives' checking power contained in Article 9 has been made consistent with the power to verify contained previously in the 1994 law and now in the 2008 legislation. As regards the power's scope, it may be emphasized that assessment of all the threats to workers' health and safety (including those involving groups of workers exposed to particular risks) is, as we have seen, now explicitly relevant to the monitoring of the health and safety measures' application and not just as the result of an (albeit sustainable) extensive interpretation of Article 2087 of the Civil Code.[64] In this respect, 'particular risks' include those linked to work-related stress (in accordance with the provisions of the European Framework Agreement dated 8 October 2004), those involving female workers during pregnancy (in accordance with the provisions of Legislative Decree no. 151/2001), those linked to differences in gender, age and provenance from other countries and those linked to the specific type of contract under which the work is carried out (Article 28(1)).

If to this are added all the considerations regarding the new definition of workers' health as a 'state of complete physical, mental and social well-being and not merely the absence of disease and infirmity' (Article 2(1)(o)), it may be said that the workers' power to check and report through their safety representatives (hereinafter also WSR) to an undertaking's management and, in the case of its inertia, to the competent bodies, extends well beyond the traditional concept of psycho-physical integrity contained in Article 9. Indeed, it extends to the risks linked to work-related stress and also (and perhaps primarily) to those attaching to differences in gender, age, nationality and type of contract, in relation to which frequent (and often fruitless) efforts have been made to identify a distinguishing characteristic.[65]

In this perspective, to the dimension just mentioned must be added another, concerning the threat to the health and safety of female and male workers deriving

62. G. Perone, *Lo statuto dei Lavoratori* (Turin: UTET, 1997), 89.
63. E. Ales, 'L'articolo 9 statuto dei lavoratori alla luce della legislazione più recente in materia di salute e sicurezza: Partecipazione o controllo?', *Rivista italiana di diritto del lavoro*, no. 1 (2011): 57.
64. L. Montuschi, *Diritto alla salute e organizzazione del lavoro* (Milan: Giuffrè, 1980), 157 et seq.
65. E. Ales, 'L'articolo 9 statuto dei lavoratori alla luce della legislazione più recente in materia di salute e sicurezza: Partecipazione o controllo?', *Rivista italiana di diritto del lavoro*, no. 1 (2011): 57

from discriminatory behaviour or actions (including harassment, obviously[66]) on the part of employers, managers, supervisors or workers' own colleagues.[67] This includes behaviour or actions based on discriminatory motives other than those expressly referred to by Article 28(1), by virtue of the reference made there to 'all the risks for workers' health and safety'.

In the new regulatory framework, therefore, extension of the WSRs' monitoring power provides a solution to the problems linked to the only potential existence (in the private sector) of equal opportunities bodies that may serve to identify and report discrimination.

[B] The Italian Path to Balanced Participation and the Compulsory Establishment of a Workers' Representative Body for Health and Safety at Work

In implementation of the Framework Directive, the national legislator has established a distinctive representative body with specific occupational health and safety duties: the Workers' Safety Representative, to be entrusted with the tasks outlined by the Community legislator.[68]

There can be no doubt that one of the most important novelties introduced by Legislative Decree no. 81/2008 is its new formulation of the rules on workers' collective representation.[69] These are set out in section VII (entitled 'Consultation and Participation of Workers' Representatives') and contain some directions that show the legislator's clearly intentional emphasis on the necessity for and mandatory nature of the representation in question.

Article 47 of Legislative Decree no. 81/2008 makes such provision, breaking with the line pursued in Article 18 of Legislative Decree no. 626/94. This discontinuity is

---

66. See Judgement *Corte di Cassazione (Penale, III Sezione)*, no. 12738/2008. This recognized the *locus standi* of a trade union (of which a female worker who had been the victim of sexual violence was a member) to bring a civil claim within the criminal proceedings, on the assumption that Legislative Decree no. 626/1994 extended the concept of health to include not only physical but also psychic integrity.
67. P. Albi, *Adempimento dell'obbligo di sicurezza e tutela della persona. Art. 2087 c.c.*, Vol. of *Il Codice Civile. Commentario. Fondato da P. Schlesinger*, ed. F.D. Busnelli (Milan: Giuffré, 2008), 305 et seq.; A. Lassandari, *Le discriminazioni nel lavoro. Nozioni, interessi, tutele*, Vol. LVI of *Trattato di Diritto Commerciale e di Diritto Pubblico dell'Economia*, ed. F. Galgano (Padua: Cedam, 2010).
68. L. Galantino, 'Il contenuto dell'obbligo di sicurezza', in *La sicurezza del lavoro*, ed. L. Galantino (Milan: Giuffrè, 1996), 8; G. Natullo, *La tutela dell'ambiente di lavoro* (Turin: Utet, 1995), 254 et seq.; P. Campanella, 'I rappresentanti dei lavoratori per la sicurezza', in *Le nuove regole per la salute e la sicurezza dei lavoratori. Commentario al D.Lgs. 9 aprile 2008, n. 81 aggiornato al D.Lgs. 3 agosto 2009, n. 106*, ed. L. Zoppoli, P. Pascucci & G. Natullo (Milan, Ipsoa, 2010), p. 471 et seq.; M. Ricci, 'Sicurezza sul lavoro: controllo e partecipazione sindacale tra iure condito e de iure condendo', *Il lavoro nella giurisprudenza*, no. 2 (2008): 120 et seq.; A. Tampieri, *Sicurezza sul lavoro e modelli di rappresentanza* (Turin: Giappichelli, 1999), 115 et seq; L. Miranda, *La dimensión colectiva de la tutela del entorno de trabajo en Europa: perfiles comparativos entre Italia, Francia y Eslovenia* (Albacete: Bomarzo, 2012), 125 et seq.
69. L. Miranda, *La dimensión colectiva de la tutela del entorno de trabajo en Europa: perfiles comparativos entre Italia, Francia y Eslovenia* (Albacete: Bomarzo, 2012), 135 et seq.

revealed not so much in terms of the overlapping of the WSR's functions with those of the workers' general representatives (which remain) as, rather, in the legislator's specific opting in favour of a system clearly marked by the attempt to exploit the participatory prerogatives of workers' representatives to the full and to guarantee their actual presence in all workplaces by institutionalizing the territorial representative bodies and developing the powers and competences devolved to them.[70]

The innovative force of Legislative Decree no. 81/2008 immediately becomes clear upon a simple reading of Article 47(1), which states, 'The workers' safety representative shall be established at a territorial or sectoral level, at an undertaking level and at a production site level.'

Indeed, the use of the expression *shall be established* is an explicit statement of the national legislator's intention: the outlining of three levels of representation unequivocally communicates the obligatory nature of such representation in all productive activities.[71] The territorial level is correspondingly accorded pride of place amongst the levels of representation, which fact makes it possible to observe its generally subrogating nature when compared to the undertaking and the production site levels.

If, in this respect, the legislator's choice in favour of the mandatory presence of safety representatives is evident, Article 47 of Legislative Decree no. 81/2008 tackles and clarifies some of the most important provisions on the subject previously identified by Legislative Decree no. 626/94.

In the first place, Article 47(2) of Legislative Decree no. 81/2008 re-proposes the formulation according to which the Workers' Safety Representative is to be elected or appointed in all undertakings or production units. The said provision nevertheless assumes a more stringent connotation if read in the light of the provision relating to the establishment of the WSR, set out in the first section.

Dealing with the actual election or designation of the WSR, however, the subsequent sections (3) and (4) re-propose the distinction between undertaking or production units with more or less than fifteen employees that was formerly contained in Article 18 of Legislative Decree no. 626/94: the effects of the distinction from a practical/applicative point of view clearly change.[72]

With reference to undertakings or production units employing up to fifteen employees, in fact, Article 47(3) of Legislative Decree no. 81/2008 provides that the

70. M. Lai, 'Il ruolo delle parti sociali', in *Il testo unico della salute e sicurezza nei luoghi di lavoro,* ed. M. Tiraboschi (Milan:Giuffrè, 2008), 519 et seq.; A. Baldassarre, 'Le rappresentanze dei lavoratori per la sicurezza e il rilancio della filosofia partecipativa', in *Il testo unico della salute e sicurezza nei luoghi di lavoro,* ed. M. Tiraboschi (Milan:Giuffrè, 2008), 533 et seq.

71. See, L. Miranda, *La dimensión colectiva de la tutela del entorno de trabajo en Europa: perfiles comparativos entre Italia, Francia y Eslovenia* (Albacete: Bomarzo, 2012), 136.

72. P. Pascucci, 'Dopo la legge n. 123 del 2007. Titolo I del d.lgs. 9 aprile 2008, n. 81 in materia di tutela della salute e della sicurezza nei luoghi di lavoro', http://csdle.lex.unict.it/Archive/WP/WP%20CSDLE%20M%20DAntona/WP%20CSDLE%20M%20DAntona-IT/20111025-121117_pascucci_n73-2008itpdf.pdf>, vol. 73, 140 et seq.; M. Lai, 'Il ruolo delle parti sociali', in *Il testo unico della salute e sicurezza nei luoghi di lavoro,* ed. M. Tiraboschi (Milan:Giuffrè, 2008), 519 et seq.

WSR must, 'as a rule'[73] be elected directly by the workers from amongst themselves or must be identified, in relation to more than one undertaking in the territory or sector of production, in accordance with Article 48 of Legislative Decree no. 81/2008.

Article 47(4) of Legislative Decree no. 81/2008, however, confirms the provision contained in Article 18(3) of Legislative Decree no. 626/94 that 'the workers' safety representative shall be elected or designated by the workers in the context of the trades union representation in the undertaking', without prejudice to the fact that 'in the absence of such representation, the representative shall be elected by the undertaking's workers from amongst themselves'. Thus the legislator's preference for the overlapping of internal trade unions representation with safety representation that had already characterized Legislative Decree no. 626/94 is confirmed.

A difference of primary importance may nevertheless be identified in terms of the concrete enforcement measures, should there be no election or designation of the WSR either in undertaking with fewer than fifteen workers or – and this constitutes a novelty that is particularly in line with the legislator's goal – in those with more than fifteen workers. Indeed, in such cases, the legislator expressly provides (in Article 47(8) and 48(1) of Legislative Decree no. 81/2008) for the subrogating intervention of the territorial workers' safety representative, establishing that the latter shall exercise the powers of the WSR with reference to all the undertakings or production units in the territory or sector of competence *in which the workers' safety representative has not been elected or designated* and not just *in undertakings that employ up to 15 employees*, as was previously the case, pursuant to Article 18(2) of Legislative Decree no. 626/94.

The novelty of this provision is to be considered fundamental also in terms of the actual presence of territorial representation within undertaking, insofar as Legislative Decree no. 81/2008, unlike Article 18(2) of Legislative Decree no. 626/94, is no longer couched in terms of the merely potential subrogating intervention of the territorial representation but, rather, imposes it in cases where the undertaking's WSR is not elected or designated, whatever the size of the undertaking may be.[74]

As regards the powers of the Safety Representatives at every level, however, the Italian path to implementing the measures fostering the balanced participation referred to in Framework Directive is twofold in its nature: basically a-conflictive, with reference to the powers regarding training, information and consultation, and conflictive when, in the absence of agreement between the employer and the workers' specialist representatives in the first instance, the latter can take action directly.[75]

Under the new legislative framework, therefore, consultation with the WSR is characterized by its timeliness and priority with regard to the most important of the employer's decision-making processes concerning the protection of workplace safety.

---

73. P. Pascucci, 'Dopo la legge n. 123 del 2007. Titolo I del d.lgs. 9 aprile 2008, n. 81 in materia di tutela della salute e della sicurezza nei luoghi di lavoro', http://csdle.lex.unict.it/Archive/WP/WP%20CSDLE%20M%20DAntona/WP%20CSDLE%20M%20DAntona-IT/20111025-121117_pascucci_n73-2008itpdf.pdf>, vol. 73, 142.
74. See, L. Miranda, *La dimensión colectiva de la tutela del entorno de trabajo en Europa: perfiles comparativos entre Italia, Francia y Eslovenia* (Albacete: Bomarzo, 2012), 140 et seq.
75. L. Miranda, *La dimensión colectiva de la tutela del entorno de trabajo en Europa: perfiles comparativos entre Italia, Francia y Eslovenia* (Albacete: Bomarzo, 2012), 183 et seq.

Whilst it cannot be deemed binding on the employer, it is an essential prerequisite, enabling representatives to participate in the employer's decision-making in a technical sense, making proposals and checking that the safety measures are implemented.

In such a perspective, Legislative Decree no. 81/2008 has sought to solve one of the most critical aspects of Legislative Decree no. 626/1994. To the WSR's powers under Article 50, it has added the principle that consultation must take place regarding two employer's decisions of primary importance: designation of the occupational health physician and, more importantly, the organization of training in general (and, therefore, no longer just the training for the officers responsible for emergencies[76]).

If consultation takes place regarding the most important health and safety topics and encompasses the very great majority of employer's decisions potentially affecting the management of health and safety, the problematic side to the provision therefore lies in the fact that the opinion of the WSR is not binding. More precisely, where the views on the workplace safety measures that have been or are to be adopted are not identical, the naturally conflictive tendency of Italian industrial relations therefore reappears.[77] Safety representatives are consequently recognized as enjoying, *in primis*, the right of access to workplaces and, second, the right to present observations during inspection and monitoring visits made by public authorities. These rights are completed by the right/duty to alert the person responsible in the undertaking to the risks the representative has identified during the course of his or her work and the fundamental right to have recourse to the competent authorities, inspection services and ordinary courts, should the safety measures adopted by the employer not be deemed appropriate.

Implementation of the balanced participation provisions is also enhanced by the provision of further, specific powers for safety representatives. These are a consequence of representatives exercising their information and consultation rights and are characterized by their priority and complementarity, if compared to the same representatives' checking duties and reporting power (regarding which, see Article 3(a) above). More specifically, they are the rights identified under letters (h) and (m) of Article 50(1) of Legislative Decree no. 81/2008, which guarantees representatives the right to promote the formulation, specification and implementation of suitable preventive measures for protecting the health and physical integrity of workers and to propose forms of preventive action.

Indeed, although the said powers constitute a mere consequence of WSR being informed and consulted and are, therefore, to be considered as subsequent to such an

76. P. Pascucci, 'Dopo la legge n. 123 del 2007. Titolo I del d.lgs. 9 aprile 2008, n. 81 in materia di tutela della salute e della sicurezza nei luoghi di lavoro', http://csdle.lex.unict.it/Archive/WP/WP%20CSDLE%20M%20DAntona/WP%20CSDLE%20M%20DAntona-IT/20111025-121117_pascucci_n73-2008itpdf.pdf>, vol. 73, 148.

77. See, L. Zoppoli, 'Rappresentanza collettiva dei lavoratori e diritti di partecipazione alla gestione delle imprese', *Giornale di diritto del lavoro e di relazioni industriali*, no. 2 (2005): 432 et seq.; E. Ales, 'Informazione e consultazione nell'impresa: diritto dei lavoratori o dovere del datore di lavoro? Un'analisi comparata', *Rivista italiana di diritto del lavoro*, no. 1 (2009): 239; L. Montuschi, 'La sicurezza nei luoghi di lavoro ovvero l'arte del possibile', *Lavoro e diritto*, no. 2 (1995): 413 et seq.; G. Ferraro & M. Lamberti, La sicurezza sul lavoro nel decreto legislativo attuativo delle direttive CEE, *Rivista giuridica del lavoro*, no. 1 (1995): 50 et seq.

event, they do at the same time constitute a measure that naturally precedes and complements the monitoring and reporting rights attributed to the same representatives.[78]

Lastly, an analysis of what has been defined as the third level of representation leads to some considerations regarding the introduction of the workers' safety representative for production sites. This figure is identified in relation to specific production contexts characterized by the simultaneous presence of more than one firm. It therefore has a considerable importance in the building sector.

The establishment of the production site representative is left to the initiative of the individual WSRs, who choose him or her from among the WSR within the undertakings operating on the production site.

The legislation first identifies the production context in which the WSR for production sites may be identified. This context is characterized by the simultaneous presence of more than one undertaking or work project. The legislator's goal is therefore to promote co-operation and the co-ordination of measures monitoring and promoting safety between the various undertakings present on the same worksite and exposed to the so-called risks deriving from interference; a situation linked to the duty that, under Article 26(3) of Legislative Decree no. 81/2008, requires the commissioning entity to draw up 'a consolidated risk-assessment document indicating the measures adopted to eliminate or, where that is impossible, reduce the risks deriving from interference to the minimum'.

The establishment of the WSR for production sites therefore meets the need to tackle the significant danger to the safety of workers operating in complex business structures. This in view of the simultaneous presence of specific risks inherent to individual undertakings' activities and risks deriving from the concomitant presence of different undertakings' employees and therefore caused by reciprocal interference between the various activities present in the same theatre of work.[79]

[C] The Role of the Joint Committees and their Relevance in the Building Sector

The Italian legal order assigns a fundamental role in occupational health and safety management to Joint Committees (hereinafter JC). These are identified by Article 2(ee) of Legislative Decree no. 81/2008 as bodies founded on the initiative of one or more of the employer/workers comparatively most representative organization at national level.[80] They are given the task of participating further, at a territorial level.

78. L. Miranda, *La dimensión colectiva de la tutela del entorno de trabajo en Europa: perfiles comparativos entre Italia, Francia y Eslovenia* (Albacete: Bomarzo, 2012), 191.
79. See *Corte di Cassazione* (*Sezione Lavoro*), Judgement no. 45/2009, reported in *Argomenti di diritto del lavoro* no. 3 (2009): 800 et seq.
80. P. Passalacqua, *voce Enti Bilaterali*, vol. IV of *Digesto delle discipline privatistiche - Sezione commerciale* (Turin: Utet, 2008), 236 et seq.; L. Miranda, *La dimensión colectiva de la tutela del entorno de trabajo en Europa: perfiles comparativos entre Italia, Francia y Eslovenia* (Albacete: Bomarzo, 2012), 210 et seq.

It is the defining framework contained in Article 2(ee) of Legislative Decree no. 81/2008 that states this, specifying that the said bodies are privileged places for programming training activities and formulating and consolidating good preventive practices, developing actions connected with health and safety at work, assisting undertakings in fulfilling their health and safety duties and every other activity or function assigned them by law or by the collective agreements of reference.

Over and above the essential function of resolving disputes concerning application of the rights to representation, information and training, in accordance with all that was previously indicated by Article 20 of Legislative Decree no. 626/1994, the legislator therefore accords the JC the fundamental role of privileged places for adopting measures actively geared to completing the participatory framework outlined by the legislation. The task of orienting and promoting training assumes a particular importance.

In completion of their specific functions, the same JC are also given the fundamental task of supporting undertakings in the identification of technical and organizational solutions aimed at guaranteeing and improving the protection of health and safety at work.[81] To such end, provided that they have staff with the relevant specific technical competences, the JC are entrusted with a particularly important task since they are given the power to carry out on-site inspections in workplaces falling within the territory and sector of production for which they have competence.[82] In such a sense, therefore, JC support undertakings in identifying specific shortcomings in health and safety protection and promote the interventions needed to make workplaces safe.

Lastly, the JC's role and functions have been enhanced by Article 30 of Legislative Decree no. 106/2009. This amended Article 51 of Legislative Decree no. 81/2008 by inserting sections 3 *bis* and 3 *ter* which attribute to (truth to tell) not better specified 'technically competent joint commissions' (to be set up within the joint committees) the important functions of:

> certifying performance of the activities and services supporting undertakings' systems, including 'asseverazione' (i.e. certification by professional experts) of the adoption and effective implementation of the models of safety organization and management referred to in Article 30, of which the supervisory bodies can take account for the purposes of programming their activities.[83]

81. As Art. 51(3) of Legislative Decree no. 81/2008 puts it.
82. Sections 6 and 7 of Art. 51 of Legislative Decree no. 81/2008 complete the said power by providing for the duty on bilateral bodies to send an annual report on their activities to the Regional Committee co-ordinating preventive and monitoring activities referred to in Art. 7.
83. The joint committees system has an important role in the building sector, where the social partners (through collective bargaining) accord the territorial joint committees and the body that unites them at a national level, the CNCPT – *Commissione Nazionale dei Comitati Paritetici Territoriali* – fundamental policy-setting, monitoring and co-ordination functions as well as that of liaising with other national bodies with competence for safety and prevention. Thus structured at a national level, the System has a network of ninety-eight bilateral committees distributed throughout the provincial territories. In accordance with legal requirements, each territorial body has the goal of studying general and specific problems concerning accident prevention, work hygiene and, generally, improving the workplace, formulating proposals and suggestions and promoting or participating in suitable initiatives. In particular, the national

As stated earlier, Article 51(3)(*ter*) of Legislative Decree no. 81/2008 attributes important functions to technically competent joint commissions, to be set up within joint committees. These are the functions of certifying performance of the activities and services supporting undertakings' systems, including *asseverazione* of the adoption and effective implementation of the models of safety organization and management referred to in Article 30, of which the supervisory bodies can take account for the purposes of planning their activities.[84]

Such fact guarantees a double advantage to entrepreneurs. On the one hand, they can profit from a checking that, if not less rigorous, is at least programmed by the supervisory bodies since it is performed periodically under the compliance auditing procedures carried out by the JC. On the other, they have the advantage that the *asseverazione* regarding adoption of a particular model of business management and organization exempts legal entities, companies and associations (including the unincorporated associations referred to in Legislative Decree no. 231 dated 8 June 2001) from administrative liability.[85]

Confirming that the interpretation of the provisions on *asseverazione* set out above is the only one consistent with the national system of accident prevention (in terms of compatibility both with domestic constitutional law and with the European Union (EU) provisions contained in Framework Directive), the European Commission opened infringement proceedings against the Italian government on 30 September 2011. The government has been asked to explain, *inter alia*, the effects of an *asseverazione* in relation to an employer's supervisory duty.

More specifically, the European Commission is challenging the Italian government regarding the incompatibility of a preventive finding of correct supervision on the part of the employer and those whom s/he delegates (that would emerge from Article 16(3) of Legislative Decree no. 81/2008 and that can be certified by way of an *asseverazione* to the effect that the company's model of health and safety management has been drawn up correctly) with the provisions of Article 5(3) of the Framework Directive, which state: 'The workers' obligations in the field of safety and health at work shall not affect the principle of the responsibility of the employer.'[86]

Except that, in the national legislative and case law context described, the so-called principle of employer's reliability has a very limited role. This by virtue of the consideration (constantly affirmed by the constitutional case law) that such principle is not operative in situations where there exists a guarantee (such as the employer's

---

and territorial committees propose initiatives directed at running prevention courses for the persons in charge of implementing both the accident-prevention legislation and the health and safety information and training interventions for construction workers, workers' safety representatives, the person responsible for the Preventive and Protective Service and for the safety co-ordinators. In addition, they provide assistance and consultations to businesses for the optimization of worksite safety standards.

84. L. Miranda, *La dimensión colectiva de la tutela del entorno de trabajo en Europa: perfiles comparativos entre Italia, Francia y Eslovenia* (Albacete: Bomarzo, 2012), 216 et seq.
85. A. Vallebona (ed.), 'Colloqui giuridici sul lavoro', no. 1 (2009), *Supplemento del Massimario di giurisprudenza del lavoro*, no. 12 (2009).
86. L. Miranda, *La dimensión colectiva de la tutela del entorno de trabajo en Europa: perfiles comparativos entre Italia, Francia y Eslovenia* (Albacete: Bomarzo, 2012), 220 et seq.

guarantee to the company's employees) which continues even in the face of a formal scrutiny of compliance with the legislative provisions conducted by the administrative supervisory bodies, given that such compliance relates to a particular, 'frozen' moment.

That is to say that the *asseverazione* of a theoretically effective model of business management and organization does not free employers a priori from their potential liability in the case of subsequent accidents. Indeed, if an accident occurs, the existence of a direct connection between the employer's act or omission, even *in vigilando*, and the event occurring must be assessed by the court.[87] This by virtue of the responsibility for safety that the legal order generally attributes to employers in principle and, more particularly, under Articles 32 and 41(2) of the Constitution and 2087 of the Civil Code. Thus entrepreneurs have to show at trial that their protective measures complied with the legal provisions, without prejudice to the court's exclusive competence to assess whether a direct connection existed between the fault and the accident at the moment when the damage occurred.

This in spite of provisions (such as the one being contested by the European Commission contained in Article 16(3) of Legislative Decree no. 81/2008) which, rather apodictically, provide for discharge of the employer's duty to supervise those to whom s/he delegates duties in those cases where the company's model of health and safety management has been adopted and effectively implemented.

That said, *asseverazione* of the adoption and effective implementation of the models of safety organization and management referred to in Article 30 of Legislative Decree no. 81/2008 does afford a partial alleviation of entrepreneurial liability in the said legislative context.

Indeed, since it constitutes a formal scrutiny of the compliance and efficiency of an undertaking's management and organizational model, it ought to facilitate the 'division of competences' between employers and the employees they have deputed to carry out specific actions to protect the workplace.

§7.07 CONSEQUENCES FOR THE PARTIES AND REMEDIES FOR BREACH OF HEALTH AND SAFETY PROVISIONS

[A] Limits on Employers' Liability

Passing now to analyse the consequences of health and safety violations and the remedies provided by the Italian legal order, it should be noted that the aim of the reform was to tackle the previous grave lack of clarity created by the co-existence of different penalty systems in the various provisions governing the subject.[88]

The legislator's intention had already been set out in Article 1(2)(f) of Act no. 123/2007. This expressly authorized the government to modify the administrative and

---

87. P. Pascucci, 'Delega delle funzioni su doppio binario. Cambia la responsabilità dei "vertici"', *Guida al Diritto – Il Sole 24 Ore, Speciale 'Sicurezza Lavoro'*, September 2009, 26 et seq.
88. P. Rausei, 'Il procedimento sanzionatorio', in *Il testo unico della salute e sicurezza nei luoghi di lavoro*, ed. M. Tiraboschi (Milan:Giuffrè, 2008), 721 et seq.

criminal penalty systems, taking particular account of the responsibilities and functions of each party under a legal duty.

The tools outlined by the legislator for achieving such an objective are essentially the following: modulation of the penalties to fit the risk; criminal law penalties of imprisonment or fines; the graduation of prohibition orders to fit the gravity of the particular provision breached and, lastly, recognition of the possibility, in certain cases, for trade union organizations and associations of victims' families to exercise the rights and powers accorded injured parties.[89]

It has already been stated how neither the principle of workers' involvement nor the possibility for employers to delegate almost all their health and safety management duties can exclude their liability in the case of a worker's accident. This, whether the accident is caused by a failure to adopt appropriate protective measures or whether it may be imputed to the entrepreneur's failure to monitor his or her employees' application of the said measures. Thus, alongside the employer's duty to organize concrete instruments of protection and prevention, there exists a further duty to monitor and carry out checks.

The main problems have arisen, however, in cases involving accidents marked by a worker's contributory negligence. In such circumstances, the case law has rigidly upheld the employer's liability, repeatedly stating that, 'the worker's possible contributory negligence cannot in any way exempt the employer'.[90] Thus the employer's duty has been further extended to include even the accidents caused by a worker's inattention, incompetence, negligence or imprudence.[91] It is therefore evident that employers have a duty to foresee and prevent their own workers' negligence.[92] It follows that the employer's liability is only excluded 'where the conduct of the worker presents features that are exceptional, abnormal or excessive when compared to the work procedure and the precise organizational directions received, such conduct being wholly unforeseeable or unimaginable'.[93]

As specified by the case law, 'abnormal conduct' is to be understood as that form of behaviour that 'by its strangeness and unpredictability, places itself outside every possible form of control on the part of the persons supervising application of the measures for preventing accidents at work'.[94] Abnormal behaviour is therefore something that contrasts with a worker's 'normal' conduct, where normality is to be

89. P. Pascucci, 'Dopo la legge n. 123 del 2007. Titolo I del d.lgs. 9 aprile 2008, n. 81 in materia di tutela della salute e della sicurezza nei luoghi di lavoro', http://csdle.lex.unict.it/Archive/WP/WP%20CSDLE%20M%20DAntona/WP%20CSDLE%20M%20DAntona-IT/20111025-121117_pascucci_n73-2008itpdf.pdf>, vol. 73, 158; P. Rausei, 'Il procedimento sanzionatorio', in *Il testo unico della salute e sicurezza nei luoghi di lavoro*, ed. M. Tiraboschi (Milan:Giuffrè, 2008), 768 et seq.

90. See, in the case law, *ex plurimis, Corte di Cassazione (Sezione Lavoro)*, Judgement nos. 7454/2002 and 16253/2004.

91. See *Corte di Cassazione (Sezione Lavoro)*, Judgements no. 7328/2004, no. 5920/2004, no. 7454/2002, no. 13690/2000 and no. 1687/1998 and *Corte di Cassazione (Penale, IV Sezione)*, Judgement no. 7735/2008.

92. S. Bellumat, 'Responsabilità dell'imprenditore e concorso di colpa del lavoratore in materia di danno da infortunio', *Rivista italiana di diritto del lavoro*, no. 1 (2005): 103 et seq.

93. So held the *Corte di Cassazione (Penale, IV sezione)* in its Judgement no. 7735/2008.

94. So held the *Corte di Cassazione (Penale, IV sezione)* in its Judgement no. 14440/2009.

considered commensurate with the given ordinary working practices. It must therefore be understood in terms of whether or not it is typical of such practices.[95]

From a criminal law perspective, furthermore, the majority of offences regarding safety at work are offences of 'gross negligence through omission'. Indeed, their essence lies in the violation of a legal requirement to take action to avoid certain consequences. The legal requirement may be identified in the law but may also be found in regulations, administrative measures and contracts. For this reason, criminal liability regarding occupational accident prevention has many broad points of contact with private law and labour law, in particular. The said offences therefore crystallize not in a pre-defined criminal action but through conduct that is characterized by a failure to observe laws, regulations or rules of conduct.

Precisely in this respect, it is of fundamental importance to emphasize how a recent judgment of the Turin Criminal Court (delivered on 16 April 2011, in a very famous reported case involving failure to take precautionary accident-prevention measures: Tyssen-Krupp) led, in a wholly innovative manner, to the conviction of the employing business's managing director for grossly negligent manslaughter. The judgment therefore establishes that employers will be liable for their grossly negligent failure to organize precautionary accident-prevention measures when, accepting the risk that the accident may occur, they do not conscientiously introduce all the safety measures needed to prevent such an event.

[B] Forms of Liability for Breach of Health and Safety Provisions

As far as its penalty system is concerned, Italian health and safety law is characterized by a double plurality: the plurality of parties that may be held liable and the plurality of categories of liability. As regards the latter aspect, breach of the legal provisions gives rise not only to a criminal liability but also an administrative one, in addition, obviously, to the civil law liability to pay compensation.

The first two forms of liability extend to all the parties involved in the system of prevention and protection. Indeed, employers, managers, supervisors, manufacturers, project designers, fitters, the occupational health physician and workers are all answerable, both in criminal and in administrative law. As regards civil law liability, however, it is fairly clear that this attaches only to the employer (commissioning entity included).

Another aspect concerns the contractual or non-contractual nature of the civil liability. The choice has important consequences since in Italy the law governing the two forms of liability is different. On the one hand, civil liability, in this domain, must be traced back to a contractual one, since the duty created by Article 2087 of the Civil Code is based on the work relationship which is, precisely, contractual in nature, at least according to the majority of legal scholars.[96] Workers may therefore quite

95. See *Corte di Cassazione (Sezione Lavoro)*, Judgement no. 5920/2004.
96. *Contra* R. Scognamiglio, 'La natura non contrattuale del rapporto di lavoro', *Rivista italiana di diritto del lavoro*, no. 1 (2007): 379 et seq.

legitimately bring typical contract law actions, including actions for specific performance, pursuant to Article 1453(1) of the Civil Code, or for partial exemption from performance, pursuant to Article 1460 of the Civil Code, which permits workers to refuse to work in a situation of imminent danger to their health.

The issue merits a more detailed analysis, however, if one considers that, under Article 32 of the Constitution, health is, above all, a fundamental personal right that transcends an individual's qualification as a worker, since it is recognized by everyone, whether or not a work relationship exists. Freed from contractual ties, the right to health would find its protection in Article 2043 of the Civil Code, under which the violation of the right to health is classified as a tort. Faced with this possibility, the case law recognizes that a concurrence of the two types of liability may arise, should the employer's conduct simultaneously injure the right to health and prejudice the worker's psycho-physical integrity.[97]

The choice between one or the other kind of action would be left to the worker. Some legal scholars are, however, opposed to admitting the possibility of a concurrence of the two forms of liability, although they acknowledge that the safety duty pertains to rights that transcend the work relationship.[98] Indeed, that this aspect of the safety duty is compatible with the regime of contractual liability appears plausible.

Another distinctive feature of the Italian system for managing occupational health and safety concerns the penalties system. As we have seen, by making the majority of the breaches a criminal offence and a considerable number of them administrative law violations, Legislative Decree no. 81/2008 outlines a particularly stringent system of penalties for employers, managers, supervisors, project designers, manufacturers, suppliers and fitters, the occupational health physician and both employed and self-employed workers.

It therefore follows that, in the case of specific breaches of the preventive provisions, the authors may be liable to criminal law penalties, on the one hand, and administrative law ones on the other. This is naturally in addition to their civil law compensatory liability for 'differential damages' i.e., damages exceeding the legally defined maximum covered by INAIL, should the said breaches have caused occupational accidents or disease. This situation is capable of generating the phenomenon known as a 'material accumulation of penalties', whereby a separate penalty is applied for every breach committed *(tot crimina tot poenae)*.

It is the grounds for protecting the constitutionally guaranteed right to health that justify such a penalties system and legitimize consideration of the administrative and criminal 'wrongs' breaching health and safety law as distinct and independent of each other in the abstract, even though they occur simultaneously.

The general principle informing the adoption of sanctions for health and safety offences is therefore to include a separate head for each breach or omission or, where appropriate, in relation to every victim of the same.

97. The hypothesis is one of 'multiple-offence conduct', as outlined (along with others) in *Corte di Cassazione (Sezione Lavoro)*, Judgement no. 2569/2001.
98. L. Montuschi, 'L'incerto cammino della sicurezza del lavoro fra esigenze di tutela, onerosità e disordine normativo', *Rivista giuridica del lavoro*, no. 2 (2001): 508 et seq.

Only in residual cases does the opposite principle apply, according to which the penalty may be applicable only once, on the assumption that a single wrong has been committed. This provided that the employer can demonstrate the unitary nature of his or her behaviour, even if it has had the consequence of failing to observe more than one rule.

[C] Types of Penalty and Consequences in Case of Breach of Health and Safety Regulations

As far as the criminal law context is concerned, Legislative Decree no. 81/2008 provides for imprisonment and fines whereas, from a civil law point of view, it establishes the duty to pay damages.

As regards the first aspect, the cases exclusively punishable with imprisonment are very limited when compared to the original intention of the Parent Act (Act no. 123/2007) to introduce highly repressive legislation. Breaches of particular gravity are punishable with up to three years' imprisonment.[99] Far more frequently, however, the two criminal law penalties of imprisonment and fines are combined.

From the civil law point of view, only physical damage (i.e., the damage referred to under Article 2043 of the Civil Code) was considered compensable in the past. The fact that new categories of damage (such as psychic injury) have subsequently emerged has caused the legal scholars to ask themselves whether the traditional distinction between pecuniary damage and non-pecuniary damage may be left behind in order to achieve a general and all-inclusive protection of the person.

The courts have intervened in this area on several occasions. An important judgment of the Constitutional Court[100] confirmed the bipartition of the two categories but interpreted Article 2059 of the Civil Code more broadly, including pain and suffering within the category of non-pecuniary damage.

Having clarified this point, it must be asked whether injury of a constitutionally recognized right, such as the right to health, may give rise to compensation not just for physical injury but also for psychic injury. The answer is certainly in the affirmative. So much so that reference has been made to a 'new season of non-pecuniary damage'.[101]

In the light of these considerations, breach of the provisions protecting workers' health and safety will therefore give rise to actions for both pecuniary and non-pecuniary damages. The judge will quantify the compensation equitably, taking the injured party's level of remuneration as a point of reference.

---

99. The only three categories punished exclusively with imprisonment are: failure to assess or incomplete assessment of the risks in businesses exposed to particular types of risk (Art. 55(2)); imprudent assignment of tasks within certain businesses (Art. 55(4)(c)) and lastly, failure to observe a suspension order (Art. 14(10)). It is clear that the first two are really special aggravating factors whereas the last category is an infringement of a general nature. P. Rausei, 'L'arresto', in *Il testo unico della salute e sicurezza nei luoghi di lavoro*, ed. M. Tiraboschi (Milan:Giuffrè, 2008), 737 et seq.
100. See *Corte Costituzionale*, Judgement no. 233/2003.
101. M. C. Cimaglia, 'Il nuovo danno non patrimoniale fa il suo ingresso nel diritto del lavoro (Commentary on the Judgement issued by the Court of Tempio Pausiana (sez. lav.) 10.07.2003)', *Rivista italiana di diritto del lavoro*, no. 1 (2004): 311 et seq.

Alongside the right to damages, workers keep their right to have their job held open for them for as long as is necessary to make a full recovery from an occupational accident.

To such end, in fact, the general rule contained in Article 2110 of the Civil Code provides that, in cases of occupational accident or disease (in addition to pregnancy or breast feeding), the duty to perform is to be suspended and the employer cannot dismiss the worker before expiry of the period of job preservation (the so-called grace period) specifically identified in collective agreements.

This general rule is, nevertheless, subject to a strict qualification in situations generated by a breach on the part of the employer. Indeed, where it is ascertained that the occupational accident or disease is to be imputed to a breach of the employer's, the days required to return to a state fit for work are not included in the calculation of the grace period.

This insofar as the impossibility of working, in such a case, may be imputed to the conduct of the same party that is destined to benefit from the said work, namely, the employer.

Analogously, the employer is under a duty, where possible, to reinstate the worker in the same position that s/he previously occupied, once s/he has recovered. Where the worker's physical integrity is incompatible with the duties previously carried out, s/he must be guaranteed reinstatement in another position that enjoys equal professional status and remuneration.

Whilst confirming the pre-eminence classically accorded to criminal law wrongs over administrative law ones, Legislative Decree no. 81/2008 equally provides for a considerable body of administrative law fines, as well as suspension of the undertaking's activity.[102]

Having analysed the general types of penalty identified by the legislation, it is appropriate to trace a general outline of the breaches typified by Legislative Decree no. 81/2008. Following the scheme applied by Legislative Decree no. 626/1994, these may be sub-divided into two macro areas.

Indeed, on the one hand, there are the forms of conduct and omissions of a general nature constituting breaches of the safety measures required by Title I of Legislative Decree no. 81/2008. In this context, the breaches that most stand out are those deriving from: failure to draw up or adopt the risk-assessment document; failure to draw up the Emergency and Evacuation Plan; failure to provide general and practical training; failure to set up the Preventive and Protective Service and appoint the person responsible for it, and lastly, failure to appoint the occupational health physician.

On the other, there are penalties that relate to the specific risks present in the workplace as it is organized by the employer. These are set out in the other Titles into which Legislative Decree no. 81/2008 is sub-divided. In their detail, they integrate the general provisions, taking care of the breaches involving: workplaces; use of working apparatus and personal protective equipment; temporary or mobile worksites; occupational health and safety signs and notices; the manual handling of heavy loads;

102. P. Rausei, 'Le contravvenzioni e le sanzioni amministrative', in *Il testo unico della salute e sicurezza nei luoghi di lavoro*, ed. M. Tiraboschi (Milan: Giuffrè, 2008), 729 et seq.

apparatus equipped with visual display units; physical agents; dangerous substances (chemical agents, carcinogenic and mutagenic agents; exposure to asbestos); exposure to biological agents and, lastly, protection from explosive atmospheres.

Precisely the sub-division of breaches into the categories typified in Title I of Legislative Decree no. 81/2008 and those provided for by the specific provisions in the law's other titles has required the legislator to adopt the so-called principle of a breach's specificity in Article 298. According to this principle, when one and the same fact is punishable under a provision contained in Title I and by one or more provisions contained in the other titles of Legislative Decree no. 81/2008, the judge is bound to apply the specific provision.

Lastly, regarding specific violations, the formulation of the administrative law penalty of suspension of an undertaking's activity is particularly important. This is contained in Article 14 of Legislative Decree no. 81/2008, which has been heavily amended by Legislative Decree no. 106/2009.[103] The provision in question permits the supervisory bodies at the Ministry of Labour and Social Security to issue orders suspending entrepreneurial activity in the occurrence of certain forms of conduct on the part of the employer.

In its original formulation, Article 14 identified three categories of conduct punishable with a suspension order. In the first place, the employment of staff not appearing on the books or in other mandatory documentation and representing 20% or more of the workers present in the workplace. In the second place, serious and repeated breaches of provisions protecting workers' health and, lastly, repeated breaches of the law governing overtime and both daily and weekly rest periods.

Precisely in relation to this last category, the original formulation of Article 14 referred to:

> repeated breaches of the law governing overtime and both daily and weekly rest periods referred to in Articles 4, 7 and 9 of Legislative Decree no. 66, dated 8 April, as subsequently modified, considering the gravity of the exposure to the risk of an accident.[104]

In this way, the negative impact on workers' health and safety caused by breaches of the working time regulations was emphasized and consequently punished, given the legislator's express reference to the increased risk of an accident. Moreover, the suspension order was added to the administrative law penalties already established for breach of the working time provisions by Legislative Decree no. 66/2003.

Article 14's current formulation (introduced when Legislative Decree no. 106/2009 came into force) maintains the first two (possible) penalties but eliminates the reference to repeated breaches of the working time provisions. Thus the supervisory bodies' power to issue the related suspension order is excluded. As a consequence, what could have been interpreted as a sign of particular attention towards working time, in terms of its impact on workers' health and safety, no longer exists today. This

103. P. Pascucci, 'Delega delle funzioni su doppio binario. Cambia la responsabilità dei "vertici"', *Guida al Diritto – Il Sole 24 Ore, Speciale 'Sicurezza Lavoro'*, September 2009, 24 et seq.

104. Article 4 concerns the maximum working time. Art. 7 covers daily rest and Art. 9 contains the law governing weekly rest.

has had a negative impact on that 'safety culture' that would appear to be the mainstay of Legislative Decree no. 81/2008.

[D] Adoption of a Proper System of Occupational Health and Safety Management

One of the most important innovations in occupational health and safety is to be found in Article 30 of Legislative Decree no. 81/2008.

This provision establishes the requisites for the organization and management models capable of averting liability for the administrative officers of legal entities, companies and associations, including unincorporated associations as defined by Legislative Decree no. 231/2001.[105] It ascribes to the organization model adopted by undertakings the fundamental functions of collaborating in risk-assessment, preventing risks, programming the training and monitoring activities and, lastly, managing emergencies.[106]

With the intention of permitting undertakings to manage health and safety, production quality and their environmental impact management (already regulated by Legislative Decree no. 231/2001) in an integrated manner, Article 30 of Legislative Decree no. 81/2008 identifies the Occupational Health and Safety Management System (hereinafter also referred to as WHSMS) as an organizational model that is geared to achieving the business's health and safety objectives but planned in such a way as to observe the most appropriate costs-benefits ratio.

Article 30 further provides that, during initial compliance with Legislative Decree no. 81/2008, organizational models conforming to the UNI-INAIL (*Principi Uniformi Istituto Nazionale Infortuni sul Lavoro*) Guidelines dated 28 September 2001 and British Standard OHSAS 18001:2007 shall be deemed to comply with the legal requirements and therefore create a legal presumption of exoneration from criminal liability for administrative officers. The presumption is clearly rebuttable and operates exclusively for the purposes of recognizing the exonerating 'suitability' requisite. A doubt remains, however. One based on the fact that a generalization is thus made, a priori, without checking whether the models in question actually function effectively.[107]

A particularly important grey area of Article 30 is whether application of the related organizational models is only pertinent to the administrative and criminal

105. A. Lanzi & P. Aldrovandi. 'L'applicazione del D.Lgs. n. 231/2001 sulla responsabilità amministrativa degli Enti per fatto di reato, ai delitti di omicidio colposo e lesioni personali colpose commessi con violazione delle norme antinfortunistiche. Commento all'art. 9', in *Legge 3 agosto 2007, n. 123. Commentario alla sicurezza del lavoro. Misure in tema di tutela della salute e della sicurezza sul lavoro e delega al governo per il riassetto e la riforma della normativa in materia*, ed. F. Bacchini (Milan: Ipsoa-Indicitalia, 2008), 225 et seq.

106. L. Golzio & A. Montefusco, 'Modelli di organizzazione e gestione', in *Il testo unico della salute e sicurezza nei luoghi di lavoro,* ed. M. Tiraboschi (Milan: Giuffrè, 2008), 391 et seq.; L. Miranda, 'Organizzazione del lavoro e sicurezza negli accordi del Gruppo Fiat-Chrysler. Verso un comune modello di rappresentanza paritetica per la sicurezza?', *Diritti lavori mercati,* no. 1 (2012): 101 et seq.

107. P. Tullini, 'I sistemi di gestione della prevenzione e della sicurezza sul lavoro', *Giornale di diritto del lavoro e relazioni industriali,* no. 3 (2010): 407 et seq.

liability of the employer as a legal entity or whether the Article has a wider-ranging impact. Indeed, some legal scholars argue that adoption of the Article 30 models may also exclude the employer's criminal liability in those cases where s/he delegates his or her functions. In fact, by virtue of the corrective Decree no. 106/2009, the employer's monitoring duty in cases of delegation 'shall be deemed to have been performed where the checking and inspection model referred to in Article 30(4) has been adopted and effectively implemented'. It would thus be possible to speak of an exempting effect in favour of the employer not only as a juristic person but also as a natural person.[108] This interpretation cannot be supported. Indeed, reading the provision's wording in the light of the case law it has produced, it can be observed that the criminal liability for workplace safety offences is characterized by a double negligence. Generic, in the case of breach of the detailed preventive rules, and specific, for breach of Articles 589 and 590 of the Criminal Code, insofar as the former are offences involving danger and mere conduct, whilst the latter are offences causing harm and having consequences.[109]

Analogously, civil law liability has been characterized by the need for the event to be referable to the employer's fault: for breach or failure involving conduct-related duties imposed by law or required by technological developments, provided that they are concretely identified.[110]

In this legislative and case law context, the so-called principle of 'employer's reliability' has always played a very limited part.[111] This by virtue of the consideration (constantly asserted in the case law[112]) that the said principle is not operative in situations where a protected position (such as that of employees vis à vis their employer) exists.

Such fact also resolves the problem of whether adoption of the organizational models in question may also influence evaluations of an employer's civil liability. Indeed, it should be noted that management systems do not only affect business organization but also imply a standardization of the employer's duty under Article 2087 of the Civil Code. They therefore become instruments for measuring the adequacy of contractual performance.[113] If that is true, it is equally appropriate to separate the relationship between the undertaking and its organizational model, on the one hand, from that between employer and worker, on the other. That is to say, discharge of the safety duty under Article 2087 of the Civil Code in the ways prescribed by Article 30 of Legislative Decree no. 81/2008 will only be significant under the first profile and therefore satisfy the legality test with regard to the public monitoring bodies. It does not affect the safety duty the employer owes to his or her workers, who are the only parties

108. A. De Vita & M. Esposito, *La sicurezza sui luoghi di lavoro. Profili della responsabilità datoriale*, (Naples: ESI, 2009), 73 et seq.
109. See, *ex plurimis*, *Corte di Cassazione (Penale, IV Sezione)*, Judgement no. 35773/2001.
110. See, *ex plurimis*, *Corte di Cassazione (Sezione Lavoro)*, Judgement no. 12863/2004.
111. See, *Corte di Cassazione (Penale IV Sezione)*, Judgement no. 18290/2003.
112. See, *ex plurimis*, *Corte di Cassazione (Penale, IV Sezione)*, Judgement no. 7032/1999.
113. P. Tullini, 'I sistemi di gestione della prevenzione e della sicurezza sul lavoro', *Giornale di diritto del lavoro e relazioni industriali*, no. 3 (2010): 407 et seq.; L. Miranda, 'Organizzazione del lavoro e sicurezza negli accordi del Gruppo Fiat-Chrysler. Verso un comune modello di rappresentanza paritetica per la sicurezza?', *Diritti lavori mercati*, no. 1 (2012): 101 et seq.

having the right to health.[114] The adoption of particular models of business organization therefore cannot erode the responsibility principle underpinning the regulation of health and safety at work.

[E] The Interaction between Health and Safety Regulation and the Social Security System

Alongside what is described above regarding the criteria for employers' liabilities, in the Italian legal system, under the profile of compensation for damages suffered by the worker, there is a close link between the health and safety preventive legislation and the social security system.[115]

Health and safety of the workers, in fact, is guaranteed not only through the application of the accident-prevention rules, but also through the economic protection of workers who have suffered damages to their health; such a function is attributed to INAIL, through the compulsory insurance against occupational accidents and diseases.[116]

The compulsory insurance, established in 1898, provides the economic protection of the workers injured during a job activity or affected by an occupational disease. It is currently regulated by the so-called *Testo Unico*, approved by the Presidential Decree of the 30 June 1965, no. 1124, as lastly amended by Act no. 493/1999 and by Legislative Decree no. 38 of 23 February 2000. The first amendment has extended the insurance protection to familiar and unpaid activities, the second and much more important one, has extended the social insurance coverage to damages not previously protected, such as psycho-physical damages.

The system is built on the compulsory insurance stipulated by the employer with INAIL and it is financed by employers' contributions, periodically re-evaluated and commensurate with the risk of the job performances and the number of workers employed.

INAIL operates according to a typical insurance-based logic by virtue of the correspondence between contributions paid and risks linked to work, the compensatory nature of the economic treatment guaranteed to the workers and the exclusion of the liability of the employer.

The insurance, therefore, protects workers in case of accident or occupational disease, generated in reason of the work which has caused a psycho-physical impairment to the worker, through the recognition of both the free health-care and the economic benefits, predetermined in an amount not directly equivalent to the damages suffered by the worker (so-called indemnity) and variable in duration according to the permanent or temporary character of the inability.

---

114. P. Albi, *Adempimento dell'obbligo di sicurezza e tutela della persona. Art. 2087 c.c.*, Vol. of *Il Codice Civile. Commentario. Fondato da P. Schlesinger*, ed. F.D. Busnelli (Milan: Giuffrè, 2008), 45 et seq.
115. See, Gaeta L., *Infortuni sul lavoro e responsabilità civile. Alle origini del diritto del lavoro* (Naples: ESI, 1986).
116. See Cinelli M., *Diritto della previdenza sociale* (Turin: Giappichelli, 2012), 429 et seq.

The insurance protection, on the one hand, operates regardless of employer's fault and even if the accident has been caused directly by the employee, if originated from the job occasion; on the other hand, it does not cover injuries and illnesses leading to permanent or temporary inability less than 6% of the full work capacity with respect to psycho-physical injury and less than 16% with reference to the reduction of general aptitude to work.

This because, at the origin of the accident insurance system, there is the concept of occupational risk that is founded on the overcoming of the civil law liability principle, by introducing an alternative criterion, based, within predetermined limits, on the imputation to the employer of the objective liability related to the exercise of economic activities, still risky for workers.[117]

The adoption of such criterion has implied that the employer, within the limits of the insurance obligation, is held responsible not only for the damages suffered by the worker that may be charged to him or her as a fault-based liability, but also of those that are a consequence of fortuitous events, force majeure or fault of the worker, with the only exception for worker's purely arbitrary acts.

The mandatory insurance achieves, therefore, the so-called transactive[118] effect, since employers' sphere of responsibility is extended, but they respond within the limits of the insurance, while workers, in exchange of greater protection, receive only a partial compensation for the damage suffered, because the social security protection neither intervene for minor injuries nor guarantees the economic equivalence between the full extent of the damage and the economic indemnity paid to the worker.

Alongside the extension of liabilities to events not directly attributable to the employer, in favour of the latter operates the rule of exemption from civil liability,[119] disciplined by Article 10 of Presidential Decree no. 1124/1965, according to which the employer is not liable for damages suffered by the employee if covered by the INAIL allowance, but only for the so-called differential damages, i.e., damages exceeding the legally defined maximum amount covered by INAIL, against which must be ascertained the responsibility of the employer, according to the civil law rules, as mentioned in the above.

Pursuant to Article 10, paragraph 2, of Presidential Decree no. 1124/1965, the exemption rule does not apply if the fact constitutes a criminal offence, intentional or negligent, prosecutable at the initiative of the public jurisdictional authorities. In this case, INAIL has the right to act against the employer, who will be called to refund the amount paid by INAIL as insurance benefit to the employee.[120]

Formulated in these terms, the exemption rule of employers' liability has created, in practice, no few problems, mainly due to the - once exclusive - compensability of the material damages suffered by a worker because of an occupational accident, with the exclusion of damage not susceptible of an economic evaluation, pursuant to the

117. Così, da ultimo, Cass. Civ., Sez. Lav., no. 6002/2012.
118. Cinelli M., *Diritto della previdenza sociale* (Turin: Giappichelli, 2012), 429 et seq.
119. See, Corte Cost. 18 luglio 1991, n. 356, edited in *Foro It.*, 1991, 2967.
120. See, *ex multis*, Cass. Civ., Sez. Lav., no. 856/2012.

provisions of Article 2059 of the Civil Code, which limits the compensation of non-pecuniary damages to the sole cases provided by law.

As already stressed in the above, over the years, such a system has been substantially dismantled by the jurisprudence of the Constitutional Court, in order to protect the right to health of workers, guaranteed by the Article 32 of the Italian Constitution, which includes the recognition of non-pecuniary damages if perceived as socially relevant.

By Judgment no. 184/1986 the concept of 'biological damage' has been recognized as a damage to health, consequent to a psycho-physical damage capable of producing both economical and not-economical consequences, that find its protection in Article 2043 of the Civil Code, according to which the violation of the right to health can be classified as a tort.[121]

Nevertheless, the compensation of biological damages deriving from accidents at work, has been recognized only later on, following three judgments of the Constitutional Court in 1991 that, by overcoming the limitation of the insurance indemnity to pecuniary damages suffered by the worker, laid down the basis for the intervention of the Italian legislator.

In its Judgment no. 87 of 15 February 1991, the Constitutional Court, while not declaring unconstitutional the exclusion of INAIL coverage for non-pecuniary damages suffered by the worker, underlined the need to distinguish between the damage to the working ability and the damage to health that, as a constitutional right, shall find a protection similar to that one provided by the Presidential Decree no. 1124/1965.

However, it is only with the decisions no. 356 of 18 July 1991 and no. 485 of 27 December 1991 of the Constitutional Court that an instrument of jurisdictional enforcement in favour of workers has been found. The decisions stated, in fact, on the one hand, that the exemption of employers liability according to Article 10 of Presidential Decree no. 1124/1965 must be interpreted in the sense that it operates only with reference to the damages linked to a reduction of working ability, thus not extending up to deny the compensation of health damages suffered by the worker, according to the general rules of the Civil Code and, specifically, to Articles 2043 and 2087;[122] on the other hand, that INAIL claims against the employer cannot be extended to those damages, not covered by the accident insurance.

To such a demand of social protection, the legislator responded through Legislative Decree no. 38 of 23 February 2000 that, in Article 13, extends accident insurance compensation to biological damages, defined as the psycho-physical impairment of the employee detected by medical-legal evaluation, in addition to the pecuniary damage resulting from the same event.

The general provision contained in Article 13, Legislative Decree no. 38 of 23 February 2000, has allowed, however, over the years, the extension of the insurance

121. See, Pedrazzoli M. (ed), *Danno biologico e oltre* (Turin: Giappichelli: 1995); Busnelli F.D., *Il danno biologico. Dal 'diritto vivente' al 'diritto vigente'* (Turin: Giappichelli, 2001); V. Luciani, 'La responsabilità civile e il danno biologico', in *Ambiente e sicurezza del lavoro*, eds. M. Rusciano & G. Natullo (Turin: Utet, 2007), 485 – 502.

122. See, in general, R. Del Punta, 'Diritti della persona e contratto di lavoro', *Giornale di diritto del lavoro e di relazioni industriali*, no. 1 (2006): 195-268.

protection also against psychic damages, understood as impairment which affect the attitude of workers to participate in normal social life, such as stress, anxiety, depression, frustration, phobias etc., actually covered by the accident insurance.

It has derived the full protection of situations of damage not only attached to the merely risky operations of the productive cycle, but also linked to the organization of employer's activities.

Therefore, in the current regulatory framework of the accident insurance, mental disorders and impairments can be considered of occupational origin if caused by specific and particular conditions of the work activity and by the organization of work - situations that occur in the presence of incongruity in employers' organizational choices (the so-called *costrittività organizzativa*).

A last profile of strict interaction between the health and safety management and the Social Security System, is that of the efficacy of an entrepreneur's protective actions and the economic incidence of the corresponding duties regarding INAIL contributions. This is by way of the tried and tested *Bonus-Malus* scheme under which a low level of accidents corresponds to a lower level of contributions and vice versa.

To such end, the Ministerial Decree of 12 December 2000 (governing the new premium rates for insurance against work accidents and occupational diseases) provides, in Articles 19 and 22 respectively, for an oscillation of the average rate during the first two years of an undertaking's activity and an oscillation of the average rate on the basis of the accident track record after the first two years of activity.

In the first case, a fixed reduction or increase by 15% in the average contribution rate may be applied to an undertaking during the first two years of its activity, on the basis of the company's observance of the rules on accident prevention and occupational health and safety.

In the second case, at the end of the two years running from the start of the undertaking's activity, INAIL may, in connection both with the interventions carried out to improve workplace safety and hygiene (in implementation of the provisions of Legislative Decree no. 81/2008) and with the actual and certified realization of a valid system for managing occupational health and safety, apply a fixed reduction of 5% or 10% in the average contribution rate (according to the number of workers for the year in question). This is for those employers who have complied with the accident prevention and workplace hygiene requirements and their social security and insurance duties.

Should it become evident, at any stage, that the conditions for entitlement to the abovementioned reduction have not been satisfied, however, INAIL will annul it and request integration of the premiums due, as well as application of the civil and administrative penalties in force.

§7.08 CONCLUSIONS

As we have seen in the above, the Italian health and safety system was characterized for more than a decade by a highly fragmented regulation that risked seriously compromising both certainty as to the law and full regulatory adaptation to important new categories of risk.

Such problems, combined with persistently worrying statistics both for occupational accidents and diseases, pushed the government to introduce by Legislative Decree no. 81/2008 an innovative regulatory framework with a view of rationalizing health and safety law both at general and at sectoral level.

If an impact assessment of Legislative Decree no. 81/2008 is still impossible, in absence of relevant case law, it is already possible to underline some innovative features that characterize the new health and safety management in practice, showing the intention of the legislator to protect not only health and safety, but even well-being at work.

These features can be divided into five main profiles, each one closely related to the other, such as: (a) the application of health and safety provisions to all sectors of production and all types of risk; (b) the innovative all-inclusive concept of 'worker'; (c) the new risk-assessment provisions; (d) the new requirements for the organization and management health and safety models; (e) rules on workers' representation.

As stated above, one of the most important features of the new national preventive legislation is its application to all sectors of activity, whether private or public, and to all types of risk, without distinctions based on the nature (i.e., entrepreneurial or non-entrepreneurial) or the type of activity pursued or, indeed, the dimensions of a business.

Under this profile, the novelties tend to guarantee a global approach to the problem of protecting safety at work, closely linked to the innovative all-inclusive concept of 'worker', that not only covers employees (regardless of the various types of contract) but also workers that, albeit not employed by the employer, are nevertheless subject to his or her managerial authority (agency workers), or self-employed, in relation to certain specific duties.

Both aspects are strictly related to the emphasis the law put on the employer's primary duty to assess risks in the workplace. In such a perspective, in fact, the new risk-assessment cannot be considered, as before, a static document that analyses only the technical risks deriving from the productive process, but it must cover all the risks to workers' health and safety potentially present in the workplace, even those linked to the organization of work and its impact on health and safety, those regarding pregnant workers and those linked to gender differences, age and provenance from other countries, regardless of the type of work contract.

In view of ensuring the maximum diffusion of the preventive measures and the certainty of responsibility in the undertakings, a very important role is played by Article 30 of the Legislative Decree no. 81/2008. It identifies the requisites for the organization and management of health and safety models, ascribing to the organization model adopted by undertakings the fundamental functions of collaborating in risk-assessment, preventing risks, programming the training and monitoring activities and, lastly, managing emergencies, attempting to ensure proper and a priori identification of positions of responsibility, the division of powers within the company, as well as the full functionality of the health and safety management in the working environment.

Similarly, the new rules on collective representation seem to be characterized by the intention to assign a more important role to the workers' specialist representatives (WSR) in the entrepreneur's organizational decisions that could affect the health and safety of all the workers present at workplace.

This happens, not only with the identification of new WSR's powers, but also, above all, through the recognition of the all-inclusive concept of 'worker' even for the purposes of applying the provisions on the participation and consultation of workers' representatives, that has the effect to facilitate, at least at the formal level and in the absence of any other legislative specification, the representation of workers not typically covered by trade union membership.

Therefore, in conclusion, the new system seems to be aimed at guaranteeing a global approach to health and safety at work, by establishing an action programme that is truly centred on a culture of accident prevention and on a constant and planned protection at work.

BIBLIOGRAPHY

Adinolfi, F. & R. Bortone. 'Commento alla direttiva europea sulla tutela della maternità'. *Giornale di diritto del lavoro e relazioni industriali*, no. 1 (1994): 372–395.

Ales, E. 'Informazione e consultazione nell'impresa: diritto dei lavoratori o dovere del datore di lavoro? Un'analisi comparata'. *Rivista italiana di diritto del lavoro*, no. 1 (2009): 239–260.

Ales, E. 'L'articolo 9 statuto dei lavoratori alla luce della legislazione più recente in materia di salute e sicurezza: Partecipazione o controllo?'. *Rivista italiana di diritto del lavoro*, no. 1 (2011): 53–67.

Albi, P. *Adempimento dell'obbligo di sicurezza e tutela della persona. Art. 2087 c.c.*, Vol. of *Il Codice Civile. Commentario. Fondato da P. Schlesinger*, edited by F.D. Busnelli. Milan: Giuffré, 2008.

Aparicio Tovar, J. 'Sicurezza sul lavoro'. In *Dizionario di diritto del lavoro comunitario*, edited by A. Baylos Grau, B. Caruso, M. D'Antona & S. Sciarra. Bologna: Monduzzi, 1996.

Bacchini, F. 'Committenti e appaltatori'. In *Il testo unico della salute e sicurezza nei luoghi di lavoro*, edited by M. Tiraboschi. Milan: Giuffrè, 2008.

Balandi, G. 'Individuale e collettivo nella tutela della salute nei luoghi di lavoro: l'art. 9 dello Statuto'. *Lavoro e Diritto*, no. 1 (1990): 219–230.

Baldassarre, A. 'Le rappresentanze dei lavoratori per la sicurezza e il rilancio della filosofia partecipativa'. In *Il testo unico della salute e sicurezza nei luoghi di lavoro*, edited by M. Tiraboschi. Milan: Giuffrè, 2008.

Biagi, M. (ed.). *Tutela dell'ambiente di lavoro e direttive CEE*. Rimini: Maggioli, 1991.

Bellumat, S. 'Responsabilità dell'imprenditore e concorso di colpa del lavoratore in materia di danno da infortunio'. *Rivista italiana di diritto del lavoro*, no. 1 (2005): 103–115.

Borgogelli, F. *Il lavoro femminile tra legge e contrattazione*. Milan: Franco Angeli, 1987.

Bortone, R. 'Il servizio di prevenzione'. In *Ambiente, salute e sicurezza,* edited by L. Montuschi. Turin: Giappichelli, 1998.

Calafà, L. 'Le lavoratrici madri'. In *Ambiente e sicurezza del lavoro,* edited by M. Rusciano & G. Natullo. Turin: Utet, 2007.

Campanella, P. 'I rappresentanti dei lavoratori per la sicurezza'. In *Le nuove regole per la salute e la sicurezza dei lavoratori. Commentario al D.Lgs. 9 aprile 2008, n. 81 aggiornato al D.Lgs. 3 agosto 2009, n. 106,* edited by L. Zoppoli, P. Pascucci & G. Natullo. Milan, Ipsoa, 2010.

Carollo, L. 'Informazione e formazione dei lavoratori'. In *Il testo unico della salute e sicurezza nei luoghi di lavoro,* edited by M. Tiraboschi. Milan: Giuffrè, 2008.

Caruso, B. 'L'Europa, il diritto alla salute e l'ambiente di lavoro'. In *Ambiente, salute e sicurezza,* edited by L. Montuschi. Turin: Giappichelli, 1998.

Cimaglia, M. C. 'Il nuovo danno non patrimoniale fa il suo ingresso nel diritto del lavoro (Commentary on the Judgement issued by the Court of Tempio Pausiana (sez. lav.) 10.07.2003)'. *Rivista italiana di diritto del lavoro,* no. 1 (2004): 311–316.

Cinelli, M. 'Il progetto di "testo unico" della sicurezza sul lavoro: alcune osservazioni "a margine"'. In *Il testo unico sulla sicurezza del lavoro,* edited by P. Pascucci, Conference Proceedings Urbino, 4 May 2007. Rome: Ispesl, 2007.

Corrias, M. *Sicurezza e obblighi del lavoratore.* Turin: Giappichelli, 2008.

De Vita, A. & M. Esposito. *La sicurezza sui luoghi di lavoro. Profili della responsabilità datoriale.* Naples: ESI, 2009.

De Vita, P. 'Formazione dei lavoratori e dei responsabili/addetti'. In *Il testo unico della salute e sicurezza nei luoghi di lavoro,* edited by M. Tiraboschi. Milan: Giuffrè, 2008.

Di Monte, O. 'L'informazione e la formazione dei lavoratori'. In *La sicurezza sul lavoro,* edited by M. Ricci. Bari: Cacucci, 1999.

Emiliani, S. P. 'Il dovere di sicurezza del lavoratore verso se stesso alla luce della normativa comunitaria'. *Argomenti di diritto del lavoro,* no. 1 (2009): 104–124.

Ferraro, G. & M. Lamberti. 'La sicurezza sul lavoro nel decreto legislativo attuativo delle direttive CEE'. *Rivista giuridica del lavoro,* no. 1 (1995): 50–75.

Galantino, L. (ed.). *La sicurezza del lavoro.* Milan: Giuffrè, 1996.

Galantino, L. 'Il contenuto dell'obbligo di sicurezza'. In *La sicurezza del lavoro,* edited by L. Galantino. Milan: Giuffrè, 1996.

Gentile, M. R. 'I dirigenti e i preposti'. In *Il testo unico della salute e sicurezza nei luoghi di lavoro,* edited by M. Tiraboschi. Milan: Giuffrè, 2008.

Ghezzi, G. 'Art. 9 (Tutela della salute e dell'integrità fisica)'. In *Statuto dei diritti dei lavoratori. Art. 1-13,* 2nd edn, edited by U. Romagnoli et al. Bologna: Zanichelli, 1979.

Golzio, L. & A. Montefusco. 'Modelli di organizzazione e gestione'. In *Il testo unico della salute e sicurezza nei luoghi di lavoro,* edited by M. Tiraboschi. Milan: Giuffrè, 2008.

Lai, M. 'Tutela della maternità nella legislazione europea'. *Diritto e pratica del lavoro,* no. 2 (1993): 77–85.

Lai, M. *La sicurezza del lavoro tra legge e contrattazione collettiva*. Turin: Giappichelli, 2002.

Lai, M. 'Il ruolo delle parti sociali'. In *Il testo unico della salute e sicurezza nei luoghi di lavoro,* edited by M. Tiraboschi. Milan:Giuffrè, 2008.

Lanzi, A. & P. Aldrovandi. 'L'applicazione del D.Lgs. n. 231/2001 sulla responsabilità amministrativa degli Enti per fatto di reato, ai delitti di omicidio colposo e lesioni personali colpose commessi con violazione delle norme antinfortunistiche. Commento all'art. 9'. In *Legge 3 agosto 2007, n. 123. Commentario alla sicurezza del lavoro. Misure in tema di tutela della salute e della sicurezza sul lavoro e delega al governo per il riassetto e la riforma della normativa in materia,* edited by F. Bacchini. Milan: Ipsoa-Indicitalia, 2008.

Lassandari, A. *Le discriminazioni nel lavoro. Nozioni, interessi, tutele,* Vol. LVI of *Trattato di Diritto Commerciale e di Diritto Pubblico dell'Economia,* edited by F. Galgano. Padua: Cedam, 2010.

Lo Faro, A. 'Azione collettiva e tutela dell'ambiente di lavoro in Europa'. *Giornale di diritto del lavoro e di relazioni industriali,* no. 1 (1991): 162–185.

Masi, M. 'Il rischio: le ragioni di una disciplina in materia di sicurezza'. In *Il testo unico della salute e sicurezza nei luoghi di lavoro,* edited by M. Tiraboschi. Milan:Giuffrè, 2008.

Miranda, L. 'Organizzazione del lavoro e sicurezza negli accordi del Gruppo Fiat-Chrysler. Verso un comune modello di rappresentanza paritetica per la sicurezza?'. *Diritti lavori mercati,* no. 1 (2012): 93–113.

Miranda, L. *La dimensión colectiva de la tutela del entorno de trabajo en Europa: perfiles comparativos entre Italia, Francia y Eslovenia.* Albacete: Editorial Bomarzo S.L., 2012.

Monda, G. M. 'La valutazione dei rischi per la sicurezza e la salute dei lavoratori'. In *Le nuove regole per la salute e la sicurezza dei lavoratori,* edited by L. Zoppoli, P. Pascucci & G. Natullo. Milan: Kluwer - Ipsoa, 2008.

Montuschi, L. *Diritto alla salute e organizzazione del lavoro*. Milan: Giuffrè, 1980.

Montuschi, L. 'La sicurezza nei luoghi di lavoro ovvero l'arte del possibile'. *Lavoro e diritto,* no. 2 (1995): 413–445.

Montuschi, L. (ed.). *Ambiente, salute e sicurezza.* Turin: Giappichelli, 1998.

Montuschi, L. 'L'incerto cammino della sicurezza del lavoro fra esigenze di tutela, onerosità e disordine normativo'. *Rivista giuridica del lavoro,* no. 2 (2001): 508–538.

Montuschi, L. 'Verso il testo unico sulla sicurezza del lavoro'. *Giornale di diritto del lavoro e di relazioni industriali,* no. 2 (2007): 799–815.

Montuschi, L. (ed.). *La nuova sicurezza sul lavoro. D.lgs. 9 aprile 2008, n. 81 e successive modifiche.* Bologna: Zanichelli, 2011.

Natullo, G. *La tutela dell'ambiente di lavoro.* Turin: Utet, 1995.

Pascucci, P. 'Dopo la legge n. 123 del 2007. Titolo I del d.lgs. 9 aprile 2008, n. 81 in materia di tutela della salute e della sicurezza nei luoghi di lavoro'. http://csdle.lex.unict.it/Archive/WP/WP%20CSDLE%20M%20DAntona/WP%20CSDLE%20M%20DAntona-IT/20111025-121117_pascucci_n73-2008itpdf.pdf, vol. 73.

Pascucci, P. 'Il campo di applicazione soggettivo delle nuove regole sulla salute e sicurezza nei luoghi di lavoro'. *Diritti lavori mercati*, no. 2 (2008): 295–306.

Pascucci, P. 'Delega delle funzioni su doppio binario. Cambia la responsabilità dei "vertici"'. *Guida al Diritto – Il Sole 24 Ore, Speciale 'Sicurezza Lavoro'*, September 2009.

Pascucci, P., 'Salute e sicurezza: dalle rappresentanze dell'art. 9 ai rappresentanti del d.lgs. 81/08', *Diritti lavori mercati*, no. 4 (2010): 670-678;

Pascucci, P. 'La nuova disciplina della sicurezza sul lavoro del 2008/2009: una rapsodia su novità e conferme'. http://olympus.uniurb.it/images/wpo/wpo1-2011.pdf, vol. 1

Passalacqua, P. *voce Enti Bilaterali*, vol. IV of *Digesto delle discipline privatistiche – Sezione commerciale.* Turin: Utet, 2008.

Perone, G. *Lo statuto dei Lavoratori.* Turin: UTET, 1997.

Persiani, M. (ed.). *Il nuovo diritto della sicurezza sul lavoro.* Turin: Utet, 2012.

Pinna, A. 'Obblighi, compiti e diritti dei lavoratori'. In *La tutela della salute nei luoghi di lavoro*, edited by G. Loy. Padua: Cedam, 1996.

Prosperetti, U. 'Commentary on the Judgement issued by the Milanese Pretura, 28 July 1971'. *Massimario di giurisprudenza del lavoro*, no. 1 (1972): 20–24.

Rausei, P. 'Le contravvenzioni e le sanzioni amministrative'. In *Il testo unico della salute e sicurezza nei luoghi di lavoro*, edited by M. Tiraboschi. Milan: Giuffrè, 2008.

Rausei, P. 'Il procedimento sanzionatorio'. In *Il testo unico della salute e sicurezza nei luoghi di lavoro*, edited by M. Tiraboschi. Milan:Giuffrè, 2008.

Rausei, P. 'Le contravvenzioni e le sanzioni amministrative'. In *Il testo unico della salute e sicurezza nei luoghi di lavoro*, edited by M. Tiraboschi. Milan:Giuffrè, 2008.

Ricci, M. 'Sicurezza sul lavoro: controllo e partecipazione sindacale tra iure condito e de iure condendo'. *Il lavoro nella giurisprudenza*, no. 2 (2008): 113–118.

Rusciano, M. & G. Natullo (eds.). *Ambiente e sicurezza del lavoro.* Turin: Utet, 2007.

Russo, A. 'La delega di funzioni e gli obblighi del datore di lavoro non delegabili'. In *Il testo unico della salute e sicurezza nei luoghi di lavoro*, edited by M. Tiraboschi. Milan: Giuffrè, 2008.

Scognamiglio, R. 'La natura non contrattuale del rapporto di lavoro'. *Rivista italiana di diritto del lavoro*, no. 1 (2007): 379–405.

Smuraglia, C. 'Sicurezza del lavoro e obblighi comunitari. I ritardi dell'Italia nell'adempimento e le vie per uscirne'. *Rivista italiana di diritto del lavoro*, no. 1 (2002): 221–240.

Tiraboschi, M. (ed.). *Il testo unico della salute e sicurezza nei luoghi di lavoro. Commentario al decreto legislativo 9 aprile 2008, n. 81.* Milan: Giuffrè, 2008.

Tampieri, A. *Sicurezza sul lavoro e modelli di rappresentanza.* Turin: Giappichelli, 1999.

Tullini, P. 'I sistemi di gestione della prevenzione e della sicurezza sul lavoro'. *Giornale di diritto del lavoro e relazioni industriali*, no. 3 (2010): 407–418.

Vallebona, A. (ed.). 'Colloqui giuridici sul lavoro', no. 1 (2009); *Supplemento del Massimario di giurisprudenza del lavoro*, no. 12 (2009).

Venturi, D. 'Lavoratore: definizione e obblighi'. In *Il testo unico della salute e sicurezza nei luoghi di lavoro*, edited by M. Tiraboschi. Milan: Giuffrè, 2008.

Zoli, C. 'Sicurezza del lavoro: contrattazione e partecipazione'. *Rivista Giuridica del Lavoro*, no. 3 (2000): 614–628.

Zoppoli, L. 'Rappresentanza collettiva dei lavoratori e diritti di partecipazione alla gestione delle imprese'. *Giornale di diritto del lavoro e di relazioni industriali*, no. 2 (2005): 326–452.

Zoppoli, L., P. Pascucci & G. Natullo (eds.). *Le nuove regole per la salute e la sicurezza dei lavoratori, Commentario al D.lgs. 9 April 2008, no. 81*. Milan: Ipsoa-Wolter Kluwer, 2008.

CHAPTER 8

Occupational Health and Safety in Spain: A Formidable Array of Employers' Liabilities

Joaquín Aparicio Tovar

§8.01 BACKGROUND, SOURCES AND BASIC CONCEPTS

The substantive provisions on occupational health and safety in Spain are, in common with other Member States, mainly derived from European Union (EU) Law. However, an important feature of the Spanish system is the part played by the labour authorities in supervising and controlling compliance with health and safety law. The first health and safety Acts, passed in 1868, involved mining safety, although they were not designed exclusively for the protection of the workers, but to ensure the general safety of mines. In these Acts preference was given to the use of administrative penalties to punish infringements over the penal ones chosen by the first *Factory Acts* passed in Britain. This penchant for administrative liability was reasserted in the first Spanish social Acts passed in the early 1900s[1] especially when the Labour Inspection, the main means of imposing administrative penalties, was created in 1906.[2] The Labour Inspection authority is charged with overseeing compliance with social law, particularly the *Ley de Accidentes de Trabajo* de 1900 (Occupational Accidents Act), *Ley del*

1. A. Martin Valverde, 'La formación del Derecho del Trabajo en España. Estudio preliminar', in *Legislación social en la historia de España. De la revolución liberal a 1936*, ed. A. Martin Valverde et al. (Madrid: Congreso de los Diputados, 1987), LV.
2. The government's power to impose penalties in Spain is indelibly linked with 'the power to regulate and enforce government policies in certain areas'. A. Nieto, *Derecho Administrativo Sancionador*, 4rd ed. (Madrid: Tecnos, 2005), 96. Government intervention has played an important part in social matters and continues to do so today. Spain has ratified conventions 81 and 129 of the ILO related to maintaining a system of labour inspection in industrial, commercial and agricultural workplaces.

Trabajo de Mujeres y Menores (Women and Minor Workers Act), also dating from 1900, and the *Ley del Descanso Dominical* of 1904 (Sunday Rest Act). The Ministry of Labour was created in 1920, and since then the Labour Inspection Department has been part of it. With the decentralization of the Spanish State that followed the Constitution of 1978, the Labour Inspection departments now act within the autonomous communities which are empowered to apply penalties as part of the enforcement of labour law. In Spain, the Labour Inspection authority has a general mandate, meaning that it oversees compliance with all labour and social security law, not only occupational health and safety.

In the early 1980s, in response to social concern about the number of occupational accidents, acts were passed to reinforce criminal liability by creating specific offences for endangering occupational health and safety.[3] The relationship between administrative and criminal penalties will be examined later. When existing civil and social security liabilities are added to the above, the result is a formidable and complex array of liabilities, which are a further feature of the Spanish system.

The *Ley de Accidentes de Trabajo* of 1900 was very important: first, because, in line with other European legislation, it set aside the requirement of wrongful conduct or negligence on the part of the employer by introducing the hitherto almost non-existent idea of *occupational risk*,[4] aside from any criminal liability, because it was limited to general offences or misdemeanours involving injury or homicide. Second, because it established the definition of an occupational accident as being injury caused to the worker while engaged in, or as a result of his or her work. This is basically the same definition used in today's General Social Security Law (including accidents going to and returning from work), and has been frequently and widely interpreted by the courts, allowing work-related diseases to be included in the scope of the law. Nevertheless, the Act did not make it compulsory for employers to take out civil liability insurance; this would have to wait till 1932, when it was brought in as a result of the social activism of the Second Republic. Third, for the first time the law expressly enforced measures to prevent occupational risks by making it compulsory for employers to apply health and safety protection measures in the premises and on the machinery under their control.

The main sources governing these issues are, in order of importance: first, the Constitution (Articles 15, 40.2 and 43); second, numerous International Labour Organization (ILO) conventions on the subject, ratified by Spain, particularly Convention 155. Since ILO conventions are international treaties, pursuant to Article 96 of the

---

3. J. Terradillos Basoco, *La siniestralidad laboral como delito* (Albacete: Bomarzo, 2006), 14. The life and safety of workers are relevant legal rights which, if endangered, are subject to criminal liability. This is mainly a result of the recognition of economic and social rights enforced in Spain following ratification of the International Covenant on Civil and Political Rights (1997) and the International Covenant on Economic, Social and Cultural Rights (1977), together with recognition of the constitutional right to life, physical and moral safety (Art. 15), and the obligation of public authorities to promote safety (Art. 40.2) and health (Art. 43) in the workplace.
4. L.E. De la Villa, 'La influencia de la Ley de Accidentes de Trabajo de 1900 en la construcción del ordenamiento laboral español', in *Cien años de Seguridad Social,* ed. L. Gonzalo & M. Nogueira (Madrid: La Fraternidad, 2000), 5.

Spanish Constitution they become part of Spanish law from the moment they are ratified. As established in Spanish jurisprudence and Article 27 of the Treaty of Vienna, they overrule ordinary Spanish legislation in the event of conflict; third, a considerable number of statutes, including *Ley* 31/1995, of 8 November, of *Prevención de Riesgos Laborales* (Prevention of Occupational Risks Act, hereinafter, LPRL). This Act implemented Directive 89/391/EC, of 12 June 1989 (hereinafter Framework Directive) in Spain. The *Ley de Infracciones y Sanciones del Orden Social* (Infringement and Penalties in the Social Sphere, hereinafter, LISOS) of 4 August 2000 regulates administrative liability, and the Penal Code of 1995 (Articles 316, 317 and 318 specifically) regulates criminal liability. Another Act to consider is Act 7/2009, of 23 November, *Libre acceso a las actividades de servicios* (on unrestricted access to and engagement in services), which implemented in Spain Directive 2006/123/EC, regarding services in the domestic market. Fourth, there are many administrative regulations, usually highly technical, governing the use of certain working materials, instruments and processes. Some of these standards derive from specific directives referred to in Article 16 of the Framework Directive.[5] The building sector has specific rules, in addition to general acts, one of the most important of which is *Ley 32/2006, de subcontratación en el sector de la construcción* of 18 October, regulating subcontracting in this sector.

Legal regulations are always 'the mandatory minimum' enabling collective bargaining to improve and adapt preventive measures to the industry or to businesses (Article 2.2 LPRL). Nevertheless, some aspects, such as the scope of the employer's overall obligations, or the regulation of administrative infringements and penalties are to a certain extent 'unsuitable' for collective bargaining[6] which, meanwhile, has a wide scope as far as workers' involvement[7] and other instrumental obligation are concerned.

Health and safety law cannot be understood without referring to other rules that mainly aim at compensating the consequences of occupational accidents and diseases, such as the Civil Code provisions on contractual (Articles 1101 - 1112 sub) and extra-contractual (Article 1902 - 1910) liability, and particularly the General Social Security Act of 1994.

The LPRL contains basic preventive concepts, which do not include health and safety, as these are general concepts. Many of these are taken from the Framework Directive. The aim of the LPRL is to promote safety in the workplace by means of various actions taken by the authorities, employers and workers, which take the form of risk prevention measures (Article 2.1). Prevention implies all measures adopted or planned at all stages of professional activity in order to prevent or reduce job-based

5. The Framework Directive established the deadline by which Member States had to implement these standards in their domestic law. Spain did not meet this deadline, and this caused delays in the implementation of a considerable number of specific directives, and for which Spain was found guilty of breach of EC Law (as it was known) by the Court of Justice. See ECJ, 26 September 1996, Case C-79/95, *Commission v. Kingdom of Spain.*
6. J. Aparicio Tovar, 'El papel de la negociación colectiva en materia de salud y seguridad en el trabajo', in *La nueva función de la negociación colectiva y los acuerdos interconfederales en el empleo* (Madrid: Ministerio de Trabajo, 1998), 109.
7. N. Mendoza Navas, *Prevención de riesgos laborales: la regulación convencional* (Albacete: Bomarzo, 2007), 107 et seq.

risks. Risk means the probability of a worker suffering injury derived from his or her job. The extent of the risk depends on the greater or lesser probability of the injury occurring, and the seriousness of the same (Article 4). The aim of Spanish legislation, by imposition of the Framework Directive (Article 6. 2c), is to eliminate risk at its source, and to this end general measures are both preferred and put in place; individual protection of workers is a secondary measure.

[A] Emergent Risks and New Concepts

Psychosocial risks are often considered emerging risks affecting the mental health of the person and often motivated by circumstances related to the workplace. The prevention of these risks relate more to the idea of achieving wellness and quality of personal and social life at work than the classical approach to prevent occupational accidents and diseases,[8] but we must not fail to note that the relationship between 'classical' risk factors (e.g., noise) and health is clearer today than the relationship between psychosocial risk factors and health. The risk factor in these cases would be the workload, i.e., 'the set of psycho-sociological and physical requirements, as are the physical, mental workload, fatigue and job dissatisfaction, to which the worker is subjected throughout their working day'.[9]

The study of these risks from the fields of psychiatry, medicine and psychology are abundant while labour standard treatment from preventive or social protection is still insufficient. The legislation does not regulate specifically psychosocial risks, except those arising from gender discrimination, in particular sexual harassment and harassment based on sex.[10] In fact, Act 3/2007 on equality between men and women rate the sexual harassment and gender harassment as direct discrimination (Article 7) and establishes the need to organize measures and procedures for prevention in the private sector (Article 48) and public administration (Article 62). Companies should promote working conditions that prevent sexual harassment and gender harassment and have procedures to prevent these risks. Also, the workers' representatives should be proactive on prevention through awareness and information. Prevention in the Public Administration requires the development of a protocol negotiated with employee representatives, including a complaint procedure in case of harassment.

There is thus a lack of specific regulation for risks such as stress, bullying or so-called burn out, which must be prevented in accordance with the general obligation to ensure the safety and health of workers adopting all the necessary measures for this (Article 14 LPRL).[11] Bullying or harassments are distinct from harassment based on sex

8. M.R. Martinez Barroso, *Riesgo psicosocial en el sistema de protección social* (Sevilla: Laborum, 2007), 191.
9. M.C. Díaz Descalzo, 'Los riesgos psicosociales en el trabajo: el estrés laboral, el síndrome del quemado y el acoso laboral', *Revista de Derecho Social*, no. 17 (2002): 184.
10. *Ley 3/2007, de igualdad entre hombres y mujeres* (of equality between men and women) rates the sexual harassment and gender harassment as direct discrimination (Article 7).
11. G. Fabregat Monfort, *La obligación empresarial de tutela frente al acoso Laboral en la empresa* (Albacete: Bomarzo, 2011), 71 et seq.

or sexual harassment, but they are also a form of discrimination.[12] It is not necessarily a physical injury to the victim. The main point is the context of violence.[13] The employer must implement measures aimed at preventing psychosocial risks after evaluation of its incidence. This will be planned taking into account 'work organization, working conditions, social relationships and the influence of environmental factors at work' (Article 15 LPRL). In application of the LPRL, prevention of psychosocial risks should also include the obligation to train workers in relation to these risks and the obligation of health surveillance to detect mental illness.[14]

Psychosocial risks can and should be prevented but it is difficult to apply automatically the general principles of preventive action. Therefore, the National Institute of Occupational Safety and Health has issued several Technical Notes on Prevention to facilitate the assessment and prevention of these risks. Technical Notes are useful guidelines or criteria for risk assessment to enable the employer to adapt the workplace to avoid or prevent psychosocial risks. These are similar to the British Code of Practice issued by the Health and Safety Executive. These Technical Notes are not legal standards and therefore do not create obligations. However, when patterns of these rules are applied in the enterprise they can be used in court proceedings as evidence to hold the employer free of liability. Also some collective agreements take action in this matter in adapting safety patterns to the enterprise.[15]

Some of these Technical Notes address how to assess psychosocial risks, their conceptual basis or their characteristics. Thus, the Technical Note nº 703 identify as 'psychosocial risk factors' the characteristics of working conditions and, above all, its organization, which affect the health of people through psychological and physiological mechanisms. It is usually considered therefore that the causes of these risks are varied, being necessary to include the organization of work as a subjective or personal characteristic of the individual.[16]

§8.02 PERSONAL SCOPE

Pursuant to Article 2.1 of the Framework Directive, the LPRL (Article 3) extends the scope of health and safety rules in the workplace beyond the traditional boundaries of Labour Law, and includes labour relations in the public sector, regardless of whether

12. M.F. Fernández López, 'El acoso moral en el Trabajo ¿una nueva forma de discriminación?', *Revista de Derecho Social*, no. 19 (2002): 19.
13. M.J. Romero Rodenas, *La protección frente al acoso moral en el trabajo* (Albacete: Bomarzo, 2004), 28.
14. F. Cavas, 'Breves consideraciones sobre la prevención de los riesgos psicosociales', in *Los riesgos psicosociales: Teoría y práctica*, ed. C. Sanchez Trigueros (Pamplona: Aranzadi, 2009), 114.
15. B. Valdés de la Vega, 'El tratamiento del acoso moral desde la autotutela colectiva', in *Los medios de tutela frente al acoso laboral*, ed. B. Valdés de la Vega (Granada: Comares, 2007), 25. The *European Framework agreement on harassment and violence at work*, 26 Apr. 2007 in Spain was incorporated to *The Interconfederal Agreement on Collective Bargaining 2008*, but before there were not many agreements with provisions in this matter.
16. S. Olarte & M. Jiménez, 'La regulación de los riesgos psicosociales en España: experiencia y prospectiva', *Revista de Prevención de Riesgos Psicosociales y Bienestar en el trabajo*, no. 1 (2010): 18.

they are under an administrative contract or public service agreement. Also included are members of co-operatives, military personnel including civilians and people employed in military establishments, prison employees and self-employed workers.

Nevertheless, it should be pointed out that these rules are not applied systematically; rather, 'exemplary' or ordinary application is limited to relationships regulated in the *Estatuto de los Trabajadores* (ET) (Workers Statute). The remaining relationships included contain specific activities that exclude them, to a greater or lesser extent, from the common system. Exceptions, however, are not standardized either: some are closer to the common system, which are those applied to government employees who are not among the excluded groups, while others have special features.

The determining factor is the status of employee, irrespective of whether the employer is a public or private organization, with administrative or labour relationships. A good example of this is the definition of the risk as 'occupational' made by the LPRL (Article 4.2), references made by the same to businesses (Article 4.1), and those made to 'job-related injuries' (Article 4.3), and finally, the phrase 'working condition' (Article 4.7).

The concepts of worker and employer are adapted to civil servants, and the LPRL in general terms, equates government with the employer, and people at its service with the worker. Act 7/2007, of 12 April, the Basic Statute for Civil Servants, made it mandatory for public administrations, as employers, to provide civil servants with effective protective measures to ensure health and safety in the workplace. However, some differences exist, for example, with reference to Article 31.1 LPRL, on Protective and Preventive Services, to Article 34.3 LPRL, on the right to participate, and to Article 45 LPRL, regarding administrative infringements. Behind this overall alignment lurks a different system for civil servants[17] that, following the enactment of RD 1488/1988, of 11 September, is evident in four issues: the organization of Protective and Preventive Services, workers' representation, control and auditing systems and the system of liabilities.[18]

Health and safety rules introduce special activities that further divorce them from the common system in two specific public service situations, namely, in military establishments and prisons.

Military centres and establishments are not excluded from overall health and safety regulations; rather, these rules have been adapted to these situations through the enactment of RD 1488/1998. The military centre or establishment, not the service relationship with the military authorities, is taken as a point of reference for the application of the LPRL. The concept of military establishment is the subject of RD 2205/1980, of 13 June, declared constitutional in 1991.[19] Therefore it is necessary to understand what 'military administration centres, bodies, units, premises and organisms' (Article 1 (b)) represent. However, public owned companies or militarized

17. S. Gonzalez Ortega & J. Aparicio Tovar, *Comentarios a la Ley 31/1995, de Prevención de Riesgos Laborales* (Madrid: Trotta, 1996), 60.
18. M.D. García Villaverde, 'El ámbito de aplicación de la Ley de Prevención de Riesgos Laborales', in *Derecho de la Prevención de Riesgos Laborales*, ed. A. Garrigues (Albacete: Bomarzo, 2009), 129.
19. STC 101/1991 of 13 May.

industries supplying the armed forces are not considered military establishments. The important point is that the labour inspection authorities have no competence in military centres and establishments, as can be seen from Articles 77.4 and 75.4 of the aforementioned RD 2205/1980. Act 42/1997, organizing Labour Inspections, has not addressed this deficit. The labour section attached to each army, navy and air force headquarters is in charge of inspecting the health and safety and working conditions.

Prisons, unlike military establishments, do not constitute a sphere where health and safety rules have to be adapted to suit the particular characteristics of the centre. In prisons, health and safety regulations are only adapted in the case of activities that require special rules, when the risk involved is considered unique. An important factor to bear in mind is that these adaptations are established in the collective bargaining processes applied to civil servants.

Since the enactment of Act 20/2007, of 11 July, the Self-Employed Worker's Statute, steps have clearly been taken to align the health and safety regulations applied to these workers with the 'common' regulations applied to ordinary employees.[20] It is clear that all workers, whether self-employed or employed by a third party, are subject to the same occupational risks, although the 'liable parties' are different. Self-employed have the right/obligation to protect themselves, but when they are involved in relationships of co-operation and collaboration with other business owners they are acquired a number of rights from the main employer.[21] There are two important points here: (a) the obligation of companies contracting self-employed to carry out jobs or services within their field of activity to ensure that said workers comply with health and safety rules; (b) the obligation of companies that have failed to comply with their health and safety commitments to compensate self-employed for any injury they may suffer.

The scope of application of the LPRL does not include the activities of the police, customs officials, civil protection services and forensic experts involved in situations of extreme risk, catastrophes and disasters, such as the armed forces, and military activity carried out by the Civil Guards. These are legally enforced exclusions, because if they were not singled out, they would become part of the scope of health and safety regulations. They are all inspired to Article 2.2 of the Framework Directive, and are restrictive, inasmuch as a series of circumstance must arise for the exclusion to occur. First, the activities must involve public service, irrespective of the relationship (administrative or labour) maintained by the personnel employed by the authority charged with rendering the service, and they must fall within the category of the functions enumerated above. This is a closed list. Second, they must be activities in which risk is an essential component (police directly in contact with delinquents, fire-fighters. etc.). Not all activities carried out within these services are excluded, and those that are not intrinsically 'risky' are included in the scope of these regulations (law enforcement officials engaged in office work, for example). The same is true for the

20. B. Alonso-Olea García, 'La prevención de riesgos laborales del trabajador autónomo', *Aranzadi Social*, no. 16 (2008): 6.
21. S. Olarte, *Prevención de riesgos laborales de los trabajadores autónomos* (Granada: Comares, 2009), 47 et seq.

armed forces and Civil Guards. Only activities that are intrinsically risky are excluded, the rest are subject to the aforementioned special activities carried out in military establishments.

The LPRL does not include domestic services rendered in the private residence, although the head of the household is obliged to ensure that the work is carried out under suitable health and safety conditions. This concept is not devoid of meaning.

§8.03 DUTIES OF THE PARTIES

[A] The Employer as the Party Obliged to Guarantee a Safe and Healthy Working Environment

The employers' essential obligation[22] is to guarantee a healthy and safe working environment. This obligation derives from their position in the employment contract, because by making use of their management skills, employers organize and oversee the productive process, as described in Article 1.1. of the Workers Statute. Employers must exercise their managerial powers in a safe and healthy workplace. When employers enter into an employment contract they acquire a commitment to ensure the safety of each employee at their service, making them safety creditors.[23] This role of the employer as the party obliged to guarantee safety in the workplace is enforced in Article 14 LPRL, which simply incorporates into Spanish law the provisions of Articles 5 and 6 of the Framework Directive. The employer's obligation to ensure safety is mirrored by the workers' corresponding right to receive this protection. The employer is subject to a number of obligations: first, a general obligation to ensure safety, and second, a series of specific obligations, some of which are instrumental to the overall obligation.

These instrumental obligations were one of the most important new rules included in the LPRL, and help make health and safety part of company management.[24] The general obligation obliges employers to 'guarantee the health and safety of their workers in all aspects of their work'. To this end, they must adopt 'as many measures as needed', engaging in 'ongoing activities aimed at improving existing levels of protection' (Article 14.2 LPRL). This obligation is classified as a standard of conduct which, if complied with, will provide the truly 'effective protection' to which workers are entitled (Article 14.1 LPRL). The standard is expressed using the example of the careful employer, it is objective and each individual employer must adapt their conduct to comply with the norm.[25] According to this standard, employers should first be aware

22. M. Alonso Olea & M.E. Casas Baamonde, *Derecho del Trabajo*, 26rd ed (Madrid: Civitas, 2009), 345.
23. P. Rivas Vallejo, 'El empleador y su deber de prevención (I). Fundamento y obligaciones generales', in *Derecho de la Prevención de Riesgos Laborales*, ed. A. Garrigues (Albacete: Bomarzo, 2009), 159.
24. M.L. Martin Hernandez, *El derecho de los trabajadores a la seguridad y salud en el trabajo*, (Madrid: CES, 2006), 201.
25. J. Aparicio Tovar, 'Las obligaciones del empresario de garantizar la salud y la seguridad en el trabajo', *Revista Española de Derecho del Trabajo*, no. 49 (1991): 723.

of both the risks involved in their particular production process and the measures that should be taken to prevent said risks, and second, be willing to effectively put these measures in place. As with all standards, that of the careful employer is open-ended and can only be met when adapted to circumstances of time and place.[26] Awareness of risks and willingness to prevent them is an ongoing process insomuch as both elements change in line with new scientific and technological developments. Because of this, their overall obligation forces employers to engage in 'permanent surveillance of preventive actions [...] and provide the necessary means to adapt health and safety measures' (Article 14.2 LPRL).

Generally speaking, specific obligations are instrumental in ensuring suitable compliance with general obligations. Many of these obligations are highly technical in nature, and in most cases they are established in regulations designed to implement specific directives (data screens, noise, asbestos, equipment, etc.). Many of these obligations are established without exception, meaning that they must be complied with to the letter of the law, with no scope for interpretation depending on time and place. Highly technical obligations are applicable to certain productive activities or to the use of identified procedures or specific products. Nevertheless, other specific and instrumental obligations are general in nature and are established in the LPRL to implement provisions of the Framework Directive, such as drawing up a health and safety plan, risk assessment, information, consultation, participation and training of workers, designing drills to be used in situations of serious and imminent risk and emergency, monitoring the health of the workers employed and providing the services and means needed to correctly organize health and safety measure, particularly so-called preventive services. Employers are also under the obligation to maintain and correctly fill in health and safety forms and to notify the labour authorities of any injuries to their workers. The combination of specific and general obligations make up what could be succinctly called the employer's obligation to ensure safety. Compliance with this obligation enables employers to provide a risk-free workplace that does not endanger their workers' safety. Under this perspective, the obligation to ensure safety is not one of means but of results;[27] and this forces employers to use the best possible risk-preventions resources, consistent with the provisions in place since 1889 and with Article 1105 of the Civil Code, which only releases employers from their liability where occurrences are due to unforeseen or unavoidable circumstances.

One of the most important employer's instrumental obligations is that of establishing Protective and Preventive Services provided with the technical resources required to ensure correct compliance with safety obligations. This obligation is expressly established in Article 7 of the Framework Directive, which has been implemented in Articles 30 and 31 LPRL, and enforced through several rules. These Articles have been modified several times, the last amendment taking place in Act 25/2009, of 22 December, passed when the provisions of Directive 2006/123/EC on

26. R. Pound, *Law and Morals* 57 (Oxford U. Press - U. North Caroline Press 1926).
27. M. Gonzalez Labrada, *Seguridad y salud en el trabajo y responsabilidad contractual del empresario* (Barcelona: Cedecs, 1996), 357; F. Calvo Gallego, *La obligación general de Prevención y la responsabilidad civil o contractual* (Pamplona: Aranzadi, 1998), 55 et seq.

services in the domestic market were incorporated into Spanish law. Although this was not strictly necessary because the Directive does not apply to labour law, Spanish legislators chose to include it. Constituting or enlisting the services of a technical consultancy to organize health and safety measures does not imply the transfer of employers' safety obligations. Employers are not released from their safety obligations, as expressly established in Articles 5 and 6 of the aforementioned Framework Directive. Article 31.2 LPRL, meanwhile, establishes that these resources or services should 'advise and assist' employers, and not replace them in this role. This instrumental obligation implies that employers are under the obligation of knowing the risks involved in their productive processes and of using the best means to prevent them, in line with scientific and technical advances. However, employers are not expected to be also chemists, doctors, engineers, etc., and for this reason must enlist the services of experts.

In small companies (with up to ten workers), employers can undertake these duties themselves provided they have the expertise needed for the activity undertaken. The LPRL gives employers the option of appointing one or more workers to these posts, creating their own health and safety service or enlisting the services of third party health and safety experts. A feature of the Spanish system is that *Mutuas de Accidentes de Trabajo y Enfermedades Profesionales* (Occupational Accidents and Disease Mutual Associations), or *Mutuas*, can carry out health and safety services for their member companies.[28] Health and safety services, be they internal or outsourced, are organized and structured units,[29] that must be made up of specialists in at least four occupational medicine specialities, occupational safety, ergonomics, industrial hygiene and applied psychology. Employers can enlist the services of more than one health and safety consultant. Many Spanish companies, whether large or small, choose to outsource their health and safety services,[30] and this is detrimental to the process of integrating health and safety attitudes in company management bodies. Any organization wishing to act as a health and safety service provider must be accredited by the labour authorities. The aim of Act 25/2009 and its implementing provisions is to standardize, simplify, but at the same time reinforce accreditation guarantees across Spain. The need for government intervention in this sphere is not incompatible with the free provision of services in the domestic market, because protecting the health and safety of workers is an overriding requirement of general interest, as set forth in Article 4.8 of Directive 2006/123/EC.

28. *Mutuas* are employers' associations classified as social security agents and strictly controlled by the Ministry of Labour. The assets of these associations are held by the General Treasury of the Social Security but managed by the *Mutua*. *Mutuas* are non-profit-making organizations and cannot distribute dividends to their members. They were formed to insure against occupational accidents and diseases.
29. A. Garrigues, 'El empleador y su deber de prevención (III). Obligaciones instrumentales: Servicios de prevención y otras modalidades de organización técnico-preventiva', in *Derecho de la Prevención de Riesgos Laborales*, ed. A. Garrigues (Albacete: Bomarzo, 2009), 279.
30. A. Garrigues, 'El combinado Ley ómnibus-RD 337/2010: Un paso adelante (*ma non troppo*) para la racionalización de las modalidades de organización técnico-preventiva y la integración de la prevención en la gestión empresarial', *Revista de Derecho Social*, no. 50 (2010): 93.

[1] The Adaptation of Work to the Individual and Specific Risk Groups

All these obligations are subject to the principle of adapting the job to the worker (Article 15.1.d. LPRL) established in Article 123 LGSS (General Social Security Act). This means that the personal circumstances of a given worker, such as a disability, language barriers, age, etc. must be taken into account when assigning jobs and assessing risks. Preventive measures must take into account any distractions or carelessness on the part of the worker (Article 15 LPRL).

Adapting the job to the person means taking into consideration the particular sensitivity of some workers to specific occupational risks that may affect them more than other workers. This principle has a clear realization in Article 25 LPRL by protecting workers especially sensitive to certain risks. By 'especially sensitive worker' can be understood 'any worker whose personal characteristics lead to a condition or physical, mental or sensory impairment, which makes it particularly vulnerable to occupational hazards'.[31] That is, the subjective factor of the worker takes precedence over the conditions or circumstances of the job or the work itself. Occupational hazards are manifested or worsen due to the personal conditions of an 'especially sensitive worker'. From this broad definition, Articles 25, 26 and 27 LPRL raise the standard of prevention for certain groups of workers. These are minor workers, working women who are pregnant, after childbirth or are breastfeeding and disabled workers. In such cases the general obligation of safety and health of the employer involves the adoption of measures with a more specific and intense character.

Employers must carefully assess any risks that could affect the reproductive functions of both men and women (Article 25. 2 LPRL). RD 1802/2008, of 3 November, on hazardous substances, adapted existing Spanish legislation to Regulation (EC) 1907/2006 of the European Parliament and of the Council (the REACH Regulation) of 18 December 2006, which includes protection against substances harmful to reproductive health.

Protection of pregnancy is dealt with specifically in Article 26 LPRL, which implements the mandates of Directive 92/85/EC, of 19 October 1992. Regulations on this issue are further enhanced by the provisions of Act 39/1999, of 5 November, promoting Conciliation of Family and Working Life, and Organic Act 3/2007, of 22 March, on effective equality of men and women. Pregnant women are under no obligation to inform their employer of their pregnancy insomuch as this falls within the scope of their private life, according to a ruling handed down by both the EU Court of Justice and the Spanish Constitutional Court.[32] Article 55.5 (b) of the Worker's Statute invalidated, on the grounds of discrimination, the dismissal of pregnant women, even when the employer has not been informed of the pregnancy. Antidiscrimination safeguards, however, cannot be applied in the same way in health and safety rules

31. A. Moreno, *La prevención de riesgos laborales de los trabajadores especialmente sensibles* (Valencia: Tirant lo Blanch, 2010), 57.
32. ECJ, 4 Oct. 2001 (C-438/99), *Jimenez Melgar*, and 27 Feb. 2003 (C-320/99), *Busch*, and STC 92/2008, of 21 July.

because the employer must be informed of the worker's pregnancy in order to effectively implement some protective measures.[33]

The first level of protection involves evaluating any risks the company might pose in hypothetical situations of pregnancy, recent birth or breastfeeding. The second level involves avoiding the exposure of women to harmful agents or situations, and this could mean that the adaptation of the conditions or working time of the worker concerned is two-fold.[34] On the one hand, it would be the physical adaptation of the characteristics of the job, on the other the adaptation of other working conditions normally associated with working time and rest periods, night work, shift work, etc. Where adaptation is not possible the next level would consist in changing to a job without risk. In practice, this measure is the most used and since 2007 a reduction has been established in social security contributions for the substitute worker. In cases where, because of risk during pregnancy or risk during breastfeeding, the worker is destined for a different job or function, a reduction of 50% of the employer's social security contribution is applied. Finally, if it is not possible to change jobs, the last level of protection is implemented, and the worker is entitled to a suspension of contract, pursuant to the provisions of Articles 45 and 48 of the Workers Statute.

This suspension is not the same as the sixteen weeks maternity leave. At this time the protection provided under Social Security Law (Articles 134 and 135 LGSS and implementing provisions) comes into play by guaranteeing a cash benefit (in additional to, naturally, health care). However, this situation can only be authorized if the health system medical services accredit that the job entails health risks that could be harmful to the pregnancy or to breast feeding. Workers are entitled to take time off work, without this being deducted from their salary, to attend prenatal examinations and courses.

Young workers constitute another particularly sensitive group, and a series of measures for their protection have been drawn up in Article 27 LPRL, echoing the provisions of Directive 94/33/EC. The protection measures are based on an assessment of the risks to which these workers are exposed, which must take into consideration their lack of experience in the job and their immaturity. Underage workers for labour purposes, and therefore protected by the regulations of prevention of occupational hazards, are those who are over 16 and under 18 years. The only exception that permits the work of minors under 16 years, is the public entertainment activity which requires authorization of the labour authority for preventive purposes. Employers are also under the obligation of informing workers of what their job entails. This reporting obligation covers the risks to which the child is exposed and the protective measures. The information must be sufficient and appropriate which means taking into account the subjective circumstances of the person – underage worker – receiving such information. Both risk assessment and the information is configured as fundamental

33. M. Purcalla, 'El empleador y su deber de prevención (II): Obligaciones específicas (colectivos especialmente sensibles)', in *Derecho de la Prevención de Riesgos Laborales*, ed. A. Garrigues (Albacete: Bomarzo, 2009), 228.
34. A. Moreno, *La prevención de riesgos laborales de los trabajadores especialmente sensibles'* (Valencia: Tirant lo Blanch, 2010), 221.

obligations that in case of breach, in addition to constituting an offence, affects the graduation of the penalty if the minor suffers an occupational accident.

The Workers Statute establishes the limits of the working day and bars these workers from working at night or from doing overtime. They are also barred from carrying our certain jobs deemed hazardous in a previous (albeit current)[35] standard, namely the Decree of 26 July 1957.

Disabled workers are the third particularly sensitive group to which higher level of protection is set. Although there is no normative definition of 'disabled worker especially sensitive', the most characteristic elements can be pointed out. It would be the worker who has a physical, mental or sensory handicap recognized with some degree of disability that would hinder his or her participation in active working life for an indefinite period, thus excluding circumstantial situations covered by temporary disability. In addition, the disability must be in connection with the concrete work in doing which the disability impairs or affects their ability and qualifications for the job, making the worker therefore more vulnerable to occupational hazards. Risk assessment is especially important in these cases, something that happens to all particularly sensitive workers and requires special attention to the type of disability, degree of involvement and the circumstances of the job. Therefore, it seems necessary that the employer has knowledge of the disability situation, whether physical, psychic or sensory and its expression has been supervening or not, temporary or permanent.

Unlike the regulation provided for women in Article 26 LPRL, Article 25 LPRL has not expressly provided a specific obligation to adapt the work to the disabled worker. Nevertheless, the general principle of adapting work to especially sensitive workers entails its realization to special cases. If it were not possible to take measures to adapt or change the workstation, and occupational risks come together to the disabled worker, the prohibition of work shall proceed. Article 25 LPRL states that 'workers will not be employed in those jobs where, because of [...] their physical, mental or sensory disability duly recognized' themselves or others are put at risk. However, the situation of disabled workers is supplemented by the provisions of labour law on employment of disabled persons and by the public subsidy of the cost of the material adaptation of the workplace.

[2] Collaboration among Employers in Cases of Outsourcing

Earlier in this text we have repeatedly described the employer as the main party liable for ensuring safety, and the recipient of the obligations described in this context. However, the general practice of outsourcing has turned individual employers into complex employers. The individual employer, be they a natural or legal person, organizes and manages the productive process and receives the *utilitas* derived from the work. However, in the case of the complex employer, there is a division between the party organizing the work and that obtaining the *utilitas*, and this is evident in

35. I. García-Perrote, 'Protección de los trabajadores especialmente sensibles a determinados riesgos', in *Seguridad y salud en el trabajo. El nuevo derecho de prevención de riesgos laborales*, ed. M.E. Casas & C. Palomeque (Madrid: La Ley, 1997), 48.

various ways. When using the freedom of enterprise (Article 38 Spanish Constitution), employers have resource to various forms of partnership,[36] some of which do not create subordinate associations between companies (commission or agency contracts, etc.), while others create subordinate associations similar to vertical integration structures (franchises, groups of companies). These bodies are officially beyond the scope of employment relationships; however, they determine the worker's performance by greatly affecting the obligations to ensure safety in the workplace insofar as they are an obstacle to accountability and therefore compliance with obligations. This being the situation, it would be necessary to introduce the notion of the employer as 'party' to the contract, identifying the structured and organized interests between different parties.[37]

One of the reasons for the high rate of accidents in Spain is the extensive use of outsourcing, particularly in the excessive use of contractors or subcontractors as a means of bringing down corporate costs. The companies at the end of the subcontractor chain are usually small enterprises with few financial resources that do not invest in health and safety.[38] This situation calls for strengthening the means of co-ordinating business activities set forth in Article 24 LPRL, an issue which was address in RD 171/2004, of 30 January.

The law identifies various situations: the first being that in which workers from different companies, which are not legally associated, come together in the same workplace. In this case, two separate circumstances are identified. The first circumstance is in which none of the employers owns the workplace in question. In this context, the workplace is merely the place where the services are rendered, and is not intended in the strictest sense described in Article 1.5 ET, i.e., as a productive, specifically organized unit registered as such with the labour authorities.[39] The second circumstance occurs when there is a main employer owing the workplace where workers employed by other companies, contracted or subcontracted by the main company carry out their jobs. The law takes into account the different legal relations between employers in order to impose comprehensive obligations.[40] Obligations are arranged on a scale, and are accumulative. The highest level is occupied by the employer who owns the workplace contracted and subcontracted to other employers. The next level corresponds to the employer who owns a workplace where workers from other companies, with which s/he has no contractual association, carry out their

36. M.A. Alcalá, 'Aspectos económicos e instrumentos jurídicos-mercantiles de descentralización empresarial (outsourcing)', *Revista de Derecho Social,* no. 23 (2003): 81 et seq.
37. M.F. Fernández López, 'El empresario como parte del contrato de trabajo: Una aproximación preliminar', in *El empresario, contrato de trabajo y colaboración entre empresas,* ed. M.F. Fernández López (Madrid: Trotta, 2004), 45.
38. M. Miñarro Yanini, 'Prevención de riesgos laborales en los supuestos de coordinación de actividades empresariales', in *Derecho de la Prevención de Riesgos Laborales,* ed. A. Garrigues (Albacete: Bomarzo, 2009), 340.
39. J. Aparicio Tovar, 'Los sujetos obligados a garantizar un medio de trabajo seguro y saludable. Especial referencia a la cooperación entre empresarios', in *Estudios jurídicos sobre siniestralidad laboral,* ed. J. Terradillos & M. Acale (Sevilla: Junta de Andalucia, 2006), 35.
40. J.M. Goerlich, 'Centros de trabajo compartidos. Contratas y subcontratas. Régimen de responsabilidades', in en VEGA, J. (ed.) *Responsabilidades y responsables en materia de prevención de riesgos laborales,* ed. J. Vega (Gran Canaria: Gobierno de Canarias, 2004), 143.

work. This is followed by employers, with no mutual contractual association who carry out their activity in the same workplace where none is superior to any other because none owns or controls the work premises. Self-employed can fall into any of the above three categories. The new notion introduced by Article 8 of the LETA (Ley del Estatuto del Trabajo Autonomo) contemplates the case in which a self-employed is under contract to the employer who owns the workplace where the former carries out his or her specific services. In this case, the employer is under the obligation of ensuring that the self-employed complies with health and safety rules in the same way as his or her employees.[41]

In the case where several employers not legally associated merely coincide in the same workplace, all are obliged to establish the *required* means of co-operation. This is an open-ended concept meaning that each employer must take into consideration the involvement of workers belonging to other companies when assessing their risks. All employers are obliged to inform each other of the risks involved in their activity, in writing in the case of serious risk, and to immediately inform the other employers of any emergency situation. Each employer must also inform their own workers.

The same obligations are imposed on the employer who owns the workplace and who, moreover, must notify the other employers of any risk intrinsic to the workplace before they start their activity, and indicate what emergency measures must the taken. This employer must likewise be informed by each employer of the risks associated with their activity, and communicate this to the other employers. In cases involving particular danger or which are particularly difficult to co-ordinate, this task must be entrusted to one or more persons qualified in and familiar with health and safety practices. The owner of the workplace must discuss organizing the work involving different companies with the workers' representatives.

In addition to all the above, employers who own the workplace and have contracted or subcontracted other companies are obliged to ensure that the contractors and subcontractors engaged by them are in compliance with health and safety rules. However, a certain confusion arises in Article 24.3 LPRL, which makes the main employer liable for health and safety at the workplace within the limits of his or her 'own activity'. The concept of 'own activity' has been introduced by Article 42 ET in order to limit the liability of the main employer in the event that contractors or subcontractors breach their wage or social security obligations to their workers. The *Tribunal Supremo* (Social Chamber 4), after some hesitation, has adopted a restrictive concept, defining 'own activity' as any task inherent to the production of the goods or services that the main employer's company sells, instead of including all tasks essential for the productive activity. Thus, contractor and subcontractor activities such as cleaning (related to hygiene) or security would be beyond the scope of Article 24.3, and this is absurd.[42] This interpretation would suggest an inconsistency between Articles 24.1 and 24.2 LPRL, which refer health and safety obligations to the workplace, and

41. S. Olarte, *Prevención de riesgos laborales de los trabajadores autónomos* (Granada: Comares, 2009), 63.
42. F. Navarro, *Coordinación de actividades empresariales y prevención de riesgos laborales* (Albacete: Bomarzo, 2005), 45.

Article 24.3 LPRL, which refers them to the activity performed. Indeed, the workplace shall be considered the determining factor.[43] The only possible conclusion is that there is no single way of interpreting the concept of 'own activity'. In the sphere of health and safety this is wider and includes all tasks needed for carrying out the productive activity.

Last but not least, the obligation of the main employer to ensure safety is, according to RD 17/2004, limited to contractors or subcontractors that carry out their activity at the main employer's workplace, which excludes the liability of the main employer in case activities are carried out outside his or her workplace.[44]

The obligation to ensure compliance with health and safety rules also falls to the employer when workers employed by contractors or subcontractors do not carry out their activity in the workplace but use machinery, equipment or products supplied by the main company.

[B] Duties and *Ius Resistentiae* of Workers

The aim of employer's obligations is to guarantee a healthy and safe working environment. As we have discussed above, as the employer owns the means of production, s/he is capable of organizing and running his or her company and can implement the health and safety measures demanded by the law. The position of the worker is completely different. S/he must obey the orders given by the employer correctly exercising his or her managerial duties and has no say in the company, except in very exceptional circumstances. However, this does not mean that the worker has no health and safety obligations.

The first obligation of the worker is to comply with the health and safety measures put in place by the employer and with their instructions in this regard, as established in Article 29 LPRL and Article 13 of the Framework Directive. Workers have the obligation of diligence, and must ensure their own safety and that of other workers who may be affected by their activity. The level of diligence is related to the level of qualification of the worker and to his or her personal circumstances. Workers must make appropriate use of instruments and resources associated with their job. However, to do so, the employer must supply them with the correct instructions and information. Workers are also under the obligation of notifying their superiors, health and safety supervisors or other workers of any possible risk they discover while discharging their duties. They are also obliged to collaborate with the company and authorities to ensure that their working conditions are safe.

Failure on the part of the worker to comply with these obligations entitles employers to impose any penalties within their power. The extent of the penalty will

43. A. Olmo Gascón, *Los derechos laborales del trabajo en contratas* (Granada: Comares, 2002), 439.
44. M. Miñarro Yanini, 'Prevención de riesgos laborales en los supuestos de coordinación de actividades empresariales', in *Derecho de la Prevención de Riesgos Laborales*, ed. A. Garrigues (Albacete: Bomarzo, 2009), 357.

depend on the extent of the breach, which must, furthermore, be attributed to the worker.

Workers are entitled to refuse to carry out orders given by their employer. They may exercise this right by interrupting their activity and leaving the workplace when they discover a serious and immediate risk to their life or health (Article 21.2 LPRL). They cannot be penalized for taking this action. If a serious or imminent risk occurs, it is because the employer has failed to comply with their obligation, or because an unforeseeable situation has arisen. In both cases, the worker cannot be penalized for leaving the workplace. In the sphere of health and safety, the principle of good faith is two-fold: on the one hand, workers must not interrupt their work without a serious or justified cause, and on the other the employer, when making their a posteriori evaluation of the worker's conduct must limit themselves to considering the circumstances at that particular time and place. The personal situation of the worker must also be taken into account.

Workers' representatives may, exceptionally, take over the powers of the employer and order activity to be suspended if a serious or imminent risk arises that affects all or a particular group of workers (Article 21.3 LPRL). The decision to interrupt production must be taken when the employer does not take these measures in a situation of imminent risk, or does not allow others to take them. This decision, which is a temporary expropriation of the managerial powers of the employer, must be taken by the majority of the workers' representatives sitting on the Works Council, or by the Health and safety representatives if the Works Council cannot be convened. It must be immediately communicated to the labour authorities who, within twenty-four hours, must ratify or overrule the decision. The extent and imminence of the risk must be weighed up on the basis of the exact circumstances and in the context of the moment at which the decision was taken. Workers' representative may not be penalized for any decision taken in good faith and workers may freely decide whether or not to follow the order to interrupt their activity. Any worker interrupting their activity retains their entitlement to receive their salary.

§8.04 WORKERS' REPRESENTATIVES

Workers have a right to participate and be represented in the company in all health and safety issues. The division between participation and representation is not well defined, but there are some differences. The first implies workers' contribution to decisions taken within the company, while the second has more to do with channelling the interests of the workers themselves. This distinction is addressed in Directives 2001/86/EC, of 8 October, supplementing the statute for a European company and 2003/3/72, of 22 July, supplementing the Statute for a European Cooperative Society, that introduce the term *involvement* of workers instead of participation.[45]

45. C. Agut, 'Implicación de los trabajadores en la prevención de riesgos laborales en la empresa', in *Derecho de la Prevención de Riesgos Laborales*, ed. A. Garrigues (Albacete: Bomarzo, 2009), 308.

The distinction between participation (or involvement) in health and safety measures and representation of worker' interests can be seen in Spanish law from the time works councils and trade union representatives were empowered to 'defend the workers' interests in matters of health and safety' (Article 34. 2 LPRL), while also conferring the right to participation, which take a different form.

Spain is one of the European countries in which workers are represented through two channels. The first is the uniform, elected body that represents the workers within the company, the Works Council (or Staff Representative in the case of companies with less than fifty workers), and the second is the trade unions, which represent the members of each trade union active in the company.

The legal system of representation in health and safety matters bypasses the trade unions and is organized through the Works Council. The Works Council choose which of its members will have specific health and safety functions, and these are called Health and Safety Representatives. This means that Health and Safety Representatives must be members of the Works Council (Article 35 LPRL). The number of Representatives is determined by the size of the company. As already stressed, the Works Council, or the Health and Safety Representatives if the Council cannot be convened, can decide to suspend production in the event of serious or imminent risk. Health and Safety Representatives are charged with overseeing and ensuring compliance with health and safety rules; receiving information from the employer regarding health and safety plans; and encouraging workers to co-operate with health and safety measures. Generally speaking, they receive information from the employer, and the latter consults them on health and safety matters. Health and Safety Representative receive general information, but this does not release the employer from his or her obligation to inform each worker individually about the risks involved in their specific job. Health and Safety Representatives are under the obligation to treat any information received confidentially, and are given the same guarantees and rights as other members of the Works Council, who cannot be penalized for carrying out their duties.

This system is not compulsory. The law provides for the option of collective bargaining in order to implement a different system for representing workers' interests. Nevertheless, an examination of many collective agreements has shown that the legally recommended system is applied in most cases. Collective agreements only bring in some small differences, mostly related to appointing health and safety representatives.[46] Most collective agreements in Spain are applied *erga omnes*.

The Health and Safety Committee, however, is a participatory body where all issues related to health and safety are periodically discussed (Article 38 LPRL). It is a joint committee formed of health and safety representatives and an equivalent number of management representatives. It must be constituted in all companies or workplaces with more than fifty workers. In addition to promoting initiatives to improve health and safety, the Committee must take part in drawing up and implementing health and safety plans.

46. N. Mendoza Navas, *Prevención de riesgos laborales: la regulación convencional* (Albacete, Bomarzo, 2007), 113.

Workers' representatives cannot be held liable for violations of health and safety rules as it is the employer's obligation to guarantee a safe working environment. If they do not carry out their supervision duties they may be removed from their post by the workers.

§8.05 CONSEQUENCES AND REMEDIES IN CASES OF VIOLATION OF HEALTH AND SAFETY PROVISIONS

[A] Liabilities of the Employer

The formidable array of liabilities imposed on employers if they fail to comply with safety obligations in the workplace is one of the features of Spanish law. Without liability, obligations would have little weight. Liability plays an important role in health and safety because it deters those liable from neglecting their duties. All Spanish regulations must necessarily include liabilities for violations of safety obligations, although there may be several different types of liability, as set forth in Article 5.4 of the Framework Directive, which allows Member States to freely organize their system of liabilities.

In Spain, liabilities may be administrative, criminal, civil or of a social security nature, as established in Article 42 LPRL. The common basis for all liabilities is a breach of the health and safety obligation incumbent on employers, although each follows a different legal system and pursues different ends. The aim of administrative liability is to defend general public interest (Article 103.1 of the Spanish Constitution), as peaceful relations would be jeopardized if the rules guaranteeing a safe and healthy workplace are not respected. The aim of criminal liability is to protect the life, health and safety of workers and is considered particularly relevant for the general public, and for the same reason, results in the State's *ius puniendi*, imposing a penalty on the offenders. They are, therefore, very different, although the purpose of both is to defend public assets, and both are forward-looking, their aim being to dissuade possible future offenders. Civil liability, however, focuses on the individual and is retrospective, its aim being to compensate for damages suffered. Social security liability is one of a wide range of social policies, and aims to provide cash benefits to guarantee a certain level of income for those who would not have been the victims of an occupational accident or disease if the employer had discharged his or her obligations correctly.

[1] Administrative

The task of monitoring and ensuring compliance with health and safety regulations cannot be left solely in the hands of private interests (employers and workers) by implementing the common actions laid down by the law for all those who feel that their rights or interests have been jeopardized. The authorities also have a constitutional duty to ensure standards of health and safety in the workplace (Article 40.2 of the Spanish Constitution). The essence of administrative liability does not reside so much in the State's *ius puniendi* but in the government's management responsibilities, because 'a risk society demands the presence of a risk management State' capable of eliminating these risks.[47] However, violations by employers of their obligation to ensure safety may lead to sanctions, i.e., actions taken by the authorities against those under their control as a result of their conduct. This, in a constitutional state, must be carried out according to the law, a principle laid down years ago by the Constitutional Court.[48] According to this principle, criminal conduct must be predetermined in legislation.[49] In other words, criminal conduct must be clearly defined (typified) within a legal rule, although the legal rule (statute) can also refer to other, lower ranking rules and thus form a body of law. This is what Article 5.2 of the LISOS achieves by establishing that health and safety violations are actions or omissions that breach legal or regulatory standards or the provisions of collective agreements.

Administrative infringements are typified by Articles 11,12 and 13 LISOS. Violations can be minor, serious or very serious, although it is hard to understand the description of the unlawful conduct provided in these articles without reference to the substantive legal rules, regulations or collective agreements governing the matter. It is in these that the employer's obligations are laid down, although it should be stressed that, in order to impose a penalty, violations of the substantive rules must necessarily lead to one of the types set forth in the above articles. Any activity that cannot be classified under the types set forth in the aforementioned articles cannot be penalized.

Violations, being classified as minor, serious and very serious, have their corresponding scale of penalties. Minor violations include, for example, lack of cleanliness in the workplace that does not endanger the health or safety of the workers, or failure to report minor accidents to the labour authorities in the established timeframe and form. Examples of serious violations include failure to carry out risk assessment, failure to report serious occupational accidents to the authorities, or assigning workers to jobs that are incompatible with their personal characteristics. Very serious violations include, for example, inappropriate assignation of workers to jobs that can lead to a serious or immanent risk to their health, failure to observe specific health and safety rules for women during pregnancy or breast feeding, or failure to adopt co-operation measures when several companies coincide in a single workplace.

Following the classification of violations into minor, serious or very serious, the corresponding penalties imposed are likewise classified as minimum, medium or maximum. This classification is based on criteria that include the degree of danger associated with the activity of the company, whether the violation is permanent or occasional, the seriousness of any injury that could be caused, the number of workers affected, or the failure to heed prior warnings. Fines range from Euros (EUR) 40 to EUR 405 for minimum penalties for minor violations to EUR 819,780 for maximum penalties

47. A. Nieto, *Derecho Administrativo Sancionador,* 4rd ed. (Madrid: Tecnos, 2005), 183.
48. STC 25/1984, of 23 February.
49. J. Mercader & C. Tolosa, *Derecho Administrativo Laboral* (Valencia: Tirant lo Blanch, 2000), 566.

for the most serious violations. A recurrence of the violation can double the amount of the fine.

Furthermore, when there is evidence of serious or imminent risk, labour inspectors may issue an order to suspend productive activity. Suspension is maintained until the cause of the same has been rectified, and it enforces action that can be very effective. The contracts of the workers affected by the order are not suspended, but they are in a condition in which they cannot carry out their work for causes beyond their control, and they therefore continue to be entitled to receive their wages and are not obliged to make up for the work lost with any other work done at any other time (Article 30 ET).

In the event of a health and safety violation of exceptional seriousness, the government (or the corresponding autonomous community authorities) may agree to suspend work for a certain period or to close the establishment. This can occur in situations where violations are widespread, and the action is not a sanction but a precautionary measure.

It is important to note that the mere breach of health and safety rules, albeit with no associated injury, can be considered an administrative violation. Injury can also be involved, but it is not an integral part of the violation.

[2] Criminal

The classification of harmful effects on the health and safety of workers originating in the workplace as a criminal offence is not a necessary outcome of the constitutional imperative of Article 40 of the Spanish Constitution. However, experience has shown the need for rules drawn up to protect the safety of workers to include criminal liabilities. On this point, the Spanish system follows other European systems by establishing criminal liabilities in certain violations of the obligation to guarantee the health and safety of workers in the workplace. In view of the scale of administrative liability, the problem lies in justifying and using these measures. The importance of the legal asset protected by these regulations, i.e., the life and safety of workers exposed to risk situations over and above those affecting other members of the public, justifies the use of criminal law[50] over and above administrative proceedings. Neglect of other kinds of liabilities, and the increased exposure to accidents of workers in unstable situations, immigrants, young people or women[51] justify criminal proceedings because the threat of sentencing 'has an educational function' and is 'a deterrent'.[52]

There are, however, many points on which the law on administrative and criminal penalties coincide. First, because there are no health and safety violations specific to penal law, it means that the scope of action of both bodies of laws overlap

50. A. Baylos & J. Terradillos, *Derecho penal del trabajo*, 2nd ed. (Madrid: Trotta, 1997), 113.
51. J. Terradillos, 'Introducción al estudio de la incidencia de las variables de género, inmigración y edad en las cifras sobre siniestralidad laboral', *La siniestralidad laboral. Incidencia de las variables de 'género', 'inmigración' y 'edad'*, ed. J. Terradillos & M. Acale (Albacete: Bomarzo, 2009), 10.
52. J. Terradillos, *La siniestralidad laboral como delito* (Albacete: Bomarzo, 2006), 32.

to a great extent. Preference should be established on the grounds of the principle of *ne bis in idem*. Second, it should be noted the principles of legality and incrimination are required in both penalty laws by Article 25 of the Spanish Constitution. The difference is that in its interpretation criminal law is stricter in demanding these principles. However, criminal rules, both for the construction of the elements of a particular offence and for determining the identity of the perpetrator of the same, must include substantive health and safety regulations.

The Penal Code typifies offences against the life and health of workers (Article 316) by imposing punishments of between six months and three years imprisonment for those who 'in violation of health and safety rules, and being legally bound by them, do not provide the resources needed to enable workers to carry out their activity with adequate health and safety measures' and seriously endanger the life, health or safety of said workers. The criminal offence consists of creating danger when 'it is an objectively ideal medium for affecting life or health',[53] without the need for injury to actually occur. When this same offence is committed through serious negligence, the lower penalty on the scale is applied (Article 317 of the Penal Code). In the case of corporations, criminal liability falls on the directors or departmental managers (Article 318 of the Penal Code). The perpetrator of the offence has to be the party subject to the obligation imposed by health and safety regulations to guarantee a safe working environment, in other words, the employer. When the employer delegates health and safety duties s/he is not criminally liable, and the parties to whom said duties have been delegated are the 'legally obliged' parties referred to in Article 316 of the Penal Code.

General offences of homicide and reckless injury are applied when the worker is effectively the victim of the same. The principle of the individual nature of the punishment prevents corporations from being held criminally liable.

[3] Civil

The immediate aim of civil liability is to redress the consequences of the damage suffered by the worker who falls victim to an occupational accident or disease due to the employer's failure to comply with safety measures. Civil liability consists of compensation, and damage is an integral part of civil liability. Without damage there is no liability.[54] A cause and effect relationship between the unlawful activity and the existence of damage must be established in order to identify the party liable. This is not so much a purely causal relationship but more a legal assessment. Compensation must be complete, covering all damages, although, being a financial restitution, it introduces the serious problem of using financial resources to compensate damages to non-financial assets, namely, personal damages. Courts always have great difficulty in determining the amount of compensation needed to redress the death or personal damage suffered by an individual. Financial restitution, however, has the positive,

53. J. Terradillos, *La siniestralidad laboral como delito* (Albacete: Bomarzo, 2006), 49.
54. J. Aparicio Tovar, 'Sobre la responsabilidad civil del empresario por infracción de las obligaciones de seguridad y salubridad en el trabajo', *Relaciones Laborales*, no. 12 (1994): 12.

practical aspect of allowing corporations to be declared legally liable, because the essence of civil liability is the allocation of financial resources to the consequences of an action, as has long been established in the law.[55]

The above characteristic effects the relaxation of the requirement of fault as the main element of civil liability. Taking fault as the be all and end all of civil liability could mean that the needs of the victim are not adequately addressed.[56] In Spain, civil liability is regulated in Article 1101 and subs. of the Civil Code, which refers to contractual liability, and 1902 and subs., dealing with extra-contractual liability. There is a growing tendency to unify both fields of civil liability and their corresponding rules, and this would lead to a gradual move towards ensuring that liability is objective. However, it must be remembered that according to the letter of the law, there can be no liability for risks without including the actions of the debtor,[57] namely, the employer. The truth is that the level of diligence demanded is very high, as we have seen when discussing the overall obligation that includes an objective standard obliging the employer to apply the best possible measures. The combined interpretation of Articles 1104 and 1105 of the Civil Code leads us to the conclusion that employers must be liable for any foreseeable occurrences and all situations that can be avoided using up to date scientific and technical resources. If an accident occurs, the burden of proof is not reversed, however, the principle of *res ipsa loquitor* (the thing speaks for itself) is on the side of the plaintiff (the worker or their beneficiaries). This obliges the employer to prove that s/he acted based on foreseeable circumstance and prevented all avoidable risks, in other words, in compliance with the provisions of Occupational Health and Safety Law. In this situation, the difficulty lies not in supplying proof, but in what has to be proved.[58]

Employers are directly liable for the damage caused by their workers in the exercise of their duties (Article 1903 CC).

The courts with competence to hear these cases must be Social Courts responsible for hearing disputes stemming from employment contracts. However, civil courts also have competence derived from *Aquilian* or extra-contractual liability. This double jurisdiction at times can result in each court delivering different rulings (specifically, there are discrepancies between Social Chamber 4 and Civil Chamber 1, both belonging to the *Tribunal Supremo*).

[4] Social

Article 115 of the General Social Security Act defines an occupational accident as 'any physical lesion suffered by the worker during, or as a consequence of, his or her job as an employee'. The social security authorities provide the financial benefits established

55. J. Rubio, *La responsabilidad civil del empresario* (Madrid: Academia de Jurisprudencia y Legislación, 1971), 28.
56. G. Viney, *Traité de droit civil. Introdution à la responsabilité*, 2rd ed. (Paris: LGDJ, 1996), 20.
57. F. Calvo Gallego, *La obligación general de Prevención y la responsabilidad civil o contractual* (Pamplona: Aranzadi, 1998), 62.
58. M. Alonso Olea, *La responsabilidad del empresario frente a terceros por actos del trabajador a su servicio* (Madrid: Civitas, 1990), 83.

for insured workers in the event of an occupational accident or disease, irrespective of whether this has been the result of insufficient safety measures. The financial benefits established by law are guaranteed for all insured persons in any event, because the aim of social security is to guarantee the welfare of all individuals with no income, thus giving tangible effect to their status as citizens.

When the accident occurs as a result of insufficient safety measures, Article 123 of the General Social Security Act obliges the employer to pay in addition between 30% and 50% of the benefit. In these cases, the benefit is divided into two parts: one being the ordinary benefit supplied by the social security system, and the other being the extra benefit. In order for the extra benefit to be authorized there must be a strict causal relationship between the lack of safety measures and the accident. The decision to grant this benefit lies with the social security authorities.

Much has been discussed about the nature of this extra benefit, particularly whether it is an administrative penalty or a civil compensation, but no single definition has been reached. The legal principles under which Chamber 4 of the Supreme Court operates is an example of this dichotomy, because after having at times opted for considering the benefit to be a compensation, it then considered it a sanction (STS 22 April 2004, Tribunal Supremo. Sala de lo Social), and finally recognized the difficulty involved in adapting it to typified legal concepts, including sanctions.

The offending employer liable for paying the extra benefit or civil compensation might not be the same as that with whom the worker has a contractual association. This situation can arise in cases of co-operation between employers.[59]

[B] Compatibility of Liabilities

Articles 42.1 and 3 of the LPRL state that failure on the part of the employer to comply with their obligations will give rise to administrative liability and, as the case may be, criminal and civil liabilities. These liabilities are declared compatible with the extra social security benefits and with civil compensation.

In Article 25 of the Spanish Constitution, recognition of the principle of *ne bis in idem* as a result of the principle of legality prevents the imposition of administrative and criminal penalties together. Article 3 of the LISOS clearly establishes this incompatibility when the subject, the offence and the grounds are the same. Preference is given to criminal penalties, as can be seen from STC 2/2006, of 16 January, and as such, proceedings to apply an administrative penalty must be suspended if another, criminal proceeding has also been initiated (Article 52.3 LISOS). However, acquittal in the criminal court does not prevent an administrative penalty from being imposed. Identification of the fact, which prevents a double penalty, enables minor negligence to be liable to administrative penalty, while serious offences are tried under criminal

59. C. Alfonso Mellado, 'Manifestaciones de la responsabilidad empresarial en el derecho del trabajo y la Seguridad Social', in *La responsabilidad laboral del empresario: siniestralidad laboral*, ed. N. Pumar (Albacete, Bomarzo, 2006), 38.

law.[60] Therefore, an activity might not merit criminal prosecution, but could be liable for administrative proceedings.

Identifying the extra social security benefit with an administrative penalty, on the one hand, and civil compensation on the other, is a very controversial issue. If the extra benefit is a penalty, the principle of *ne bis in idem*, prohibiting a two-fold sanction, would be applicable. If, on the other hand, it is considered a compensation, it would have to be deducted from the amount awarded as civil compensation. The two-fold penalty cannot be prohibited on the grounds of incompatibility with an administrative penalty because there is no shared basis; the extra benefit is awarded to meet the needs of the victim, and this does not occur in the case of an administrative penalty.

The sum of the social security benefits, including the extra benefit, is considered an additional civil compensation. However, the problem arises when calculating the amount of civil compensation. Even jurisdiction issues are in dispute: Civil Chamber number 1 of *Tribunal Supremo* continues to uphold its competence in this matter,[61] as does Social Chamber 4 of the same Tribunal. The Civil Chamber tends to consider social security benefits and additional compensation awarded for moral damages or *pecunia doloris* as independent rather than complementary, and as a result, usually awards higher compensation. Chamber 4 adheres to the principle that social security benefits derived from the occupational accident are an insurance paid by the employer to cover his or her liability, and for this reason the compensation paid in the form of benefits should be deducted from the total amount of compensation awarded.[62]

However, as according to the law, benefits and compensation are compatible, the Social Chamber 4ª considers them to be complementary. To calculate the total amount to be paid in damages, net of the social security benefits, it has used as a guide the scale included in the Motor Vehicle Insurance Law. After its decision of 17 July 2007, the Social Chamber has gradually refined its position and modified this particular criterion. Therefore, it cannot be said that a definitive solution has been reached. In any event, the notion of social security benefits paid for occupational accidents being the result of an employer's liability insurance is open to criticism. Indeed, social security benefits, including those derived from occupational accidents, have their origin in a 'readjustment of the contingency of a occupational accident'.[63]

§8.06 CONCLUSIONS: SOME SPECIAL FEATURES OF THE SPANISH CONSTRUCTION SECTOR

The construction sector has had an enormous influence on the Spanish economy up to 2008 when, due to the effect of the economic and financial crisis, it fell sharply. In spite of this, it is still an important sector, full of striking contrasts, containing both large

60. J. Terradillos, *La siniestralidad laboral como delito* (Albacete: Bomarzo, 2006), 128.
61. STS, 1ª, 15 January 2008.
62. A. Desdentado, 'La compatibilidad de responsabilidades y el cálculo de la indemnización por daños', in *La responsabilidad laboral del empresario: siniestralidad laboral*, ed. N. Pumar, (Albacete, Bomarzo, 2006), 79.
63. M. F. Fernández López, 'Accidente de trabajo: prestaciones e indemnizaciones. Un nuevo paso en el "vaivén" judicial', *Revista de Derecho Social*, no. 45 (2009): 69.

international Spanish companies involved in large infrastructure project (highways, railway lines, large dams) and building all over the world, and thousands of tiny companies and self-employed. This contrast does much to explain the widespread use of subcontractors. Most of the workers in the construction sector have temporary contracts and there is a significant number of foreign workers, all of which, when combined with a potentially dangerous profession, both leads to and explains the high accident rate. For this reason, several industry-specific provisions have been implemented.

Some of these rules have their origin in Directive 92/57/EEC, of 24 June, on temporary or mobile construction sites, specifically RD 1627/1997, of 24 October, incorporating the aforementioned Directive into Spanish law while at the same time implementing Article 24 LPRL on collaboration between employers. The important Act 32/2006 on subcontracting in the construction industry stemmed from a popular legislative initiative put forward by the Construction Federation, part of the trade union *Comisiones Obreras*.[64]

RD 1627/1977 distinguishes various parties forming the productive cycle of this sector: the developed, the co-ordinator, the contractor, the subcontractor and the self-employed worker. The developer, being the owner of the project, although not directly involved in the building, is in charge of enlisting the services of the contractor and is affected by the co-ordination duties described in Article 24 LPRL. The process can be separated into two phases, the project and the execution. The project phase must include a safety analysis that will subsequently be incorporated into the execution stage. At the execution stage this safety analysis is used to draw up a health and safety plan for each contractor, which must be approved by the co-ordinator.[65] The co-ordinator is the key to all construction projects involving more than one company. The co-ordinator must be technically qualified, and is appointed by the developer. The contractors and subcontractors must comply with the health and safety measures and make sure that all their employees comply with health and safety regulations. The main company, together with the contractors and subcontractors, are jointly liable for their obligations, as long as the violation occurs in the workplace (Article 42.3 LISOS). Self-employed, in addition to complying with their safety obligation in a similar way as employees, must obey the instructions of the co-ordinator.

The aim of Act 32/2006, regulating subcontracting in the construction industry, is to protect the health and ensure the safety of workers. The nature of the Act is clearly to prevent[66] the special risks to which construction workers are exposed. It follows the same line as the Community Health and Safety at Work Strategy (2007–2012)[67] that

64. More than 500,000 people signed the initiative that the Spanish Lower Chamber, with a Popular Party majority, did not act on. Two years later, with the Lower Chamber under a different majority, it was made into an Act as a result of public outrage following a construction accident in which seven workers lost their lives.
65. J. López Gandia, M.J. Romero, D. Toscani, *La prevención de riesgos laborales en el sector de la construcción* (Albacete: Bomarzo, 2008), 13.
66. A. Merino, *Comentarios a la Ley 32/2006, de 18 de octubre, reguladora de la subcontratación en el sector de la construcción* (Albacete: Bomarzo, 2006), 29.
67. COM (2007) 62 final.

identifies increased workplace risks as a result of the spread of subcontracting. To achieve its health and safety objectives, the Act first demands that companies entering the construction industry subcontractor market must have a genuine productive organization, i.e., that they are not fake, manpower trafficking companies. This demand that the companies involved are genuine is further strengthened by the obligation that 30% of company employees should have permanent contracts, a requirement that improves the quality of work and encourages vocational training. The company must prove compliance with legal requirements by registering with a Register of Accredited Companies. All contractors or subcontractors must always keep a Subcontractor's register in the workplace in which they record all the subcontracting agreements entered into. The register must be open to inspection by the labour authorities, other subcontractors and the workers' representative.

One of the most important new provisions introduced has been to limit vertical subcontracting to only three levels (except in special cases), and to ban self-employed or business owners with no employees from forming part of the various subcontracting levels. The developer or main contractor may subcontract several activities on the first level (horizontal contracting), and so it is common to find a number of different companies working on the same construction site.

Obviously, there are limits to freedom of enterprise, and these limits must be contrasted with the freedom of establishment and freedom to provide services described in Articles 49 and 56 TFEU. The requirement that companies must have a genuine productive organization, and limiting the number of vertical subcontractors to three, seems not to be in contrast with EU Law unfair because the aim is first to guarantee the health and safety of workers, and second to 'streamline the market'.[68]

The obligation to register in the Registry of Accredited Companies is in line with the interpretation of this issue made by the European Court of Justice.[69] Under Spanish law registry can be made on the basis of a statement made by the company's representative, a situation that clearly devaluates the ultimate aim of ensuring protection.[70] A similar conclusion can be made regarding the requirement of keeping a Subcontractor register. This can be subject to mutual recognition, provided the law of the country of origin of the subcontractor entering the Spanish market demand that the Book should contain all the information required under Spanish law. However, it should be borne in mind that the Book must be kept in the workplace. Neither does EU Law stand in the way of the requirement that subcontractors must have a minimum number of workers with a permanent job contract. This requirement does not specify a minimum period of employment before entering into the contract agreement, which could be problematic.

Ensuring health and safety in the workplace, associated with protecting the life and dignity of the individual, is part of the Charter of Fundamental Rights of the EU, the

68. M. Llobera Vila, *Límites a la subcontratación en el sector de la construcción y libre prestación de servicios* (Albacete: Bomarzo, 2010), 67.
69. ECJ, 21 Oct. 2004, C-45/03, *Commission v. Luxemburg* and of 19 Jan. 2006, C-244/04, *Commission v. Germany*.
70. M.F. Fernández López, 'El derecho de las contratas entre la licitud y la prevención', *Revista de Derecho Social*, no. 36 (2006): 54.

European Social Charter and of the constitutions of most Member States. It could, therefore, form the basis for justifying a 'weak interpretation'[71] of economic freedom, as established in the decision of the ECJ of 14 October 2004, C-36/02, *Omega*.

BIBLIOGRAPHY

Agut, C. 'Implicación de los trabajadores en la prevención de riesgos laborales en la empresa'. In *Derecho de la prevención de riesgos laborales,* edited by A. Garrigues. Albacete: Bomarzo, 2009.

Alcalá, M.A. 'Aspectos económicos e instrumentos jurídicos-mercantiles de descentralización empresarial (outsourcing)'. *Revista de Derecho Social* 23 (2003).

Alfonso Mellado, C. 'Manifestaciones de la responsabilidad empresarial en el derecho del trabajo y la Seguridad Social'. In *La responsabilidad laboral del empresario: siniestralidad laboral,* edited by N. Pumar. Albacete: Bomarzo, 2006.

Alonso Olea, M. *La responsabilidad del empresario frente a terceros por actos del trabajador a su servicio.* Madrid: Civitas, 1990.

Alonso Olea, M. & M. E. Casas Baamonde. *Derecho del Trabajo.* 26th edn. Madrid: Civitas, 2009.

Alonso-Olea García, B. 'La prevención de riesgos laborales del trabajador autónomo'. *Aranzadi Social* 16 (2008).

Aparicio Tovar, J. 'Las obligaciones del empresario de garantizar la salud y la seguridad en el trabajo'. *Revista Española de Derecho del Trabajo* 49 (1991).

Aparicio Tovar, J. 'Sobre la responsabilidad civil del empresario por infracción de las obligaciones de seguridad y salubridad en el trabajo', *Relaciones Laborales* 12 (1994).

Aparicio Tovar, J. 'El papel de la negociación colectiva en materia de salud y seguridad en el trabajo'. In *La nueva función de la negociación colectiva y los acuerdos interconfederales sobre el empleo.* Madrid: Ministerio de Trabajo, 1998.

Aparicio Tovar, J. 'Los sujetos obligados a garantizar un medio de trabajo seguro y saludable. Especial referencia a la cooperación entre empresarios'. In *Estudios jurídicos sobre siniestralidad laboral,* edited by J. Terradillos & M. Acale. Sevilla: Junta de Andalucía, 2006.

Baylos, A. & Terradillos, J. *Derecho penal del trabajo,* 2rd. edn. Madrid: Trotta. 1997.

Calvo Gallego, F. *La obligación general de Prevención y la responsabilidad civil o contractual.* Pamplona: Aranzadi, 1998.

Cavas, F. 'Breves consideraciones sobre la prevención de los riesgos psicosociales'. In *Los Riesgos Psicosociales: Teoría y Práctica,* edited by C. Sanchez Trigueros. Pamplona: Aranzadi, 2009.

Desdentado, A. 'La compatibilidad de responsabilidades y el cálculo de la indemnización por daños'. In *La responsabilidad laboral del empresario: siniestralidad laboral,* edited by N. Pumar. Albacete: Bomarzo, 2006.

71. M. Llobera Vila, *Límites a la subcontratación en el sector de la construcción y libre prestación de servicios* (Albacete: Bomarzo, 2010), 67.

Diaz Descaalzo, M.C. 'Los riesgos piscosociales en el trabajo: el estrés laboral, el síndrome del quemado y el acoso laboral. Su consideración como accidente de trabajo'. *Revista de Derecho Social* 17 (2002).

Fabregat Monfort, G. *La obligación empresarial de tutela frente al acoso laboral en la empresa.* Albacete: Bomarzo. 2011.

Frenandez López, M. F. 'El acoso moral en el trabajo ¿una nueva forma de discriminación?'. *Revista de Derecho Social* 19 (2002).

Fernandez López, M.F. 'El empresario como parte del contrato de trabajo: Una aproximación preliminar'. In *El empresario, contrato de trabajo y colaboración entre empresas,* edited by M.F. Fernández. Madrid: Trotta, 2004.

Fernandez López, M.F. 'El derecho de las contratas entre la licitud y la prevención'. *Revista de Derecho Social* 36 (2006).

Fernandez López, M.F. 'Accidente de trabajo: prestaciones e indemnizaciones. Un nuevo paso en el "vaivén" judicial'. *Revista de Derecho Social* 45 (2009).

García Villaverde, M.D. 'El ámbito de aplicación de la Ley de Prevención de Riesgos Laborales'. In *Derecho de la prevención de riesgos laborales,* edited by A. Garrigues. Albacete: Bomarzo, 2009.

García-Perrote, I. 'Protección de los trabajadores especialmente sensibles a determinados riesgos'. In *Seguridad y salud en el trabajo. El nuevo derecho de prevención de riesgos laborales,* edited by M.E. Casas & C. Palomeque. Madrid: La Ley, 1997.

Garrigues, A. 'El empleador y su deber de prevención (III). Obligaciones instrumental. Servicios de prevención y otras modalidades de organización técnico-preventiva'. In *Derecho de la prevención de riesgos laborales,* edited by A. Garrigues. Albacete: Bomarzo, 2009.

Garrigues, A. 'El combinado Ley ómnibus-RD 337/2010: Un paso adelante (*ma non troppo*) para la racionalización de las modalidades de organización técnico-preventiva y la integración de la prevención en la gestión empresarial'. *Revista de Derecho Social* 50 (2010).

Goerlich, J.M. 'Centros de trabajo compartidos. Contratas y subcontratas. Régimen de responsabilidades'. In *Responsabilidades y responsables en materia de prevención de riesgos laborale,* edited by J. Vega. Gran Canaria: Gobierno de Canarias. 2004.

Gonzalez Labrada, M. *Seguridad y salud en el trabajo y responsabilidad contractual del empresario.* Barcelona: Cedecs, 1996.

Gonzalez Ortega, S. & Aparicio Tovar, J. *Comentarios a la Ley 31/1995, de Prevención de Riesgos Laborales.* Madrid: Trotta, 1996.

Lopéz Gandía, J. & M.J., Romero & Toscani, D. *La prevención de riesgos laborales en el sector de la construcción.* Albacete: Bomarzo, 2008.

Llobera Vila, M. *Límites a la subcontratación en el sector de la construcción y libre prestación de servicios.* Albacete: Bomarzo, 2010.

Martin Hernández, M.L. *El derecho de los trabajadores a la seguridad y salud en el trabajo.* Madrid: CES, 2006.

Martin Valverde, A. 'La formación del Derecho del Trabajo en España. Estudio preliminar'. In *Legislación social en la historia de España. De la revolución liberal a 1936*, edited by A. Martin Valverde et al. Madrid: Congreso de los Diputados, 1987.

Martinez Barroso, M.R. *Riesgo psicosocial en el sistema de protección social.* Sevilla: Laborum, 2007.

Mendoza Navas, N. *Prevención de riesgos laborales: la regulación convencional.* Albacete: Bomarzo, 2007.

Mercader, J. & Tolosa, C. *Derecho Administrativo Laboral.* Valencia: Tirant lo Blanch, 2000.

Merino, A. *Comentarios a la Ley 32/2006, de 18 de octubre, reguladora de la subcontratación en el sector de la construcción.* Albacete: Bomarzo, 2006.

Miñarro Yanini, M. 'Prevención de riesgos laborales en los supuestos de coordinación de actividades empresariales'. In *Derecho de la prevención de riesgos laborales,* edited by A. Garrigues. Albacete: Bomarzo, 2009.

Moreno, A. *La prevención de riesgos laborales de los trabajadores especialmente sensibles'.* Valencia: Tirant lo Blanch, 2010.

Navarro Nieto, F. *Coordinación de actividades empresariales y prevención de riesgos laborales.* Albacete: Bomarzo, 2005.

Nieto, A. *Derecho Administrativo Sancionador.* 4rd. edn. Madrid: Tecnos, 2005.

Olarte, S. *Prevención de riesgos laborales de los trabajadores autónomos.* Granada: Comares, 2009.

Olarte, S. & García, M. 'La regulación de los riesgos psicosociales en España: experiencia y prospectiva'. *Revista de Prevención de Riesgos Psicosociales Bienestar en el Trabajo* 1 (2010).

Olmo Gascón, M. *Los derechos laborales del trabajo en contratas.* Granada: Comares, 2002.

Pound, R. *Law and Morals.* Chapel Hill: Oxford University Press-The University North Caroline Press, 1926.

Purcalla, M. 'El empleador y su deber de prevención (II): Obligaciones específicas (colectivos especialmente sensibles)'. In *Derecho de la prevención de riesgos laborales,* edited by A. Garrigues. Albacete: Bomarzo, 2009.

Rivas Vallejo, P. 'El empleador y su deber de prevención (I). Fundamento y obligaciones generales'. In *Derecho de la prevención de riesgos laborales,* edited by A. Garrigues. Albacete: Bomarzo, 2009.

Romero Rodenas, M.J. *La protección frente al acoso moral en el trabajo.* Albacete: Bomarzo, 2004.

Rubio, J. *La responsabilidad civil del empresario.* Madrid: Academia de Jurisprudencia y Legislación, 1971.

Terradillos, J. *La siniestralidad laboral como delito.* Albacete: Bomarzo, 2006.

Terradillos, J. 'Introducción al estudio de la incidencia de las variables de género, inmigración y edad en las cifras sobre siniestralidad laboral'. In *La siniestralidad laboral. Incidencia de las variables de 'género', 'inmigración' y 'edad'*, edited by J. Terradillos & M. Acale, Albacete: Bomarzo, 2009.

Valdés de la Vega, B. 'El tratamiento del acoso moral desde la autotuela colectiva'. In *Los medios de tutela frente al acoso laboral,* edited by Valdés de la Vega, Granada: Comares, 2007.

Villa, L.E. De La 'La influencia de la Ley de Accidentes de Trabajo de 1900 en la construcción del ordenamiento laboral español'. In *Cien años de Seguridad Social,* edited by B. Gonzalo & M. Nogueira, Madrid: La Fraternidad, 2000.

Viney, G. *Traité de droit civil. Introdution à la responsabilité*. 2rd. edn., Paris: LGDJ, 1996.

CHAPTER 9

Occupational Health and Safety in a Diverse, Post-Industrial Society: A Swedish Dilemma

Maria Steinberg

§9.01 INTRODUCTION AND HISTORICAL BACKGROUND

On the whole, it can be said that Swedish work environment legislation is a product of industrial society and the problems it has generated over time.[1] The Swedish work environment legislation has its roots in a male-dominated, homogeneous industrial work force. The legislation of today is therefore not always well suited for a workforce where most women want to work full-time, where every fifth person has his roots in a country outside Sweden, and where the population are growing older. Industrial workers are fewer every year and civil servants and other professionals are becoming a larger part of the work force.[2] One can say that present problems have more to do with stress than with physical dangers on the job.

The work environment question returned to the public agenda at the beginning of 2012, when the prime minister proposed that Swedish workers would have to postpone retirement to age 75 in order to preserve the welfare state in its present form.[3] At present, the retirement age in Sweden is flexible, ranging from 61–67 years.[4] The Chairman of the Swedish Confederation of Professionals (SACO)'s reply to the prime

1. Håkan Hydén, 'Hur får vi en bättre arbetsmiljö? Om strategier i arbetsmiljölagstiftningen över tid och vad vi kan lära av det?', in *Framtidens arbetsmiljö-och tillsynsarbete*, ed. Bo Johansson, Kaj Frick, Jan Johansson (Lund: Studentlitteratur, 2004), 127.
2. Swedish Government Official Reports series, 'Arbetsmiljön och utanförskapet-en tankeram för den framtida arbetsmiljöpolitiken' (Stockholm: DS 2008:16), 25.
3. Ewa Stenberg, 'Fler måste kunna ändra karriär mitt i livet', *Dagens Nyheter*, 7 February 2012, sec. News.
4. Sweden http://www.pensionsmyndigheten.se/AnsokOmPension.html, 3 May 2008.

minister's suggestion was that there was a need for an increase in working time and that this increase could only be feasible if work environments became more worker-friendly.[5]

Problems on the agenda today concern the lack of legal protection against stress, unlimited working time for many professionals, harassment, the protection of 'whistle blowers', and the use and abuse of social media and no discussion of the rate of suicides as part of a bad work environment. The largest trade union for white collar workers (*Unionen*) claims that 20% of its members work overtime.[6] The chairman of TCO (Swedish Confederation of Professional Employees), with 1.2 million members, considers stress the worst work environment problem of this decade.[7]

About half of Sweden's 9 million inhabitants are in the workforce. The majority of adults work full-time. Ninety percent of working men with children at home work full-time; 70% of working women work full-time – though that figure drops to 66% for women with pre-school children at home.[8] The Swedish workforce is sharply segregated by sex, the female part being largely concentrated in the 'caring' sector: e.g., in hospitals, schools, and institutions for the elderly.[9] More women than men suffer harm from their work environment generally, though men are more often the victim of accidents on the job. With women it is often a matter of ergonomic injury or of stress or other psychosocially debilitating condition.[10]

Over the past hundred years, the majority of Swedish governments have been headed by the Social Democrats. For the past six years, however, the Alliance – a coalition of the Conservative (*Moderaterna*), Liberal (*Folkpartiet*), Central (*Centerpartiet*), and Christian (*Kristdemokraterna*) parties have been in power. The biggest change brought by the Alliance was cutting the budget for the Work Environment Authority (hereinafter WE Authority) and closing down the Working Life research centre (*Arbetslivsinstitutet*).

In 2008, the Government announced a framework for discussion about the work environment. This framework was published as part of a governmental investigation in which it was pointed out that the Alliance was aware of the fact that the workforce needed to be more flexible. It was estimated that by 2020 21% of the population would be over 65 years of age. Meanwhile, in 2008, some 20%–25% of the workforce in response to inquiries said they suffered from some kind of disability. The most frequent

5. Göran Arrius, chairman of SACO http://www.saco.se/Nyheter--debatt/Aktuellt-fran-Saco/Debatt/hasardinslagen-i-ppm-frater-pa-hela-pensionssystemet/. This was a press release from SACO, published in *Dagens Nyheter*, Sec. Debates, 11 Feb. 2012.
6. Åsa Johansson and Daniel Gullstrand, 'Alltid uppkopplad –Aldrig avkopplad', in *Undersökning om det gränslösa arbetslivet* (Stockholm: Unionen, 2011), 8.
7. Eva Nordmark, chairman of TCO comments in a press release, 2 Mar. 2012, on a study made by Roger Mörtvik and Jana Fromm, *TCO granskar stressbarometern 2012, Dags för åtgärder 1/12* (Stockholm: TCO, 2012). http://www.tco.se/Templates/Page1____238.aspx?DataID=12476 accessed on 3 May 2008.
8. Most small children in Sweden stay home with one of their parents the first one and a half years. After that, they go to day care which is subsidized by the Local Government.
9. Helene Sandmark (ed.), *Perspektiv på kvinnors hälsa i arbetslivet* (Lund: Studentlitteratur, 2012),11.
10. Kerstin Fredriksson, 'Det ojämlika arbetslivet', in *Perspektiv på kvinnors hälsa i arbetslivet,* ed. Helene Sandmark (Lund: Studentlitteratur, 2012), 21-22.

work-related illnesses today derive from ergonomic factors. The second most common are those traceable to organizational or social factors. The problem of unlimited working hours in certain sectors was discussed, as was the WE Authority and its role as agent of control and source of information. One of the Alliance's key aims was to create a work environment which would include as many people as possible. They also stressed the need to achieve a better balance between the State and the Social Partners (i.e., the representatives of both the confederations of employers and the trade unions).[11] This may mean that the Social Partners will have to take more responsibility for the work environment.

In 2009, the Government presented a new investigation under the title *Inclusive Working Life* (SOU 2009:93). The idea was to develop 'more and better jobs' - as a matter of survival for Sweden and Europe.[12]

The Government realized there is a need for improvement in non-traditional areas of work. In the spring of 2011, the Government asked the WE Authority to develop and put into effect inspections that would reduce the number of women being driven out of the workforce by difficulties related to their work environment. The focus was to be on ergonomics, the source of most problems that cause women to go on sick leave. Women in Sweden take 45% more sick days than men do. And since more women than men are forced to quit work before retirement age, they tend to have smaller pensions and a less secure retirement.[13]

In July 2011, the Government launched an investigation aimed at *A Better Work Environment through More Effective Sanctions* (SOU 2011:57).[14] The idea is to replace some of the criminal sanctions for less serious offences with administrative fines up to approximately Euros (EUR) 100,000. All prohibitions and injunctions from the WE Authority shall include conditional fines. Today these can lead to criminal sanctions.

The Government is aware of psychosocial problems such as stress and harassment. In December 2011, they invited the Social Partners to a seminar on psychosocial problems arising from the work environment.[15] At present, there is no special Provision focused on psychosocial issues, though several proposals for such a Provision have been made since 1980. The employers' confederations are against such a Provision, as it would, they say, impinge on the employers' right to manage their employees. Because of this basic principle, the WE Authority, will not issue such a Provision without the consent of all parties.

The aim of this article is to describe and analyse Swedish legislation concerning Occupational Health and Safety from a labour law perspective. The approach will be as follows.

---

11. Swedish Government Official Reports series, 'Arbetsmiljön och utanförskapet-en tankeram för den framtida arbetsmiljöpolitiken' (Stockholm: DS 2008:16), 33, 36,54-56, 136-138.
12. Swedish Government Official Reports series, 'Inkluderande arbetsliv' (Stockholm: DS 2009:93), 16.
13. Sweden Government Decision 2011-05-12. (A2011/2209/ARM) and http://www.av.se/dokument/aktuellt/Artikel_B_Callermo.pdf. 3 May 2012.
14. Swedish Government Official Reports series, En bättre arbetsmiljö genom effektivare sanktioner (Stockholm: SOU 2011:57).
15. Swedish Ministry of Labour, 'Invitation to a Conference on the psychosocial work environment' Stockholm, 5 Dec. 2011.

First, the historical background will be sketched out. Then will follow a presentation of the basic concepts of Occupational Health and Safety Law. Something will then be said about personal scope and application, duties of the parties involved, compliance and control, the key role of workers' representatives in health and safety management, the consequences for the parties concerned, remedies for breaches of health and safety provisions, and some of the features of the Swedish health and safety legal system, which will include some final comments.

Sweden started to become an industrial nation in the middle of the nineteenth century. As in other European countries, industrial workers in Sweden began to form trade unions in response to their dissatisfaction with working conditions. This resulted in many strikes. The Swedish Government realized they must do something about this and, inspired by other European countries, they proposed a preventive law and State supervision.

This first Swedish National Health and Safety Act (*Yrkesfarelagen* 1889:19), was limited to protecting workers in the industrial and construction sectors. Employers were made responsible for the lives and health of their employees. This early legislation was accompanied by advice on how to prevent accidents in the course of working with certain tools and machines and how to render work environments safe from disease caused by bad ventilation, dust, smoke, and steam. Recognition was given to the need for training in dangerous jobs and the risks associated with moving heavy objects. Employees were obligated to be careful and cooperative, and a State agency, the Labour Inspection, was created to further these aims. The Inspection had the responsibility, through its inspectors, of informing employers of potential dangers and advising them on how to protect their employees from harm. In cases of acute danger, they had the power to order a work stoppage. Employers, furthermore, faced the possibility of being hit with heavy fines for neglecting safety precautions.

In 1900, a special law was passed for the protection of minors and women: *The Act on the Use of Minors and Women Working in Industrial Occupations* (1900:75). A woman was not allowed to be employed within four weeks of giving birth unless she had the dispensation of a special medical certificate. Women were not allowed to work in mines. As a result of this law, new mothers were deprived of an income for four weeks and women who had been working in the mines lost their jobs. In 1909, women were prohibited from working at night in certain industries (1909:131). The prohibition of night work for women had been advocated by the *International Society for Legal Safety at Work*, formed in Paris in 1900.[16] Work by minors was also limited.

Swedish workers were still not happy with the conditions in industry. In 1909 there were frequent and large strikes. This prompted the Government to initiate a revision of the law. Representatives of the workers and the trade unions were canvassed for their opinions. Some thought it was strange that the officials coming from the Labour Inspection only met with employers in the course of their inspections; they requested that the inspectors should also have contact with them. After all, they

16. Maria Steinberg, 'Är arbetsmiljölagen sexistisk?', in *Perspektiv på kvinnors hälsa i arbetslivet*, ed. Helene Sandmark (Lund: Studentlitteratur, 2012), 224–226.

argued, it was the workers who actually used the machinery and breathed the air of the factory floor; it was they who knew best what the dangers were that they faced.[17]

The revision led to the Work Protection Act of 1912 (1912:206). This new Act embraced all workers in commercial enterprises industrial or otherwise. It included the special restrictions on work by women and children. It gave workers the right to elect a spokesman for their interests (Swedish *ombud,* as in *ombudsman*), and it required inspectors to confer with these officials. It also imposed stiffer penalties: imprisonment for up to six months could be the consequence to an employer who violated a prohibition issued as a result of an inspection. Workers could be fined for creating a hazard by removing protective apparatus.

Both old and new restriction on women's work became part of the 1912 Act: women were forbidden to work in environments where there was a considerable risk of accidents, or at any job that was very strenuous or might be dangerous to their health. Furthermore, they were to be kept out of work environments where they might be exposed to 'moral' danger. A special Inspector for Female Employees was appointed by the Labour Inspection Board, whose duty was not only to inspect women's working conditions, but also to seek ways to improve their dietary habits, living conditions, and health insurance arrangements. The Special Inspector was also expected to work for the 'moral and intellectual' improvement of his charges.[18]

The Act was revised twice during the 1930s. In 1931, the power of the *Ombud* was considerably expanded. Employers were now required to consult the *Ombud* concerning health and safety issues in the workplace, while the *Ombud* was required to collect information about conditions affecting health and safety there. S/he was also expected to be proactive in promoting health and safety, and one of the *Ombud's* duties was to persuade his or her fellow employees to participate in this activity. If the *Ombud* received a complaint or became aware of a problem, it became his or her responsibility to engage the employer or employer's representative in an effort to rectify the situation. The employer was obliged to report back to the *Ombud* within a reasonable time (*skälig tid*), taking note of the problem and suggesting means for solving it. If the employer failed to solve the problem, the *Ombud* could turn to the Labour Inspection for help. This regulation was influenced by International Labour Organization (ILO) recommendation number 131 of 1929.[19]

In 1938, there was a further revision. More co-operation between employer and employees was mandated. The *Ombud,* now called the Safety Delegate (hereinafter

17. Maria Steinberg, *Skyddsombud i allas intresse* (Stockholm: Norstedts Juridik, Dr of law, diss. University of Stockholm, 2004), 67–68.
18. Sweden, Government Bill 1912:104 Förslag till lag om arbetarskydd, 59-60, 134. The Swedish Courts study Bills when making decisions.
19. Maria Steinberg, *Skyddsombud i allas intresse* (Stockholm: Norstedts Juridik, Dr of law, diss. University of Stockholm, 2004), 74-77. See also Annika Åkerblom, *Arbetarskydd för kvinnor, Kvinnlig yrkesinspektris i Sverige 1913-1949* (Uppsala: Uppsala Studies in Economic History, 43, 1998), Ph.D. diss.

'SD'), could also be elected by the trade union. Security Committees were proposed for 'larger' workplaces – the word 'larger', however, not being defined.[20]

In 1942, the first health and safety collective agreement was signed between the Employees confederation (SAF) representing the private sector, and the blue collar union (LO). There was hope on the part of the employers that the State would be less insistent in regulating health and safety issues if collective agreements were in effect.[21]

In 1949, a new Work Safety Act (1949:1) was ratified. A National Board of Occupational Safety and Health (NBOSH) took over the main responsibilities regarding health and safety. Formerly these had been under the supervision of the National Insurance Bureau (or, in the case of women, that of the Board of Social Affairs). The new Act further included farm workers as well as pupils in trade schools, and soldiers still in training. It also covered inmates of institutions, alcoholics, and the mentally disabled (if these performed any kind of work). Further special restrictions on women workers were incorporated into the 1949 Work Safety Act (1949:1), but the prohibition against women working in environments where there might be 'moral' danger was dropped. Meanwhile in 1942, inspired by the collective agreement between SAF and LO, a special chapter entitled Cooperation Between Employer and Employees ('Employees' here referring principally to SD) was added to the Act. Employees were now permitted to appoint a SD if there were at least five persons employed on the premises (previously the requirement had been ten persons). Nevertheless there remained the problem of SD for employees in small businesses, especially where they changed frequently, as in the construction industry. To solve this problem, the trade unions were allowed to appoint Regional SD with the same powers as those appointed for specific workplaces, but these had to be initially approved by the NBOSH. Anyone, including an employee, who harassed a SD can be made to pay compensation for their damage.

It was now recommended that Safety Committee (hereinafter 'SC') consisting of employers and SD be formed wherever there were fifty or more employees.

The 1950s and 1960s saw the development of collective agreements concerning the work environment. This began in the private sector, where the setting up of the Occupational Health Service (hereinafter OHS) was proposed for all who were covered by such agreements.[22] Meanwhile, the most important reform enacted during this time was the 1963 addition to the Work Safety Act (1963:245), which made the Act applicable to all public employees.

In the late 1960s, there was a serious epidemic of strikes in Sweden. One result of this was an increased willingness to revise and improve labour law, chiefly in the

20. Sweden Government Bill '1930:71 med anhållan om riksdagens yttrande rörande vissa av den internationella arbetsorganisationens konferens 1929 fattade beslut', 78 and Bill '1938:265 med förslag till lag angående ändringar i vissa delar av lagen den 29 juni 1912 (nr 206) om arbetarskydd m m', 55-56.
21. Maria Steinberg *Skyddsombud i allas intresse* (Stockholm: Norstedts Juridik, Dr of law, diss. University of Stockholm, 2004), 80-85.
22. Maria Steinberg *Skyddsombud i allas intresse* (Stockholm: Norstedts Juridik, Dr of law, diss. University of Stockholm, 2004), 90-93.

direction of affirming the right of employees to have a say in determining their working conditions.[23]

In 1972, came the Private Board Representation Act (1972:829), and in 1974 the Employment Protection Act (1974:12) and the Labour Union Representation Act. These were followed in 1976 by the Co-determination in Workplace Act (1976:580) and, in 1977, the Work Environment Act (1977:1160). In 1979 came the Equal Treatment of Women and Men in Employment Act (1979:503), the nation's first law aimed at combating sex discrimination.[24] All these were responses to the demands of employees for a more democratic workplace.

From 1973 on, reform was very much oriented towards affirming and protecting the right of employees to influence their working conditions. Trade unions, very strong at the time, were now given priority in the election of SD if they had a collective agreement with the employers in question, a privilege they had been seeking since 1912. Furthermore, SD were now given the right to participate in employers' planning of changes and to call a stop to work in cases of immediate and grave danger. The tasks of SC were expanded. Recognition was given to the idea that costs incurred in improving a work environment should be counted as part of a firm's costs of production.[25]

In 1976, the Social Partners in the private sector agreed on an extensive collective agreement concerning work environments. This agreement dealt with the role of SD, SC, and the OHS. SD were to be given forty-hours of instruction on the subject of work environments. This need was met by an extensive training programme called 'Improving the Work Environment'. For many years this programme was heavily subsidized by the government. The subsidy has now been taken away.[26]

In 1977, the Work Environment Act (hereinafter 'WE Act') replaced the 1949 Act. Both the new Act and its predecessors had been greatly influenced by the ILO, of which Sweden became a member in 1920.[27] The new Act was to cover all kinds of work including even that of self-employed persons using technical apparatus. The Act covered pupils from grade 7 up, persons employed in the defence sector (including future members of the armed forces in training), and persons in the caring professions. The special chapter about women was omitted, though a woman's right to nurse her children remained in the Act. One of the aims of the new WE Act is to make as little distinction as possible between the sexes. Employers were simply required to adjust work environments to employees' physical and mental needs. This was presumed sufficient to cover any special needs of women.

The highlighted innovation of the new Act was its dealing with psychosocial issues. The individual employee now had a legally guaranteed right to take part in the establishment of his or her work environment. S/he also had the right to take the

23. Tore Sigeman, *Labour Law* 18 (Norstedts Juridik 2010).
24. Laura Carlson, *The Fundamentals of Swedish Law* 363–364 (Studentlitteratur 2009).
25. Sweden Government Bill 'Åtgärder för bättre arbetsmiljö' 1973:30, 148.
26. Maria Steinberg *Skyddsombud i allas intresse* (Stockholm: Norstedts Juridik, Dr of law, diss. University of Stockholm, 2004), 94-97.
27. Swedish Government Official Reports series, 'Internationella konventioner inom arbetarskyddet' Appendix 3 (Stockholm: SOU 1976:4), 23.

unilateral decision to stop working in certain situations without the risk of being held financially responsible for any resulting damage. Employers, moreover, were required to consider the special needs of immigrants and persons with various disabilities in their workforce. SD were authorized to call a halt to solitary work in the case of insufficient protection for the worker. Employers were, moreover, obliged to inform SD of any plans to make changes that might affect the work environment. SD were given the right to appoint a Senior SD, and this person was authorized to appeal decisions from the Labour Inspection to the NBOSH, and thence to the Government. During 1979–1981, several collective agreements concerning work environment were signed by the Social Partners in the public sector.[28]

In the early 1980s, there was a widespread debate over whether the OHS should be mandatory for everyone. In 1986 an addition to the WE Act stipulated that employers must arrange the OHS wherever necessary (Chapter 3, section 2). It was further stipulated that every SC must discuss an organization for work adjustments and rehabilitation (Chapter 6, section 9).

In 1990, the WE Act was expanded to cover all pupils and students from first grade up to 7 and including high school. Pupils and students above grade 6 had a right to appoint their own SD with special rights.[29]

In 1991, there were major revisions of the 1977 WE Act. LO, the central union for blue collar workers, had demanded harsher penalties and more stringent fixing of responsibility in cases where an employee was injured as a result of his or her employer breaking the law. A further motive for revision was the need to make clearer just who was chiefly responsible for work environments, for many employers had been handing over work environment problems to SC instead of solving them themselves. The most important result of the 1991 revision was that employers were obliged to carry out systematic work environment management, while their employees were given an increased say in the establishment and maintenance of the work environment; and work organization as such was now a recognized concern of the WE Act. Employers were to be held responsible for the organization of work adjustment and, when necessary, for worker rehabilitation. SD were now authorized to participate in any changes made in the work organization and systematic management of the work environment.

A new concept in criminal law was defined: A person could be charged with a Work Environmental Offence if his or her failure to observe a requirement of the WE Act led to death, bodily injury, sickness, or the placing of another person in danger (Swedish Criminal Code *Brottsbalken* 3:7-9). Other technical stipulations – many inspired by European Union (EU) regulations – were added to the Act at this time, such as the requirement to give employees technical information and information about dangerous substances.[30]

28. Maria Steinberg *Skyddsombud i allas intresse* (Stockholm: Norstedts Juridik, Dr of law, diss. University of Stockholm, 2004), 96-97.
29. Swedish Government Bill 'Om elevers medverkan i arbetsmiljöarbetet'(Stockholm: 1989/90:61).
30. Swedish Government Bill 'Arbetsmiljö och rehabilitering' (Stockholm: 1990/91:140), 31-56, 108-116.

The Social Partners were very active as this revision was being drawn up. All the collective agreements concerned with the work environment - which at that time covered about 90% of the labour market - were revised or abolished. In the mid 1990s there were further revisions.[31] In 1994, a special responsibility for temporarily hired persons was included in the Act. Here, as elsewhere, Sweden had been greatly influenced by the European Community and its directives, an influence going back even before Sweden's adherence to the 1994 EES agreement and its full membership of the EU in 1995.

In 1999, employers were made responsible for maintaining OHS. This requirement, however, is considered vague and in need of being followed up by collective agreements to be effective. In 2000–2012, the WE Act underwent several alterations: in 2001 the Labour Inspection and the NBOSH were merged into a single body we have already mentioned - the WE Authority. One aim of this was bringing implementation by the two organizations into line with each other.

Starting in 2008, appeals from the new Authority were to be addressed to a County Administrative Court, instead of to the national Government as previously. This was in conformity with EU usage. In 2008, the WE Act incorporated regulation 1907/2006/EC on chemicals known as 'Reach'.[32] The EU's latest Machinery directive 2006/42/EC was also incorporated into the WE Act.[33] The applicability of the Act was now extended to work in an employer's household. The following year, 2009, several sections were added to the Act in order to satisfy the requirements of the EU, e.g., the Council directive 92/57/EU concerned with building sites.[34] The potential influence of SD on 'foreign' worksites was increased, as was student's control over their 'work' environment.[35] In July 2011, the Government launched an investigation aimed at procuring 'A better work environment through more effective sanctions.' The investigators proposed stricter penalties for breaking the rules of the WE Act and the Working Time Act, and they imposed sanction fees for breaking certain Provisions. Today, the highest fine is EUR 10,000. It is suggested that the highest fine should be EUR 100,000.[36]

31. Maria Steinberg *Skyddsombud i allas intresse* (Stockholm: Norstedts Juridik, Dr of law, diss. University of Stockholm, 2004), 99.
32. Commission Regulation (EU) No 125/2012 of 14 Feb. 2012 amending Annex XIV to Regulation (EC) No 1907/2006 of the European Parliament and of the Council on the Registration, Evaluation, Authorisation and Restriction of Chemicals (REACH).
33. Corrigendum to the Commission communication in the framework of the implementation of Directive 2006/42/EC of the European Parliament and of the Council of 17 May 2006 on machinery, and amending Directive 95/16/EC (OJ C 87, 23 Mar. 2012).
34. Council Directive 92/57/EEC of 24 Jun. 1992 on the implementation of minimum safety and health requirements at temporary or mobile constructions sites.
35. Swedish Government Bill 2008/09:138. 'Elevers och studerandes medverkan i arbetsmiljöarbetet, m.m.'
36. Swedish Government Official Reports series, 'En bättre arbetsmiljö genom effektivare sanktioner' (Stockholm: SOU 2011:57).

§9.02 BASIC CONCEPTS

[A] Risk Prevention and Protection from Risk

The Swedish WE Act is not characterized by strict definitions. It is rather a framework of very general expressions filled out with over 100 detailed Provisions issued by the WE Authority. Only in a few recent Provisions does one find exact definitions. The development of the Swedish work environment legislation has gone from rule-oriented legislation to a more goal-oriented one, according to Hydén.[37] The WE Act is on the whole action oriented; it will tell persons responsible what they must do. The word 'prevention' first appears in the Swedish version of the Act when the purpose of the Act is explained. 'The purpose of the Act is to prevent occupational accidents and diseases and generally to achieve a good working environment' (Chapter 1, section 1). The second occurrence of 'prevent[ive]' in the Act is in explaining to employers their general obligations, viz.:

'The employer shall take all the precautions necessary to prevent the employee from being exposed to health hazards or accident risks' (Chapter 3, section 2).

When the Act says 'all the precautions necessary', it refers back to the text in the Bill that preceded the Act. Here it is stated that the action the employer shall take must not be disproportionate in relation to the intended result. For example, the WE Authority must not demand a very costly investment to prevent a very minor contingency. The law puts higher demands on the work environment in new building than in old ones.[38] But in every case it is the WE Authority who shall make the first judgment (of course, it can be very difficult for anyone to know in advance where to draw the line).

The word 'prevent[ion]' occurs again in relation to the OHS. The Act states that this service 'shall in particular work for the prevention and elimination of health risks at workplaces ... ' (Chapter 3, section 2c). The word 'preventive' is used again in reference to demands made on employees. Every employee must follow the Provisions 'and exercise such other precautions as are needed for the prevention of occupational accident and diseases' (Chapter 3, section 4) – here 'prevention' refers to part of a demand which is not precisely defined.

The word is used again, in relation to the construction industry:

> 'The building work environment co-ordinator ... shall see to co-ordinating the work with preventing risks for occupational accidents and diseases ... ' (Ch. 3, Sec. 7b).
> 'The person who is responsible for the co-ordination of work ... shall make sure that
> 1. the work of preventing risk for occupational accidents and diseases is co-ordinated at the common worksite' (Ch. 3, Sec. 7d).
> The Swedish equivalents of 'protection' and 'prevention' (Sw. *skydd*) occur in the original text dealing with the protection of employees from certain injuries.

37. Håkan Hydén et al. *Den Nordiska modellen, Fackföreningarna och arbetsrätten i Norden –nu och i framtiden* (Malmö: Liber, 1990), 186.
38. Swedish Government Bill 1976/77:149, 'Arbetsmiljölag', 253.

> 'Adequate safety (*säkerhets*) precautions shall be taken to prevent injuries being caused by falls, collapses, fire, explosion, electric current or other comparable factors' (Ch. 2, Sec. 4).

The term 'protect[ive]' is also used in connection with personal equipment:

> Personal protective equipment shall be used when adequate security from occupational accidents and diseases cannot be achieved by other means (Ch. 2, Sec. 7).

Chapter 2 of the WE Act uses similar words when discussing planning for work. This should 'be carried out in a healthy and safe surroundings' (section 2). Work should be carried out in a 'suitable working environment' (section 3). Touching questions of occupational hygiene the word 'satisfactory' is used (section 4). The expression 'adequate safeguards against occupational accidents and diseases' is used in connection with machinery (section 5). The phrase 'adequate security' is used in the discussion of substances that can cause occupational accidents and diseases (section 6). More precise demands for the work environment are spelled out in the Provisions.

[B] The Concept of 'Health', 'Danger' and 'Risk'

The words 'health' and 'danger' are rarely used in the WE Act, but occur in reference to the OHS. The Act states that this service shall work specifically for 'the prevention and elimination of health risks at workplaces ...' (Chapter 3, section 2c). The word 'health' is used in connection with the employee's right to stop work if it 'entails an immediate and serious danger to life or health' (Chapter 3, section 4). The expression 'ill health' is used many times together with the word 'risk'. The employer shall bear in mind the basic principle that 'everything capable of leading to occupational accidents or diseases has to be altered or replaced in such a way that the risk of occupational accidents or diseases is eliminated' (Chapter 3, section 2). There are no definitions of words given in the Act.

The word 'risk' is used twice in the Act: 'The employer shall consider the special risk of occupational accidents or diseases which can be entailed by an employee working alone' (Chapter 3, section 2). 'An employer ... shall ... continually investigate the hazards of the activity ...' (Chapter 3, section 2b). See also section 2c above concerning the OHS. 'The employer shall make sure that the employee...knows what measures shall be taken for the avoidance of risks in the work' (Chapter 3, section 3).

The WE Act tells what the requirements are without giving definitions.

The general requirements for the work environment are mainly found in Chapter 2 of the WE Act and in the Provisions issued by the WE Authority. The demands on the work environment as stipulated in section one are that:

> The working environment shall be satisfactory with regard to the nature of the work and social and technical progress in the community.

Acknowledged here is the fact that certain kinds of work will by their nature always be dangerous, such as that of fire-fighters, police officers, and those dealing with violent patients in psychiatric institutions. Conflicts sometimes arise between different laws.

For example, between society's need to put out a fire, capture a terrorist, or care for a violent psychopath and the need of fire-fighters, police officers, and psychiatric care workers to do their job in a decent work environment. Employers have to balance one set of needs against another.[39]

Employers are furthermore expected to move with the times as social and technical changes take place in the larger community. This refers to, among other things, the fact that solitary work has become more dangerous in recent years because workplaces are troubled by increasing amounts of violence: e.g., in the case of persons working alone in small shops open late at night, or that of social workers visiting their clients' homes. However, safety in such situations has been increased by the advent of the cell-phone.

The Section further specifies:

> Working conditions shall be adapted to people's differing physical and mental aptitudes.

This requirement acknowledges the fact that people are different physically and mentally, such differences often being related to differences in age, gender, or work experience. However, employees cannot expect every part of their job to be adjusted to their personal requirements.[40] The physical layout of workplaces should nevertheless, as far as possible, accommodate personal differences. For example, the design of tables, chairs, and mechanical equipment and the composition of chemical substances should, as far as possible be adjusted to individual employers' needs. This kind of adjustment can affect the organization and content of work. Respect and considerations must be given to mental health. Stress, monotony, and isolation on the job should be minimized.[41]

The section also stipulates:

> The employee shall be given the opportunity of participating in the design of his own working situation and in processes of change and development affecting his own work.

Employee participation is seen as an essential element in promoting psychological well-being in the workplace. In the public sector there are special collective agreements that give employees the opportunity to participate in decisions affecting them. This will be discussed later. The same section stipulates that:

> Technology, work organisation, and job content shall be designed in such a way that the employee is not subjected to physical or mental strains which can lead to ill health or accidents.

This part of the section leaves room for a broad interpretation. Essentially it means that work organizations must not be designed in any way that can be mentally harmful. Further the section requires that:

39. Swedish Government Bill 1976/77: 149 'Arbetsmiljölag', 220.
40. Swedish Government Bill 1990/91:140 'Arbetsmiljö och rehabilitering', 44-45.
41. Maria Steinberg, *The Prevention of Psychosocial risks in the workplace: Swedish Law* (unpublished ms, Örebro U. 2012).

> Forms of remuneration and the distribution of working hours shall also be taken into account in this connection.

Forms of remuneration include pay by piecework, which (though not forbidden) can be stressful.

Employers should be aware that remuneration linked to the amount of production rather than determined by a set and regular salary, can be accompanied by a higher rate of accidents and diseases. In the distribution of working hours, moreover, employers must give their employees sufficient break-time. These demands are also covered by the Working Time Act, which in many workplaces is superseded by collective agreements by the Social Partners. The same section further requires that:

> Closely controlled or restricted work shall be avoided or limited.

It was the assembly line that was in focus when this section was incorporated in the WE Act. Today many other forms of work are also recognized as restrictive, e.g., work in call-centres, work as a personal assistant, or work done alone with persons of extremely limited mobility. Section One finally requires that:

> Efforts shall be made to ensure that work provides opportunities of variety, social contact, and co-operation, as well as coherence between different tasks. Furthermore, efforts shall be made to ensure that working conditions provide opportunities for personal and vocational development, as well as for self-determination and the exercise of professional responsibility.

These requirements, however, are advisory and not legally binding.

This section of the Act is the only one that recognizes that mental stress and bad organization can have a deleterious effect on the work environment and should be avoided. Some Provisions emanating from this section are concerned with mental effects. Here are some examples:

> *Care in Private Homes* (AFS 1990:18). Many employees, for the most part women, work in facilities for the elderly or in the care of people with disabilities. The employees take care of these persons' personal needs at their homes. Most of the time this is managed by local authorities, but increasingly private agencies are taking over. Under either form of management the work is subsidized by taxes. The provision states that when planning, organising, and carrying out care activities, consideration shall be given to all circumstances that may have an effect on physical or mental health. It further states that mentally strenuous work shall as far as possible be avoided, or at least limited. When doing mentally strenuous work employees must receive continual special supervision and support. Jobs requiring great mental strain must not be assigned as solitary work. Work defined as mentally strenuous may involve violence, alcohol or drugs, or care of the mentally ill.

Another Provision that consider mental issues is:

> *Violence and Threats in the Working Environment* (AFS 1993:2). This provision focuses on situations where there is a risk of violence or the threat of violence. In such situations the employer must take several preventive measures such as investigating the risks and doing whatever is necessary in the arrangement of

> working conditions to avert violence or the threat of violence. In addition, employers must establish special security routines. If a job is dangerous, employees engaged in it must be given sufficient information, instruction, and hands-on training to be able to perform it with maximum safety. Every employee is entitled to support and guidance. Employers must ensure that fire alarms and other safety devices are installed as needed and fully functional. Very dangerous work should never be performed in solitude. Employees who have been subjected to violence or threats must be given prompt assistance and support to prevent or alleviate any injury, physical or mental.

This Provision is almost twenty years old and is still an effective instrument, if used properly.

It is extremely rare for violence at work to be dealt with by the courts and to become a work environment offence. One unusual case came before the Court of Appeal for Blekinge and Skåne:

> An assistant at a State treatment home was killed by one of the inmates while working alone. The Appellate Court found the director and vice director of the home guilty of a work environmental offence. Both received a conditional prison sentence and a personal fine: equivalent to 3,100 euros in one case and to 2,000 euros in the other.[42]

There is a Provision called *Victimization at Work* (AFS 1993:17). This includes victimization, discrimination, and harassment. The consequences can be that the employer will give notice of termination of employment to the employee who has discriminated against someone. In these cases the Employment Protection Act (1982:80) guarantees protection against unfair dismissals.

The term 'victimization' is defined as 'the recurrent directing of offensive acts against an individual with the intention of hurting that person and the likely result of excluding him or her from acceptance by the workplace community'. Employers are required to plan and organize work in ways that will discourage victimization. It is their responsibility to make clear to employees that no kind of victimization will be accepted in the workplace. Routines must be established for discovering the early signs of victimization; and if it appears, counter-measures must be taken immediately. Employees who have been victimized must be given immediate help and support. The WE Authority is invested with legal powers to assure that this Provision is respected, but it can only exercise them preventatively: e.g., by demanding injunctions against victimization and by advising employers concerning strategies to prevent it. The Authority does not deal with individual cases and therefore has no power to determine compensation for damages in such cases. According to Chapter 6 of the School Act (2010:800) pupils can receive individual compensation if the school they attend has failed to take sufficient action to prevent their victimization.

Sexual harassment is an example of victimization, which is dealt with in the Discrimination Act (2008:567). Protection against sexual harassment was first established in the Equal Treatment of Men and Women Act of 1991. Today this protection is

42. The Court of Appeal over Skåne and Blekinge, Dep. 3, Case B 2098-10, 2011-07-14.

found in the Discrimination Act. The definition of sexual harassment according to that Act is the following:

> Conduct of a sexual nature that violates someone's dignity (Ch. 1, Sec. 4, p. 4).

The definition specifies things like unfavourable treatment, sexually suggestive rudeness, and unwanted licentious behaviour. According to the proposal, such behaviour may include verbal expressions of a wish to engage in sexual relations or physical offences such as rude touching, caressing, pinching, or rubbing against another person's body. Displaying pornographic pictures or objects can also been seen as violations of a person's dignity.[43]

Each employer must be active in general to prevent sexual harassment. If an employer knows that sexual harassment has occurred s/he must act on that knowledge. An employer is not permitted to reprimand a person for rejecting sexual advances. The employer's obligation to prevent sexual harassment in general (Chapter 3, section 6) means that s/he must take overall action. This can include adopting a policy stating clearly that sexual harassment is forbidden. It can also include issuing an action plan that describes what kind of procedures are to be followed if someone is harassed in this way. It is recommended that supervisors and trade union representatives be instructed in identifying this kind of harassment. Such a policy may include specification of the punitive consequences an act of sexual harassment can lead to: e.g., transfer to another post or notice of termination of employment. The Commentary of Fransson and Stüber notes that it is unclear whether employers are required or merely advised to have a policy and an action plan relative to this matter. They conclude, however, that some form of written documentation is required.[44]

An employer must investigate and take action if sexual harassment has occurred (Chapter 2, section 3). The obligation to investigate and take action falls on employers and former employers. To come under the Act harassment must have some connection with the employee's work. Therefore, an employer is responsible for investigating harassment alleged to have occurred on a business trip, at a conference, or at a staff party.[45] If an employer fails to make an investigation, s/he may have to pay discrimination damages (Chapter 5, section 1).

In all cases it is the employer who has to pay such damages, even if it was an employee who committed the harassment (Chapter 5, section 2). The obligation to investigate starts as soon as the employer or a supervisor learns about the allegation of harassment. The employer may have been informed by a trade union representative or by the alleged victim him or herself. Any suspicion of sexual harassment obliges an employer to make inquiry concerning it. When the employer has obtained sufficient

---

43. Susanne Fransson & Eberhard Stüber, *Diskrimineringslagen En kommentar* (Stockholm: Norstedts Gula bibliotek, 2010), 92-93 and Sweden Bill 1997/98:55 'Kvinnofrid', 139. Sweden Bill 2007/08:95 'Ett starkare skydd mot diskriminering' 494.
44. Susanne Fransson & Eberhard Stüber, *Diskrimineringslagen En kommentar* (Stockholm: Norstedts Gula bibliotek, 2010), 399-401, Sweden Bill 2007/08:95, 536, Sweden Bill 1997/98:55,138.
45. Susanne Fransson & Eberhard Stüber, *Diskrimineringslagen En kommentar* (Stockholm: Norstedts Gula bibliotek, 2010), 240-241.

information relative to the case, s/he must commence an investigation, if possible speaking first with the alleged victim and then with the one or ones accused of the harassment.

Sexual harassment can have serious consequences for the employment status of the person accused. S/he may be transferred to another job or be given a notice of termination of the employment. The precise consequences may be determined by collective agreement. Of course, a very grave act of sexual harassment may lead to the pressing of criminal charges. In any case employers must always attempt to curtail such harassment and do everything reasonably possible to prevent its recurrence. Victims must be given appropriate support and have the right to be informed what kind of action, if any, their employer intends to take.[46]

Some examples from the Labour Court:

> **Continual harassment**[47] An employer managing a casino was informed that one of his female employees had been sexually harassed by one of the supervisors. The court found that the employer had not investigated the charge of sexual harassment with sufficient thoroughness and found the employer's contacts with the man (M) who had offended the woman (W) to have been culpably superficial. Furthermore the employer had done nothing to support W; she and M had continued to be assigned to work together. The court found that the employer had broken the law and was therefore obliged to pay compensation for general damage to W owing to the employer's failure to conduct a sufficient investigation and to take action afterwards to support the victim and protect her from further harassment. The employer had to pay W the equivalent of 8,000 euro.
>
> **Rape, outside the work place**[48] A man (M) had tried to rape a woman (W) outside working hours. The employer should have kept W informed of what actions were planned to prevent further occurrences of rape or sexual harassment. The court found that the employer maintained no regular contact with W. The Court determined that the employer had said that the man who had committed rape and sexual harassment would be transferred to another worksite when W returned to work (she was presently on sick leave). The employer had talked to the accused (M) about his behaviour towards co-workers. Everyone in the company had been informed that sexual harassment in the workplace was unacceptable. W had been assigned a contact person to follow-up on the case. Therefore the Court could not find the employer negligent of the obligation to take action. Concerning the right to receive compensation for damages, the court referred to the stipulation in the Provision that there is no right to claim damages if the harassment ceased after the employer was informed. However, the Court found that the employer neglected the obligation to investigate the allegation of rape outside the workplace and the question of how M treated W afterwards. The employer had not been prompt enough in investigating the allegation of sexual harassment in the workplace, and because of this failure was required to pay damages to W. At the same time it was acknowledged that the employer had given W immediate support and had eventually conducted an investigation. The damage that the employer had to pay W amounted to 5,000 euro.

46. Susanne Fransson & Eberhard Stüber, *Diskrimineringslagen En kommentar* (Stockholm: Norstedts Gula bibliotek, 2010), 240-244.
47. Labour Court 2002/102.
48. Labour Court 2005/22.

One can summarize the preventative legal requirements concerning sexual harassment in the workplace as follows. An employer's obligation is: (1) to conduct a serious investigation; this includes interviews with both the alleged victim and alleged perpetrator. (2) The employer must maintain some form of contact with the alleged victim and give him or her moral and emotional support. (3) The employer must take action to prevent harassment in the future; such action might include closer surveillance of the work situation and, perhaps, transfer of the perpetrator to another workplace.

Another Provision that deals with the protection of mental health is:

> *First Aid and Support in Crisis* (AFS 1999:7) The term 'support in crisis' refers to the counselling and other forms of care needed to mitigate psychic trauma caused by accidents or sudden emergencies, and panicky reactions to the same. The employer's obligation is to plan, organize, and maintain a support system for dealing with such crises. In every workplace there should be planned routines in place for providing support in the eventuality of a crisis. Employers and supervisors are to have sufficient knowledge of how support works in time of crisis. This Provision came into use when many Swedish holidaymakers lost their lives as a result of the tsunami in Thailand.

These are the major Provisions for the prevention or mitigation of psychosocial and mental strain in the workplace.

A great many Provisions deal with dangerous substances such as asbestos and lead, and many of these are aimed at particular fields of employment such as mining or work with animals. There are, of course, Provisions addressing the use of particular tools and types of machinery. One of the most important is *Ergonomics for the Prevention of Musculoskeletal Disorders* (AFS 2012:02). The importance of these Provisions is confirmed by statistics covering a large number of work-related ergonomic injuries. Many of the Provisions have been added or revised under the influence of EU directives.

Employers must always take the work environment into consideration when planning a job according to the WE Act (Chapter 2, section 2). The premises shall be suitable (section 3). There is a special Provision that describes the requirements for a workplace. The WE Act indicates specific factors that must be considered, such as air quality, lighting, noise, vibrations. It mandates the safe use of machinery, high-tech devices, and substances of every kind (sections 4–6). Special rules cover the use of personal protective equipment. Such equipment shall only be used when safety in the work environment cannot be attained in other ways. If used, such equipment must be supplied by and at the expense of the employer (section 7).

The employer is also responsible for providing facilities such as toilets and, if the work is dirty, showers; also a place where employees can warm or cook their food and eat it; and lockers for clothes and the safe deposit of valuables (section 8). All these rules are developed further in specific Provisions.

§9.03 PERSONAL SCOPE OF APPLICATION

[A] The National Legislation's Application to All Sectors of Production and All Types of Risk

The Act applies to all forms of work, public and private. It covers everything from industry and farm work, to work in offices. It covers non-profit institutions.

The Act applies when a worker carries out work for an employer. 'Workers' here includes trainees and interns, according to the Act.[49]

An 'employer' can be a private person, a company, or a local or county government or employed by the State (whose employees include, for example, the police). The Act also applies, to some extent, to the armed forces in Sweden and abroad and to the civil defence. Students from the age of 6 (pre-schoolers) including university students are protected by the Act. Persons who carry out work in certain institutions such as inmates in prisons are covered by the Act. These groups do not have the same but similar rights as workers to appoint SD and members of the SC.

The Act applies to those who are employed by the employer and working in that person's home, e.g., personal assistants.

The Act applies for most part for participants in employment market policy programmes and in work placement or other professional development activities which have been assigned by the social welfare board. It also applies to a certain extent for those who are working, when seeking asylum in Sweden. The Act furthermore applies to those who work on ships and, in part, to the self-employed. A self-employed person must try to live up to the goal of the Act and must follow the demands in Chapter 2, which regulates technical devices and dangerous chemicals. The self-employed must follow Provisions, if those have included the self-employed.[50] The WE Authority can make inspections on self-employed persons' premises, if there is a need for it or if the self- employed person has asked for it, (WE Ordinance, section 15).

[B] Specific Legislation Regarding Pregnant Women and Young Workers

[1] Women

There are in general today no special laws protecting or limiting the activities of female workers as such. However, Sweden has implemented the Council Directive 92/85/EEC on pregnant workers and workers who have recently given birth or are breast

49. Hans Gullberg & Karl-Ingvar Rundqvist, *Arbetsmiljölagen med kommentarer och författningar* (Stockholm: Norstedts Juridik, 2010), 48.
50. Hans Gullberg & Karl-Ingvar Rundqvist, *Arbetsmiljölagen med kommentarer och författningar* (Stockholm: Norstedts Juridik, 2010), 374.

feeding.[51] The WE Authority has issued a special Provision, *Pregnant and nursing employees* (AFS 1994:34, revised 2007:5). The Provision makes very high demands on employers to adapt work environments to the needs of pregnant or nursing woman. The downside is that these demands can make an employer hesitant to employ young women.

Sweden has a very good system of parental leave based on the idea of the two parents sharing equally in its benefits. Since 1974 Swedish parents have been able to obtain both parental leave and cash benefits. In 1976 Sweden passed its first act regulating parents' absence from work; it covered a period of seven months. Today there is a new parental leave act, dating from 1995, which allows eighteen months of full-time parental leave, which can be taken out in portions. There is a right to work part-time until one's children are 8 years old. Parents can share their parental leave and spread it out during the child's first 8 years. It is obligatory for the mother to be on parental leave two weeks before and two weeks after giving birth. The cash amount is the same as for sick leave. Fathers will receive ten additional days of these benefits after the child is born, if both stay home with the newborn.[52] The parental leave cash benefits are paid for a maximum of 240 days and amount to approximately 80% of the ordinary salary, up to a certain amount. All but sixty of the days (which must go to the mother) may be shared.

Most women workers are employed in the public sector, e.g., in schools, hospitals, and other caring institutions, where their employment falls ultimately under the responsibility of politicians. There are no special rules for politicians as employers - as there are, for example, for employers in the construction industry. Most Labour Inspection injunctions and prohibitions are aimed at the private sector and male-dominated occupations.

The WE Act can be considered discriminatory towards women in many ways. The Act and its Provisions are chiefly concerned with safety in male-dominated occupations. There are no Provisions dealing with school or hospital environments. Employer obligations are spelled out mainly for sectors chiefly employing men, such as the construction industry.[53]

[2] Young Workers

In general one cannot say there is much child labour in Sweden. Most young persons up to 18 years go to school. Quite a few young persons from 16–18 years old will work

51. Directive 92/85/EEC – pregnant workers of 19 Oct. 1992 on the introduction of measures to encourage improvements in the safety and health at work of pregnant workers and workers who have recently given birth or are breastfeeding.
52. Laura Carlsson, *Searching for Equality, Sex Discrimination, Parental leave and the Swedish Model with Comparisons to EU, UK and US Law* 116,118,135,141 (dr of Law, dissertation, Justus Förlag 2006).
53. Maria Steinberg, *Is the Swedish Work Environment Law Sexist? Perspective of Women's Health,* in *Working Life: An Anthology* 241–248 (Helene Sandmark ed., Studentlitteratur, 2012).

in the summer while school is over. Some also have jobs after school. Statistics show that minors are more at risk than adults in work situations.[54]

Sweden has had special protection for minors since 1852. This legislation was renewed and revised in 1881. Today there is a special chapter in the WE Act covering this area, along with a special Provision concerning minors. The legislation is largely influenced by ILO conventions 138 and 182 and EU directives 94/33/EEC concerning young persons at work.[55]

As defined by the WE Act, a minor is any person under 18 years of age. Minors under 16 are generally forbidden to work, but there are exceptions: e.g., that minors between 13–16 years can be employed at certain forms of light work that are not harmful to health, development, and schooling. But minors must not be employed in work that is excessively strenuous or where there is considerable risk of accident.

There is a rather complicated Provision entitled *Minors* (AFS 2012:03). In general, its restrictions are designed to provide protection against physical, psychological, and social risk. Training for a job, for example, must be adapted to the young person's age and maturity. A minor must work under the supervision of a suitable mature person. SD must be kept informed of the employment of minors. Special rules deal with dangerous work. In certain circumstances a properly trained minor may be allowed to do potentially hazardous work under qualified supervision.

Other exceptions to the prohibition of work for minors are made if they are part of a training programme where there is a qualified instructor in supervision. Other work that is absolutely forbidden to minors are diving, participation in certain kinds of cancer research, and work involving the risk of contact with infected blood.

Other special rules govern minors' working hours. Those over 16 but under 18 years of age may work up to eight hours a day but may not work between 10 pm and 6 am – or alternatively between 11 pm and 7 am, and they must always be allowed at least twelve consecutive hours of rest during every twenty-four hours. Minors under 13 years may be hired to do light gardening. They may also sell Christmas magazines door to door – a traditionally popular activity for children. The WE Authority is authorized in special circumstances to make exceptions to the rules for individuals. Any employer who hires a minor in opposition to the rules is subject to a fine. The Provision is under revision.

[C] The Distinctive Concepts of 'Worker', 'Employer' and 'Commissioning Entity'

All workers are covered by the WE Act. There are no special rules for part-time workers or those with fixed-term employment. They are included in the WE Act. Thanks to

54. Unga i arbete, *Korta arbetsskadefakta Nr* 2/2012, http://www.av.se/dokument/statistik/sf/Af_2012_02.pdf accessed on May, 2012.
55. Directive 94/33/EC – young workers of 22 Jun. 1994 on the protection of young people at work.

Directive 97/81/EC, concerning part-time workers,[56] and to Directive 99/70/EC about fixed-term employment, there is a special law to protect them, the Anti-Discrimination Act (2002:293).

Those are also protected by the Employment Protection Act.[57]

There are no special laws for distance workers. There are frameworks for distance work issued by the Social Partners, but these are not binding in the public sector. Distance work is covered by collective agreements[58] which are covered by the Labour laws and the WE Act. The situation for distance workers is presently a bit unclear. The WE Authority gives guidelines for this kind of work. An employer who chooses to have a person work at home has more responsibility in this regard than one whose employee made that choice independently. The WE Act includes protection for persons employed on ships and in the armed forces, but there are special laws as well for these groups. Construction workers are protected by special regulations in the WE Act and by special Provisions. There are also Provisions affording special protection to divers, longshoremen, mine workers, and pregnant women. There is a special Discrimination Act to further protect the disabled but moral harassments are not included.

Sweden is now in the process of implementing the Directive 2008/104/EC on agency work.[59] Seconded workers are now protected both in the WE Act Chapter 3, section 12 and in the Work Environment Management (AFS 2001:1). The employer is the main person responsible (section 2). Chapter 3 of the WE Act takes up the general obligations. There is no definition in the Act of who is an employer. One may say that it is someone who supervises workers. Self-employed persons must in principle follow the WE Act (section 5). Employers who share a common worksite must consult with each other (section 7g).

The person who is in charge of a common worksite for several activities is to co-ordinate preventative measures for the whole worksite, scheduling jobs so that there is no risk of injury owing to one job conflicting with another. This person is also responsible for co-ordinating sanitary facilities and making sure that there is adequate personal space for everybody. These responsibilities can be transferred to anyone else engaged in activity at the worksite (section 7d). However, those involved in manufacturing or providing or delivering machinery or other equipment to the worksite have only limited responsibilities (section 8). Those who use agency workers to supplement the regular workforce are responsible for taking all needed safety measures with regard to them (section 12).

The Act regards some other categories as employees: e.g., persons undergoing training or education. Pre-school children and school children receiving instruction in recreation centres, however, are not covered. As most parents in Sweden work,

56. Council Directive 97/81/EC of 15 Dec. 1997 concerning the Framework Agreement on part-time working concluded by UNICE, CEEP and the ETUC. Council Directive 1999/70/EC of 28 Jun. 1999 concerning the framework agreement on fixed-term work concluded by ETUC, UNICE and CEEP.
57. Birgitta Nyström, *EU och arbetsrätten* (Stockholm: Norstedt Juridik, 2011), 314-317.
58. Birgitta Nyström, *EU och arbetsrätten* (Stockholm: Norstedt Juridik, 2011), 319.
59. Swedish Government Official Reports, 'Bemanningsdirektivets genomförande i Sverige' (SOU 2011:5). *

Swedish children normally begin pre-school at age one and a half. Up to the age of 12 many attend activities at recreation centres after school. The recreation centres are normally run by local government. There has been a request by the Ombudsman for Children that pre-schoolers and children at recreational centres should be covered by the Act, on the grounds that pre-schools and recreation centres are part of the educational system. An investigation was made but the Government did not accept this proposal.[60]

§9.04 DUTIES OF THE PARTIES INVOLVED

[A] The Employers' Main Duty: Risk Assessment

The main liability for fulfilling the requirements stipulated in Chapter 2, WE Act and in the Provisions falls on the employer. The employees and employer must co-operate in pursuing the goals of the Act (Chapter 3, section 1a). The employer is required to take all reasonable precautions against the occurrence of occupational accidents and diseases, this to include injuries to mental health. This is chiefly to be done by trying to alter or replace circumstances potentially conducive to ill health or accidents. Special attention is given to work done in solitude (section 2):

> *Solitary Work* (1982:3) was the first Provision to take mental health issues into account. The Provision states that employers must give special attention to an employee's mental condition when assigning him or her to solitary work. The employer must, for example, make every effort to allow the employee opportunities for contact with colleagues and others. This can be done by giving the employee frequent breaks or the possibility of alternating solitary jobs with non-solitary ones. The SD must be informed whenever solitary work is initiated.

Every Swedish employer is obliged to follow an action plan of Systematic Work Environment Management. This includes maintaining a work environment policy, making risk assessments, allocating tasks with regard to health and safety, and drawing up and following a work environment action plan including follow-up routines. All of this should be set out in writing. Employers shall also investigate all injuries occurred at work and constantly monitor possible hazards in the work environment and take measures to neutralize them. When measures cannot be taken immediately, the employer must draw up an action plan including a prepared scheme of job adaptation and rehabilitation, in compliance with the Social Insurance Code (section 2a) and Chapter 3, section 2a WE Act. Employers with less than ten employees are exempted.

There are no special Provisions regarding stress, but prevention of stress-induced injury may be effected through Systematic Work Environment Management. Furthermore, there are no financial incentives to induce an employer to develop a health and safety system, except for the economic loss a bad reputation can lead to.

60. Swedish Government Official Reports series, 'Bättre arbetsmiljöregler 1. Samverkan, utbildning och avtal m.m.' (Stockholm: SOU 2006:44), 77-134.

More specific demands are laid down in the Provision *Systematic Work Environment Management* (AFS 2001:1). The Swedish Provision was at first (1992) called *Internal control*, but its name was changed in 2001. The idea behind this Provision comes from Norway. The Swedish Provision is also based on the Framework Directive.

The basic idea of the Provision is that employers must regard questions of the work environment as part of their day-to-day responsibilities and observance of public law. It requires every employer to make inventories and assessments of risk on a regular basis, giving special attention to those which might lead to physical or psychosocial harm. Employers must give their employees, SD, and Student and Pupil SD the opportunity to participate in this form of review. The Provisions require that every employer must have a written:

> **Risk assessments.** Employers must regularly and frequently examine working conditions and make written assessments of risks they entail. If an employee has been in an accident or suffers an injury because of his or her work, the employer must investigate the causes of the same for the purpose of attempting to prevent occupational accidents or diseases resulting from the same conditions in the future. A risk assessment must then be made with an outline of the employer's plan to make changes.
>
> **Action plan.** This is to deal with such risks as have not been immediately removed in consequence of the risk assessment. These might be psychosocial problems or matter relating to air quality or noise. The employer and the SD must together evaluate the risks. Suggestions as to how they shall be dealt with are then to be incorporated in the plan. The plan must state who is responsible for dealing with the risks and include a time-table for accomplishing that operation. It should further be stated when there is to be a follow-up control to monitor the effect of the measures.

Every employer with ten or more employees shall present a written:

> **Work Environment Policy and written routines** for carrying out systematic work environment management, and an **Allocation of tasks**.

[B] The Employer and the Management: The Delegation of Duties

According to the already mentioned Provision *Systematic Work Environment Management*, the following has to be highlighted:

> **Allocations of tasks** (sometimes called 'Delegation'). One or more managers, supervisors, or even employees, may be given specific tasks aimed at minimizing risks at work. There should be enough of such persons with sufficient authority and resources to meet this responsibility. These tasks shall be described as clearly as possible. The persons to whom they have been assigned, shall carry on work environment management as a routine part of their daily activities in connection with decision-making and work supervision. The tasks are to be assigned to specific individuals, personally or by position, and deputised to other specific individuals, if the appointed persons are absent sick or on leave. At any given moment everyone should know what to do in relation to the work environment. Final liability, however, always remains with the employer. 'Authority' here means the power to decide and take action. 'Resources' means money, staff,

equipment, facilities in general, time and knowledge. If the person receiving an assignment is aware of an insufficiency of any of these, s/he must apply to his or her superior for a change, and if that is impossible, the person must formally resign from the duties assigned. This should not put him or her at risk of being held responsible in case of an accident. (Sec. 6 AFS 2001:1).

Competence/knowledge. Employers are responsible to make sure that those who have been assigned these specific tasks have sufficient knowledge of the work environment rules and understanding of physical, psychological, and social conditions that could result in occupational accidents or diseases. They should know what constitutes a good work environment and how to exercise work environment management competently. In a commentary on this section it is explained that managers and supervisory personnel need to have a good knowledge of the WE Act and the Work Environment Ordinance and Provisions and of how they are meant to be applied. They are also expected to have 'good insight into human reactions'. They should understand the impact on employee health of 'heavy workloads, overtime, victimization, threats and violence, and substance abuse.' (Sec. 7 AFS 2001:1).

Follow-ups. The employer shall make an annual follow-up inspection to observe and record the effects of work environment management.

Job adaptation and rehabilitation. It is the employer's duty to organise job adaptation and rehabilitation as part of the systematic work environment management.

It has been questioned if Sweden was complying with the Framework Directive. The Commission claimed that Sweden had neglected to fulfil some of the requirements of Article 7.8. of the Framework Directive. The Commission found that Sweden had unclear rules about allocation of tasks and was lacking in specific rules about requirements of knowledge and resources of the environment from those who were in charge of the work environment. The European Court of Justice (ECJ) did not find that Sweden had been negligent in the way that the Commission had claimed.[61]

However, further to this decision, the WE Authority has improved the Provision about Systematic Work Environment Management in these areas.

In the period 1991–2004, the Supreme Court tried five cases of work environment offences related to the allocation of tasks.

The first case concerned whether a foreman could be held responsible for someone's death:

> A person working with concrete was crushed to death by a transport vehicle whose safety mechanism failed to function. The Supreme Court found that the foreman could not be held responsible because he had not been formally issued with an allocation of tasks or job description mentioning health and safety.[62]

Here one can see that a foreman could not be held responsible without any instructions or allocation of tasks, for s/he cannot automatically be assumed to have the knowledge that is needed in such cases.

61. ECJ, 15 Jun. 2006, C-459/04 *The Commission v. Sweden*.
62. Nytt Juridiskt Arkiv (NJA) 1991 P. 247.

The second case is concerned with whether a superior manager was responsible for the negligence of a subordinate manager who had sufficient knowledge of the work environment:

> A worker employed by the Street Office at a Local Government was injured during a re-building project. He had fallen as a result of there being no adequate safety equipment in use. The Supreme Court found the superior Manager of the Street Office was not responsible for the injury. There was no written allocation of tasks. The Court found that responsibility lay entirely with the subordinate Department Manager in charge of the re-building works, who, they said, was fully qualified to bear that responsibility. In the judgement of the Court (though it was not a unanimous decision), the Manager of the Street Office was justified in assuming that the subordinate Department Manager had undertaken full responsibility.[63]

In short, a superior manager who had not allocated the task of the work environment to his subordinate manager was not held responsible because the subordinate was deemed qualified to understand the risks involved.

The third case dealt with a military drill instructor's responsibility for those under his command:

> In the course of a field exercise, an army trainee was drowned while attempting to cross a body of water under the weight of his pack and other gear. The water had proved deeper than the drill instructor had anticipated. The Supreme Court absolved the instructor of responsibility on the grounds that he had simply been conducting the exercise in the routine way in which he had conducted it many times before. While this meant he was not exactly following the written safety instructions, those instructions, according to the Court, were badly written and hard to understand. The fault was attributed to the bad instructions.[64]

Does this mean that an instructor who follows bad routines or unclear instructions will not be held responsible for accidents that happen as a result? One wonders, does the responsibility really lie on the heads of those who wrote up the instructions?

The fourth case was concerned with whether a Principal of a school could be held responsible for a student's accident:

> A fourteen-year-old high school student seriously injured his hand in a machine during a lesson. An experienced teacher was in charge of the lesson and was present. The Supreme Court stated that a teacher is always responsible for the immediate supervision of his or her students, with or without an explicit allocation of task. The principal, of course, also has responsibilities. But in this case the machine had been used for many years without anyone ever being hurt by it. The teacher, moreover, was well experienced. The principal could not have anticipated the accident and therefore was not deemed responsible.[65]

One can read between the lines that, in the opinion of the Court, the experienced teacher had responsibilities in the case. But the prosecutor did not accuse the teacher of any crime, and so his responsibility was not tried in court.

63. NJA 1993, P. 245.
64. NJA 2004, P. 34.
65. NJA 2004, P. 80.

The fifth case dealt with the question of how inactive the manager of a premise can be after a work accident:

> A worker was hurt very badly while cleaning a machine. She had received no safety instructions. A similar accident had occurred six months earlier. The manager of the premises had not investigated the first accident nor done anything to prevent its recurrence. The manager claimed he had not been officially charged by the Head manager with any responsibility for the work environment. He had also assumed that the technical manager under him took care of all safety issues. The Supreme Court declared that allocation of safety tasks must be made very clear. The Court found the manager of the premises responsible for the accident because of his negligence. He had known about the earlier accident and had not made any investigation or taken any action to prevent a similar accident from happening.[66]

The Court looks seriously at any case where an employer has not taken active measures after a work accident to avert the risk of the same accident happening again.

One of the conclusions one can draw is that a foreman without an allocation of tasks and a job description that does not include health and safety issues will not be held responsible for a subordinate's death. But in some cases a superior manager can expect a subordinate manager to assume some health and safety responsibilities. Especially when the manager is qualified to do so. Sometimes the fault of an accident lies with bad routines and instructions. A principal is not likely to be held responsible if there is a competent teacher supervising students. A manager's responsibility for the work environment is all the greater if s/he has not been active after an accident to prevent its recurrence and the same kind of accident happens again.

[C] Working Time: Also a Work Environment Issue

Working time is another part of the employer's obligations. The first Working Time Act of 1919 only covered industrial workers. In 1970 Sweden passed a new general Working Time Act, which mandated the limit of a forty-hours week.

The present Act dates from 1982. The Social Partners can supersede any of the terms of the Act by a collective agreement, apart from the so-called EU locks. As 70% of the Swedish workforce belongs to a trade union, employees' working hours in Sweden are for the most part regulated by such agreements. The Swedish Working Time Act reflects the influence of conventions formed by the ILO and has been adapted to Directive 2000/34/EC and to Directive 2003/88/ EC concerned with certain aspects of the organization of working time.

Both employers and trade unions have been unhappy with the EU directives prescribing limitations on certain night work and the need for eleven hours rest during every twenty-four hours. These requirements, they feel, have tied down their hands. Previously both sides could find common ground on working hours through freely negotiated collective agreements; now negotiations are constrained by 'EU locks'.

66. NJA2004, P. 126.

The Working Time Act (1982: 637) applies to all activities in which employees work for an employer (section 1). This does not, however, include circumstances in which the employer has no supervisory responsibility, such as the work of a consultant hired ad hoc or that of an upper level manager, or to work performed in the employer's home. Nor does the Act apply to work on shipboard or in connection with certain forms of land and air transport. Protection in these latter three cases comes under special laws. The Act applies to the armed forces, police, and emergency services, if this does not conflict with the provisions of Directive 2003/88/EC (section 2).

Exemptions from this Act can be made by collective agreement approved by a central employee's organization. The Government can make separate rules for those engaged in the armed forces and defence industries (sections 3–4).

Normally working hours should not exceed forty per week averaged out over a four-week period (section 5). 'Call-hours', hours in which the employee is at the employer's disposal, shall not number more than forty-eight averaged over a four-week period, or fifty hours averaged over a calendar month. Any time an employee works is counted as working hours (section 6). Overtime, if necessary, is allowed of up to 48 hours in a four-week period, 50 hours in a calendar month, or 200 hours in a calendar year (section 8). Under the pressure of special circumstances, extra overtime may be allowed up to 150 hours in a year (section 8a). In calculating compensation for overtime, lawful leave of any kind is to be counted together with working hours. An employer may require emergency overtime in case of natural disaster or other extreme accident, and this may be prolonged as long as the extraordinary need persists (section 9).

There are special rules for part-time employment. An employer is allowed to request 200 additional hours up to full-time employment. If there are special reasons, an employer can request another 150 hours from a part-time worker (section 10a). This means that a part-time worker can work 350 additional hours and then another 350 hours as overtime in the course of a year. One limitation on this is that the additional hours may not exceed forty-eight hours on average over four weeks or fifty hours in the course of a month (section 10a).

No employee shall work more than forty-eight hours on average in a seven-day period over a period of four months. Average working hours are to be reckoned to include annual leave and other lawful absences, e.g., for sickness (section 10b). The employer shall keep a register of overtime, call-hours, and additional working hours of part-time workers. Employees, SD, and trade union representatives are allowed to take part of these records. The WE Authority has issued a Provision prescribing how to keep such records (section 11).

Although rest from work should be for a period no shorter than eleven consecutive hours out of every twenty-four, temporary departures from this rule for special reasons are permitted. Exceptions can be made, for example, in cases where it could not have been possible for the employer to anticipate the extra need for work. Work between midnight and 5 am is normally forbidden except where unavoidable to meet public needs or in other special circumstances (section 13). In practice this may include the needs of certain industries to run night shifts.

Pursuant to the Working Time Directive, new sections have been added to the Working Time Act. One new section forbids night work that exceeds eight hours in the course of a twenty-four hour period averaged out over four months. And if the night work involves 'particular risks or great physical or mental effort', the eight-hour limitation must be respected. Deviations may be made temporarily if there are circumstances that could not be foreseen. The worker should later receive corresponding compensatory leave. A 'night worker' is a person who normally works at least three hours between 10 pm and 6 am (section 13a).

Every employee has the right to a weekly rest of thirty-six consecutive hours during a seven-day period. This rest period must not include stand-by hours. The rest should, if possible, take place during the weekend. Temporary exceptions can be made (section 14). The Act recognizes three occasions during the day when an employer must allow his or her workers time off from work. Every employee has a right to rest after five consecutive hours in the course of a day. This rest does not count as working hours, and the employee is free to leave the workplace while taking it. If the employee cannot leave the work place, his or her meal breaks will be counted as working hours. The time spent on meal breaks is included in working time. Employers are responsible for giving their employees breaks during the day, which shall be included as working time (sections 15–18).

The WE Authority can authorize exemptions from the requirements of certain sections in cases where the Social Partners are unable to achieve a collective agreement and there are special reasons for such exemptions. This is apt to occur, for example, with regard to regular working hours, on-call hours, additional working hours for part-time workers, and the disposition of working hours and rest intervals. The possibility of exemption is, however, subject to the requirements of Directive 2003/88/EC.

A new section was inserted into the Act in 2011 giving SD the right to approach the employer with a demand for needed improvements. If the employer does not take action to meet such a demand within a certain time, the SD can make a request to the WE Authority, which then must take a decision whether or not to endorse the demand. Such a demand may concern such things as extra overtime, emergency overtime, and overtime in connection with part-time employment (section 19a). If a collective agreement covers stipulations concerned with extra overtime, extra additional working hours, or emergency overtime, the WE Authority cannot interfere.

Supervision of the Working Time Act is the responsibility of the WE Authority, which can issue both positive commands and prohibitions. Non-compliance with any of these can lead to a contingent fine (see below). Non-compliance with the Act, whether intentional or because of negligence, can result in the imposition of a personal fine. The same penalty is incurred if an employer gives the WE Authority false information (sections 20–25). Moreover, contravention of a collective agreement can lead to a liability for damages.

There are special charges for overtime. These can amount to over EUR 40 for every overtime hour taken in infringement of the overtime regulations. But this liability only rests on employers who are not subscribed to a collective agreement, and the charges can in any case be reduced or waived in special circumstances (section 26).

Decisions of the WE Authority can be appealed to an Administrative Court. Any Senior SD can make such an appeal; if no one holds that office, the right devolves on any SD, and if there are none of these it devolves on the trade union - provided the issue concerns one of its members and has been raised previously (sections 29-32).

[D] The Role of Occupational Health Services

The status of the OHS has been a problematic issue in Sweden. It started as an integrated service in big companies in which the company doctor and safety engineer started to work together.[67] In 1967 the first collective agreement on the issue was signed in the private sector. In the 1970s the public sector joined with their own collective agreements.[68]

Today the WE Act includes a section dealing with OHS, requiring employers to provide it as appropriate according to working conditions. The OHS shall be 'an independent expert resource' in the field of 'working environment and rehabilitation'. It must 'work for the prevention and elimination of health risks at workplaces and shall have the competence to identify and describe connections between the working environment, organization, productivity and health' (Chapter 3, section 2c). Provision 2001:1 concerning Systematic Work Environment Management includes a section on OHS. The WE Authority has found it difficult to enforce this section. The problematic side of it, however, is that a company can simply pay for OHS in order to comply with the law, but is not obliged by the law to follow the recommendations of the OHS. Employers can choose someone else to help them if needed.

The different public sectors have different collective agreements to provide for OHS. Parts of the private sector also have collective agreements. Such agreements are considered to be the best means for securing effective OHS. Today OHS consists of specially educated occupational health doctors, nurses, physiotherapists, safety engineers, and psychologists or persons trained in the behavioural sciences.

[E] The Training Duty

A very important section in Chapter 3 of the WE Act concerns information provided to employees about risks in their work environment. Every employer is obliged to inform his or her employees about risks associated with their work and how to avoid them. Employers must take note of their employees' training, and only those who have received adequate preparation should be permitted access to areas where there is a serious risk of occupational accidents or diseases. Employers may need to modify working conditions in view of the fact that different people have different susceptibilities (section 3).

67. Lars Ettarp and Eva Kvarfordt, *Arbetsmiljöboken* (Stockholm, Publica, 1981), 47.
68. Maria Steinberg, *Skyddsombud i allas intresse* (Stockholm: Norstedts Juridik, Dr of law, diss. University of Stockholm, 2004), 90.

[F] Duties on Contractors and Subcontractors with Specific References to the Building Sector

There are special, very detailed responsibilities for the managers and supervisors of every kind of construction work. These have been drafted in accordance with EC's building site directives 92/57/EEC. Today a special building work environment co-ordinator (hereinafter BWC) must be appointed to oversee all aspects of the planning and execution of a building project. The BWC can, but need not, be from one of the companies engaged in the construction project. This co-ordinator retains his or her responsibility until it is transferred in writing to – and accepted in writing by – someone else. These rules are complemented by a special Provision, *Building and Civil Engineering Work* (AFS 1999:03), which has been revised several times. Architects and design engineers must make sure that work environment considerations are taken into account, when applicable, during both construction and future use of the building. The BWC must co-ordinate work environment issues, including co-ordination of rules, activities, etc. (sections 6-7c).

[G] Workers' Duties

Employees are also responsible for the work environment. Every employee has a duty to participate in activities and discussions concerning his or her work environment and, by active participation, to assist in the implementation of measures necessary to achieving a good working environment. Every employee must conform to the provisions of the WE Act, use required safety devices, and take the prescribed precautions. Of course, an employee always has the right to stop working in the face of immediate danger to life or limb and cannot be held liable for any damages resulting from such an action. This, however, presupposes that there was no possibility of informing the employer or SD before stopping work. In such cases the employee who has stopped work must contact his or her employer or the SD as soon as possible (section 4).

§9.05 COMPLIANCE AND CONTROL

As already said in the above, the WE Authority was established in 2001, when the NBOSH and ten regional WE Authorities merged into a single Authority. The Authority's duty is to ensure compliance with the WE Act and the Working Time Act – and, in certain respects with the Tobacco Act and the Environmental Code and the Provisions issued by the WE Authority. The Authority is the liaison office for information for posted workers according to section 9 of the Act (1999:678) of posted employees. The Act has implemented the Council's directive on posted workers 96/71/EC.

The WE Authority is run by an Authority's Director General, as chair. S/he shall install a special committee for co-operation. This committee shall consist of representatives from the labour market representing the trade unions and the employers' confederations. When issuing new Provisions, the WE Authority shall again co-operate with the Social Partners. There shall also be a special committee for transparency.[69] This committee consisted today of three members of the Parliament and a work psychologist.[70]

The WE Authority issues Provisions and manages the three regional Inspection Offices. Under these offices are ten Inspection Districts with work environment inspectors responsible for supervision by inspection. The WE Authority issues prohibitions and injunctions.

§9.06 WORKERS' REPRESENTATIVES

Cooperation between employers and employees is, of course, essential to achieving a safe work environment. The WE Act is based on the presumption of this kind of co-operation. This includes collective agreements and the participation of SD (Safety delegates). Collective agreements concerning the work environment shall never include preventative rules less stringent than those contained in the relevant parts of the Act or its Provisions.

Chapter 6 WE Act deals mainly with SC (Safety Committee), ordinary SD, and Pupil and Student SD. The chapter is complemented by regulations in the WE Ordinance.

The public sector is covered by two different work environment agreements. The local and regional governments have an agreement dating back to 2005 called *Renewal, Work Environment and Cooperation* (FAS 05)[71] which establishes the degrees of influence on an individual basis with annual conferences with the employer, monthly worksite meetings, and regularly held meetings with combined SC for work environment questions and negotiations according to the Co-determination Act. The State has a somewhat similar agreement dating from 1997.[72] The private sector can have many different work environment agreements or none at all.[73]

A SD must be appointed wherever there are five or more persons employed. It is the responsibility of the local trade union having a collective agreement with the employer to make this appointment. In the absence of such an agreement, the employees themselves may appoint the SD. Where there is more than one SD, a Senior SD must be appointed from among them and will be responsible for co-ordinating their activities.

Regional SD (hereinafter RSD) are appointed by trade unions. At least one member of the union must be present among the employees on a worksite for the RSD

---

69. The Social Partners on the Labour Market means representative from the employers' confederations and employees trade unions. Instruction for the WE Authority (2007:913). Sec. 6, 8–10.
70. Hans Gullberg & Karl-Ingvar Rundqvist *Arbetsmiljölagen i lydelse den 1 januari 2010* (Stockholm, Norstedts Juridik, 2010), 32.
71. Förnyelse, arbetsmiljö och samverkan (FAS 05). Samverkansavtal för landsting och kommuner.
72. 'Samverkan för utveckling' (1997), Kollektivavtal inom staten.
73. Maria Steinberg, *Skyddsombud i allas intresse* (Stockholm: Norstedts Juridik, Dr of law, diss. University of Stockholm, 2004), 125, 249.

to have access thereto. The RSD have the same rights as other SD and are expected to help employees appoint SD and organize safety-oriented activities. Today many RSD help employers draw up programmes of Systematic Work Environment Management.

All SD represent employees in matters concerning their work environment. They have the right to participate in the planning of new or alteration of existing premises, equipment, and working methods and techniques, as well as in the organization of work and the use of chemical and other substances. An SD takes part in the preparation of action plans, that being a function of Systematic Work Environment Management (section 4).

The following unique and much discussed case was tried by a District Court:

> A SD was charged with negligence leading to a work environment offence. The prosecutor claimed the SD had some special legal responsibilities according to Ch. 6, Sec. 4 of the WE Act. The section stipulated that the SD shall 'oversee safeguards against occupational accidents and diseases'. The SD had failed to do this adequately. He was acquitted on the grounds that his negligence had been less than criminal in extent.[74]

The general opinion among lawyers in the field is that an SD does not have a special legal liability as a representative for the employees. Such a person will always have a legal liability as an employee to follow the WE Act and its Provisions.

Employers are required to give SD paid leaves for training and for time spent in carrying out their duties (section 5).

Employers must inform SD whenever changes are planned which will affect the work environment. SD have the right to inspect all relevant documents and to be given any other information relevant to their concern with the work environment (section 6).

Since the SD and members of the SC have very extensive rights to information, they have certain obligations not to spread the information. These obligations are found in Chapter 7, section 13 of the WE Act:

> A person who has been appointed to serve as a SD or as a SC member or who has taken part in job adaptation and rehabilitation activities pursuant to this Act, may not improperly divulge or utilize facts which have come to his or her knowledge in the course of his or her duties concerning professional secrets, working procedures, business dealings, the personal circumstances of individuals or matters relating to national defence.

If anyone breaks this rule, they can be made to pay a personal fine or it will be at the most one year imprisonment according to the Criminal Code (Chapter 20, section 3).

SD have a right to demand that employers conduct an investigation, and if necessary a correction, any time they deem the work environment unsatisfactory. If an employer does not comply with such a demand, the SD have the right to appeal to the WE Authority, which is then required to conduct an inspection and, if appropriate, take legal action and issue a prohibition or injunction. SD mandate also extends to the protection of agency workers but only in relation with the SD's employer or employees working for contractors (section 6a). If an SD is not content with the resulting decision,

74. Umeå District Court, Case B 2683-08 4 Jun. 2009.

the Senior SD for the workplace can appeal the issue to an Administrative Court (Chapter 9, section 3).

SD have the authority to order a work stoppage in situations where there is an immediate and serious danger to life or health and no help from the employer forthcoming. The SD can order work to stop if considerations of health and safety in solitary work so demand. Authority to call for a work stoppage also exists in cases where an employer has failed to respect a prohibition issued by the WE Authority. SD cannot be held liable for any damage that occurs as a result of action taken by them under the terms of the law.

A study published in 2004 showed that SD were able to obtain some kind of support from their employers and/or the WE Authority in more than 90% of the cases connected with their exercise of the above mentioned rights. The same study showed that in 20% of all cases the activity of the SD resulted in improvement of environments in respect of mental health.[75]

An SD who has been harassed and who has been appointed by the trade union can, with the support of the trade union, bring the case to the Labour Court. There is only one Labour Court in Sweden. Several conditions must be met for a dispute to come before the Labour Court. One is that it must come from an employer organization, an employee organization or an employer engaged in a collective agreement on an individual basis. A non-unionized SD can take a harassment to a District Court and then appeal to the Labour Court. Interference with the activity of an SD (if a union member) is protected under the WE Act, the Labour Union Representative Act, and the Co-Determination Act. SD may be compensated for harassment by as much as the equivalent of EUR 3,500 if their employer has not allowed them to participate in planning of the workplace.[76]

Pupils and students undergoing education shall participate in safety activities. Pupils from grade 7 through high school are encouraged to appoint Pupil SDs. These have a right to be kept informed on health and safety matters. Two of them have a right to participate in the SC meetings. Student SD can receive compensation for damages and have the right to take problems to the WE Authority on the same terms as SD (sections 6a, 17–18).

The Labour Court has tried several cases involving SD. Most of these cases have had to do with SD right to call a halt to work in accordance with Chapter 6 section 7 of the WE Act.[77] In each case, the claim was for compensation, but so far no SD has been awarded compensation in such a case; invariably the Labour Court found that the SD had not followed correct procedures in stopping work.[78]

A recent case dealt with the SD right to participate in an employer's plans for the future concerning the work environment:

75. Maria Steinberg, *Skyddsombud i allas intresse* (Stockholm: Norstedts Juridik, Dr of law, diss. University of Stockholm, 2004), 272-291, 303-310.
76. Maria Steinberg, *Skyddsombudsrätt* (Stockholm: Norstedts Juridik, 2006), 107-108.
77. Labour Court 1979/164, 1979/165, 1984/58, 2000/92, 2002/98.
78. Labour Court, 2000/92, 2002/98,2008/77. See also Maria Steinberg, *Skyddsombud i allas intresse* (Stockholm: Norstedts Juridik, Dr of law, diss. University of Stockholm, 2004), 198.

> A SD was not invited to participate in an employer's planning of a new staff room. This constituted a violation of Ch. 6 Sec. 4 of the WE Act. The employer had to pay damages to both the SD and the trade union amounting to the equivalent of 3,500 euro each.[79]

This is a good example of how important the court considers SD participation in the planning of work environments.

SC are formed whenever there are at least fifty employees on a worksite. An SC must consist of both employers and union representatives and must include at least one SD. If there are students (any persons receiving post-high school education) and pupils from grade 7 two Student SD must be appointed; the same goes for Pupil SD at schools. The SC has certain tasks laid down in the WE Act. The Committee serves in an advisory capacity to the employer and is expected to participate in planning of the work environment. This is to include plans related to OHS and action plans etc., and encompasses planning changes in work organization. It further includes dispensing information and imparting instruction about the work environment. The Committee shall discuss job adaption and rehabilitation at the worksite (sections 8–9). If any member of the SC, including a Student SD, disagrees with a decision, the question can be taken to the WE Authority, insofar as it falls under that authority (section 9 WE Ordinance). There are no special joint committees for the building sector. The different employers shall co-operate and the SD has the right to participate in the planning according to Chapter 6, section 4 WE Act. When a worksite gets established all the employers and SD could have a joint meeting.

There is no special protection for 'whistle blowers'.

§9.07 CONSEQUENCES FOR THE PARTIES AND REMEDIES FOR BREACH OF HEALTH AND SAFETY PROVISIONS

If an employer does not comply with the WE Act or a Provision, the WE Authority shall give that employer a written notice of inspection accompanied by a written set of demands and the positing of a certain time by which these must be met. Most employers in Sweden readily accede to the demands of an inspection notice. If the employer does not comply the WE Authority will, after issuing a warning, impose a prohibition or injunction (Chapter 7, section 7). The WE Authority can also impose a prohibition immediately if it considers there to be a serious danger (Chapter 9, section 5). Most of the time, the WE Authority will suggest a contingent fine in connection with their injunction or prohibition. A contingent fine is an administrative sanction in the form of a sum of money which the WE Authority specifies in an injunction or prohibition. The contingent fine will be imposed by an Administrative Court, if the injunction or prohibition is not complied with. Contingent fines are usually directed against legal persons, such as a company. It can also be directed against a regional or local government acting as an employer. It can be directed at natural persons as well. The level of a contingent fine depends among other things upon the cost of the

79. Labour Court, 2007/70.

measures to be taken and on how big a fine is needed to 'persuade' the employer to take them.[80] A contingent fine can be adjusted by the Administrative Court, if the employer has lived up to some of the demands from the WE Authority. It can be adjusted to zero, if the employer has complied totally with the requirements of the WE Authority. The money goes to the State.

The Administrative Court handles contingent fines issued by the WE Authority, as well as appeals by employers or Senior SD. The General Courts deal with work environment offences, cases where no contingent fine has been issued, and infractions against certain sections of the WE Act or its Provisions. Cases of neglect by employers in connection with sexual harassment, and terminations of employment on grounds of misconduct or reduced capacity to work are dealt with primarily by the Labour Court.

At present, a contingent fine cannot be levied on the State as an employer - e.g., in the management of universities, or the police force. A Government investigation has been proposed dealing with sanctions. It discusses the possible elimination of personal criminal fines and liability to imprisonment in preventive cases. It suggests that the WE Authority shall be able to impose a contingent fine on the State in its role as employer. Sanction up to EUR 100,000 are suggested when employers are not complying with the WE Act. Sanction shall be pointed out in the legislation.[81]

Most of the time when the Authority intervenes and suggests a contingent fine, the purpose is to improve the physical work environment in a male-dominated workplace. Only 30% of the Authority's injunctions and prohibitions from 2007 to 2010 dealt with workplaces, such as schools and hospitals and other caring institutions, where the majority of the workforce is female.[82] Few of its prohibitions and injunctions concern psychosocial problems, such as violence, unjust forms of organization, or harassment. In a study of the 1,476 decisions made in 2010 and 2011, only 10% were found to deal primarily with psychosocial issues.[83]

Violation of health and safety regulations can result in a contingent fine, a personal fine, imprisonment, and/or, in some cases, payment of damages.

At the maximum, failure to comply with an injunction or prohibition without a conditional fine can bring an employer to prison for up to one year. But this has never happened. At most employers have had to pay a personal fine, or in place of it, a contingent fine proposed by the WE Authority (Chapter 8, section 1). Such fines can be levied on the employer as a company or on the local municipality. In some cases it can also be levied on a physical person.

A personal fine may be levied directly, without injunction or prohibition, on a person who intentionally or by negligence unlawfully hires minors, makes criminal misrepresentations to the WE Authority, or unlawfully removes safety devices (Chapter 8, section 2).

80. http://www.av.se/inenglish/inspections/contingent.aspx.
81. Swedish Government Official Reports series, 'En bättre arbetsmiljö genom effektivare sanktioner' (Stockholm: SOU 2011:57), 11-20.
82. Maria Steinberg, *Is the Swedish Work Environment Law Sexist? Perspective of Women's Health*, in *Working Life: An Anthology* 243 (Helene Sandmark ed., Studentlitteratur 2012).
83. Maria Steinberg, *The Prevention of Psychosocial Risk in the Workplace: Swedish Law* (unpublished ms, 2012).

Violation of a rule stated in a Provision can lead to the imposition of a pecuniary charge, which can be from EUR 100 to EUR 10,000. An employer might have to pay this, for example, for failure to make an annual check-up of a container or a lift (Chapter 8, section 5).

If an employer or a Senior SD is dissatisfied with a ruling of the WE Authority, they can appeal to an Administrative Court. Both have then the right to appeal to the Administrative Court of Appeal and, if necessary, to the Supreme Administrative Court (Chapter 9). The two last courts will decide whether the appeal will be tried or not.

In 2009, there were twenty-nine cases appealed from the WE Authority to the Administrative Court. The highest contingent fine imposed was the equivalent of EUR 50,000. This happened when a county government failed to correct a serious defect in the ventilation system of a school building.[84] In 2010, the Administrative Courts imposed a contingent fine in only six cases. The largest contingent fine was for the equivalent of approximately EUR 25,000 in a case where a county government had failed to correct violations in the areas reserved for the staff of an ambulance service.[85] The most debated conditional fines were imposed on county governments as the employers in charge of (the majority of) Swedish hospitals. The overcrowded conditions in many hospitals result in ergonomic distress for their staff. In an attempt to prevent ergonomic injuries the WE Authority has imposed several contingent fines. A frequently violated requirement is that no hospital bed must have less than 80 cm of free space on either side. The contingent fine for violations of this requirement has been set at the equivalent of EUR 3,000.[86]

Work environment disputes are also dealt with by the General Courts on other grounds. An occupational accident, injury, or serious danger can count as a Work Environmental Offence in accordance with Chapter 3 section 10 of the Criminal Code. Such an offence will have taken place if carelessness was the cause of death or serious bodily injury or illness, or if it created circumstances of serious danger. For charges to be brought someone needs to have carelessly neglected duties prescribed in the WE Act or the Provisions; this could be the carelessness of an employer or employer's representative who has been charged with responsibility by a written task allocation or ex officio. Conviction for such offences can result in imprisonment for up to six years in very serious cases. In several cases the person found guilty can receive a probationary sentence (*villkorlig dom*), which means the next time s/he is convicted of a similar offence a prison sentence might be mandatory. However, so far no one has actually gone to prison; matters have always been settled by the payment of a personal fine and/or probationary sentence. An example:

The Supreme Court tried the case of an employer who had failed to follow a Provision.

The question was whether this failure could be deemed the cause of a consequent death on the job:

84. Administrative Court of Appeal in Stockholm. Dep. 4. Case 9675-08, 25 Feb. 2009.
85. Administrative Court of Stockholm, Dep. 11. Case 10503-10, 26 Feb. 2010.
86. Administrative Court of Stockholm, Dep. 555, Case 24237-08, 12 Jun. 2009.

> A man was killed while unloading cargo from a ship. The employer had not followed a Provision which required that a special 'signal person' accompany the stevedores unloading a ship. The signal person's duty is to give the stevedores a signal to indicate no one is present in the area, while cargo is being loaded or unloaded. The Court found there was a causal relationship between the accident and the failure of a signal person to be present, as prescribed by the Provision. The executive manager was found guilty.[87]

Provisions must be followed. Violation of a Provision resulting in an accident can be deemed a criminal offence.

The Court may decide that a company shall pay a 'corporate fine'. Today this kind of fine will amount to the equivalent of something between EUR 500 and EUR 1,000,000, according to the Criminal Code (Chapter 36, sections 7-10a). If the corporate fine is no more than the equivalent of EUR 50,000, the company can accept a summary punishment and pay the fine without going to court.[88]

Infraction of certain sections in the WE Act or Provisions may be dealt with by the General Courts. This may either be because no conditional fine is set or because the Act or Provision so requires - which can be the case if someone illicitly hires a minor, removes safety devices, or lies to the WE Authority (Chapter 8, section 2 WE Act).

Action on work environmental infractions may be initiated by the WE Authority or the police. Usually it takes a serious accident to lead to the imputation of such an offence. In the course of 2010 the General Courts heard cases dealing with sixty-five work environmental infractions, almost 25% of which went to appeal. The gravamen of the charge in thirty-four of the sixty-five cases was that a worker had been injured by a tool or piece of machinery. The next most frequent (twenty of the sixty-five) were cases in which someone was injured as the result of a fall. Six of the cases that came before the Court involved death. In two cases, the employer in question had to pay a corporate fine:[89]

> One person was killed and two seriously wounded when a concrete block fell on them. The company had to pay a corporate fine equivalent to 150,000 euro. The person responsible for the block received a conditional prison sentence and had to pay a personal fine equivalent to 1,500 euro.[90]

In the year 2010, the General Courts tried eighty-three preventative cases in which there had been a failure to comply with the WE Act or certain Provisions. In most instances - sixty-seven out of eighty-three cases - there was no trial as the accused agreed to pay a corporate fine. Over 20% of the cases concerned the lack of a building plan for the work environment. Ten per cent were concerned with the lack of a (required) environmental action plan, and another 10% with failure to notify the WE Authority of an injury, accident, or other serious mishap. The average corporate fine was on the order of EUR 2,500; in two cases principals of schools had to pay a personal fine equivalent to EUR 1,020 and EUR 1,350, respectively, because they had not

87. NJA 2007, p. 369.
88. Swedish Government Bill. 2005/06:59 'Corporate fine'.
89. www.av.se/lagoch ratt/domar/2010.
90. Solna District Court, Dep. 2, Case B 7473-08, 2010-01-15.

informed the WE Authority of serious violence committed against a student, in one case, and of a serious accident in the school's workshop, in the other.[91] The General Courts deal mostly with factory and construction site accidents; they hear few cases involving psychological injury. In the period 2006–2010, the General Courts delivered 446 verdicts concerning the work environment, only nine of which (2%) related to psychosocial issues. Interestingly, six dealt with an employer's failure to report a serious accident or injury to the WE Authority. These had to do with incidents of robbery and assault.

During 2007–2009, the General Courts tried 155 cases related to the work environment. They found seventy-eight employers guilty of violations under the WE Act, of whom forty-six were convicted of work environment offences. Almost all the cases arose from events in industries (e.g., construction) where the bulk of the workforce is male.[92]

An individual worker hardly ever pays a fine for removal of a safety device or other work environment offence under the criminal act. However, an individual worker is sometimes given a notice of dismissal or termination of employment. This can happen if a worker has committed sexual harassment, in which case s/he is likely to be summoned before the Labour Court.

Here are a few examples of cases in which an employer issued a notice of termination of employment in these kind of cases:

> A Vice-director[93] who had sexually harassed a woman gave her notice of termination. The woman complained to the Court and was awarded the equivalent of 4,000 euro in harassment damages.
> A gym teacher[94] used very rude verbal expressions of a sexual nature towards his young students, co-workers, and (female) principal. Upon receiving a sharp warning from his employer he desisted from this bad behaviour. Nevertheless, his employer terminated his employment. The court found this totally wrong: there was not even grounds for a warning of termination since the teacher had thoroughly ceased to commit the offence.
> A male officer[95] in a job centre was in the habit of commenting on the looks of female job seekers. His colleagues reproached him for this behaviour and was given a notice of termination. The court found that there was no justification for dismissing him and that his employer should instead have simply had a talk with him and admonished him about his behaviour.

A conclusion that can be drawn from these cases is that normally an employer has no right to give a notice of termination without previous warning. Nor is a notice of termination justified if the employee in question, having been warned, has thenceforth desisted from his or her bad behaviour.

In three other cases, however, the Labour Court accepted a termination of employment without previous warning given:

91. www.av.se lag och rätt, domar 2010.
92. Maria Steinberg, *Is the Swedish Work Environment Law Sexist? Perspective of Women's Health*, in *Working Life: An Anthology* 247–248 (Helene Sandmark ed., Studentlitteratur 2012).
93. Labour Court 1991/65.
94. Labour Court 1996/82.
95. Labour Court 2008/91.

An employee[96] had downloaded pornographic movies into his employer's computer. The Court found that this was against 'Good Labour Market Practice' and accepted the termination.
A male employee[97] had sexually harassed a much younger co-worker physically at least 15 times, and as often mentally by using offensive language. The Labour Court accepted the termination of employment because of the serious nature of the harassment in this case. According to the then-in-effect Equal Treatment of Men and Women Act, the employer was required to conduct an investigation. But the Court waived that obligation in this case owing to the seriousness of the offence, which involved repeated physical violence, and the fact that the victim was much younger than the perpetrator and was relatively new at her job.
The head manager[98] of an association devoted to helping former drug addicts was accused of sexual harassing three women. His employment was terminated without notice. The court found the incidents of harassment serious enough to justify such termination of employment. The Court considered the man's behaviour more than usually offensive because he expressed himself in a very obnoxious manner. Moreover, two of his three victims had a history of suffering sexual abuse and consequent traumatisation, and this was known to the accused, who should therefore have been especially careful not to commit this kind of verbal injury. He was also the closest person in the organization to all three. All of which circumstances made his behaviour more than usually culpable. The Labour Court has pronounced that in cases of very grievous sexual harassment immediate termination of employment is justified and no further investigation need be made.

A person who sexually harassed another person cannot be dismissed if s/he ceases his or her behaviour, unless the harassment has been very violent, in which case the employer can terminate the employment immediately.

§9.08 THE INTERACTION BETWEEN HEALTH AND SAFETY PROVISION AND THE SOCIAL SECURITY SYSTEM

The interaction between health and safety provisions and the social security system takes place in several different ways in Swedish Law. The problem begins when someone gets sick and cannot work. There is a compensatory side, a need for job adjustment and rehabilitation. At the same time, a person with limited work capacity is sometimes discriminated against and may even be given notice of termination of employment.

[A] The Compensatory Side as Part of the Social Insurance Code

National Social Insurance is well developed in Sweden and is linked with rehabilitation programmes and compensation for on the job injuries.

96. Labour Court 2007/62.
97. Labour Court 2006/54.
98. Labour Court 2011/84.

When an employee in Sweden cannot work because of any kind of illness, whether professional or not, s/he will receive compensation from the employer in accordance with the Sick Pay Act (1991:1047). Compensation is normally set at about 80% of the person's salary. There is no compensation for the first day of absence on sick leave, but after that the employer is obliged to pay compensation for thirteen calendar days. The employee must present a medical certificate to justify absences of more than one week. If s/he can show that, the employer must pay compensation according to the Sick Pay Act. After that the National Social Insurance will compensate up to maximum EUR 25,300 per month.

Employees do not take all of their salaries. Part of it is paid through the employer's taxes (*arbetsgivareavgift*) towards the social security system. A sick employee will receive 80% of compensation from the social security system after the first thirteen calendar days. The person remains employed while receiving compensation.

Sweden has a National Social Insurance whose regulations are laid down in the Code of Social Insurance. These deal with sickness benefits, rehabilitation benefits, and benefits in compensation for job-related injury. The system has recently been very complicated. Sickness benefits are accorded to employees who are insured and whose ability to work has been limited to one-fourth or less of full-time performance (Chapter 27, section 2). The amount of compensation is set at 80% of full salary, up to a certain amount.

This amount is related to the national price-based index. The insured are entitled to this kind of benefit for up to twelve months. In special cases, their benefits can be prolonged, at 75% of salary, for up to a maximum of 550 days. This kind of limitation is not set in the case of occupational accidents.[99] Compensation is accompanied by a continual review of the receiver's ability to return to work.

The employer is required to investigate whether a sick employee is fit to be given an alternative form of work, e.g., a lighter job in the same department. The aim is to help the sick person get back to work as soon as possible. After ninety days, it is the responsibility of the Social Insurance Bureau to determine whether the employee is well enough to be transferred to other duties under the same employer. The National Insurance Bureau has seldom time to be active in this field. If the results are negative after 180 days, there must be another investigation to see whether the insured person, who is unable to work at any job under his or her present employer, is fit to undertake employment elsewhere on the labour market (Chapter 27, sections 46–48). The Social Insurance Bureau may, however, make humanitarian exceptions to the rigour of this requirement, e.g., in the case of a cancer patient.[100]

Employees on sick leave are required to undergo whatever rehabilitation is necessary in order to return to work. Such rehabilitation normally consists of some kind of training or re-training under a supervisor at the work place. During this time, the employee is entitled to rehabilitation benefits based on the results of an Insurance

---

99. Udén, Cecilia, National Social Insurance Bureau, 27 Mar. 2012.
100. Lotti Ryberg-Welander, *Socialförsäkringsrätt* (Stockholm, Norstedts Juridik, 2011), 139, 142, 147.

Officer's investigation of his or her needs. The employer is expected to take an active part in this rehabilitation. The insured person must also co-operate in the procedure or have the benefits withdrawn. It is the responsibility of the Social Insurance Bureau to make sure that the employee receives the right kind of rehabilitation and that it is properly conducted.[101]

A new Occupational Accident Act was passed by the Parliament in 1976. It was an innovative legal construction, aware of the problem for the single worker to prove his or her injury being caused by the work, in particular the non-accidental injuries. The burden of proof was thus put on the Public Insurance Authority (*Försäkringkassan*). If there was a known, medical acknowledged, risk within the work carried by the employee (chemical substances, monotonic physical movements, psychological pressure etc.), it was for the authority to show that there was another cause in the case of the single employee. The Act also presumed every employee to be insured in his or her actual condition and not to be compared to some kind of ideal person. This was important in particular for female industry workers, handling tools made for male workers.

The presumption of the actual condition is still remaining, but today the burden of proof is different. There is to be enough (medical) evidence for the injury being caused by the work in each single case, which of course is far more difficult for the single worker to achieve. The worker has to show evidence that the injury is a result of the heavy work carried out, and not of leisure activities or other circumstances.[102]

In addition to work injury compensation, approximately 80% of Swedish workers are covered by extra insurance thanks to collective agreements.[103] Victimization at work may count as an occupational injury but is very hard to prove as the burden of evidence falling on the employee who claims it is very heavy. There have been a few cases in which the highest Administrative Courts have deemed victimization an occupational injury, and the vindication of such claims did become a little easier after 1993, when a court deemed a union representative's suicide after a hard series of negotiations at his workplace to be an occupational injury. In this case, the court accepted that the workplace circumstances were sufficiently detrimental, and the union representative's family received some compensatory benefits.[104]

There is also a civil liability system for employers and employees. This can be used if an employee is not covered by the National Social Insurance. This can be the case when an employer has not paid taxes for the employee's work.

[B] Work Adaption and Rehabilitation as Part of the WE Act

The WE Act demands that the employer must adjust the work environment to the workers (Chapter 2, section 1 and Chapter 3, section 3). There is a special Provision dealing with job adjustment and rehabilitation.

101. Lotti Ryberg-Welander, *Socialförsäkringsrätt* (Stockholm, Norstedts Juridik, 2011), 159-176.
102. Lotti Ryberg-Welander, *Socialförsäkringsrätt* (Stockholm, Norstedts Juridik, 2011), 159-176.
103. www.afa.se.
104. Regeringsrätten 1249-1996. Today it is called the Supreme Administrative Court.

The employer has a general duty to adapt work to the employees and to have an organization in place both for this purpose and for effecting rehabilitation. These duties are specified in the Act itself and in the Provision, *Work Adjustment and Rehabilitation* (AFS 1994:1). The WE Authority can impose a conditional fine if the demands of the Act and Provision are not met. The Act places several demands on the employer in this area, stipulating that:

> Working conditions shall be adapted to people's differing physical and mental aptitudes (Ch. 2, Sec. 1).

This means that the employer must adapt work environments to the average worker, male or female, according to age and level of experience. The Act stipulates that every workplace must have:

> a suitably organised scheme of job adaption and rehabilitation... (Ch. 3, Sec. 2a).

Such schemes must fulfil the requirements of the WE Act and of Chapter 30 of the Code of Social Insurance. The employer must also consider an:

> employee's specific aptitude for a given job by modifying working conditions and taking other appropriate measures if necessary.

In planning a job, an employer must also take into consideration the fact that:

> 'individual persons have differing aptitudes for the tasks assigned them' (Ch. 3 Sec. 3).

This Section prescribes the making of individual adjustments by employers. For example, chairs and desks are to be adjusted for the general comfort of employees – though employers, however, are not responsible for adapting every work situation to every individual.[105] It is the duty of the SC to discuss the organization of job adaptation and rehabilitation (Chapter 6, section 9).

The Provision *Work Adaption and Rehabilitation,* requires both physical and mental job adaptation and rehabilitation. The employer must have set objectives for this and be constantly aware of changing needs in relation to it. Individual job adaptation may involve such things as choice of tasks, use of technical equipment, adjustment and layout of the work setting, various kinds of work aids, changes in organization, changes in work tasks and procedures, and individual support and coaching.

Employees must be informed how these forms of adaptation are to be carried out. The responsibility for organization for adaptation and rehabilitation can be assigned to a particular individual. This person must have the competence and be given the authority, resources, and knowledge needed for the job. Adaptation and rehabilitation are to be carried out in close contact with the persons involved and with their SD. The principle is that there shall be co-operation between the Social Insurance Office and other authorities involved. These might include a hospital, social agency, or OHS. The

105. Swedish Government Bill 1990/91:140 '*Arbetsmiljö och rehabilitering*' 45-46.

employer is to conduct an annual review of the organization and make any appropriate changes. The intended objectives, the allocation of tasks, and the routines for performing these tasks, along with the results of an annual revision, must be set forth in written documents. Special attention is to be given to employees with any kind of disability. Efforts to combat alcohol and drug abuse come under the rehabilitation scheme. If the employer does not comply with such demands, the WE Authority may issue an injunction.

[C] Protection of the Disabled

A person who becomes ill while in employment and does not recuperate fully may have to change jobs within the workplace or be given a notice of termination.

According to the WE Act, employers have a duty to adapt work assignments to the individual worker. In this respect, the WE Authority can only compel the employer to take preventative action. But an employer who is unwilling to adapt the work environment to the needs of a disabled employee might be in violation of the Discrimination Act.

Disabled persons are protected by the Discrimination Act, which defines their condition as:

> 'a permanent physical, mental, or intellectual limitation of a person's functional capacity which, as a consequence of injury or illness, has existed since birth, has arisen subsequently, or can be expected to arise (Ch. 1. Sec. 5). An employer is not allowed to discriminate, directly or indirectly or harass an employee because the latter suffers from any kind of disability' (Ch. 1, Sec. 4). A disabled person who has been denied employment may suspect discrimination and bring charges against the employer who refused to hire him or her. When the court decides such a case they are to compare the claimant's situation with that of an equivalent person who is not disabled. In order to qualify for such a comparison the disabled person must be able to carry out 'the most central work tasks'. An employer who hires a disabled person must give that person all reasonable support (Ch. 2, Sec 1).[106]

Only a few cases of this sort have come before the Labour Court. One, in 2003, concerned diabetes; another, in 2005, was about multiple sclerosis; and another, in 2006, had to do with food allergies.[107] Most recently, in 2010, a partially blind person[108] applied for work with the National Insurance Bureau. The job she sought dealt with the disbursement of sick pay. She was denied the job on the grounds of her impaired vision. The question was whether the employer should have hired her, adapted the work to her condition, and generally provided sufficient support for her to be able to perform the job efficiently. The support she would have needed included an extra

106. Susanne Fransson & Eberhard Stüber, *The Discrimination Act, A Commentary* 76, 148 (Norstedts Yellow Library 2010).
107. Labour Court 2003/47, 2005/32 and 2006/97.
108. Labour Court 2010/13. See also Susanne Fransson & Eberhard Stüber, *The Discrimination Act, A Commentary* 154 (Norstedts Yellow Library 2010).

person to help her. The Court found the employer innocent of the charge of discrimination, for he was within his rights in declining to undergo the extra cost of providing that kind of support.

There are several cases in the Labour Court that deal with the question of the employment of persons with an impaired capacity to work. The issue usually revolves around whether the employer has tried to provide an alternative kind of suitable work before giving notice of termination. An employee may justifiably be given notice of termination if s/he 'cannot perform work of any value'.[109] According to case law, an employer must undertake an investigation to determine a disabled person's capacity to work. When necessary the employer must offer some kind of work-related rehabilitation to enable the employee to perform his or her job. That failing, the employer must attempt to find another, more suitable job which the disabled person could perform before deciding that his or her employment should be terminated. The employee must co-operate by, for example, giving the employer sufficient information to enable the latter to help rectify the situation.

Some principles have been laid down by the Labour Court, one of which is that the first responsibility of the employer in such situations is to attempt to get the employee back to work at his or her former workplace. At the same time the action the employer takes must not be harmful to other employees. Nor must fulfilment of this duty compel an employer to expand work activity on the premises:[110]

> An employee[111] had failed to inform his employer about his psychological troubles. Therefore the Labour Court found that his employer was not responsible for rehabilitating him and was allowed to give him notice of termination of employment.

An employer is expected to remain actively engaged when an employee is undergoing medical treatment with the aim of coming back to work:

> A bus driver[112] was no longer able to perform his duties. The bus company he worked for dismissed him. The Court ruled that they had no right to do this. The company should have kept itself informed of his progress in therapy and given him other kinds of work till he was healed. As it was a very large company, it should have been able, with some adaptation, to provide other, temporary work which the patient would be capable of performing while he was undergoing therapy.

In sum, a person with limited work capacity must co-operate with his or her employer in efforts to continue the employment. An employer has the right to dismiss a person who has no work capacity of any value.

---

109. Labour Court 1997/115.
110. Westregård, Annamaria 'De nya sjukpenningsreglerna i ett arbetsrättsligt perspektiv', *Arbetsmarknad och Arbetsliv*, no. 4 (2009).
111. Labour Court 2001/55 and Jakob Broman, Bo Ericson & Carolina Öhrn, *Rehabilitation responsibilities* 73–74 (Norstedts Juridik 2011).
112. Labour Court 2003/44.

§9.09 CONCLUSIONS

The Swedish WE Act came into being to a large extent as a product of the needs of industry, including the construction industry. Much of the content of its related legislation and Provisions was designed to meet those needs. Today a great many employees in Sweden are civil servants and workers in the service sector, and many of these are women. Civil servants and those who primarily work with people - e.g., in schools, hospitals, and the caring sector generally - have needs that largely differ from those of industrial workers. Therefore, there is a need of a Provision that deals with psychosocial problems such as stress-induced injury. Nearly a third of the Swedish workforce is publicly employed and therefore ultimately under the management of politicians. There is need for more precise definitions and more transparent regulations governing the responsibility of politicians in this capacity.

Maybe there is a need for a better balance of focus in order to prevent not only death caused by falling down from a building but also from committing suicide because of bad work environment.

The present Act has been influenced by the conventions and recommendations of the ILO, and today also involves the implementation of directives from the EU. There has been both an upside and a downside to this.

Part of the upside has been that rules have often been stricter than they might otherwise have been. For example, after introducing systematic work environment management, in accordance with Framework Directive, Sweden was accused by the European Commission of not living up to the requirements of that directive. This led to a decision in favour of Sweden by the European Court.[113] Nevertheless, Sweden has re-formulated its regulations for greater precision and clarity concerning the allocation of tasks and the need for a thorough knowledge of the work environment. One downside, from the point of view of the traditional Swedish bias towards treating the sexes equally, has been the necessity, imposed by the EU, for a Provision specially dealing with women who are pregnant or nursing.

Another strain on Swedish habits comes from the EU 'locks' in the Working Time Act. Both the caring and industrial sectors in Sweden have been used to negotiating over lengthy shifts and night work on the basis of collective agreements, and the Social Partners remain unhappy with the constraints imposed by the EU 'locks'. Again, the European Social Partners have issued a 'Framework on Stress', which, though approved by the Swedish Social Partners at the central level, have not yet been implemented on branch level, where decisions are reached by local collective agreement. And herein lies a serious dilemma.

There is a widespread opinion that the WE Act and its Provisions are more concerned with injuries due to accidents than with injuries due to stress and other psychosocial circumstances. Male workers are more subject to the former and female to the latter. The details of the Act and its Provisions seem still to be developing more in regard to the industrial than to the caring sector. This is not to say there is no need

113. ECJ, 15 Jun. 2006, C-459/04, *Commission v. Sweden*.

for the health and safety regulation in male-dominated workplaces, but there does seem to be some discrimination – or at least, lack of balance – in the protection the law provides for the two sexes. In general one can say that the most important regulations in the Swedish WE Act are those dealing with Systematic Work Environment Management and the rights and functions of SD; and those in the Act and its Provisions which give full weight to ergonomic and psychosocial factors.

Sweden today is a diverse society, and this means it has a great need for clear preventative Provisions with regard to persons who do not have Swedish (or English) as their first language. Other problems on the agenda today include the need for better protection for 'whistle blowers' and protection against moral harassment (especially through the burgeoning social-network media). The problems of posted workers, those who come from other countries are much discussed.

One pressing problem at the moment is the cost and difficulties connected with rehabilitation, and the strict requirement of the National Insurance Bureau to pay out insurance benefits for up to six months, after which time the sick employee who still has the capacity to work may have to seek employment elsewhere in labour market. There has been debate as to whether a person in those circumstances must seek work anywhere in Sweden or only locally. Formerly many who were unable to find other work locally were entitled to receive what was called a 'pre-pension'. That option is no longer available in the same way.

The ageing of the Swedish population is making it necessary to increase working hours overall in order to meet the costs of the welfare state. An investigation is pending to discuss protection in employment for those who want to work until the age of 69.In conclusion, we are in a situation where work environment issues play, and will continue to play, an important part in the political agenda.

BIBLIOGRAPHY

Åkerblom, A. *Arbetarskydd för kvinnor, Kvinnlig yrkesinspektris i Sverige 1913-1949.* Uppsala: Uppsala Studies in Economic History, 43, 1998, diss.

Arrius, G. president of SACO press release, published in *Dagens Nyheter*, Sec. Debates. http://www.saco.se/Nyheter - debatt/Aktuellt-fran-Saco/Debatt/hasardinslagen-i-ppm-frater-pa-hela-pensionssystemet/, 11 February, 2012.

Broman, Jakob, Ericson, Bo & Öhrn, Carolina. *Rehabilitation responsibilities.* Stockholm: Norstedts Juridik, 2011.

Carlsson, L. *Searching for Equality, Sex Discrimination, Parental leave and the Swedish Model with Comparisons to EU, UK and US Law.* Uppsala: Justus Förlag, 2006.

Carlsson, L. *The Fundamentals of Swedish Law.* Lund: Studentlitteratur, 2009.

Ettarp, Lars & Eva Kvarfordt. *Arbetsmiljöboken.* Stockholm: Publica, 1981.

Fransson, Susanne & Eberhard Stüber. *Diskrimineringslagen. En kommentar.* Stockholm: Norstedts Gula bibliotek, 2010.

Fredriksson, K. 'Det ojämlika arbetslivet'. In *Perspektiv på kvinnors hälsa i arbetslivet,* edited by H. Sandmark. Lund: Studentlitteratur, 2012.

Gullberg, Hans & Karl-Ingvar Rundqvist. *Arbetsmiljölagen med kommentarer och författningar.* Stockholm: Norstedts Juridik, 2010.
Hydén, Håkan et al. *Den Nordiska modellen, Fackföreningarna och arbetsrätten i Norden –nu och i framtiden.* Malmö: Liber, 1990.
Johansson, Åsa & Daniel Gullstrand. *Alltid uppkopplad –Aldrig avkopplad, En undersökning om det gränslösa arbetslivet.* Stockholm: Unionen, 2011.
Nordmark, Ewa, chairman of TCO comments in a press release, 2012-03-02, on a study made by Roger Mörtvik and Jana Fromm, *TCO granskar stressbarometern 2012, Dags för åtgärder 1/12,* Stockholm: TCO, 2012. http://www.tco.se/Templates/Page1____238.aspx?DataID = 12476, 3 maj 2012.
Nyström, B. *EU och arbetsrätten.* Stockholm: Norstedt Juridik, 2011.
Ryberg-Welander, L. *Socialförsäkringsrätt.* Stockholm: Norstedts Juridik, 2011.
Sandmark, H. (ed.). *Perspektiv på kvinnors hälsa i arbetslivet.* Lund: Studentlitteratur, 2012.
Sigeman, T. *Labour Law.* Stockholm: Norstedts Juridik, 2010.
Stenberg, Ewa. 'More must be able to change careers in the middle of life'. *Dagens Nyheter,* 7 February 2012, sec. News.
Steinberg, M. *Skyddsombud i allas intresse.* Stockholm: Norstedts Juridik, 2004.
Steinberg, M. *Skyddsombudsrätt.* Stockholm: Norstedts Juridik, 2006.
Steinberg, M. 'Är arbetsmiljölagen sexistisk?'. In *Perspektiv på kvinnors hälsa i arbetslivet,* edited by Helene Sandmark. Lund: Studentlitteratur, 2012.
Steinberg, M. *The Prevention of Psychosocial risks in the workplace: Swedish Law.* Örebro: Örebro University, 2012, not yet published.
Sweden, http://www.pensionsmyndigheten.se/AnsokOmPension.html, 3 May 2012.
Sweden Government Decision 2011-05-12. (A2011/2209/ARM) and http://www.av.se/dokument/aktuellt/Artikel_B_Callermo.pdf. 3 May, 2012.
Sweden Ministry of Labour. ' Invitation to a Conference on the psychosocial work environment' Stockholm, December 5th, 2011.
Swedish Government Bill '1912:104 förslag till lag om arbetarskydd'.
Swedish Government Bill '1930:71 med anhållan om riksdagens yttrande rörande vissa av den internationella arbetsorganisationens konferens 1929 fattade beslut'.
Swedish Government Bill '1938:265 med förslag till lag angående ändringar i vissa delar av lagen den 29 juni 1912 (nr 206) om arbetarskydd m m'.
Swedish Government Bill '1973:130 angående ändringar i arbetarskyddslagstiftningen och andra åtgärder för bättre arbetsmiljö'.
Swedish Government Bill '1976/77: 149 Arbetsmiljölag'.
Swedish Government Bill '1989/90:61 Om elevers medverkan i arbetsmiljöarbetet'.
Swedish Government Bill '1990/91:140 Arbetsmiljö och rehabilitering'.
Swedish Government Bill '1997/98:55 Kvinnofrid'.
Swedish Government Bill '2005/06:59 Företagsbot'.
Swedish Government Bill '2007/08:95 Ett starkare skydd mot diskriminering'.
Swedish Government Bill '2008/09:138 Elevers och studerandes medverkan i arbetsmiljöarbetet, m.m.'.
Swedish Government Official Reports series. 'Internationella konventioner inom arbetarskyddet': Appendix 3, SOU 1976:4, Stockholm.

Swedish Government Official Report series. 'Bättre arbetsmiljöregler 1. Samverkan, utbildning och avtal m.m.': SOU 2006:44, Stockholm.

Swedish Government Official Reports series. 'Arbetsmiljön och utanförskapet-en tankeram för den framtida arbetsmiljöpolitiken': DS 2008:16, Stockholm.

Swedish Government Official Reports series. 'Inkluderande arbetsliv': Ds 2009:93, Stockholm.

Swedish Government Official Reports series. 'En bättre arbetsmiljö genom effektivare sanktioner': SOU 2011:57, Stockholm.

Udén, C. National Social Insurance Bureau, Stockholm: 27 Mar. 2012.

Unga i arbete, *Korta arbetsskadefakta Nr* 1/2011 http://www.av.se/dokument/statistik/sf/Af_2011_01.pdf

Unga i arbete, *Korta arbetsskadefakta Nr* 2/2012, http://www.av.se/dokument/statistik/sf/Af_2012_02.pdf

Westregård, A. 'De nya sjukpenningsreglerna i ett arbetsrättsligt perspektiv' *Arbetsmarknad och Arbetsliv*, 2009 nr. 4.

CHAPTER 10

Occupational Health and Safety in the Netherlands: A Shift of Responsibilities

Teun Jaspers and Frans Pennings

§10.01 HISTORICAL AND LEGAL BACKGROUND

[A] Introduction

The shift of terminology by the legislature in 1980 is a sign of a changing approach to the issue of health and safety at work. In the 1980 newly presented law, the government was using the term 'work environment' replacing the traditional terms of health and safety. It was the result of a discussion in which, apart from the academic world, the main advising body of the country in the field of social and economic policies – the Social Economic Council, played a decisive role.[1]

The discussion has been started more than a decade before under the heading of: humanization of labour.[2] It was more than a change of wording. It puts the issue in a broader perspective. Following the opinion of a British governmental committee, the Robens Committee, the general view was that 'the primary responsibility (for healthy and safe working conditions) has to lie with those who create the risk and those who work with them'. That should imply a retreat of the government and the central legislator from this field by offering more room to the parties directly involved. The role of the government should 'predominantly [be] concerned with influencing attitudes

1. Cf. A.J.C.M. Geers, *Recht en Humanisering van de Arbeid* (Law and the Humanisation of Labour) (Kluwer: Deventer, 1988).
2. The sociologist H.J. van Zuthem introduced this term in 1967 in his study: *Arbeidsbeleid in de onderneming* (Assen, 1967). He stated that the humanity of labour should be the core objective of human resources management.

and creating a framework for better health and safety organization and action by industry itself'.[3]

Alongside the United Kingdom, the *Robens Report* influenced significantly the legislatures in, *inter alia*, some Scandinavian countries and the Netherlands. According to this modern concept everything was revolving around the 'duty to care' of the employer and the obligation of the employee to co-operate as reasonably could be required from him or her. Mutuality was introduced in a field where in the past the top-down approach was the usual one. This can be seen as a shift of approach in more than one sense. Apart from the redistribution of responsibilities between the government, the employer and the employee and the introduction of self-regulation instead of state regulation, the change can also be seen in terms of (modern industrial) citizenship. The legislature leaves it to the parties involved to regulate by themselves what they think is necessary. The law sets the goals to be achieved and leaves it to the parties themselves to choose the instruments to attain the goals. They are assumed to be able and capable to do so. They can even do it better than the legislature. Tailor made rules are more effective and will be more easily accepted by those who are affected.

Although the main and primary task lies with the employer, s/he and the employee have been made more jointly responsible for the work environment, each with his or her specific duties and obligations. This is in line with the ideology of the recent governments of treating employers as well as employees (and their representatives) as self-responsible actors and not as solely addressees of top-down state regulation. In this approach the Dutch legislator is obviously showing the importance of the idea of citizenship as leading principle.

Hereafter we will demonstrate how these thoughts, advocated as new ideas, have been inserted in the Dutch law and practice. As will become clear, that does not mean that this approach is solving all the problems easily. New risks, in particular psychosocial risks, which are often connected to the behaviour, the attitudes of the employee and to the way the employee is acting as employee and in his or her private sphere, are raising new questions also as to the approach chosen. This new development we are facing in all developed economies, is challenging when we approach it from the joint responsibility perspective.

[B] Historical Background

The first Dutch statutory provisions on health and safety at work were adopted at the end of the nineteenth century. Since then the concept of health and safety has been changed over time.[4]

3. Robens Committee, Safety and Health at Work, *Report of the Committee 1970–1972* 7 (Her Majesty's Stationery Office 1972).
4. Cf. A.J.C.M. Geers, *Recht en Humanisering van de Arbeid* (Law and the Humanisation of Labour) (Kluwer: Deventer, 1988). Cf. J. Popma, M.H. Schaapman & T. Wilthagen, *The Netherlands: Implementation within wider Regulatory Reform*, in *Regulating Health and Safety Management in the European Union. A Study of the Dynamics of Change* 183 ff (D. Walters ed., 2002) (Peter Lang: New York, Bern, Berlin, Bruxelles, Frankfurt am Main, Oxford, Wien).

The legislative interference with health and safety started with an Act to protect young workers, the *Kinderwetje* (the Child (Labour) Act), which was based on a bill proposed by a member of Parliament and meant to combat the excessive working time and terrible working conditions of children.[5] This Act was succeeded by the *Arbeidswet* (Labour Act), which was meant to ensure the protection of women and (again) young workers.[6] The latter Act also created the possibility to set up a labour inspectorate. Before the start of the twentieth century a third Act was adopted, a (framework) Act on health and safety at work.[7]

After World War I, as a reaction to the revolutionary threat following from the developments in Russia and Germany, the Labour Act was renewed and strengthened, more in particular as regards working time. The Act limited the number of working hours to eight per day.[8] In the light of later developments it is interesting to note that from the beginning of the interference of the legislator the target was not only the improvement of the health and safety of the workers at the workplace. Moreover, it has to serve the well-being of the employees. The regulation of the working and rest hours as part of the policy on health and safety has to contribute also to 'the happiness and the mental well-being' of the person.[9]

The Health and Safety Act, amended in 1934, introduced health and safety committees, which were mostly to be established on a voluntary basis. This Act was a reaction to a proposal of trade unions to introduce effective participation of employees in health and safety matters in the enterprises, which proposal was rejected.

The Health and Safety Act of 1959 brought a change of approach, which can be identified in the obligation to establish an industrial medical service on the company level with the explicit objective to promote preventive medical care in the enterprise.[10]

A new approach was adopted by the Act of 1980, the *Arbeidsomstandighedenwet* (hereinafter Arbo Act or ARBO[11]). This Act not only adopted prevention as its main approach, but it introduced the well-being of the employees, or in other words, the humanization of labour, as a separate objective. This new approach did not focus only

5. The Child (Labour) Act of 1874
6. The Labour Act of 1889. This act was the result of a parliamentary inquiry on the application and functioning of the Child Labour Act and more generally on the working circumstances of workers in factories and workshops in view of their safety, health and well-being. The inquiry contained more recommendations than were adopted in the Labour Act. Cf. A.J.C.M. Geers, *Recht en Humanisering van de Arbeid* (Law and the Humanisation of work) (Kluwer: Deventer, 1988), 38 ff.
7. The Health and Safety Act of 1895. By ministerial decrees more detailed regulations could be issued. This legislative technique (framework Act and decrees) allowed for an easier way of making and changing rules; for this purpose decrees were made on the basis of the Act. In some sectors specific legislation has been adopted, such as mining, phosphormatches, quarrying and stevedores. The 1895 Act was maintained until the 1980s, while it was several times amended but these amendments did not concern the concept of health and safety.
8. It also gave a regulation for rest time in all kinds of circumstances.
9. *Handelingen TK* 1918–1919, p. 2876 (Proceedings of the Second Chamber of Parliament).
10. This new approach has been applied in some specific regulations: the Act on the use of silicosis (1951) and the Act on dangerous equipment (1952).
11. Literally: Working Environment Act.

on a decrease of the number of occupational accidents and diseases, which could be beneficial because of the reduction of costs, but it also aimed at the reduction of the absenteeism due to incapacity for work caused by occupational accidents and diseases.[12] The third objective of the new Act was to offer employees possibilities for self-fulfilment by improving their working conditions.

Next to the regulations specifically concerning health and safety of employees, two other pillars of regulations deal with health and safety issues. The oldest is the *Burgerlijk Wetboek* (Civil Code – CC, which dates back to 1907), in particular the part on the employment contract. Article 7:658 CC lays the duty on the employer to ensure healthy and safe working conditions. If s/he infringes this obligation, s/he is liable for the damages suffered by his or her employees resulting from his or her lack of care. At the time when this obligation was inserted in the Civil Code, this provision was mainly aimed at the financial compensation of the damages suffered by the employee caused by occupational accidents. Later, as from the end of the 1960s, this provision lost part of its ground (and application) when the damages suffered were compensated by the social security law provisions made for this purpose.[13] Indeed, social security legislation is the third bunch of legislation relevant to health and safety issues. We will deal with all these three parts below.

A main characteristic of the approach of the Dutch government and legislature is the deliberate choice for *self-regulation* by those bodies that are most able to ensure health and safety at enterprise level. The relevant legislation (primary and secondary legislation, discussed below in section 10.01[C]) shows that policy clearly. This legislation is characterized by the use of so-called goal provisions: these provisions contain objectives (in terms of preventing or reducing risks to the health and safety of the workers) that have to be attained. The provisions do not mention the exact means by which these goals have to be reached, since the choice for these means is left to the social partners. Indeed, the health and safety policy, the regulations and the necessary measures have to be established by the employer(s) and the representative bodies of the workers, be it trade unions or works councils. They are laid down in so-called Arbo-catalogues, which are the result of negotiations by both sides of industry (the catalogues will be discussed below). At the end of the day, however, it is the (individual) employer who has to take the concrete measures to implement the relevant regulations, in particular the Arbo-catalogue.

Because of this considerable room for own responsibility of social partners to develop their standards we used the phrase 'more room for citizens' responsibility' in

12. In terms used in the social security schemes: rehabilitation.
13. Until 1967, the loss of income was compensated by the occupational accidents law and the mandatory medical care legislation. The first one has been replaced by the Disability Act (WAO) that often offered a relatively generous (i.e., until the 1980s 80%, later 70%, of the former wage in case of full disability) compensation. It covers occupational accidents and occupational diseases as well as disablement in general. Since the disability benefits have deteriorated substantially the Civil Code provisions have gained importance. In 2006, a general act on medical care has been adopted which requires all residents to buy insurance for a basic package of care.

the title of this chapter.[14] Below we will discuss the exact contents of this responsibility in more detail and we will also try to address the question as to what the effects are of this room.

A second typical Dutch element is the introduction of financial incentives, in particular for the employer, in order to encourage him or her to take the necessary measures. If the employer does not take these measures, s/he may face the obligation to pay damages to employees, which can amount to high levels.[15] Such payments may be due both in case of illness of the employee and for the payment of all other kinds of damages caused by an accident or an occupational disease. We will come back to these payments in section 10.05[B][2].

The legislature strongly believes in this approach, since it is supposed to contribute much better to the attainment of the goal of healthy and safe working conditions than public supervision of statutory obligations can do.

The concept of decent working conditions has in many countries widened to go beyond the terms of health and safety, and also includes the 'well-being' of the employee. In the Netherlands, the development of this concept has gone in a direction different from that in other European Union (EU) Member States and EU law. The notion of 'well-being' as part of the working conditions of an employee in the enterprise - a notion being part of the health and safety policy since the beginning of the twentieth century though not explicitly - was introduced in the Dutch legislation in 1980 when a large reform of the existing legislation took place.[16] The attempt to regulate humanization of work was the motor behind this change.

However, at the time when other Member States have adopted this notion in their legislation, the Dutch legislature deleted it from the Arbo Act rather recently, i.e., 2007.[17] This was because the content of this notion was considered as too vague. Therefore, it has been deleted as a separate element of the health and safety policy.

[C] The Dutch System of Regulation of Health and Safety

Since the change of approach in 1980, the Arbo Act has the character of a framework Act. It primarily contains a series of obligations for employers - and employees - targeting on improving and maintaining healthy and safe working conditions. The framework character implies that the general Act neither provides concrete provisions on the obligations of the employer nor gives concrete norms and standards to be

14. More room for citizen's responsibility is not only present on the collective side. It lies also on the individual employer and employee in the sense that they can make arrangements between themselves tailored on the particular situation. This implies that according to the approach of the Dutch legislator the individual employer and employee has to take responsibility by their own in respect to the working environment.
15. It can cover all kinds of damages actually suffered.
16. Cf. AJ.C.M. Geers, *Recht en humanisering van de arbeid* (The Law and the Humanisation of Work) (Deventer: Kluwer, 1988), 82–97.
17. The provision was repealed by the amendment of the Work Environment Act in 2007.

applied. Instead it imposes the obligation on the employer to develop and pursue a policy aiming at healthy and safe working conditions.

The main obligation is the prevention of accidents and occupational diseases. In this respect, the Dutch law is in line with the general approach of the Directive89/391/EEC (hereinafter Framework Directive). In addition, the employer has to adapt the working methods, the equipment used and the nature of the work to be done to the personal capacities of the worker 'in so far as reasonably can be required'. S/he also has to avoid monotonous and pace driven work, again subject to what reasonably can be required from him or her. The Act – again completely in accordance with the Framework Directive – prescribes as an important tool for attaining these objectives that the employer has to carry out an inventory of the risks which the workers face and that these risks have to be systematically evaluated. Another general principle is the involvement of the workers' representatives in the whole chain of health and safety policy measures. As we will see hereafter, in some cases the workers' representatives, mainly the works council, has a right to consent. Besides that the employer has to offer effective information in due time and training courses to the individual employees.

In addition, the Act defines the function and tasks of the labour inspectors in general terms. In specific circumstances, the inspectors have the competence to give orders to the employer. Such orders can include a warning, an order to observe specific obligations by the Act and the imposition of a fine or an order to stop (dangerous) work. By a more recent amendment of the law the labour inspector has been provided with a specific tool: administrative coercion (see hereafter).

A novelty introduced by the Act is the introduction of a system of certification, not required by the Framework Directive. The Act provides that in case the employer is making use of an external service – a so-called Arbo service – only certified institutions may be used to carry out the various tasks required by the law concerning health and safety at work. The employer is no (longer) obliged to contract such a service. S/he can carry out these tasks by him or herself usually by one or more of his or her employees with specific expertise and tasks in the field of health and safety. For some specific tasks, listed in the law, the employer has to rely on certified experts (individuals or services).[18] The workers' representatives have a strong position in this field. The employer needs their consent to his or her decision of organizing these tasks. If s/he fails to get it, s/he has to contract a certified Arbo service.

As was already remarked, this approach of the Arbo Act 1980 is quite similar to that of the Framework Directive. Therefore, the implementation of this directive into Dutch law has not caused many problems. Since the Arbo Act encompasses all crucial elements of the directive it was not necessary to adapt the existing Dutch law to the requirements of the directive.[19]

18. For instance, occupational medical care services, doctors or experts in the field. A special certification body has been established. We will come back to this issue.
19. It does not mean the transposition is perfect. See hereafter.

Still, the Dutch implementation of the Framework Directive has specific characteristics. Its leading principle is the distribution of responsibilities and obligations over the various parties involved. This is in accordance with the general approach in Dutch labour market and labour issues policy and legislation to 'put the government at a distance'. In other words, the social partners are the main actors to regulate themselves their mutual relationship and the set of obligations and rights adhered to these.[20] The legislature was convinced and assured that the social partners were very well able to do so.

Also in the health and safety field the government has deliberately chosen for this approach. The actors, more in particular the employer, employees and their representative bodies, have to take their responsibility in order to improve and maintain healthy and safe working conditions. This approach fits very well in a policy of creating room for tailor made regulations and measures on a decentralized level rather than on the central level. These are made not only on the sectoral level but also on company or plant level.

Thus the main principle is the (re-)allocation of responsibilities to the main actors in this field. That implies that, next to the employer and the employee, trade unions, the works councils, the labour inspectorate, social security institutions and other health and safety services, be it private or public, have a role to play. It is left to the main actors, the employers and the workers' representatives (the key parties), to shape their various tasks and obligations by means of the instruments they prefer. Various instruments are available for this purpose, varying from collective agreements to informal arrangements.

The Dutch health and safety legislation has a layered structure. Three layers belong to the public domain and one to the private sphere:

(1) the (general) *Arbeidsomstandighedenwet* (Arbo Act) as a Framework Act;
(2) the Working Environment Decree (WED), based on the Arbo Act, which provides for a flexible way of regulating the main obligations;
(3) the Working Environment Regulation (WER) containing more concrete norms and standards;
(4) arrangements by the parties, be it *collective agreements* or arrangements, laid down in *Working Conditions Catalogues* (Arbo-catalogues) as a set of concrete norms and standards.

This can be illustrated in a Figure 10.1:

20. This approach is applied in many fields of labour and supplementary social security. The collective labour agreement is the main instrument. The parties are allowed to make their own rules in order to create 'tailor made' arrangements. In some areas, they are even allowed to deviate to the detriment of workers from statutory law. Sometimes trade unions indeed agree with such deteriorations if they can gain advantages in other fields.

Figure 10.1 Dutch Law on Health and Safety

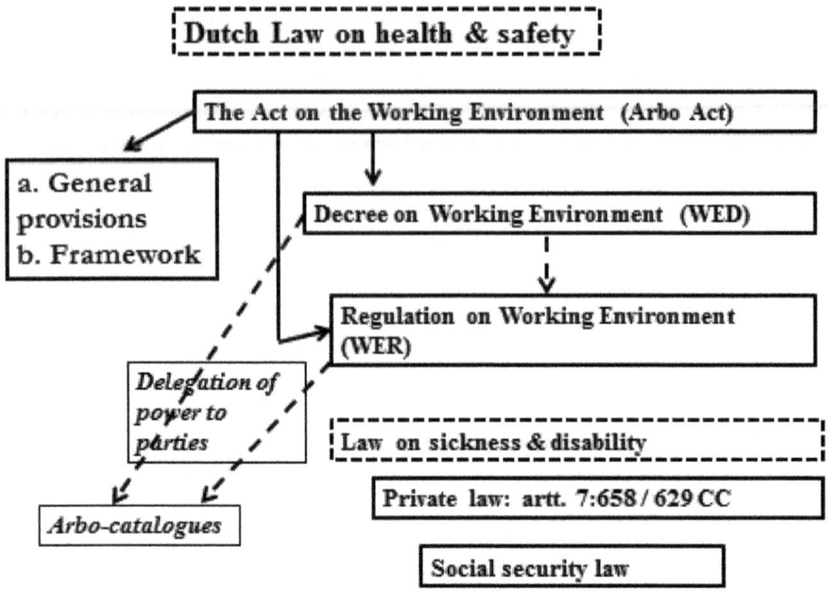

The Working Environment Decree of 1997 replaced the numerous decrees which date from the time before the introduction of the Arbo Act. By the most recent amendment of the Working Environment Decree in 2007, it was envisaged to bring the Dutch protective health and safety system at the level required by the Framework Directive. In this context, this means levelling down the level of the standards: insofar as the national standards exceeded the European level many of them were repealed or adjusted to require no more than required by EU law.[21] In this way a European level playing field has been established.

At the same time, the more detailed regulation of the older Act was replaced by more general rules setting targets to be achieved or limits to be respected.[22] It thereby created room for more tailored and more precise regulation on the lower levels (layers 3 and more in particular 4).

21. One can wonder whether the Dutch legislator has respected a so-called non-regression principle expressed in the pre-amble of the directive: 'Whereas this Directive does not justify any reduction in levels of protection already achieved in individual Member States, the Member State being committed, under the Treaty, to encouraging improvements in conditions in this area and to harmonizing conditions while maintaining the improvements made.' The directive itself does not contain an explicit provision on non-regression.

22. An example of this kind of target ruling is Article 5.2 WED: the workplace and the equipment to be used have to be shaped in a way that the health and safety of the employee will not be imperiled by the physical workload. An example of ruling by setting limits is Article 6.8 WED: a daily exposure to noise has to be limited to a certain amount of decibels (87). Another important example of putting limits is to be found in the sectors where dangerous substances are at stake.

Another instrument of the WED is what can be called 'process norms'. One of them, the most important one, obliges the employer to deploy a policy in respect to psychosocial workload, such as stress, sexual harassment and mobbing.[23]

The Regulation (WER) and the Arbo-catalogues[24] are the instruments that give more precise and detailed norms and rules. The advantage of this approach is the possibility of adapting the rules to the specific characteristics of the branch or sector of activities.

Box 10.1. The Construction Industry

This industry has a complicated form of health and safety regulation. The WED contains a paragraph dealing with general rules applicable also to this industry. Next to this regulation targeted at the health and safety of employees, another Government Decree deals with all aspects of construction activities. It mainly contains technical rules that have to be observed and respected in the construction industry. Although they are not directed to health and safety of the employees, these rules contain also safety measures to be taken or observed in relation to the construction workers. Rather recently the part of the WED mentioned above has been transferred to the Construction Decree in order to bring the rules in conformity with each other. Next to these regulations several Arbo-catalogues have been issued dealing with various aspects of the activities in the construction industry. For instance: paving, infrastructure, foundation, painting, and glazing. All these catalogues contain special rules for the various situations at work which can cause accidents, in order to prevent these accidents from occurring and to protect workers by specific means. A good example is a document dealing with breathing protection. In an extended advice several issues are tackled related to the danger of inhaling of all kinds of dangerous substances. It also contains an overview of available protection means that can or have to be used in order to diminish or exclude the risks of inhaling such substances. It does not prescribe which protection means have to be used but it gives all relevant information about the tools and means. It is up to the employer and the employees to make use of this information. Apart from physical risks the catalogue also contains issues on stress related disorders: various forms of psychological stress, overburdening, burn-out, post traumatic disorder. Various measures are recommended varying from adequate information of the employee, consultation on the quality of the work and the way of performing, regular consultation, adapting of the work to the capacities of the worker. The nature of this part of the catalogue is also an advice and not an order to be observed and explicitly sanctioned. It is on a more or less voluntary basis. That does not imply that not observing this advice is without consequences. The remedies rest in the liability of the employer in case of damages (see hereafter) either in private law or in public law by intervention of the labour inspectorate.

An interesting aspect is the danger of the use of asbestos. If asbestos has to be removed it has to be done by a certified enterprise. Since using asbestos in buildings is not allowed, the problem has become less important. The requirement of certification for enterprises dealing with asbestos is aimed at getting a grip on the danger of asbestosis. The case law on the issue meaning that the employer is fully reliable for damages caused by the use of asbestos, even

---

23. By the most recent amendment of the ARBO this issue has been introduced in the act itself; Article 1(1) sub e and f. The act defines psychosocial workload as all the factors that causes *directly as well as indirectly* stress referring explicitly to sexual harassment, aggression, violence and mobbing (emphasis by the authors). See more about this in §10.04 below.
24. At present there are more than 130 Arbo catalogues.

for damages arisen in the past, has had a substantial influence on this policy. In a series of decisions the Supreme Court constantly has decided in that way (the first one dates from 1993: HR 25 June 1993, NJ 1993, 686. Since then, *inter alia*, HR 2 October 1998, JAR 1998/228).
Another example of explicit regulation in the construction industry is dealing with the lifting and pushing of heavy things. The DWC (Dutch Working Conditions Act) contains some general provisions on this issue. As usual it is a goal provision: the employer has to take care that lifting or pushing of heavy things (physical burdens) does not cause danger to the safety and health of the employees. Furthermore, it provides for some tools or working methods to help the employer to fulfil his or her care obligation. On this general basis the Arbo-catalogue contains among other rules a maximum burden of lifting of 25 kilo per person. This shows that the Arbo-catalogue is the more detailed regulation adapted to the special features of the construction industry.

Unlike the three first mentioned instruments, the Arbo-catalogues are not made by the government but by the employer(s) and employees or their organizations (see Figure 10.2).

This approach implies the involvement of trade unions and/or works councils. They are the main actors to make concrete norms and standards on healthy and safe working conditions. Actually they are the main actors for implementing the Act by making implementation rules.

A second non-government driven instrument is called 'standardisation'. Standardization is the task of a private organization, the Dutch Normalization Institute,[25] which develops and issues the norms to be observed by enterprises in a certain branch of industry. These norms have a private law nature. Since this type of regulation - a kind of self-regulation - can respond better and faster to new technical developments and the needs of the industries, the government supports the activities of this institute, as it considers them as an interpretation of what is requested according to the 'state of technology', a legal term used in the Arbo Act. In recent decades this institute has developed sets of norms, which standards, being private norms in nature, acquire a public law status if the government adopts them in its policy measures.[26]

This practice can have some real shortcomings. Since the norms are of a private nature, accessibility, cognoscibility and legal certainty are not guaranteed. Mostly one has to pay for receiving a particular set of norms or to take part in the development of the standards. A shortcoming can also be that some parties that are involved, such as the workers and their representatives, are not allowed to participate in the development of the norms. Nevertheless, the norms can be applied as part of the set of health and safety norms to which the minister refers to in his or her policy measures.

Finally, the instrument of certification has to be mentioned. By this instrument the quality of the agents involved in the health and safety policy and practice of

25. The Dutch Normalisation Institute (NEN) was founded in 1916 by the Dutch Business and Trade Company and the National Association of Engineers. It has joined the International Organisation for Standardisation (ISO) and European institutions including the European Committee for Standardisation (CEN).
26. The Minister of Social Affairs, the competent person for health and safety issues, often refers to these NEN norms.

enterprises is checked and supposed to be secured. An independent body assesses whether the requirements for certification are fulfilled. It also periodically supervises the actions of the various bodies involved. The minister appoints the certification body and the Council for Accreditation checks the reliability and the independency of the certification body and its expertise.

Figure 10.2 The Actors

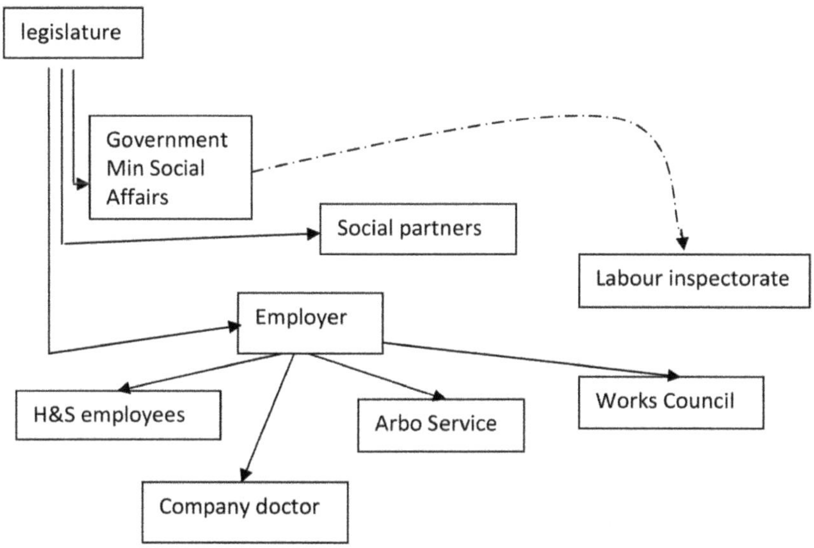

A general feature of the Dutch regulations in this field is that they are made by private parties. This private involvement is not unlimited, but beyond doubt the impact of private parties goes quite far. This can be seen from the structure of the standard setting as discussed *supra* and it corresponds to a main principle of regulation that has become quite common in Dutch labour law: retreat of the government and more room for self-regulation and self-responsibility. That phenomenon can also be seen in the approach of the financial consequences of damages due to occupational accidents and diseases.

As stated *supra*, the objective of the whole body of regulations is prevention. Only in case prevention has failed, compensation is an option. This is not a specific Dutch phenomenon, but characteristic for the Dutch system is the systematic use of a steering instrument: the financial incentive. In the last decade, financial incentives as steering instrument have become popular in Dutch policy. It is assumed that they invite or force individuals, employers and employees, to act in their own interest, but still in a way desired by the legislature and that they thus contribute to the attainment of the goals set in the law. Facing the payment of a substantial compensation for the damages suffered if the obligations have not been respected, the employer will 'voluntarily' act in conformity with the goal to be reached: healthy and safe working conditions.

This approach has reshaped the responsibilities of the parties and is put in practice in two ways: (a) by threatening with financial compensation in case of damages suffered (to be paid by the employer) and (b) on the side of the employee: a denial of a financial compensation if s/he fails to contribute to the prevention or to reduce the damages s/he might suffer or is suffering. The legislature has introduced – financial – incentives to the parties directly involved in working environment. It is supposed to be an incentive to improve and maintain healthy and safe working conditions. Protection is a responsibility once and for all of the employer(s) and the employees.

§10.02 BASIC CONCEPTS

[A] The Concept of Health and Safety and Risks

The present Dutch law neither explicitly uses nor provides for a definition of health and safety or working conditions. Even an explicit concept of health and safety is lacking.

The meaning and the content of health and safety at work may vary with the time, the state of knowledge and technical developments. For that reason, the legislature found it neither fruitful nor effective to define health and safety at work. Nor does a definition fit into the 'new' approach, also advocated at the European level.[27]

It seems more interesting and important to see whether the law indicates which risks are covered by health and safety regulations and which dangers or dangerous situations have to be effectively combated by taking protective and preventive measures. The Act takes a pragmatic approach: the employer has to develop and apply a policy in that field. In the words of Article 3 Arbo Act: 'the employer provides safe and healthy conditions concerning all aspects connected to the work to be done; he also pursues a policy that aims at working conditions that are as good as possible. To attain that goal, several obligations rest on him'. These obligations are phrased in a general way and do not contain detailed and concrete instructions. Also the – before mentioned – *Working Environment Decree* and, more importantly, the *Working Environment Regulation* (WER) do not provide instructions. They generally indicate and list the risks and the dangers to be avoided and which, if they occur, have to be taken away or remedied. The more detailed regulation has to be found in the Arbo-catalogues as mentioned above. But even in these instruments a body of detailed instructions is lacking.

[1] Legislation

The body of rules depicted above shows in the first place that physical risks are covered. These risks still belong to the usual treatment of accidents and diseases[28] and

27. Cf the Framework Directive 89/391/EEC.
28. We refer to accidents and diseases that cause physical symptoms.

form a main part of the various regulations: to prevent or to remedy injuries by dangerous, toxic substances, machines, tools, equipment etc.

Over the past few decades, 'new' risks have been introduced in the legislation. In 1993, by amendment of the ARBO of 1980, psychosocial risks were taken on board. Article 1(3)(e) ARBO refers to psychosocial workload as containing 'all factors which either directly or indirectly, including sexual harassment, aggression and violence, mobbing and heavy workload, may cause stress at work'. Stress is defined as a situation that causes negative physical, psychical or social consequences. The Act provides that these factors are relevant if they cause stress at work. Sexual harassment and aggression as such do not fall under the terms of providing safe and healthy working conditions, this is only the case if they lead to stress. However, since the term stress is rather widely defined, in practice these psychosocial risks are covered by the Act and are recognized as risks for the health of workers. So if (sexual) harassment or mobbing result in medical problems (including mental illness) it is very likely that these risks fall within the scope of the regulations. Under discussion is still whether all 'modern' diseases such as Repetitive Strain Injury (RSI), burn-out and impaired neuropsychological development fall under the concept of (new) risks covered by the regulations.

As indicated above, in this respect the Arbo-catalogues are very important sources, because they are more elaborated as to risks covered and the measures to be taken. Most of the industries issue rules in various forms, containing all kinds of means, tools and instruments dealing with physical as well as psychosocial risks and dangers. They may differ in the details of their provisions, but they all provide information and give proposals for improving the working conditions. In nearly all of them attention is paid to the psychosocial risks related to performing work in specific industries. Psychosocial risks are most often described in more specific detail in the catalogues (see the example of the construction industry, mentioned *supra*). A striking example is found in hospital care and in care in all kinds of homes (for elderly, disabled and young people).

Box 10.2. Care in nursing homes and private houses

This sector knows an extensive catalogue on this issue. The catalogue contains a detailed regulation of the physical risks related to the specific tasks to be performed in this sector such as lifting of patients, helping people who are not able to do the necessary things as to his or her external care. Next to these issues there exists an extensive body of guidelines related to psychosocial issues from harassment, overburdening, burn-out, stress. The guidelines contain checklists on which a policy of the institution can be assessed whether it is aiming at sufficient, operative and effective means to prevent physical overburdening. It can be helpful when establishing or filling in the risks inventory and assessment. It is providing all kinds of tools to attain that goal.

> In the sphere of psychosocial risks the catalogue entails a whole body of means and tools. The approach is the common one in Dutch practice. The catalogue is laying down the obligation for the employer to carry out a preventive policy in order to protect the employees against aggression, violence, harassment and mobbing. It should contain a code of behaviour, information on harassment and training of employees, a procedure of reporting and registration of incidents of harassment, reception, guidance and after-care of employees exposed to harassment.
> As to work-related stress the catalogue provides for a five steps approach. The first step is making a system of picking the signs and inserting them in the (mandatory) risks inventory and assessment. Second, if necessary carrying out a research to get a better and deeper insight the problems that could cause stress. The third step is taking effective measures in order to reduce the workload to be laid down in a concrete plan. The fourth and fifth stages are the carrying out of the measures in practice and checking the effect of the measures.

[2] The Role of the Courts

A second source, equally or even more important, is the judge. The labour law judge as well as the social security law judge play an increasingly important role in interpreting the concept of risk of health and safety. This is at least partly due to the fact that the legislation is rather vague and leaves room for the courts to interpret the content and to determine the boundaries of this term.

For instance, employers are subject to the obligation to pay damages to the extent that they can be held responsible and liable for not observing the – health and safety – obligations imposed on them. Since the law uses very wide and open norms it is to the judge to take a stand in the dispute, not only on which risks are covered, but also whether one or both actors have infringed the law on health and safety at work. In an extensive case law, courts have shed light on these problems. The legislature has deliberately chosen to leave room to the courts in order to give room for more flexibility. Since central legislation is deemed not effective and too far from the workplace and courts are considered much better equipped to assess the concrete circumstances of the case.

[B] The Risks Covered

As in all legal systems, traditionally the focus of the regulations lies on the physical risks. That was mainly due to the dominant character of economic activities, mainly of a manufacturing nature. The change of focus in the 1990s has been caused by various factors. An important one is the extended introduction of advanced technologies which has influenced the production methods substantially. The rise of the services industries connected to a decline of the traditional manufacturing has also contributed to other kinds than physical risks. Another development to be mentioned is the increasing attention to issues as (all forms of) harassment as part of a growing awareness of discriminatory practices, more in particular on enterprise level. The new risks are taken together under the heading of psychosocial risks. In the regulatory field, the state reacted by introducing state regulation. Also in the case law the courts, faced with problems connected to all kinds of psychosocial risks are dealing with it. From the statistics of the causes of accidents and occupational diseases it is evident that the psychosocial risks are taking an increasing share. Nearly 25% have a psychosocial background.

Special attention should be paid to the regulations on working time. The connection between health and safety on the one hand, and working time on the other is since the first regulations either on working time or on health and safety common. The first regulation on working time whether it concerns young workers or women, is based on health and safety of these groups of vulnerable workers. The Working Time Directive of the European Community of 1993 (Directive 93/104/EEC) was directly based on the health and safety of employees. It was (re)confirmed by the decision of the Court of Justice in 1996.[29] We will deal with this issue separately.

[1] Physical Risks

Traditionally, these risks were covered by extensive and detailed state regulations. By adopting the new approach in the 1980s, the government and the parliament have deregulated the body of laws. The detailed state regulations have been replaced by the Work Environment Decree and the Work Environment Regulation and more in particular by the various Arbo-catalogues. These instruments were perceived as more adequate because they fit better to the demands, needs and possibilities of the enterprise and were, for that reason, deemed to serve the health and safety of the employees better. The WED and the WER contain only partly concrete substantive norms, mostly minimum or maximum provisions (noise, toxic agents etc.) or prohibitions.[30] In general, the provisions of these regulations are so-called purpose provisions: aiming at achieving a pre-established objective. Leaving room for the employer to choose his or her own methods and measures provided s/he achieves what the law prescribes.

Box 10.3. In particular for the construction industry Article 3 paragraph 16 of the WED states that in order to prevent the danger of falling a safe scaffold or floor has to be built; anyhow measures against the danger of falling have to be taken in case of the existence of a working environment that entails an increased risk of falling, such as holes in a floor, or in case the danger actually exists that a fall of more than 2.5 meter can occur.
Another example is Article 5 WED that stipulates in general that work has to be organized, the production methods have to be used or protective tools have to be available in such a way that the physical burdening does not endanger the health and safety of the worker.

29. ECJ 12 Nov. 1996, Case C-84/94. (*United Kingdom of Great Britain and Northern Ireland v. Council of the European Union*), [1996] *ECR* I-5755.
30. For instance, a prohibition of the use of certain substances unless in exceptional (controlled) situations.

Another instrument that is often used in regard to physical risks is the use of limiting values. In particular it is used in industries where workers are exposed to noise and toxic agents.

> **Box 10.4.** In this field the whole range of regulatory instruments is used. The WED provides the Minister of Social Affairs with the power to issue special regulations on these issues. For the various toxic agents the regulation contains detailed limiting values the workers may be exposed to. In case the exposure is going beyond the maximum or possibly could go beyond, the employer is obliged to take the necessary and effective measures to secure the health and safety of the worker.
> An example of the construction industry: these Ministerial regulations are elaborated by Arbouw[31] in the Arbo-catalogue which provides the employer and the worker with concrete information, tools and methods how to handle situations where workers could be exposed to, inhaling toxic agents.

In case purpose provisions are not suitable, because the objective cannot be formulated in sufficient clear terms, the Arbo law is offering an alternative: process-provisions. The employer has to develop and to follow a procedure aiming at identifying, assessing and handling the risks to which the workers are exposed. This is part of the general duty to care.

Apart from the regulations by decrees, regulations and catalogues, the protection against physical risks is also an issue in the case law. It is for sure that the situation in the field of physical risks is clearer than as to psychosocial risks as we will see hereafter.

When an employee suffers damages caused by an accident while performing his or her tasks or suffers from an occupational physical disease which relates to the work done, the employer is held responsible and liable. The employee has to make plausible that the damages have occurred during performing the work; the employer has to prove that this is not the case in order to be released from his or her responsibility and liability. Actually the employer has to prove that s/he has complied with the responsibilities expressed in Article 7:658(1) CC.[32]

This provision[33] imposes obligation on the employer to take such measures as to the equipment and the maintenance of the workplaces, to the tools used for performing the work and to give orders and instructions to the employee for performing the work as may *reasonably* be deemed necessary to prevent the employee from suffering damage while doing his or her work. In Article 658(2) the employer is made liable for

31. Arbouw is the bilateral body representing Dutch employers and employees in the construction industry: www.arbouw.nl
32. This is in conformity with the wording and the meaning of the Framework Directive 391/89/EEC which contains a similar obligation.
33. This provision is part of the law on the employment contract, which in its turn is part of the Civil Code. It has to be borne in mind that this provision of liability differs from the common provision on civil liability in the sense that the burden for the employer is much heavier than in civil law for a person who caused damages to another person. The liability under common civil law is more a 'liability based on guilt', which implies that in case the person who suffered damages (caused by an accident by the other person), has (partly) contributed to the occurrence of the damages, is also liable. This can result in shared liability. Such shared liability does not exist in cases based on Art. 658 CC.

any damage which the employee suffers while carrying out his or her work, unless either the employer can prove that the obligations referred to in paragraph 1 have been respected and complied with or the damage was to a large extent the result of *intent* or *deliberate recklessness* of the employee.[34]

These obligations are elaborated by, *inter alia*, the provisions of the Health and Safety legislation (Arbo Act, the WED and WER and more in detail the Arbo-catalogues). The risks inventory and evaluation is in this respect a very important source, since the inventory and evaluation have to contain concrete measures to prevent the occurrence of the risks at the level of the enterprise. When assessing the liability of the employer, the court takes all these aspects into account. The employer has to provide information to his or her employees and must train them to make them aware of the risks. The mere existence of protective measures (equipment, tools) is not sufficient to become (released) exempted from liability.

Employers have to check periodically whether the measures are actually applied and the instructions followed by the employees, where they have to take into account that employees when doing a job on a routine basis may become inattentive as to the exact application of the rules and thus incur safety or health risks. Courts have to weigh all such aspects when deciding a case.

Special attention has to be paid to RSI. The problem here is to decide in an individual case whether this injury has been caused by the work performed. One essential element of the liability of the employer is whether the damages have arisen either while the employee performed the work; whether they are (mainly) caused by activities in the private sphere; or, as a third possibility, whether they are caused as a result of a combination of factors at work and at home. In the second situation, there is no liability of the employer, whereas in the first situation there is no doubt that s/he is liable. But even in this case the employee bears the burden of proof: s/he has to provide sufficient evidence that the damages occurred while s/he was exercising his or her tasks. The existence of a causal link between the injury and the work performed has to be proven by the employee.[35] The medical report addressing this particular issue may play an important role in this procedure.

To conclude: in cases of physical damages, the burden rests mainly upon the employer when an accident at work has caused damages or in case of an occupational disease. The employer is relieved from his or her obligations if s/he can prove that s/he has done what reasonably can be required in order to avoid the occurrence of the accident or the occupational disease or in case the employee has caused the damages intentionally or due to deliberate recklessness.

34. The obligation is not absolute: what the employer *reasonably* has to do taking into account the state of knowledge, science and technology.
35. HR 23 Jun. 2006, JAR 2006/174. The Supreme Court does not require a full proof, but the employee has to make sufficiently plausible that there exists a causal link. If he succeeds, the burden of proof shifts to the employer.

[2] Psychosocial Risks

The recognition of this type of risks causing problems of health and safety at work and for that reason to be taken seriously, took place at first in the ambit of the social security law and practicing policies. Psychical deficiencies were considered as causes for illness and disability and therefore covered by the sickness and disability schemes. The substantial attention in the health and safety at work politics for this kind of risks as causes of accidents and occupational diseases dates from the end of the twentieth century. The Arbo Act was dealing with sexual harassment by introducing in 1998 a – rather detailed – definition of this term.[36]

Since then, it belonged to the whole body of measures the employer has to take as part of his or her 'duty to care' in order to prevent (sexual) harassment could take place or to take the necessary measures by which the employees exposed to harassment were adequately and effectively protected. By a more general change of the health and safety at work law in 2007 psychosocial risks have been explicitly included in the legislation. As we have seen before, in the Arbo Act psychosocial risks have been defined as 'psychosocial workload that contains all factors that either directly or indirectly may cause stress at work'. It includes sexual harassment, aggression, violence, mobbing. In the same article, stress is defined as 'a situation that causes negative physical, psychical or social consequences'.[37] It has been elaborated in the Decree on the Work Environment (WED): if a worker has been or could be exposed to psychosocial workload, measures have to be taken in order to prevent or, if that is not possible, to reduce the psychosocial burden at work; such plans have to be based on and inserted in the mandatory inventory and evaluation of risks workers are facing or could face at work in the enterprise.[38] Similarly to the 'traditional' risks, the psychosocial risks are further elaborated in more detailed rules to be followed and aiming at the achievement of the main objective. The numerous Arbo-catalogues are providing for more extensive rulings and measures that could be taken in this respect. They all have a paragraph on psychosocial risks but the extent of detailed rules and measures to be taken varies widely. They are addressing the usual forms such as stress, burn-out, harassment, mobbing, overburdening. In some sectors, special issues are addressed such as violence, be it physical or more psychical.

The concept of psychosocial risks has been subject to court cases since the early 1990s. The first – lower – court decisions dates from 1991 to 1995 in which sexual harassment was the subject matter. Usually it was directly connected to dismissal cases.[39] Since 2003, burn-out and stress have been subject of court cases.[40] In these

36. It has been part of the implementation of the Equal Treatment legislation of the European Union. Cf. Article 1 para. 3 of the Arbo Act 1998.
37. Article 1, para. 3 sub E and F of the Arbo Act.
38. Article 2.15 WED. In para. 2 of this Article, an obligation is added for the employer to inform and to offer training to the workers about the risks (in particular the psychosocial risks).
39. Between 1991 and now, there have been more than fifty cases. Only one case was about something else.
40. The cases were based on the specific liability of the employer for the damages caused in the working environment. The number of court cases is not very high, around fifteen.

cases, the courts found that it is the responsibility of the employer in the first place to take the necessary measures to avoid, prevent and anyhow remedy situations leading to sexual harassment and stress or burn-out. If s/he fails to fulfil his or her duties, s/he is responsible and liable for the damages caused or in case of dismissal the dismissal has been declared invalid resulting usually to an increased amount of compensation.

In 2005, the Supreme Court issued a decision on the coverage of psychosocial risks.[41] It confirmed the case law of the lower courts and by that it put an end to the uncertainty, raised in practice and in literature as to whether also these risks fall under the protection of Article 7:658 CC and if so, to what extent. In its decision the Supreme Court decided that not only physical damages are covered by that provision, but also psychical damages in case it was about overburdening of the employee for several years. In the view of the Supreme Court, there is no reason that the liability of the employer being based on the principles of Article 7:658 CC - which include the obligation to care, aiming at prevention or, in case that is not possible, to reduce the occurrence of the risks covered as well as the consequences if the risks occur - is limited to physical damages. The employer is also liable when psychosocial damages have occurred because:

> it is the employer who determines in a decisive way where and how the work has been and has to be performed as well which equipment and tools are or will be used. A restrictive interpretation of Article 7:658 would lead to an arbitrary distinction, since physical and psychical welfare are closely linked. Therefore there is neither a fundamental nor a practical justification for restricting the damages to be protected by Article 7:658 CC to situations in which only physical injury or harm occurred.

After this key decision, courts have been faced with various questions on what has to be understood by psychosocial risks and whether certain forms of psychosocial damages fall under the protective umbrella of Article 7:658 CC. In a series of decisions, the scope of this article has been extended to various manifestations of psychosocial diseases. For instance, burn-out is acknowledged as an occupational disease[42] if it has been caused by doing the work. That does not imply that the employer is responsible and liable in all cases where a burn-out is at stake. It has to be sure that the burn-out is caused by or during the exercise of the tasks by the employee. A problem can arise when the burn-out is partly or mainly caused by circumstances in the private life of the employee. Generally, this problem is solved by assuming that the damages are caused by the performance of the work, unless the employer can prove that s/he has taken all the necessary measures which can reasonably be expected to prevent or to reduce the risk of becoming burnt-out.

Another key element that has been developed in the case law is whether the employer - reasonably - could have foreseen and recognized that the way the employee was performing the work, should or could lead to an overburdening of the

41. HR 11 Mar. 2005, JAR 2005/84 (ABN-AMRO/Nieuwenhuys).
42. In The Netherlands, there does not exist a closed list of industrial or occupational diseases.

employee.[43] In such cases even a pro-active approach can be demanded from the employer.[44] The same way of reasoning is used in cases in which stress is at stake.

In cases of harassment, mobbing and aggression (by other employees or clients) leading to illness or at least incapacity to work, the employer is obliged to take measures to stop and prevent further damages to the employee. To be relieved from responsibility and liability, the employer has to develop and pursue an effective policy aiming at prevention of the occurrence of these risks. In such circumstances, a pro-active approach can be required as well.

Sexual harassment is addressed in the ARBO and in the Act on Equal Treatment (ET Act). Before 2007, the ARBO explicitly mentioned and defined sexual harassment, but in 2007 the definition has been replaced by a general term: 'psychosocial workload', that explicitly includes sexual harassment.

Finally, it has to be mentioned that certain decisions or actions of employees can lead to liability of their employer when they result into the mental injury of fellow employees. In such cases, it has to established whether the employer has infringed the so-called employer's *Fürsorgepflicht*, i.e., his or her duty to care, which can include the obligation to take measures aiming at protecting the fellow employees against this kind of actions. This duty is based on a general obligation in the CC that imposes the employer to act as 'a good employer' (Article 7:611 CC). If the employee suffers damages by falling mentally ill and s/he is therefore not able to perform the tasks, s/he may be entitled to compensation.[45]

As mentioned above, in cases of a combination of causes partly within the working environment and partly in the private sphere, the problem arises whether the responsibility and the liability have to be shared by both: the employer and the employee. Since the Article of the Civil Code – Article 7:658 CC – uses the adagio 'all or nothing', it is logical that only in particular circumstances the employer and the employee share the risks. There are a few cases in which the courts decided that such situation was indeed at stake. These decisions were based on medical reports in which the medical experts give a reliable estimation of the risks borne by each of the parties. So it has to be clear that the damages are caused by the exposure to risks at work and by the behaviour of the employee him or herself.[46]

43. Supreme Court (HR) 27 Apr. 2007, JAR 2007/177. Generally speaking, the employee has to give indications to the employer that the workload is too heavy and that (s)he cannot handle it (HR 3 Apr. 2009, JAR 2009/111). The judges are rather reluctant as to the extent the employee has to inform the employer. If the employer is or can be aware of the personal characteristics of the employee that (can) contributes to the illness due to a burn-out, the employer is liable. Information by colleagues of the employee has to be taken into account as well. Recently Gerechtshof 's-Hertogenbosch (Court of Appeal) 9 Nov. 2010, JAR 2011/41.
44. Gerechtshof 's-Hertogenbosch 18 Feb. 2011.
45. S/he is entitled to the continuation of payment of the wages during the illness; see §10.04[B][2][b] below.
46. The case law has been developed in the asbestos cases: when an employee suffers cancer caused by being exposed to asbestos and by continuing smoking the court has divided the liability between the employer and the employee. HR 31 Mar. 2006, JAR 2006/100 and in nearly similar case HR 9 Jan. 2009, JAR 2009/59. This approach is criticized by labour law scholars because the estimation by doctors is at least questionable.

The issue of psychosocial risks resulting in illness occurs also as to the question whether the employee will be paid during sickness due to psychosocial causes. In this context, the question could rise whether the symptoms of an employee are recognized as illness, which is a requirement to be fulfilled in order to become eligible for payment of the wages by the employer according to Article 7:629 CC.[47] The solution of the courts is comprehensive. The judges rely on the view of the medical doctors. If they decide that the illness is caused by psychosocial problems, the judge accepts this view. There is no case law in which a court decided 'on its own' that a symptom is of a psychosocial nature or not. The law requires that the sickness has to be based on a cause objectively medically recognized as such.[48]

A related problem which has occurred in practice is that it is uncertain whether the labour inspectorate in its capacity as controller of the workplace and the methods of working, has any competence in the private sphere of the employee (including when s/he is working at home) in order to make the working situation safer and healthier. This problem does not only occur in cases of psychosocial risks but also in cases of physical accidents and diseases.[49] Since the inspector has the competence of checking and interfering only within the workplace one can wonder whether s/he can effectively operate. Until now this problem has not been solved.

[3] Working Time

Similarly as the Working Environment Act, the Working Time Act of 1996 (WHA)[50] imposes on the employer the obligation to care in terms of an obligation to pursue a policy in the enterprise in relation to the working time. Article 4, paragraph 1 of the WHA states: 'the employer pursues a policy as to working time and rest periods as appropriate as possible; he takes into account as reasonably can be required, the personal circumstances of the employees'. The WHA is directly related to the WEA where the first declares that 'the policy with regard to working hours and rest periods has to be pursued in connection to the policy on health and safety of the workers'.[51] According to the general structure of the WEA, the policy of the employer on working time has to be based on a (written) inventory and an evaluation of the risks establishing the risks the actual working time can cause. On that basis, the actual working time has to be arranged in such a way that damages to the health and safety of a worker will not occur. Also, in this field prevention is the first aim of any regulation by the employer. The act contains only minimum norms to be respected.

47. We will deal with this Article of the Code Civil hereafter in more details, in §10.08[B][2].
48. In case of a conflict between the employer and the employee resulting in a notification by the employee s/he is psychically ill caused by the conflict, it can be decided that this is not an illness according to Article 7:629 CC. Cf. S. Klosse & F.M. Noordam, *Socialezekerheidsrecht* (Kluwer, 2010), 281.
49. The most striking examples are dangerous sports, or at least sports that are risky as to causing injury; and leisure activities in which one can be exposed to dangers or risky situations.
50. Replacing the working time regulations from the (general) Labour Act of 1919 in which the working time was regulated.
51. Article 4:1 WHA refers to the WEA explicitly.

Apart from the prohibition of child work aiming at the protection of the health and safety of the child and supporting the education of the children, the Act does not contain regulation for categories of workers. The reference made in the Act to the personal circumstances of the employee provides for a tool for the employer by establishing the working time schemes for his or her employees within the enterprise to take the personal circumstances of an employee into account. Pregnancy and the period after confinement are explicitly mentioned in the law as a personal circumstance to be taken into account (Article 4:7 WHA). This fits in the policy of making the combination of work and care not impossible or too much hampered by working time arrangements.

Since 2003, the Act has specified this aspect: 'personal circumstances' implies caring of children, of (dependent) family members and relatives as well as societal responsibilities taken by the employee. It is not an absolute protection since the law is using a 'reasonability' clause implying that the interests of the individual employee have not per se priority over the interests of the enterprise. Economic interests may not endanger the health of an employee but the employer has some room to regulate the working time when weighing the interests of the enterprise and of the employee. In case law the judges are weighing these interests in the individual cases. Against the primary obligation of the employer on the employee lies the obligation of the latter to be reasonable as to the working time proposed by the employer . The judge decides on the basis of the concrete facts of a case.[52]

An important element could be and usually is whether the works council has given its consent to the proposed working time schedule. According to Article 27 of the Works Council Act, the works council has a right to give or to deny its approval to a regulation of the working time and rest periods in the enterprise. In its decision of granting or denying its approval the works council can take into account whether and if so, how the employer is taking care of the health and safety of his or her employees when establishing the working time.

The Working Time Act provides for minimum norms as to the maximum working time and the minimum rest periods. The Act is following the norms set in the EU directive 2003/88/EC.[53] Health and safety considerations are explicitly taken into account as to the regulation of night work. On this issue, the WHA is directly related to the WEA. Article 18 WEA states that employees who are working in night shifts are entitled to regular medical check-ups. In case it appears from these check-ups that the employee is suffering damages to his or her health, the employer has to adapt the working time unless it is reasonably impossible for him or her to do so.

52. See, for instance, Kantonrechter Leeuwarden 5 Apr. 2006, JAR 2006/109.
53. Including – in general terms – the case law of the CoJ as to the definition of working time in the cases of Simap (Case C-303/98), Pfeiffer (Case C-397/01 to C-403/01) and Jaeger (Case C-151/02).

§10.03 THE PERSONAL SCOPE OF APPLICATION AND THE DUTIES OF THE PARTIES

[A] The Scope in General

The Arbo Act, the decree and the regulation address employers (private as well as public employers). The regulations on sectoral and branch level cover all employers of the sector or branch concerned. If a catalogue has been established on enterprise level, it (solely) binds the enterprise concerned. As said before, if it is not a collective agreement, the catalogue does not have a legally binding effect. In such case the employers are allowed to deviate from the catalogue, but they have to meet the requirements of the Act by offering the same level of protection and solutions equivalent to those contained in the catalogue. Supervision of abiding by the Act is the responsibility of the labour inspectorate and/or the parties which adopted the catalogue.

Catalogues contain a series of duties for employers and employees. Public authorities supervise the catalogues only marginally. This concerns the way the catalogue has been established. Relevant is whether the catalogue defines for which sector or branch the catalogue applies. Also relevant is whether those who adopted the catalogue represent the employees and employers of the sector or branch concerned. Part of this supervision is that the inspectorate checks whether the catalogue is available for the employers and employees concerned. Finally, it is checked with the help of a so-called *quick scan* whether it can be assumed that by following the catalogue the goal provisions of the Act are implemented.

Catalogues which satisfy this marginal test are mentioned in a general policy rule, issued by the minister. By referring to the catalogue in a particular policy rule, the government makes clear that by following the catalogue it can be assumed, in principle, that the statutory obligations have been complied with. As a result, the catalogue as such is not binding for the individual employer, but it is 'a benchmark' for following the statutory goal obligations.

An individual company can thus deviate from a catalogue, if it demonstrates that it satisfies the goal provisions by providing an adequate safety level.[54]

On the employees' side all employees regularly working at the workplace or the place they perform the work under (formal or actual) supervision of the employer are covered by the Act, the decree and the regulation. They are also bound by the catalogue in the same way as they are by a collective agreement if the catalogue is such an agreement.

If the catalogue has a voluntary character, it depends on the employment relationship, whether the catalogue has a binding effect. If the catalogue is referred to in the individual employment contract, it has binding effect. Also if the employee claims that the employer has not adequately fulfilled his or her obligations 'as a good

54. Unless the subject has been regulated by a collective agreement to be applied to the enterprise.

employer', the standards of the catalogue may be referred to and thus become relevant for deciding the case.

Besides the regular employees, also other persons working under the supervision of the employer are covered, such as temporary agency workers, apprentices, home workers,[55] self-employed persons and employers who perform the work jointly with their employees. The user firm is responsible for all the workers carrying out the work. In some cases there is a joint responsibility and liability of the user firm and the posting firm or agency. That implies that most of the subcontractors are bound by these instruments.[56]

[B] Scope of Application and Special Groups

The Arbo Act applies to persons working on the basis of an employment contract regardless whether they have a contract of indefinite period of time, fixed term contract or a contract on call. Also civil servants are covered and persons who work under supervision of another person without an employment relationship, for instance trainees.

The act also applies to pupils and students in schools and in open areas if they perform activities that resemble work.

Also home workers are subject to this act; this is provided by the Working Environment Decree (WED).

In general, the Arbo legislation is applicable on all employees. Only for two specific groups of persons employed in an enterprise special, more protective regulations exist.

A special problem arises as to the so-called self-employed persons without staff.[57] Since they are considered as independent workers they do not have a contract of employment, even though they may be economically as dependent on a principle as employees are. They work on commission which implies they have no employer. They are neither covered by labour law nor by employees' schemes of social security.

Currently, it has been proposed that they will be covered by health and safety regulations when they work for a firm. There is also a proposal to subsume them under the (public) Disability Benefits Act.[58]

55. Unless they operate as self-employed persons. So if they are not self-employed they are covered, regardless whether they work under a contract of employment or not.
56. It has to be taken into account that catalogues are usually voluntary. That means that the responsibility of subcontractors is not so heavy.
57. The number of persons under this heading has increased since the last decade. By estimation they amount to 750,000 people. See also *The Protection of Working Relationships. A comparative study* (F. Pennings & C.Bosse eds., Kluwer L. Intl. 2011).
58. At the moment, they have to rely on a private insurance against accidents, sickness and disability.

[1] Young Workers

For persons under 18 - young workers - a special regime applies, following from the implementation of Directive 94/33/EEC on the protection of young at work.[59] This also applies to pupils and student and trainees. These persons are covered by the Arbo Act and some additional rules of the WED. These provisions require that in the risk inventory and assessment which an employer has to perform, young workers are especially taken into account. For this purpose the specific risks have to be assessed which are the result of a lack of work experience, the inability to assess the risks and a not yet completed physical and psychical development. Also the organization of the work place has to be taken into account and the nature and duration of effects of agents and physical factors. Also the training level of the young and the need for information have to be assessed. The young are entitled to medical examination if it appears from the risk inventory that they have to do work with specific risks meant in the previous sentences and the employer has to organize adequate supervision which takes account of these risks.

In relation to other specified dangerous goods or agents they are allowed to perform the work only if they work under expert supervision which can prevent the risks. There are also a number of prohibited activities, such as working with specific dangerous goods, to be exposed to toxic or biological substances, or working under particular noise levels or dangerous vibrations (Article 6.27 WED).

From the point of view of concern with the health and safety of young workers at work, the Working Time Act prohibits young workers to work (Article 3:2 paragraph 2 WHA). The Act defines young as persons less than 16 years of age. These young persons are protected for doing work based on an agreement. This may be any agreement to which not only the child, but also other persons are committed. This rule implies a broad interpretation and includes also unpaid work on a voluntary basis. For some types of work an exception has been made, related to specified age groups. For example, non-industrial work of a light nature can be done by children of 13 years and older and newspapers can be distributed outside school hours by children over 15 years of age. Another exception is provided by the WHA giving the labour inspectorate the competence to make exemptions for certain activities of young persons subject to a general clause: the work carried out may not harm the health of the child and may not collide with school hours.

[2] Women

There are no special rules for women; only in relation to pregnant women, who have recently given birth and those who are breastfeeding (hence: protected women). These follow from Directive 92/85/EEC of 19 October 1992 on the introduction of measures

59. OJ *L* 1994, 216.

to encourage improvements in the safety and health at work of pregnant workers and workers who have recently given birth or are breastfeeding.[60]

The Working Time Act provides that the work of a pregnant worker has to be organized in such a way that account is taken of her special circumstances. Furthermore, the protected women are entitled to extra breaks and the right to work in a steady and regular working pattern. In principle she cannot be obliged to perform night work (unless the employer shows that this can reasonably not be required from him) and to work more than ten hours in a service, more than fifty hours a week in a period of four consecutive weeks and more than on average forty-five hours in a period of sixteen consecutive weeks. The employer shall allow the pregnant worker to undergo the necessary medical pregnancy investigations and she remains entitled to wages if these had to take place during working time.

In addition to these rules, Article 4:6 of the Working Time Act stipulates that the employer has to organize the work is such a way that the pregnant worker does not work during the period of twenty-eight days before the presumed day of delivery and not during forty-two days after the delivery.[61] The special rules on working times which apply during pregnancy apply also six months after delivery. During the first nine months of the life of her child, the mother is entitled to interrupt her work for breastfeeding or expressing milk and to do this in an appropriate room (for at least a quarter of the working time per service).

The WED provides that the employer has to make a risk inventory (such as we discussed in relation of the young workers) if s/he employs protected women. The Decree prohibits some activities for the protected women (Article 4.109), such as when she has to work with specific goods or agents or activities such as, like in the case of young, diving and work underground (Article 6.29 Decree).

§10.04 THE DUTIES AND ROLES OF WORKERS' REPRESENTATIVES

Pursuant to the character of legislation in this field, i.e., self-regulation, the role of workers' representatives has been increased. An important development is the introduction of Arbo-catalogues in the last decade as effective instruments. As said above, the workers' representatives, be it the trade unions or the works councils, are one side of the key players by establishing these catalogues. In the process of framing the Arbo-catalogues the trade unions participate in the negotiations with the employer(s), even when they do not reach an agreement. It is the explicit objective of the legislature that the social partners make rules which fit best with the requirements of the sector involved and are thus as concrete as possible.

Since works councils operate at enterprise level, they play a different role. Their functions, rights and duties are governed by the Works Council Act (hereinafter WCA).

60. OJ *L* 1992, 348.
61. These are the periods the employer is not allowed to employ the woman; the periods she is entitled to leave are longer i.e., one month before the delivery and three months after the delivery.

The WCA itself refers in several provisions to the field of health and safety. The role of the works councils on this field has been strengthened over time. Initially the works council has been attributed with the right to be informed and consulted on health and safety issues as part of the right to give an advice to the employer according to Article 25 WCA. Also the Arbo Act contained provisions on the workers' involvement which overlapped the regulations in the WCA substantially. By an amendment to the Arbo Act in 2007, a dc duplication took place by repealing the provisions of the ARBO which doubled with the obligations following from the WCA. The aim of the government and the parliament was a reduction of the bureaucratic burden for the employer. The Arbo Act kept a general provision aiming at a good co-operation between the employer and the works council in developing and carrying out a health and safety policy in the enterprise.[62]

The most important provision in the WCA is Article 27 in which the works council is attributed with a right to consent on health and safety issues. According to that Article, the employer needs the consent of the works council when taking decisions on the establishing, the changing or the withdrawal of rules or measures concerning the health and safety of his or her employees.[63] Though the wording is rather vague, its meaning becomes clear when the view of the legislature is taken into account, since this means that all kinds of rulings in respect to health and safety can be covered by this consent requirement.[64] Since in Dutch law the responsibility for health and safety measures was conferred to the employer, the potential scope of Article 27 WCA can be wide. Actually it will depend upon the role and acting of the works council what the scope will be. Although the WCA does not refer explicitly to the Arbo Act, it is obvious that the scope of Article 27 can be determined by the various regulations on health and safety. However, it has to be kept in mind that since the works council operates at enterprise level, the influence of the works council on the decisions of the employer is restricted to those the latter actually intends to take. The link between Article 27 WCA and the health and safety regulations indicates the possible issues subject to the consent by the works council. So if the employer wants to make rules at enterprise level within the ambit of the various obligations stemming from the health and safety regulations, the works council can claim that its consent is required. The restriction is that the regulations concerned must have a general nature. That means the regulation must be generally applicable and not be restricted to a single case or employee. The main subjects for which the works council can claim that its right of consent is needed include:

---

62. Article 12 Arbo Act. Until 2007, the employer had the obligation to inform on his or her own initiative (that means the employer has to inform the works council even when it has not asked for) the works council of a requirement for exemption of obligations, a requirement to stoppage of work, a requirement for abiding the rules and for reporting on accidents to the labour inspectorate.
63. Article 27(1) sub d WCA. According to this provision the requirement of consent applies also on issues as a policy to combat absenteeism due to illness as well as a policy on reintegration of (partly) disabled employees into work.
64. In this sense, the debates in parliament in 1996 and 1997; *Kamerstukken* (Parliamentary Papers) II 1996/97, No. 24 615.

(a) a ruling on the inventory and evaluation of risks;
(b) rulings on the prevention or restriction of heavy accidents due to dangerous substances;
(c) the establishment and the way of operating of prevention services; and
(d) the choice of the external service deploying the health and safety activities to be performed.

Since Article 27 WCA contains also an obligation for the employer to pursue an effective policy[65] on the prevention of illness of his or her employees as well as reintegration of workers who are partly disabled, the works council can claim its consent on the regulation the employer intends to make.

It can be concluded that the works council can have real influence on health and safety measures to be taken by the employer, at least when they make use of the powers conferred to them by the information and consultation provisions. The works council can decide itself the degree of its involvement. It even can extend its power if it is able (and powerful enough) to widen its involvement by concluding an agreement with the employer entailing extra rights on specific issues related to health and safety.

Finally, the WCA (Article 28) attributes to the works council the task to encourage the actual application of all health and safety measures and regulations taken by the employer. This article provides the works council with the possibility to check whether the employer actually applies his or her regulations and measures.[66]

§10.05 CONSEQUENCES FOR THE PARTIES AND REMEDIES IN CASE OF VIOLATION OF HEALTH AND SAFETY PROVISIONS

[A] Consequences for the Main Actors

The main actors in the enforcement of the health and safety measures are, on the one hand, the workers' representatives and employees themselves and, on the other hand, the labour inspectors. The inspectors check whether the employer complies with the rules laid down in the Act, the decree, the regulation and the catalogues. The Act has provided the labour inspector the power to check the observance of the rules and to give orders to the employer to comply with his or her duties within a prescribed period of time. The inspector is even allowed to order to stop the work or to order that the workers have to leave the workplace.[67] If the employer does not obey an order of the inspector, the latter can enforce his or her decision by imposing a sanction on the employer. There are several types of sanctions s/he can impose.

65. That means an obligation to take measures aiming at achieving that goal.
66. Another issue is whether the works council can enforce the regulations to be applied. Actually there are some obstacles in procedural law. Since the works council does not have legal personality it can start a normal civil law procedure, although in some cases the works council can start an interim procedure.
67. In that case the employer is obliged to continue the payment of the wages.

The main obligations rest upon the employer because s/he manages the production processes. The law, EU law as well as national law, is clear on this division of responsibilities. The employee has so-called derived responsibilities. S/he is obliged to follow the instructions of the employer, to use the equipment and tools in the way as prescribed by the various regulations and by the employer (Article 11 ARBO).

Apart from this, the employees' representatives have a role on behalf of the employees and the well-functioning of the enterprise. Their roles have been described above.

The redistribution of powers, responsibilities and obligations between the public authorities and the private sector, i.e., the employers, that has started in the 1980s and was completed in the following decades, has resulted in a new structure in which various actors operate. There are several reasons for this. Apart from the employers' liability framed by private law (labour law) principles, the law provides the employer with another set of obligations: on the one hand s/he has to continue to pay the wages of his or her employee while s/he is ill and not able to work, regardless the reason of the illness. So also in case of sickness caused by accidents at work or occupational diseases the obligation to pay the wages rests on the employer. However the employer has, on the basis of particular provisions of the Civil Code and the Sickness and Disability Laws (see hereafter), to make serious and effective efforts to re-integrate ill and disabled employees in their job or at least in a suitable job within the company.[68] For the fulfilment of these multiple tasks s/he has to make use of specialized bodies or persons.

These specialized bodies or persons can be special health and safety service providers or so-called prevention persons who have special expertise in the field.[69] In order to guarantee the quality of the service provided, certificates issued by a special body can be used.[70]

In order to be certificated, the provider will be assessed on reliability, independency and expertise. Besides these general requirements, certification norms have been developed by the parties themselves for the sectors or branches in order to adjust the requirements to the needs and peculiarities of the sector. It is expected that this kind of regulation fits better than public regulation with the specific needs. If a body or person has passed the certification procedure, it is assumed to be qualified.[71]

68. According to the law, he is even obliged to look for a suitable job in other enterprises.
69. Usually the prevention persons are employed in the enterprise but they have (or must have) an independent role, independent from the employer; that means in this respect not subordinated to him or her. In case such a person fails to be present in the company it can be a service from outside the company. This external service needs to be certified by a special body competent to issue a certification.
70. These certifying institutions (of a private nature) have to be recognized by a separate independent institution: the Council for Accreditation, established by the Minister of Social Affairs. Cf. J. Popma, 'Normstelling tussen publiek en privaat', in *Het recht op veilige, gezonde en waardige arbeid*, ed. J. Popma (The right to safe, healthy and decent labour) (Geller Publishing: Nieuwerkerk aan de IJssel, 2008), ch. 5.
71. Research did not give clear evidence that this practice has improved the health and safety of and in enterprises. Most of the attention of these services seems to be devoted to guidance of employees aiming at reintegration in the job. Prevention does not have a high priority. Cf. J. Popma, 'Normstelling tussen publiek en privaat', in *Het recht op veilige, gezonde en waardige*

For norm-setting another instrument is used: standardization. A special standardization institution, a private institution, the already mentioned Dutch Normalization Institute (NEN), can issue standards for certain sectors or types of activities. The standards have the character of private law, but are adopted by public authorities. This approach is considered as the most effective, since it can quickly respond to new technological developments. Moreover, this kind of self-regulation is deemed to be advantageous, because it is supported by its users and thus it is expected that they will better abide by the rules.

As mentioned above, the labour inspectorate is the institution checking the implementation of the health and safety standards. It has been given the powers to do so by the ARBO and it is part of and supervised by the Ministry of Social Affairs.

Within this framework, the inspectorate has developed a policy how to check the implementation of the Act, since it does not have the means for an extensive and systematic check and assessment of all working situations. The labour inspection decides itself which sectors, branches or enterprises will be subjected to a checking procedure. In practice, this means that it reacts when its attention is drawn to a particular situation or when it suspects that there are problems or infringements of the rules in a particular enterprise.[72] In addition, it can organize random checks of the health and safety situation.

[B] The Arsenal of Remedies

On the health and safety laws, various remedies and sanctions are possible in case the obligations and standards are violated. One category consists of criminal sanctions. A second consists of private law remedies: this is the liability of the employer for damages (Article 7:658 CC) and the obligation of the employer to continue to pay wages to his or her ill employees for a maximum period of 104 weeks (Article 7:629 CC). Third, the health and safety law knows administrative sanctions.

[1] Criminal Remedies

The infringement of the duties and obligations of the employer laid down in the public regulations, can be sanctioned by a criminal remedy. The ARBO contains a general prohibition for the employer to act contrary to the law. Also if s/he neglects to act according to the law and s/he reasonably knows or should know how to act, s/he is considered to violate the law if his or her failure to act can cause danger to life or

arbeid, ed. J. Popma (The right to safe, healthy and decent labour) (Geller Publishing: Nieuwerkerk aan de Ijssel, 2008), ch. 5. The labour inspectorate is evaluating the practice regularly with a similar outcome.

72. Some sectors have a rather bad reputation: transport, subcontracting in the construction industry, catering and cleaning industries. The health care sector may be often a target industry, not because of its reputation but due to the fact that the work is often heavy and stressing.

serious damages to the health of his or her employees. The Act distinguishes between crimes and offences and different types of sanctions ensue for these.

A separate, general Act, i.e., the Act on Economic Crimes and Offences (*Wet economische delicten, WED*) defines a violation of the general obligations of the ARBO as a crime. The corresponding sanction is, according to the WED, imprisonment or a fine.[73] In addition, the ARBO provides explicitly which infringements have to be sanctioned as offences, punishable by a fine.[74] In case of recidivism, the fine can be increased by 50%. However, in practice the criminal approach is not the one usually chosen. This is because the labour inspector has to establish that there is a violation of the Act and second, the public prosecutor has to decide whether the employer will be prosecuted. Because of this complicated procedure and the defaming effect for the employer such prosecutions are rare.

[2] Private Law Remedies

Thus sanctions in case of non-abidance of the health and safety legislation are in practice private law sanctions. This is considered as more effective, since it provides an incentive to the main actors to act. Consequently, it is left to the employer[75] and the employee or their representatives to start a procedure by lodging claims before a civil court. It is left to the initiative of the employee (of the employer eventually) to attempt compliance with the obligations set by the law. Usually it is individual employees who go to court claiming the damages suffered by them or who require the continuation of the payment of the wage after they have become incapacitated for work. Only if the obligations of the employer rest on a collective labour agreement, the trade union is entitled to sue the employer on its own behalf, that of the employee or its member employees. Apart from procedures to claim compensation, the unions can follow summary proceedings in order to try to remedy the situation in which obligations following from health and safety law are not complied with. This kind of procedure can be used, and is actually used indeed to force the employer to fulfil his or her obligations, if necessary under pressure of a penalty. In a summary procedure the union can obtain result within a short time. Although such procedures can be successful, at least theoretically, they are rarely used. When trade unions are involved in negotiations on catalogues, they do not want to make use of the possibility of lodging a claim in a summary procedure and for individuals there are many reasons which make it unattractive to follow these procedures. Their dependent position in the enterprise is a real obstacle to go to court trying to force the employer to fulfil his or her

73. In addition to these sanctions, the employer can be sanctioned by suspension, partly or even fully, of the (activities of the) enterprise.
74. The law distinguishes between first category offences (the minor offences) with a maximum fine of Euros (EUR) 9,000 and second category fine of EUR 22,500 (the serious ones).
75. In case the employee has not fulfilled his or her obligations imposed on him or her and the employer has suffered damages. In practice, this is rare.

obligations. There is no general provision aiming at the protection of employees who are victimized; experiences of employees acting as a whistle-blower are not positive.[76]

The Dutch system of private law remedies consists of two layers. First, there is the one of compensation, according to Article 7:658 CC, if an employee suffers damages caused by accidents or occupational diseases. Second, this is the obligation of the employer to pay wages to an employee who is ill and not able to perform his or her tasks. This obligation exists for a maximum period of two years (Article 7:629 CC). We will discuss these two ways separately.

[a] Compensation by Virtue of Article 7:658 CC

As explained above, next to the enforcement procedures, the sanction of liability on the basis of Article 7:658 CC is considered as a functional and effective remedy. It is seen as an incentive for the employer to improve and to maintain safe and healthy working conditions. This approach is part of the introduction in labour law of the 'market' concept: for achieving the objectives of safe and healthy working conditions it is assumed that the parties involved, i.e., the employer and the employee, are better equipped and more challenged if they feel the (financial) consequences of their acting and not-acting. In other words, they will be pushed to act in order to pursue their own interests. The extent to which this approach is successful depends, however, on whether the statutory schemes provide for compensation in such cases and the extent to which private insurance is available to cover the employer's risks. We will deal with this separately hereafter. In case of private insurance of these kinds of risks, which is often bought, the compensation to be paid on the basis of Article 7:658 CC will actually be paid by the insurance company. This coverage may seem to substantially reduce the incentive for the employer, since now only the level of contributions to the insurance company is all what can influence the activities of the employer to realize safe and healthy working conditions. However, the actual situation depends on the terms of the insurance agreement between the employer and the insurance company, since the insurance company may itself require the employer to take measures in order to reduce claims to the insurance. The legislature has left it to the parties involved to make themselves the applicable arrangement, expecting, again, that market forces will lead to the best outcomes.

The compensation to be paid by the employer consists of the damages actually suffered, i.e., the economic damages of the employee, which will consist of the wages minus compensation received, if any. These other compensations may exist of social security benefits the employee may receive (see hereafter). The compensation due can include all kinds of regular expenses the employee is entitled to that are not included

76. Even in case their complaints were granted, the end was not very hopeful for them. They have been dismissed and got sometimes some compensation. A regulation to protect whistle-blowers against reprisals is steadily under discussion. Since some years it has been decided that this issue has to be left to the social partners to regulate in collective labour agreements. In a legislative proposal of 2004, the employee/whistle-blower will be better protected against, for instance, dismissal. Also this proposal has not been adopted yet.

in the social security benefits. Third, this compensation may consist of costs for medical treatments if and insofar as not covered by the applicable insurance schemes. Costs for adjustments to the house of the employee if necessary because of his or her incapacity and costs of aids are also to be compensated.

The judge can also decide *ex aequo et bono* to award compensation for immaterial damages.[77]

Since 1999, the protection by this provision, Article 7:658 C.C. has been extended to other groups of workers whose employment relationship is not based on a contract of employment with the employer. By amendment of the law as from 1 January 1999 also temporary agency workers or posted workers are covered. Till quite recently, the self-employed were excluded from the protection of Article 7:658 CC offered to workers who got injured by an occupational accident. Opposite to the proposal to subsume the self-employed under the Act on Health and Safety at Work (Arbo Act) resulting in the responsibility of the employer for healthy and safe conditions also to all persons who are working for the employer but not on a basis of an employment contract, a proposal to extend the (civil) liability of the employer for the damages self-employed workers suffer, is still lacking.

The Supreme Court[78] has filled that gap by deciding that this category of workers: the lonely working self-employed, will also be covered, under certain circumstances. In case a -self-employed - worker is suffering damage caused by an occupational accident and the enterprise has chosen to let perform a work by a self-employed worker instead of an employee,[79] there is no ground to treat him differently from a regular employee in respect of compensating the damages suffered subject to the fact the work belongs to the regular activities of the enterprise. The court referred to the objective of the introduction of paragraph 4 of Article 7:658 by which the liability of the employer has been extended to all persons 'who work for him under his supervision or commissioned by him'. He has to respect the *obligation to care* also to these workers. On the basis of the actual situation of the case concerned, it has to be assessed whether the employer is liable. A number of facts have to be taken into account among which prominent are: (a) the actual relationship between the employer and the person performing the work; (b) the nature of the work done; (c) to which extent the employer actually has a say over the conditions under which the work has to be carried out and the risks attached to that and (d) whether the work has been done as part of the activities the employer usually is carrying out. Again the actual situation, including the practice of the employer in the past, is decisive. This decision has resulted in an extension of the liability of the employer not only to temporary agency workers or posted workers but also to self-employed people acting in the way described. In literature this extension has been criticized as going beyond the scope of the Employment Act that covers only workers with an employment contract. In this respect the decision has been called revolutionary.

---

77. It is based on the general provisions of the Code Civil: Article 6:160 CC.
78. HR 23 Mar. 2012, JAR 2012, 110 (*Davelaar-Allspan*).
79. Usually, the employer outsources this kind of activities and hires self-employed to let him do that work.

[b] Payments by Virtue of Article 7:629 CC and Social Security Benefits

Under Dutch law, an employee who becomes ill has to rely on his or her employer for an income provision. According to the law governing the individual employment contract, the employer is obliged to continue to pay 70% of the wage for 104 weeks or at least during the first 52 weeks the full statutory minimum wage.[80]

In addition, the Sickness Benefits Act (*Ziektewet* – SBA) provides that an employee who can claim sick pay from his or her employer (see hereafter) does not receive sickness benefit on the basis of the SBA. The rationale of this division of responsibilities was that the legislature assumed that the longer a person remains ill, the more difficult it becomes to get him or her back into work, since the longer one remains inactive, the more one tends to consider oneself to be completely unable to work. For this reason it was considered necessary to make employers and employees pro-active in reducing any periods of sickness, and to ensure that this began even in the early stages of sickness.[81] As a result of the financial effects of the shift of responsibilities, employers are expected to check more carefully whether an employee is rightfully absent or not. Employers can take out private insurance to cover their risk, or they can bear it themselves.

Large companies very often choose for bearing the risks themselves as they have the means (organizationally as well as financially) to do so and are better off. As a result they are directly interested in reducing the costs, so they take their responsibility seriously. Smaller firms mainly opt for the insurance possibility.[82] The SBA serves as a safety net provision only, mainly for the case in which the employment contract is terminated during the 104 weeks' period; in this case the employee may claim sickness benefit amounting to 70% of his or her wages for the remainder of that period.

The SBA is also essential for several categories of persons who are not employees in the sense of the Civil Code, but who are covered by the SBA. Examples are persons who have been put on a par with employees, such as home workers. The Sickness Benefit is still relevant in situations where the person concerned does not have an employer (anymore) or where the legislature found it unjustified that the employer has to pay sick pay.

There are only few grounds on the basis of which an employee is excluded from the right to sick pay; these are defined by the Civil Code. One of these is that the employer does not have to pay sick pay if the employee concerned refuses to carry out suitable work for his or her employer without good reason, provided s/he is capable of doing so. In this sphere the Dutch system has a 'three stages' approach: first, the

80. See also, F. Pennings, 'The responsibility of the modern enterprise in the reduction of sickness and the promotion of reintegration of disabled workers', in *Social responsibility in labour relations* 223–238 (F. Pennings, Y. Konijn & A. Veldman eds., Wolters Kluwer 2008) and F. Pennings, *The New Dutch Disability Benefits Act: the Link between Income Provision and Participation in Work*, in *Too Sick to work?* 95–118 (S. Devetzi & S. Stendahl eds., Kluwer L. Intl. 2011).
81. For further details, see F. Pennings, *The Netherlands*, in *International Encyclopaedia of Laws, Encyclopaedia of Social Security Law* (Kluwer L. Intl, loose leaf).
82. Hereafter we come back to the issue as to whether the objective behind the approach of responsibility and liability is upheld and if so how.

employer has to re-employ the (partly) disabled person in his or her own job. If that is not possible the second stage is re-employment within the firm in a suitable job if necessary by adjusting the workplace to the capacities of the employee concerned. In case this cannot be realized, the employer has to look for suitable work with another employer. If the employee refuses to co-operate that can imply that the employee is not entitled to sick pay (see hereafter).

Illness denotes a process-wise event consisting of a disruption of physical or mental functions of a human being accompanied by a decrease in the level of power of performance. The case law further shows that a person who is not able to do his or her work owing to a lack of education or experience is not incapacitated as a result of illness or infirmity. Limitations in connection with low intelligence, slight build, advanced age or maladjusted disposition of an insured person may therefore not be considered as illness or infirmity either.

The fact that an insured person is incapacitated to work has to be assessed on the basis of 'his or her work'. In general, this is the work the insured person did lastly – i.e., before ceasing that work due to illness or infirmity – and which s/he actually did. Relevant is, therefore, not the work agreed on by contract by the insured person and his or her employer or the work that would be done after the insured person had been declared healthy.

The criteria for sickness under the Civil Code and the Sickness Benefits Act are basically the same and thus imply that a person is ill if s/he is incapable of doing his or her work as a direct and objectively medically determined consequence of illness.[83] Infirmity is also considered as illness. Relevant is, therefore, whether a person cannot do his or her work as the result of sickness. If the cause is elsewhere the person is not considered ill. Furthermore, the illness must be medically determinable. So, if doctors cannot find an objective reason for the symptoms expressed by the employee, there will probably be no illness or infirmity. Because of the outcomes of the strict conditions, there were many complaints of persons whose complaints were accepted as real by the doctors, but they were not considered ill as no medically objective reason could be found. A solution to this problem has been found by a court decision. The Central Appeals Court[84] has developed an additional criterion: if several independent doctors (meaning that they do not work for the applicant) confirm unanimously that a person has certain deficiencies, the person can be considered as ill despite the lack of objective medical data.

In current discussions on sick pay entitlement, reporting sickness due to conflicts at the workplace play a large role. For instance, if an employee has problems with his

83. There is relatively little case law on the definition of sickness in relation to sick pay due by the employer. One reason is that labour law disputes are not so easily brought before court as social security cases (as long as it is not a dismissal case). Another reason is that, before a dispute on whether a person is sick can be brought before the court, s/he has to ask the benefit administration for an expert opinion. This opinion often solves the dispute. A well-known problem in this respect is that an employee has reported ill due to a dispute at workplace, usually with his or her supervisor. As such this reason is not sufficient. By medical examination it has to be established that the person is medically ill.
84. The highest court in social security cases.

or her supervisor, s/he may report ill. For this purpose guidelines called *Werkwijzers* are made, which help doctors to reduce sickness and to encourage reintegration into work. These guidelines are not binding, but they are useful as they give guidance in their approach.[85] These guidelines go to the extreme to keep the employee out of the 'sickness area', but seeking alternative solutions, such as change of working place or supervisor.

[C] Reintegration Efforts Required of the Employer and Employee

As said before, the characteristic Dutch approach in the field of health and safety at work is the allocation of responsibility on the employer as well as the employee. Reintegration of workers suffering from the consequences of occupational accidents or diseases and therefore only partially able and capable to work, is a basic part of the health and safety policy. One of the main instruments to improve the working conditions of the employees on enterprise level is that both the employer and employee have statutory reintegration obligations. For that reason we also discuss the issue of reintegration obligations mentioned in the law.

[1] Act on the Employment Contract

These obligations can be mainly found in the *Wet Verbetering Poortwachter* (the Gatekeepers Act[86]), which requires employers and employees to undertake reintegration efforts in case of illness if incapacity for work is expected to last for a long period.[87]

Thus, if an employee is expected to be ill for more than six weeks, the employer and employee are obliged to make a plan for reintegration. The plan can entail, for instance, that the workplace of the employee is adjusted to his or her impairments and/or that training or work experience in another job have to be tried. Subsequently the employer and employee have to meet on a regular basis to see how the reintegration efforts are proceeding and if the plan has to be adjusted. One party can oblige the other to co-operate, if necessary by invoking legal means before court.

An assessment of whether the reintegration activities have been sufficient takes place three months before the employee applies for disability benefit.[88] The employee has to produce a report on the reintegration activities which have taken place. If the employer's actions are considered insufficient by the benefit administration, his or her obligation to pay wages is extended (by up to a maximum of twelve months). Note that this is in addition to the already existing obligation to pay wages during twenty-four months. If it is the employee who has not co-operated satisfactorily, s/he can be

85. The guidelines are made by Stichting STECR, Expertisecentrum Participatie. This is a centre of experts. The guideline on conflicts at work is the *Werkwijzer arbeidsconflicten*.
86. The purpose of this Act was to narrow 'the gate' to the Disability Benefits Act.
87. Of course, in most cases of illness, such as fevers, etc. no measures are necessary.
88. Actually this is after one year and nine months of continuous sickness, since the waiting period for being eligible for disablement benefit is twenty-four months.

refused disability benefit for a certain period. In serious cases of non-co-operation dismissal is possible.[89]

Thus the new system introduced a serious activation system in the period prior to the claim for disability benefit. The purpose of the introduction of this feature was not only to reduce the costs of sickness and disability benefits,[90] but it is even more important objective is to give to both employer and employee an incentive to improve the working conditions in terms of combating unhealthy and unsafe working conditions. It therefore contributes to the aim of prevention.

[2] The Disability Benefits Act

The *Wet werk and inkomen naar arbeidsvermogen* (WIA - Disability Benefits Act) makes a distinction between persons: (a) who are permanently disabled to at least a level of 80% and (b) those who are either not permanently disabled, or who are permanently disabled to a lower extent. The former group (group a) deserves, in the view of the legislature, a generous disability benefit, and activation measures are not considered relevant for them.

The second group (group b) is made subject to conditions and rules meant to reinforce their activation and therefore should contribute to the main aim: prevention of illness and disability due to unhealthy and unsafe working conditions.

To be eligible for benefit the worker has to be at least 35% disabled.[91] Persons in this category, if they satisfy qualifying conditions, first receive a wage-related benefit, the duration of which depends on their employment record. After this period (or immediately, if they do not satisfy the qualifying conditions), they receive a benefit to supplement their wage, if they are still earning an income corresponding to at least 50% of their remaining earnings capacity; if they earn less (or nothing) they receive a (low) benefit.

Relevant to whether a person is disabled is whether his or her earnings capacity is reduced. This a two-tier assessment. A claimant's remaining earnings capacity is defined as the wage which s/he can earn by doing 'generally accepted work which s/he is capable of doing with his or her strength and competence' despite his or her illness or infirmity. The determination of earnings capacity thus requires a medical assessment but also an ergonomic assessment by a job expert.

The *medical assessment* is done by a medical expert (a doctor)[92] who has to assess whether the claimant suffers from illness or infirmity. The doctor has to list the effects of the illness or infirmity on the claimant. For this, a specific scheme is used,[93]

---

89. This is an exemption to the general rule of the prohibition of a dismissal during twenty-four months of sickness (Article 7:670(1) 1 CC).
90. As a result the sickness rate has been reduced. But one has to be careful by pointing this policy as the only factor for lower sickness rates.
91. Under the previous law the percentage was much lower: 15%.
92. Working for the benefit administration.
93. This expert uses a specific system, the so-called Claim Assessment and Control System For the selection of jobs (*Claim Beoordelings- en Borgingssysteem* (CBBS). This system contains a - non-exhaustive - survey of existing jobs (about 10,000) and indicates which skills (education

standardizing the practice of assessing claimants and to reduce differences in outcomes resulting from the assessment procedures themselves. In this system, it is not the limitations of claimants to do work which are important, but their ability to work.

Six aspects are relevant: social functioning, adaptation to the environment's requirements, dynamic activities, stationary attitudes and working times. The first two categories are mainly meant to assess the abilities of people with psychical problems.[94] The findings are laid down in a list showing the capabilities of the claimant and presented to the so-called jobs expert.

Obviously, if the limitations are so serious that the person concerned is not able to work at all, no further investigations are required: s/he receives a full benefit. In all other cases the job expert, on the basis of the outcomes of the medical investigation, investigates which types of work the person in question can still do despite his or her medically indicated limitations. It does not matter in this respect whether the person in question has a real chance of being employed in this work, in other words whether there are any vacancies. This is because the law explicitly stipulates that whether the person in question can actually obtain the labour in question should not be considered. We already saw in the introduction how sensitive this issue of considering labour market factors had become.

An important factor in this respect is what has to be understood by 'the work s/he still can do'. The Act uses the expression: 'generally accepted work'. That means: all work can be taken into account for this purpose, not only suitable work (for instance, the work the person did before). From among the jobs the person can do which match these criteria, the job expert has to choose those in which the claimant would earn the highest wage. A special decree (*Schattingsbesluit* – Assessment of Work incapacity Decree) provides more specific guidelines. The concept is used in a rather wide way. There are only minor exceptions.[95]

Persons who are incapacitated at a level of less than 35% are not eligible for benefit. It was the view of the legislature that their incapacity rate is so low that they should in any case be able to remain in work. That implies that in these cases the employer remains responsible for this category. In some collective agreements provisions have also been included to keep such individuals in work, and the ability to dismiss these workers because of their disability has been limited.[96]

---

and experience) are required to hold such a position, and what mental or physical requirements the job has. The assessment is more detail regulated. On the one hand, guaranteeing that only realistic job types are selected to determine the earnings capacity and on the other hand that the scope of jobs to be used in the evaluation is not too narrow. For that reason, three different job types which the claimant can do, have to be taken. Another criterion is: those job types have to be taken which have the highest income, as this will lead to a higher earnings capacity and thus a lower benefit.

94. They concern concentration of attention, ability to show one's feelings, and ability to cope with conflicts and to co-operate.
95. An example of such an exception is jobs which hardly exist on the labour market. If there is a functional age limit for specific work on account of the law or a collective labour agreement, this work may also not be considered as 'suitable' work for this person to be taken into account.
96. B. Barentsen, *Wet werk en inkomen naar arbeidsvermogen* (Deventer: Kluwer, 2006).

[3] The Relation to Dismissal Law and Unemployment

In order to keep these reintegration measures as effective as possible, the dismissal law was adapted to this approach. The prohibition of dismissal of sick employees (during two years of sickness) is meant to encourage the employer to actively look for adaptations of the work place, to jobs available and to the organization of the work in his or her enterprise. This is considered as a contribution to better working conditions for those employees suffering from accidents at work or occupational diseases.

Some examples of how this system functions, will illustrate this. If there are, for instance, psychosocial problems, the employer has, on the basis of his or her duty of care to find a solution, for instance by reorganizing the work. S/he is obliged to do so under the general rule of the CC that he has to be 'a good employer'. Of course it depends on the situation what can be exactly required from the employer.[97] If the employer has made a 'reasonable' offer, the employee has to consider it seriously and cannot refuse without very good reasons.

As already stated, persons who are ill are, generally speaking, protected against dismissal by Article 7:670 CC. An exception applies, however, if they refuse to co-operate in the reintegration efforts or if they refuse suitable work. Dismissal is, however, allowed only if it is an adequate measure; first other measures have to be tried, such as suspension of payment of sick pay.

The public decree on dismissal (*Ontslagbesluit*) has a similar regulation: a permit to dismiss sick employees will be refused.[98] As an exemption to the general ruling a dismissal permit is granted only if reintegration into work is not deemed likely within twenty-six weeks.[99] Whether attending training would enable the person concerned to re-integrate is also considered in these cases.

The other possibility of terminating the employment contract: by dissolution of the contract by the court (Article 7:685 CC), is still open to the employer because the general prohibition of dismissal of Article 7:670 CC and the Decree on dismissal is not applicable to this way of terminating. The possibility to dissolve the employment contract is restricted in this way that the court has to make sure whether the request of dissolving is related to the prohibition to dismiss the employee. So if the request is linked to the sickness of the employee, a dissolution has to be refused.[100]

---

97. According to the Dutch employment law system, an employer has to continue to pay the full wage, in case the employee remains available for work that s/he can do, and the employer does not make use of it (Article 7:628 CC) unless the employee refuses the work offered which can be considered as 'suitable' work for this employee. The employer has adapted the work or the workplace to the extent that reasonably could have been expected from him of her.
98. In short it has to be noticed that according to this decree dismissal is only possible when the employer has received a permit to do so from the administrative authority (in case the agency that also decides about claims for unemployment and disablement benefits).
99. This is based on a special decree (Article 5.2) annexed to the Decree on dismissal. The Dutch dismissal system is complicated: although the employer receives in this case a permit he cannot give notice to the sick employee due to the general protection against dismissal of sick employees during two years. This is in line with the obligation of being obliged to continuous payment of the wages during two years.
100. If the employer has a reason for terminating the contract by the court without a connection to the illness, the court is free to dissolve the contract. Usually the employee is granted a

[D] Administrative Remedies

A main task of controlling whether the obligations have been respected by the employer – and the employee – is attributed to the labour inspectorate. As described above (section 10.04) works councils as well as trade unions can exercise a checking function as well. But apart from that possibility, the responsibility for supervision and enforcement rests upon a public body. The ARBO allocated that task to the labour inspectors who act on the responsibility of the government, i.e., the Minister of Social Affairs (Article 24 ARBO). In principle and in practice the labour inspectorate has a certain discretionary power as to the application of the means and ways of control and enforcement. The final responsibility rests on the Minister of Social Affairs. The enforcement policy is in practice determined by the developments in the various industries, actual events and signals from the various industries. Choices are made also in the light of policy directions. Next to checking the implementation of the Act the labour inspectorate has also a role to play in providing all relevant information on the policy to be pursued. The labour inspectors focus their inspection activities on enterprises and sectors with high risks and repeated or continuing violations. There are also random checks. Another field of special attention concerns enterprises that operate in the grey areas or with suspect illegal activities.

To remedy effectively a dangerous situation the labour inspectors can make use of two powers attributed to them. First, we mention the less severe one: an order to the employer to fulfil his or her obligations and to take the necessary measures. The second goes further: the inspector can order to suspend the activities of the enterprise or a part of it where the dangerous situation exists.[101]

The main line of sanctioning is either criminal or administrative. As said above the enforcement by administrative means is preferred over criminal ones. The labour inspector reports and notifies to the public prosecutor if a criminal procedure has to be started.[102] This way of enforcement is not very often used. The Government prefers enforcement by the parties themselves or by the labour inspectors via the administrative way. There is hardly any case of prosecution of the employer or the company violating the health and safety rules.[103]

The administrative lane consists of two directions. The traditional one is to sanction the offender with a fine. The severity of the sanction depends on the nature of the offence. In case of a serious offence the sanction could be imprisonment or more likely high fines. This is applied when a serious danger to the employee or other persons (clients, suppliers) has occurred and the employer reasonably could have known. A milder way of reacting is at stake when the employer has infringed the obligations that are listed in the Act (Article 33 ARBO). The fine that can be imposed

compensation set by the court. The amount can depend on the circumstances of the case. The court uses, generally speaking, a kind of an internal guideline.

101. The payment of the wages is not suspended.
102. It is up to the prosecutor to decide whether s/he proceeds to prosecute.
103. In April 2012, the public prosecutor has started a preliminary procedure accusing a company of neglecting safety rules in a chemical enterprise which has caused a fatal accident.

can vary dependent on the seriousness of the offence.[104] A second category of offences are the so-called directly fined facts. They are listed in a special regulation. A direct reaction is the normal way of acting by the labour inspectors, usually consisting of giving an order and to impose measures to be taken immediately. The third category concerns the 'other fined facts'. They can be handled in an informal way: the inspector indicates the problems raised and trusts the employer will remedy the situation voluntarily; in this case no fine is imposed. The formal way of reacting consists of two phases: a warning or an order is to be followed by a report of the offence and measures.

A final sanction has been introduced in 2007: the administrative coercion. This instrument is aiming at undoing the infringement or to prevent either the continuation or the repetition of the infringement. A penalty will be imposed on the employer when the latter does not respect an order given by the inspector. This instrument is strengthening the power of the inspector of imposing orders on the employer if the employer does not respect and does not comply with the order. An administrative coercive measure will be imposed when there is a reasonable presumption that the employees are endangered or the imposition of a fine turned out not to be effective.

When an administrative sanction has been imposed the administrative procedural law has to be applied. Since this procedural law differs from the criminal procedural law in the sense that is less severe, this way of administrative sanctioning is preferred. The law on giving evidence is less strong. A second differing characteristic is the two-phases-procedure: the procedure starts with a complaint that provides the enforcement body[105] with the possibility to reconsider the measure imposed and therefore the seriousness of the offence. When it maintains its view the offender can go to the administrative court.

§10.06 CONCLUDING REMARKS

In the introduction, we put the issue of health and safety not only in the perspective of protection of the employees but also in terms of citizenship. Citizenship involves the persons having the rights and duties to participate in society. Dutch health and safety law recognizes the primary duty of care lies with the employer, in accordance with the EU Law. But it goes further, as we have shown. The introduction of self-responsibility and consequently self-regulation by the parties involved, points to what we have called citizenship in a double sense: (a) individual citizenship: emphasizing the responsibility of the employer and the (individual) employee reflected in sets of rights and duties for both of them; and (b) collective citizenship by involving the representatives of, in

104. The amounts at present vary between EUR 9,000 and EUR 22,500. In case of recidivism the (administrative) fine can be increased by 50%. This sanction can be combined with one of the orders mentioned above: an order to take measures or an order of suspension. This will be the case when very dangerous situations have occurred, more in particular when it is a life threatening situation.

105. That is the Minister of Social Affairs. Research has shown that only a minority of the employers have started the complaint procedure: less than 10%. In around 20% of the cases the employer has won the case. The number of court cases is accordingly low.

particular, the employees in the process of norm-setting. At the end of this exercise we did in the previous paragraphs we will draw some conclusions.

In order to be able to assess whether the Dutch law on the one hand has transposed the Framework Directive in a correct way and second, whether and if so how the citizenship approach has been materialized, it is necessary to read not only the legal texts but also to look at the objective of the directive to be pursued.

Without doubt, the general approach of the Framework Directive has been taken very seriously by the Dutch legislature. Obviously, prevention is the main objective of the law on this issue. It had already opted for this approach before the Directive came into force. Second, the 'duty to care' to be fulfilled by the employer is a main characteristic (also) of the Dutch law and approach. So far so good. But apart from this more formal assessment we also have to assess the Dutch law and practice from the point of view of an effective policy and ruling of the matter at stake. This is also an important requirement of EU Law. A last criterion for assessing the correct implementation of the EU law is the enforcement procedures and their effectiveness. On this point one can really doubt whether the Dutch law is in compliance with EU law. The 'modest' role of the labour inspectorate, in qualitative as well as in quantitative terms, points to it.

In order to give a final assessment of the Dutch system we sketch the system in its specific Dutch features.

[A] Responsibilities have Shifted

In the Netherlands a shift of responsibility from the State to social partners as well as to the individual employer and the employees has taken place. That is part of the general tendency in labour law (and not only on this field) of deregulation and emphasizing the instrument of self-regulation. Clear examples of this approach are: the role of Arbo-catalogues as the detailed and more tailored regulations of health and safety at work. The acts and even the regulations on lower level have a general framework nature expressing the goals to be reached. The more concrete design has been left to the parties that are more directly involved and affected by the regulations.

In this respect, it has to be noticed that there is nothing new under the sun. The Netherlands is following the EU approach as to the use of the inventory and evaluation of risks as an instrument on enterprise level. But it went further as we have shown.

[B] The Role of the Workers' Representatives has become more Important

The shift to self-regulation corresponds to the wish of increasing the role of workers' representatives in the design and the application of effective regulations on this field. Again the Arbo-catalogues are a good example, as well as the regulations in the Arbo Act and in particular the Works Council Act on the role of works councils. More in particular, the right to consent of the works council on various topics related to a health and safety policy in the enterprise underlies this approach. Since it is not only law in the books, it appears to be effective. Survey research has to give more evidence that this is the case.

[C] Financial Incentives are Used

Another typical Dutch feature is the use of the instrument of the (financial) incentive and permeates – one could say – the whole system. It seems to be effective in the sense that it influences the behaviour of the parties involved. We have shown this influence in section 10.05 where we discussed the system of sanctions and remedies. As a consequence of this approach and based on a deliberate policy aiming at an effective reintegration of ill employees, including those suffering from the consequences of accidents at work and occupational diseases, the law imposes on the employer primarily a set of obligations that should result in better working conditions. Consequently that should contribute to the main goal of the European policy and law on this field. Although it is very difficult to prove that this system works, it seems to be successful to the extent that the sickness rate has been reduced and good practices of prevention and reintegration are known.

[D] The Criterion of Sufficient Protection

A benchmark to assess whether the current Dutch system respond sufficiently to another objective of the directive – protection of the workers against the possibility to be exposed to dangerous and unhealthy working conditions – has to be established as well. Our investigation makes clear that this is the case. That is the result of the multi-level system. More in particular, the more concrete lower level regulations – the Arbo-catalogues – and the way these regulations are applied by the parties themselves contribute to the positive result. Another ground for that conclusion is the – until now existing – confidence in the functioning of this system in society. The role which the directly involved and affected people and organizations play is a crucial element.

[E] The Concept of Health and Safety

A satisfying development is also the rather broad meaning and interpretation of the concept of health and safety. The case law of the courts as well as the rulings of the parties concerned in the Arbo-catalogues are contributing to that, although even the catalogues are – till now – rather vague in wording. Actually the courts seem to play a more important role in terms of imposing serious duties on the employer to take measures to avoid the occurrence of physical and more particularly psychosocial risks. Psychosocial risks are recognized as risks to be covered by the whole body of regulations on health and safety, although there is still more to be achieved. A special problem appears to be the multi-causal disablement due to partly accidents at work and occupational diseases and partly private circumstances. The responsibility seems to become unclear which is detrimental for the employees affected.

[F] Enforcement and Control

A last remark has to be made on the shift of responsibilities from the state to the 'market'. In order to be able to check effectively whether the rules are respected by the employers, a well-functioning labour inspectorate seems to be necessary.[106] Apart from the problem of understaffing of the Dutch labour inspectorate, at least compared to the far majority of the other EU Member States, supervision of the norms is very limited, since it is nearly fully left to private actors. One can wonder whether in this way the Netherlands meets the requirements of the EU legislation. The instrument of the administrative coercion seems promising, but it remains to be seen how this instrument will be used.

From the perspective of protection of the employees, it can be concluded that the objective is achieved in the Dutch system except perhaps when we look at the enforcement and control issues. The reliance on financial remedies and sanctions to be used by the employees as one of the most important and strongly emphasized means is innovative and challenging but at the same time risky from the point of view of the values and interests of the employees. The workers' representatives are playing a crucial role since the labour inspectorate is understaffed and therefore restricted in their task as controller directed by the state. A positive point from the effectiveness perspective is the main role that Arbo-catalogues are playing. The tailored nature of these instruments increases their effect and their effectiveness. But again this depends upon the activation capacity of the employees' representatives or (if they do not exist) of the employees themselves and of the employer(s). If the latter is neither facing nor fearing financial incentives there seems to be apart from employees' pressures a lack of means of coercion.

When looking at the Dutch system from the citizenship perspective, we have to assess whether this approach and the way the Dutch law makers are applying this concept are taken seriously and are consistent. It can be stated that in any case, from the formal aspects of the laws, employers and employees have been given a large responsibility to define the rules applicable to their enterprise. Within the framework of the goal provisions in the statute laws there are actually no limits to what they can agree. Collective (industrial) citizenship appears to be guaranteed and provided with potentially effective instruments. But that depends mainly on the operational power of the employees' representatives and the willingness of the employer to deal and wheel with them. We refer here to the role of works councils on enterprise level and trade unions as parties to the Arbo-catalogues.

If assessed in respect of the individual employees, the picture is somewhat diffuse. Certainly they have been given some instruments which enforce their position. We refer to the possibility to claim compensation and reintegration efforts to be

106. J. Popma, Inkrimping Arbeidsinspectie in strijd met ILO Verdrag 81, ARBAC, November 2011; and J. van Drongelen, 'Een nieuwe Arbeidsinspectie: tussen behoud en vernieuwing', in *Sociaal Recht: tussen behoud en vernieuwing*, eds. W. Plessen et al. (Zutphen: Paris, 2011), 143 et seq.

made by the employer. The last are the most important from the point of view of recognizing effectively the notion of citizenship: the employee has a right to reintegration and the employer is obliged to made serious efforts to make the reintegration actually possible which can include an obligation to adapt the organization of the work to the capacities and possibilities of his or her employees including the employee suffering from damages due to unhealthy or unsafe working conditions. The threat of such claims influences the deployment of general policies in this field and, more important the taking of preventive measures. However, it has to be noted that these instruments lie more in the area of compensation than in prevention.

BIBLIOGRAPHY

Barentsen, B. *Wet werk en inkomen naar arbeidsvermogen*. Deventer: Kluwer, 2006.

Drongelen, J. van. 'Een nieuwe Arbeidsinspectie: tussen behoud en vernieuwing', edited by W. Plessen, H. Van Drongelen & F. Hendrickx. Paris: Zutphen 2011, 143 ff.

Geers, A.J.C.M. *Recht en humanisering van de arbeid*. Deventer: Kluwer, 1988.

Handelingen TK 1918–1919

Jaspers, A.Ph.C.M. 'Veiligheid en gezondheid op het werk: een nieuwe benadering dankzij Europa?'. In *Sociaal recht: tussen behoud en verandering*, edited by W. Plessen, H. Van Drongelen & F. Hendrickx. Paris: Zutphen 2011.

Kamerstukken TK 1996/97, Nr. 24 615.

Klosse, S. & F.M. Noordam. *Sociale-zekerheidsrecht*. Kluwer, 2010.

Pennings, F. 'The Netherlands'. *International Encyclopaedia of Laws, Encyclopaedia of Social Security Law*. Alphen aan den Rijn: Kluwer Law International, looseleaf.

Pennings, F. & C. Bosse (eds). *The Protection of Working Relationships. A comparative study*. Alphen aan den Rijn: Kluwer Law International, 2011.

Pennings, F. 'The New Dutch Disability Benefits Act: the Link between Income Provision and Participation in Work'. In *Too Sick to work?*, edited by S. Devetzi & S. Stendahl. Alphen aan den Rijn: Kluwer Law International, 2011.

Pennings, F. 'The responsibility of the modern enterprise in the reduction of sickness and the promotion of reintegration of disabled workers'. In *Social responsibility in labour relations*, edited by F. Pennings, Y. Konijn & A. Veldman. Alphen aan den Rijn: Wolters Kluwer, 2008.

Popma, J. 'Normstelling tussen publiek en privaat'. In *Het recht op veilige, gezonde en waardige arbeid*, edited by J.R. Popma e.o. Geller Publishing: Nieuwerkerk aan de Ijssel, 2008.

Popma, J., M.H. Schaapman & T. Wilthagen. 'The Netherlands: Implementation within wider Regulatory Reform'. In *Regulating Health and Safety Management in the European Union. A Study of the Dynamics of Change*, edited by D. Walters. Peter Lang: New York, Bern, Berlin, Bruxelles, Frankfurt am Main, Oxford, Wien, 2002.

Popma, J. Inkrimping Arbeidsinspectie in strijd met ILO Verdrag 81, ARBAC, November 2011.

Robens Committee, Safety and Health at Work. *Report of the Committee 1970–1972.* London: Her Majesty's Stationery Office, 1972.

Zuthem, H.J. van. *Arbeidsbeleid in de onderneming.* Assen, 1967.

CHAPTER 11

Occupational Health and Safety in the UK: At a Crossroads?

Mark Bell

§11.01 INTRODUCTION AND LEGAL BACKGROUND

Health and safety law in the UK appears to stand at a crossroads. In recent years, there has been an increasingly critical public debate about the utility of health and safety legislation, with some parts of the media promoting the idea that this branch of law is excessively protective leading to undue restrictions on businesses. This is summed up in the popular expression: 'health and safety gone mad'.[1] The Conservative-Liberal Democrat Coalition that took office in May 2010 acted swiftly to establish a review of health and safety law in order to combat the perceived 'compensation culture' and to ensure that the law did not 'overwhelm businesses with red tape'.[2] This review (by the Conservative peer, Lord Young) concluded that 'stories of individuals suing their employers for disproportionately large sums of money for personal injury claims, often for the most trivial of reasons, are a regular feature of our newspapers'.[3] In response, Lord Young proposed an agenda for reform designed to reduce bureaucracy for business and to instil 'common sense' in the application of health and safety law. This approach assumes that the basic problem is 'too much' health and safety, but it is hotly contested by the trade union movement. The latter estimates that 'at least 20,000 people die prematurely every year because of occupational injury and disease', while

1. House of Commons Work and Pensions Committee, *The Role of the Health and Safety Commission and the Health and Safety Executive in Regulating Workplace Health and Safety*, Third Report of Session 2007–2008, vol. 1, HC 246 (Stationery Office 2008).
2. *PM Announces Review of Health and Safety Laws*, http://www.number10.gov.uk/news/latest-news/2010/06/pm-announces-review-of-health-and-safety-laws-51726, 14 Jun. 2010.
3. HM Government, *Common Sense Common Safety – A Report by Lord Young of Graffham to the Prime Minister following a Whitehall-Wide Review of the Operation of Health and Safety Laws and the Growth of Compensation Culture* 7 (Cabinet Office 2010).

around 246,000 are injured at work annually.[4] Trade unions contrast these statistics with the rapid decline in enforcement activities related to health and safety legislation during the past decade. The number of workplace inspections has fallen by more than two-thirds, alongside a 48% reduction in the number of prosecutions.[5] Consequently, trade unions are arguing for a strengthening in the law, and its enforcement, an agenda that appears to be diametrically opposed to the direction that the government is pursuing. Given this polarized backdrop, this is an appropriate juncture for an overview of the current state of occupational health and safety law.

This branch of British labour law is a complex framework of legislation and common law, incorporating elements of criminal, contract and tort law. The legislative framework is mainly founded on the Health and Safety at Work Act etc. (HSWA) 1974,[6] which is an umbrella for a wide variety of Statutory Regulations. In principle, the HSWA and its Regulations form part of criminal law;[7] breach of the legislation is a criminal offence. Yet the legislation intertwines with tort, because individuals may be able to challenge breach of the Regulations under the tort of breach of statutory duty. More broadly, an individual who suffers an occupational injury or disease might also have recourse to the tort of negligence or an action in contract law for breach of the employer's implied duty of care for the employee. This chapter will, thus, consider both the role of statute and the role of common law as regards the protection of workers' health and safety. It commences with an introduction to the main facets of the British legal framework in order to clarify some of the complexity mentioned above. It then examines core concepts in health and safety law, before turning to examine the duties imposed on various actors, such as employers and employees. The role of workers' representatives is analysed and the chapter concludes with an overview of the remedies for breach of health and safety law. In keeping with the general themes of this collection of essays, particular attention will be paid to the rules applying to the construction industry.

Modern health and safety law can be traced back to the recommendations of the 1972 Robens Committee,[8] which led to the HSWA 1974. Although it has been amended in the subsequent period, this Act remains the cornerstone of statutory health and safety law. It establishes general duties on employers towards employees, as well as any other person who may be affected by health and safety risks. It applies to employment in both the public and private sectors. Duties are also imposed on the self-employed, employees, those who control premises and the manufacturers of

4. This estimate is based on those injuries that require at least three days off work. Trades Union Congress, *The Case for Health and Safety*, 2, http://www.tuc.org.uk/workplace/tuc-18421-f0.cfm, 30 Sep. 2010.
5. S. Tombs & D. Whyte, *Regulatory Surrender: Death, Injury and the Non-Enforcement of Law* 48–51 (Inst. Empl. Rights 2010).
6. The law in Northern Ireland is governed by the Health and Safety at Work (Northern Ireland) Order 1978, SI 1978/1039 (NI 9), which is not considered further in this chapter.
7. B Barrett, *The Health and Safety (Offences) Act 2008: The Cost of Behaving Dangerously at the Workplace*, 38 Indus. L. J. 73, 78 (2009).
8. Lord Robens, *Safety and Health at Work: Report of the Committee 1970–1972*, Cmnd 5034 (HMSO 1972).

articles and substances for use at work. The Act cannot be enforced by individuals.[9] Instead, the Health and Safety Executive (HSE)[10] was created and this body enjoys powers to inspect workplaces and a range of instruments for enforcement, including via criminal prosecution. The Act includes powers for the issuing of secondary legislation ('Regulations') by the relevant minister on a wide range of matters related to health and safety. Therefore, the Act is accompanied by a large number of Regulations. Some of these deal with specific health risks, such as noise or hazardous substances, but others deal with general matters of regulation, such as mechanisms for the representation and participation of workers.[11]

The volume of Regulations adopted under the HSWA changed significantly in response to developments in EU law. The advent of the 1989 Framework Directive on the safety and health of workers[12] and its six daughter Directives led to a new generation of Regulations. The HSWA is used as the legal foundation for adapting British law to the requirements of EU Directives and commentators have noted that 'virtually all' legislative changes in Britain simply reflect the implementation of EU law.[13] This general picture was confirmed in a 2011 independent review of health and safety law. The 'Löfstedt Review' reported evidence that '41 of 65 new health and safety regulations introduced between 1997 and 2009 originated in the EU'.[14] Of particular importance are the Management of Health and Safety at Work (MHSW) Regulations 1999[15] as this is the main legislative response to the 1989 Framework Directive. In relation to construction, detailed standards are set out in the Construction (Design and Management) Regulations 2007.[16] Lord Young's review has criticized the proliferation of Regulations under the HSWA. In the short term, it has proposed codifying the large number of Regulations into a single set of Regulations,[17] but it diagnosed the underlying cause of the problem as the EU 'where the tendency has been to look first at extending prescriptive regulation rather than examining ways of ensuring risk is managed in a proportionate way'.[18] This breathed fresh life into an enduring scepticism on the part of the Conservative Party, in particular, towards EU

9. Section 47(1) HSWA 1974.
10. Originally, there were two organizations, the HSE and the Health and Safety Commission. These were merged in 2008.
11. A detailed examination of the wide variety of Regulations is provided in B. Cotter & D. Bennett eds., *Munkman on Employer's Liability* (14th ed., LexisNexis 2006).
12. Directive 89/391 on the introduction of measures to encourage improvements in the safety and health of workers at work [1989] OJ L183/1.
13. B. Cotter & D. Bennett eds., *Munkman on Employer's Liability*, 1.96 (14th ed., LexisNexis 2006); P. James & D Walters, *Regulating Health and Safety at Work: An Agenda for Change?* 10 (Inst. Empl. Rights 2005).
14. R. Löfstedt, *Reclaiming Health and Safety for All: An Independent Review of Health and Safety Legislation*, Cmnd 8219, 4 (Stationery Office 2011).
15. SI 1999/3242.
16. SI 2007/320.
17. HM Government, *Common Sense Common Safety – A Report by Lord Young of Graffham to the Prime Minister following a Whitehall-Wide Review of the Operation of Health and Safety Laws and the Growth of Compensation Culture* 39 (Cabinet Office 2010).
18. HM Government, *Common Sense Common Safety – A Report by Lord Young of Graffham to the Prime Minister following a Whitehall-wide Review of the Operation of Health and Safety Laws and the Growth of Compensation Culture* (Cabinet Office 2010).

social regulation. It dovetails with a wider trend to criticize 'gold-plating' of EU Directives;[19] this is the perception that UK regulators go beyond the minimum required to comply with EU legislation. A House of Commons inquiry in 2008 found only limited evidence of this in the field of health and safety, but it recommended an approach that focused on reducing administrative burdens for business in the future.[20] In its response to the Löfstedt Review, the government has committed itself to promoting 'a more proportionate, risk-based approach to health and safety' within EU law and policy.[21]

Although the legislative framework is founded in criminal law and enforcement by an official regulator, this must be seen alongside the possibilities for individuals who have suffered damage to their health at work to seek compensation through civil proceedings. This takes a number of forms. First, there is the possibility for civil liability to arise where damage has been caused by a breach of legislative requirements on health and safety. This is excluded in relation to the duties found within the HSWA, but, in principle, an individual may bring civil proceedings for damage caused by a breach of one of the Regulations adopted under the HSWA.[22] For example, the MHSW Regulations place a duty on the employer to undertake a risk assessment.[23] If the employer failed in this duty and it could be shown that this breach caused damage to the employee through failing to provide a safe system of work, then an action for compensation could be brought.[24] This is, though, subject to the possibility for individual Regulations to contain provisions that expressly exclude civil liability. For example, some elements of the Construction Regulations do not give rise to civil liability in respect of those who are not employees.[25] Actions for compensation for breach of the Regulations take the form of the tort of breach of statutory duty. There are four main steps to establishing that this tort has been committed: (1) the legislative provision was breached; (2) the damage suffered was of a kind that the provision was intended to prevent; (3) the person belongs to a class which the provision was intended to protect; (4) the breach of the duty caused the loss.[26] An important advantage to the litigant of bringing an action for breach of statutory duty is that normally it is not necessary to show fault on the part of the employer; in this respect, it is akin to a form of strict liability once the breach has been proven.[27] This gave rise to concern in the Löfstedt Review that strict liability led to the awarding of compensation even where the

19. On the current government's policy to avoid 'gold-plating': http://www.bis.gov.uk/policies/better-regulation/policy/european-legislation/goldplating, 8 Aug. 2011.
20. House of Commons Work and Pensions Committee, *The Role of the Health and Safety Commission and the Health and Safety Executive in Regulating Workplace Health and Safety*, Third Report of Session 2007–2008, vol. 1, HC 246, para. 43 (Stationery Office 2008).
21. Department for Work and Pensions, *The Government Response to the Löfstedt Report*, http://www.dwp.gov.uk/docs/lofstedt-report-response.pdf, Nov. 2011.
22. Section 47(2) HSWA.
23. Regulation 3(1).
24. I. Smith et al., *Health and Safety: The Modern Legal Framework* 30 (2nd ed., Butterworths 2000).
25. Regulation 45, Construction (Design and Management) Regulations 2007, SI 2007/320.
26. M. Ford & J. Clarke, *Redgrave's Health and Safety* 2.21 (5th ed., LexisNexis Butterworths 2007).
27. B. Cotter & D. Bennett eds., *Munkman on Employer's Liability* 5.31 (14th ed., LexisNexis 2006).

employer had taken all reasonably practicable steps to prevent the breach of the law.[28] In response, the government has announced that it will review legislation that applies strict liability to employers in civil proceedings.[29]

The second option for individuals is to bring an action in the tort of negligence. In order to succeed, three basic elements need to be satisfied: (1) a duty of care exists on the part of the defendant towards the class of persons to whom the claimant belongs; (2) there has been a breach in the duty of care by the defendant; (3) the defendant's conduct caused the damage to the claimant.[30] In addition, it is important to recognize that there will be no liability if the damage arising is too remote.[31] In 1937, the House of Lords held that there was a general duty of care on the part of employers towards employees with regard to their safety at work.[32] The Court identified four principal components of this duty of care: the provision of a safe workplace; safe equipment; competent and safe fellow employees; and a safe system of work. Smith et al. point out that some of these categories have since been the subject of Regulations adopted under the HSWA, such as the provision of safe equipment.[33] In these areas, it may be more likely that a claimant will bring an action for breach of statutory duty, given the stricter principles of liability that will apply. There are others, however, where legislation remains limited, such as liability arising from the negligence of fellow employees,[34] and here the tort of negligence remains central.

In the period since 1945, the courts have substantially expanded the notion of 'implied duties' on the employer arising from the contract of employment.[35] These duties take various forms, but they are unwritten obligations that the courts have interpreted as being a necessary incident of the contract. The implied contractual duties include a duty on the employer to protect the health and safety of employees. The view of the courts has been that today it is 'immaterial' whether the duty of care arises in tort or in contract due to the similarity in the substance of the duty.[36] Nevertheless, different rules regarding procedures and remedies will apply depending on whether the action is brought in tort or in contract.[37]

---

28. R Löfstedt, *Reclaiming Health and Safety for All: An Independent Review of Health and Safety Legislation*, Cmnd 8219, 6 (The Stationery Office 2011).
29. Department for Work and Pensions, *The Government Response to the Löfstedt Report*, http://www.dwp.gov.uk/docs/lofstedt-report-response.pdf, Nov. 2011, 13.
30. W.V.H. Rogers, *Winfield and Jolowicz on Tort* 150 (18th ed., Sweet & Maxwell 2010).
31. D. Wilby, *The General Principles of Negligence*, in *Munkman on Employer's Liability*, 2.03 (14th ed., B. Cotter & D. Bennett eds., LexisNexis 2006).
32. *Wilson & Clyde Coal Co v. English* [1938] AC 57 (HL).
33. I. Smith et al., *Health and Safety: The Modern Legal Framework* 29–30 (2nd ed., Butterworths 2000).
34. I. Smith et al., *Health and Safety: The Modern Legal Framework* (2nd ed., Butterworths 2000).
35. S. Deakin & G. Morris, *Labour Law*, 296 (5th ed., Hart Publg. 2009).
36. Stuart-Smith LJ, *Johnstone v. Bloomsbury Health Authority* [1991] ICR 269, 274 (CA); Rodger LJ, *Barber v. Somerset County Council* [2004] 1 WLR 1089, 1098 (HL).
37. For example, damages for loss of future earnings may be awarded in tort, but it 'will not normally be recoverable in contract': S. Deakin & G. Morris, *Labour Law* 296 (5th ed., Hart Publg. 2009).

As can be seen from this short overview, British health and safety law has two distinct branches: the statutory regulation administered by the HSE, and the possibilities arising in tort and contract law for individuals to seek compensation when their health and safety has been damaged at work. Yet the two do not exist in isolation from each other and both will be examined throughout this chapter.

§11.02 BASIC CONCEPTS

[A] Health

There is no definition of 'health' in the HSWA 1974 or the MHSW Regulations 1999. In relation to civil liability, section 47(6) HSWA defines 'damage' as including 'the death of, or injury to, any person (including any disease and any impairment of a person's physical or working condition)'. This non-exhaustive definition leaves a broad scope for interpreting the concept of 'disease'. In this vein, the work of the HSE ranges from combating risks associated with traditional industrial injuries, such as chemicals or asbestos, to providing guidance on contemporary forms of work-related disease, such as musculoskeletal disorders.[38] The Trades Union Congress (TUC) has highlighted the shift in the nature of occupational health and safety risks from the former to the latter.[39] HSE statistics show a significant decline in the rate of fatal and non-fatal injuries in the past fifteen years, across both employees and the self-employed.[40] For example, the number of fatal injuries at work was 171 in 2010/2011, compared to 236 in 2003/2004.[41] The largest causes of work-related ill-health are stress, depression or anxiety, and musculoskeletal disorders. In 2010/2011, the HSE estimated that 1.8 million people were suffering an illness that they believed was caused or made worse by their current or past work.[42] Amongst those conditions that started in the past twelve months, around 75% were either 'stress, depression or anxiety', or musculoskeletal disorders.[43] Nevertheless, the HSE estimates that there was a 21% fall in musculoskeletal disorders during the period 2006/2007 to 2010/2011, with a 14% reduction in the prevalence of stress.[44] Data from general practitioners (the local doctor who is typically the first point of contact for a patient) reveals that 55% of working days lost were due to mental ill-health and 37% were due to musculoskeletal disorders.[45]

The trend towards recognizing the role played by mental health is also reflected in the common law where a series of cases has recognized that psychiatric injuries, such as stress, depression, and mental illness, are covered by the employer's duty of

38. Musculoskeletal disorders include conditions such as back pain and repetitive strain injuries.
39. Trades Union Congress, *The Case for Health and Safety*, 2, http://www.tuc.org.uk/workplace/tuc-18421-f0.cfm, 30 Sep. 2010, 3.
40. HSE, *Statistics 2009/10*, 8–9 (HSE 2010).
41. HSE, *Annual Statistics Report 2010/11*, 1 (HSE 2011).
42. HSE, *Annual Statistics Report 2010/11*, 3 (HSE 2011).
43. HSE, *Annual Statistics Report 2010/11* (HSE 2011).
44. HSE, *Annual Statistics Report 2010/11*, 4 (HSE 2011).
45. HSE, *Annual Statistics Report 2010/11*, 7 (HSE 2011).

care.[46] It should, though, be kept in mind that the nature of occupational health risks varies considerably depending on the type of work being performed. When looking at work-related illness (which includes stress), the sectors with the highest rates of illness are health and social work, followed by public administration.[47] The picture is very different if the focus is on injuries at work, where the most dangerous occupations are agriculture, transport and construction.[48] Around one-third of fatal injuries occur in construction,[49] and the rate of fatal injuries in this sector is four times the average for all industries.[50]

[1] Working Time

The implementation of the Working Time Directive[51] in Britain has contributed to broadening the traditional understanding of occupational 'health'. Prior to the Directive, there was very little regulation of working hours. This was exemplified in the case of *Johnstone v. Bloomsbury Health Authority*,[52] which concerned a doctor who suffered stress and depression following a regular requirement to work long working hours, even over 100 hours per week. In the absence of any statutory regulation of working time, the question arose whether an employer could, as a matter of principle, include a term in the contract of employment that would require the employee to work very long hours, even where this risked damaging the employee's health. The majority of the Court of Appeal took the view that an express contractual term to this effect could not be overridden by the implied duty of the employer to protect the health of employees.[53] On the facts of the case, the Court managed to interpret the employee's contract in a manner that avoided such an unrestrained obligation, but it exposed the limits of the common law for tackling this source of risks to health.

The historical absence of regulation of working hours forms part of the explanation for the British government's initial hostility to the Working Time Directive. At the time that the Directive was adopted, there was a Conservative government and it was adamant that the EU's legal competence for regulating occupational health and safety should not extend to cover the regulation of working time. This view was ultimately rejected by the Court of Justice, which adopted a broad interpretation of the concept of health: 'a state of complete physical, mental and social well-being that does not consist

46. For example, *Walker v. Northumberland County Council* [1995] 1 All ER 737 (QB); *Barber v. Somerset County Council* [2004] 1 WLR 1089 (HL).
47. HSE, *Annual Statistics Report 2010/11*, 16 (HSE, 2011).
48. HSE, *Annual Statistics Report 2010/11* (HSE 2011).
49. House of Commons Work and Pensions Committee, *The Role of the Health and Safety Commission and the Health and Safety Executive in Regulating Workplace Health and Safety*, Third Report of Session 2007–2008, vol. 1, HC 246, 124 (Stationery Office 2008).
50. Department of Work and Pensions, *The Government Response to the Rita Donaghy Report: One Death Is Too Many – Inquiry into the Underlying Causes of Construction Fatal Accidents*, 4 (Stationery Office 2010).
51. Directive 2003/88 concerning certain aspects of the organization of working time [2003] OJ L229/9.
52. [1991] ICR 269 (CA).
53. Browne-Wilkinson V-C, *ibid.*, 283.

only in the absence of illness or infirmity'.[54] Following the judgment, the UK adopted the Working Time Regulations 1998,[55] which largely mirror the contents of the Working Time Directive. Notably, the Regulations allow individual workers to 'opt-out' of the requirement that the average working week does not exceed forty-eight hours.[56] Although they include measures to promote occupational health, the Regulations are legally distinct from other aspects of health and safety legislation. They were not adopted under the umbrella of the HSWA, but are free-standing Regulations issued under the European Communities Act 1972 (a general instrument for implementing EU law). The avenues for enforcement differ from other branches of health and safety law. Specifically, there is a larger role for enforcement via individual litigation. In respect of daily and weekly rest breaks, and paid annual holidays, these provisions can be enforced by a worker making a complaint to an Employment Tribunal.[57] In contrast, the requirements on the maximum working week and the protection of night workers are enforceable by the HSE through the normal process of investigation and prosecution.[58] This hybrid model of enforcement leaves the impression that the Regulations are perceived to be on the margins of health and safety law. This is reinforced by the HSE website. There is a long list of 'topics' accessible from its homepage, but working time is not mentioned.[59] Within the website, there is a page on working time, but this advises that all initial enquiries should be directed to the Arbitration, Conciliation and Advisory Service.[60] This suggests that regulating working time is not central to the HSE's activities. Moreover, there remains political resistance to the Working Time Directive. The Coalition's Programme for Government promises 'to work to limit the application of the Working Time Directive in the United Kingdom.'[61]

[2] Harassment, Discrimination and the Protection of Vulnerable Groups

Bullying, harassment and discrimination are not specifically forbidden within health and safety legislation, although the HSE has recognized that these phenomena can be related to occupational health, in particular as causes of stress at work.[62] Nevertheless, the main legal remedies lie elsewhere and this is generally not perceived as a matter of health and safety law. Space does not permit a detailed examination of the law in this area, however, the key instruments are the following. In relation to discrimination, the Equality Act 2010 prohibits discrimination based on the personal characteristics of age, disability, gender reassignment, marriage and civil partnership, pregnancy and maternity, race, religion or belief, sex, and sexual orientation. This applies to the workplace

54. Paragraph 15, Case C-84/94 *UK v. Council* [1996] ECR I-5755.
55. SI 1998/1833.
56. Regulation 4(1).
57. Regulation 30(1).
58. Regulation 29. On the HSE's enforcement powers, see section §11.05 of this chapter.
59. http://www.hse.gov.uk/index.htm, 8 Aug. 2011.
60. http://www.hse.gov.uk/contact/faqs/workingtimedirective.htm, 8 Aug. 2011.
61. HM Government, *The Coalition: Our Programme for Government* 19 (Cabinet Office 2010).
62. http://www.hse.gov.uk/stress/furtheradvice/bullyingharassment.htm, 8 Aug. 2011.

and it includes an express prohibition of harassment.[63] As discussed in section §11.05 of this chapter, the provisions on disability are particularly relevant to workers who have experienced occupational injury or ill-health. In addition, the Prevention of Harassment Act 1997 makes it unlawful to harass another person for any reason; it is not necessary to demonstrate a discriminatory reason under this legislation. There is no definition of harassment in the Act, but section 7 clarifies that the conduct must have occurred on at least two occasions and this includes 'alarming the person or causing the person distress'. Harassment under the Act is both a criminal offence and it can give rise to a claim in civil proceedings. Employers may be vicariously liable for harassment committed by their employees, provided that it took place in the course of their employment.[64] Finally, it should be noted that the employer's duty of care arising in the tort of negligence can extend to liability for the damage to the victim's health where the employer fails to prevent workplace bullying.[65]

Beyond the general protections outlined above, the law also recognizes that there are certain groups who face specific health risks. There are various provisions within the Working Time Regulations conferring additional protection in respect of young workers and night workers. For example, a young worker may not work more than eight hours per day or forty hours per week,[66] while night workers must receive periodic health assessments.[67] The MHSW Regulations also include special obligations on the employer in respect of young workers; certain types of work are excluded for young workers, such as those involving harmful exposure to radiation or extreme temperatures.[68] The MHSW Regulations include specific protections for new or expectant mothers. Once the employee has notified the employer that she is pregnant, has given birth in the previous six months or is breastfeeding, the employer must conduct a risk assessment. The employer is under a duty to alter working conditions and hours of work if this is necessary to avoid risks to her or her baby.[69] Indeed, if such alterations cannot lead to an avoidance of the risk, then the employer must suspend (with pay) the employee.[70]

[B] Risk

This section will introduce the main avenues for tackling risks to health and safety within British law. These will be divided into two categories. First, there are the duties found within the HSWA and Regulations adopted thereto. Second, there is the tort of negligence and its approach to attaching liability where certain risks materialize in damage to individuals. It is important to distinguish the enforcement powers of the HSE under the HSWA and associated Regulations from actions by individuals in tort. Breach

63. Section 26 Equality Act 2010.
64. *Majrowski v. Guy's and St. Thomas's NHS Trust* [2006] IRLR 695 (HL).
65. Paragraph 102, *Green v. DB Group Services (UK) Ltd* [2006] IRLR 764 (QB).
66. Regulation 5A, Working Time Regs.
67. Regulation 7.
68. Regulation 19, MHSW Regs.
69. Regulation 16.
70. Regulation 16(3), MHSW Regs and s. 68 Employment Rights Act 1996.

of the legislation can lead to prosecution by the HSE for the risk caused to health and safety, whether or not damage has been caused to any individual. Where actions are brought in tort, then there must be evidence that the risk has actually given rise to harm.

[1] Risk and the Health and Safety at Work Act

The general approach to risk is found at the outset of the HSWA: 'it shall be the duty of every employer to ensure, so far as is reasonably practicable, the health, safety and welfare at work of all his employees.'[71] The 'reasonably practicable' test indicates that an employer is not liable for *every* risk that arises in the course of employment. The phrase has its origins in earlier health and safety legislation and the commonly cited interpretation is found in *Edwards v. National Coal Board.*[72] This case concerned a worker in a coal mine who was killed after the collapse of a road. The reason for the collapse was a hidden defect that could not have been detected in advance and which could only have been addressed if all such roads were reinforced. The defendants argued that this was not 'reasonably practicable'. Asquith LJ gave this term the following interpretation:

> 'reasonably practicable' is a narrower term than 'physically possible' and seems to me to imply that a computation must be made by the owner, in which the quantum of risk is placed on one scale and the sacrifice involved in the measures necessary for averting the risk (whether in money, time or trouble) is placed in the other; and if it be shown that there is a gross disproportion between them – the risk being insignificant in relation to the sacrifice – the defendants discharge the onus on them.[73]

While the Court was not satisfied that the mine owners in the *Edwards* case had actually done all that was 'reasonably practicable' to prevent the risk of the road collapsing, the test created has been criticized by some commentators for permitting economic considerations to be taken into account in deciding the scope of the duty to ensure health and safety.[74] The European Commission shared these concerns and initiated infringement proceedings on the basis that this was not compatible with Article 5(1) of the Framework Directive, which contains no such proviso.[75] The Court of Justice, however, upheld the validity of the reasonably practicable test, concluding that it was not clear that the Directive required Member States to adopt a system of no-fault liability.[76]

The 'reasonably practicable' test remains central to determining the scope of employer's obligations and continues to generate litigation surrounding its boundaries.

---

71. Section 2(1).
72. [1949] 1 KB 704 (CA).
73. *Ibid.*, 712.
74. P. James & D Walters, *Regulating Health and Safety at Work: An Agenda for Change?* 8 (Inst. Empl. Rights 2005).
75. 'The employer shall have a duty to ensure the safety and health of workers in every aspect related to the work', Article 5(1), Directive 89/391.
76. Paragraph 51, Case C-127/05, *Commission v. UK* [2007] ECR I-4619.

For example, in *R v. HTM*,[77] two employees were killed when they were moving a mobile lighting tower and this struck electricity cables. The employer claimed to have taken all reasonable precautions and to have provided training, which included lowering the tower when it was being moved. The Court of Appeal accepted that it was relevant to assessing the scope of the employer's liability to take into account the likelihood of a particular risk materializing;[78] negligence by the employees could be relevant evidence in determining whether the employer had taken all reasonably practicable steps to avoid the risk.[79] It must also be borne in mind that the burden of proof when asserting that a certain step was *not* reasonably practicable lies on the defendant.[80] Section 40 HSWA is a rare example of a reversed burden of proof in the criminal law: 'it shall be for the accused to prove ... that it was not practicable or not reasonably practicable to do more than was in fact done to satisfy the duty or requirement'.

In addition to the HSWA, it is necessary to consider the approach to risk found within the large volume of Regulations adopted under its auspices. While it is impossible to examine all of these, the MHSW Regulations are of special significance. They place a duty on an employer to conduct a 'suitable and sufficient assessment of the risks to the health and safety of his employees to which they are exposed whilst they are at work', as well as the risks faced by other persons not employed by the employer but 'in connection with the conduct by him of his undertaking'.[81] An employer must carry out a specific risk assessment before employing anyone under the age of 18,[82] and employers with more than five employees must keep a record of such assessments.[83] The extent to which British law has been shaped by EU legal requirements is evident in the inclusion within Schedule 1 of the MHSW Regulations of 'general principles of prevention' that must be the basis for employers' protective measures. These are a reproduction of Article 6(2) of the Framework Directive and include themes such as seeking to alleviate monotonous work. The general principles of prevention are also woven into other elements of health and safety legislation. The Construction Regulations require these principles to be respected in the design, planning and preparation of a project and 'so far as is reasonably practicable' during the 'carrying out of the construction work'.[84]

Critics of health and safety argue that risk assessments are linked to the supposed tendency to apply health and safety requirements in an excessively restrictive manner. Lord Young's review argued that there was an inclination towards the *elimination* of all risks, rather than the proportionate approach reflected in the 'reasonably practicable'

77. [2007] 2 All ER 665 (CA).
78. *Ibid.*, para. 22.
79. See further, V. Howes, *Duties and Liabilities under the Health and Safety at Work Act 1974: A Step Forward?*, 38 Indus. L. J. 306, 311–312 (2009).
80. M. Ford & J. Clarke, *Redgrave's Health and Safety* 2.56 (5th ed., LexisNexis Butterworths 2007).
81. Regulation 3(1), MHSW Regs.
82. Regulation 3(4).
83. Regulation 3(6).
84. Regulation 7, Construction (Design and Management) Regulations 2007.

test.[85] There appears to be two extremes in practice: some organizations that have misinterpreted, in an overly-bureaucratic fashion, the process of risk assessment, alongside other organizations that fail to undertake even basic risk assessments.[86] Lord Young's review proposed several deregulatory steps: a simplification of risk assessments for office accommodation and 'low hazard environments, such as shops and classrooms'; exempting employers from conducting risk assessments for home-workers; and exempting self-employed people from the requirement to have written risk assessments, unless in hazardous occupations.[87] While this has been presented as mere 'common sense', it seems to reflect an outdated view of occupational health risks as those likely to lead to serious physical injury. The most common sources of work-related ill-health and absences from work are stress and musculoskeletal disorders; arguably these are risks that materialize in both hazardous and non-hazardous environments, and are likely to effect all workers, including home-workers.

[2] Risk and the Tort of Negligence

In relation to the tort of negligence and, where they overlap, the employer's implied contractual duty of care, the scope of liability is mediated by the concept of reasonable foreseeability. The defendant will only be held to have breached the duty of care if s/he failed to act upon a reasonably foreseeable risk of damage:

> negligence is the omission to do something which a reasonable man, guided upon those considerations which ordinarily regulate the conduct of human affairs would do, or doing something which a prudent and reasonable man would not do.[88]

This standard is particularly tricky in relation to employer liability for psychiatric injuries to employees. In *Walker v. Northumberland County Council*,[89] a social services manager suffered two mental breakdowns in the space of a year when working under conditions of high pressure due to limited staff resources and a growing volume of child abuse casework. Much of the case turned on whether it was reasonably foreseeable that he would have suffered this psychiatric injury. The judge drew attention to the dilemma: 'given that the professional work is intrinsically demanding and stressful, at what point is the employer's duty to take protective steps engaged?'[90] On the facts, the failure of the employer to provide adequate support on his return to work following the first mental breakdown was deemed crucial, because at this point it was reasonably

85. HM Government, *Common Sense Common Safety – A Report by Lord Young of Graffham to the Prime Minister following a Whitehall-Wide Review of the Operation of Health and Safety Laws and the Growth of Compensation Culture* 7 (Cabinet Office 2010).
86. House of Commons Work and Pensions Committee, *The Role of the Health and Safety Commission and the Health and Safety Executive in Regulating Workplace Health and Safety*, Third Report of Session 2007–2008, vol. 1, HC 246, 19 (Stationery Office 2008).
87. HM Government, *Common Sense Common Safety – A Report by Lord Young of Graffham to the Prime Minister following a Whitehall-Wide Review of the Operation of Health and Safety Laws and the Growth of Compensation Culture* 28–29 (Cabinet Office 2010).
88. W.V.H. Rogers, *Winfield and Jolowicz on Tort*, 279 (18th ed., Sweet & Maxwell 2010).
89. *Walker v. Northumberland County Council* [1995] ICR 702 (QB).
90. *Walker v. Northumberland County Council* [1995] ICR 702 (QB), 710.

foreseeable that he was exposed to a higher than normal risk of psychiatric injury. Further clarification was provided in the seminal case of *Barber v. Somerset County Council*.[91] The House of Lords approved a set of 'practical propositions' developed by Baroness Hale in the Court of Appeal. These recognize that psychiatric injuries are more difficult to foresee than physically injuries and 'an employer is usually entitled to assume that the employee can withstand the normal pressures of the job unless he knows of some particular problem or vulnerability'.[92] Consequently, there is a strong onus on the employee to tell the employer about the stressful conditions under which s/he is working in order for any subsequent injury to be viewed as reasonably foreseeable. This can give rise to difficult factual disputes as to whether the employee had provided sufficient indication to the employer about work-related stress. Nevertheless, contextual information can be taken into account. In *Hone v. Six Continents Retail Ltd*,[93] the fact that the employer knew that the employee was working well in excess of the maximum forty-eight hours working week (without having signed an agreement to opt-out of this limit) should have made the employer particularly alert to concerns raised by the employee. Prior knowledge of risks will not always be sufficient to render the employer liable for psychiatric injury. In *Johnston v. Unisys*,[94] the employer's knowledge was limited to the fact that the employee had experienced work-related stress during the mid-1980s. This was deemed to be too remote for liability to be attached to a psychiatric injury following his dismissal in 1994.

[C] Protection and Prevention

The main way in which the law seeks to protect health and safety is through the imposition of duties on a range of parties, such as employers, employees and the self-employed. These are discussed in more detail in the next section, but the most extensive duties are attached to employers. The duty to conduct risk assessments is a central feature of the law's efforts to prevent damage to health at the workplace. In addition, the MHSW Regulations require the employer to establish mechanisms for ongoing health surveillance and to appoint competent persons for providing assistance on compliance with the legal duties.[95] It is not necessary for the competent person(s) to be employed directly by the employer, however, they must have sufficient time, resources and information in order to fulfil their functions. Although the MHSW Regulations require the person to be 'competent', there is no detailed regulation of the qualifications needed in order to exercise such a role. Academic commentators and politicians have criticized the poor quality of some health and safety consultants, who may lack the necessary qualifications.[96] In response, in 2011, the HSE launched an

91. [2004] 1 WLR 1089 (HL).
92. *Walker v. Northumberland County Council* [1995] ICR 702 (QB), 1092.
93. [2005] EWCA Civ 922.
94. [2001] 2 ICR 480 (HL).
95. Regulations 6 and 7, MHSW Regs.
96. P. James & D Walters, *Regulating Health and Safety at Work: An Agenda for Change?* 43 (Inst. Empl. Rights 2005); HM Government, *Common Sense Common Safety – A Report by Lord Young of Graffham to the Prime Minister following a Whitehall-Wide Review of the Operation of Health*

'Occupational Safety and Health Consultants Register' to provide a source for identifying qualified advisors.[97]

In relation to construction, there is a special set of requirements designed to prevent accidents and injuries at work.[98] A construction project is notifiable if the construction phase will last more than thirty days or if the combined total number of working days for all those workers participating exceeds 500.[99] In such cases, a 'Construction Design and Management (CDM) Coordinator' must be appointed by the client. This person has a range of responsibilities relating to the coordination of health and safety during the pre-construction phase. The CDM Coordinator must lodge with the HSE basic information concerning the project, such as its duration and the estimated number of workers involved. In addition, the client must appoint a 'principal contractor'. This person is responsible for planning, managing and monitoring the construction phase of the project with a view to ensuring that it is carried out without risks to health and safety 'so far as is reasonably practicable'.[100] The principal contractor has duties in relation to all other contractors involved in the project and in relation to all workers, regardless of employment status.[101] Workers must receive an induction to the site and the principal contractor should ensure that they receive information and training.[102] There is a duty on the principal contractor to consult with the workers or their representatives on 'health, safety and welfare',[103] and this person must ensure sufficient welfare facilities for the workers in areas such as sanitation, washing, drinking water, and rest facilities. In 2005, the HSE published a detailed investigation of occupational health and safety, which included case studies from the construction industry. This suggested that compliance was patchy and depended on the nature of the construction site. Large sites with trade union organization seemed to have better levels of compliance.[104] Moreover, compliance levels appeared higher in relation to those who were employees, and lower for those who were sub-contracted labour or agency workers.[105]

---

and Safety Laws and the Growth of Compensation Culture 31 (Cabinet Office 2010); House of Commons Work and Pensions Committee, *The Role of the Health and Safety Commission and the Health and Safety Executive in Regulating Workplace Health and Safety*, Third Report of Session 2007–2008, vol. 1, HC 246, 21 (Stationery Office 2008).

97. HSE, *The Health and Safety Executive Annual Report and Accounts 2010/11* 9 HC1066 (Stationery Office 2011); Department for Work and Pensions, *The Government Response to the Löfstedt Report*, http://www.dwp.gov.uk/docs/lofstedt-report-response.pdf, Nov. 2011, 5.
98. F. Wright, *Law of Health and Safety at Work*, 211 (Sweet & Maxwell 1997). This has been shaped by the need to comply with Directive 92/57 on the implementation of minimum safety and health requirements at temporary or mobile work sites [1992] OJ L245/6.
99. Regulation 2(3), Construction (Design and Management) Regs.
100. Regulation 22(1), Construction (Design and Management) Regs.
101. Department of Work and Pensions, *The Government Response to the Rita Donaghy Report: One Death Is Too Many – Inquiry into the Underlying Causes of Construction Fatal Accidents* 7 (Stationery Office 2010).
102. Regulation 22(2), Construction (Design and Management) Regs.
103. Regulation 24(b), Construction (Design and Management) Regs.
104. D. Walters et al., *The Role and Effectiveness of Safety Representatives in Influencing Workplace Health and Safety*, Research Report 363, 104 (HSE 2005).
105. D. Walters et al., *The Role and Effectiveness of Safety Representatives in Influencing Workplace Health and Safety*, Research Report 363, 99 (HSE 2005).

The HSE enjoys inspection powers that seek to prevent breaches in health and safety. HSE inspectors can enter any premises at 'any reasonable time (or, in a situation which in his opinion is or may be dangerous, at any time)'.[106] If the inspector concludes that there is a contravention of any of the statutory provisions, then an 'improvement notice' can be issued. This requires the person to which it is issued to take action to remedy the contravention within a specified period of time.[107] If, however, there is a risk of 'serious personal injury', then the inspector can issue a 'prohibition notice'. This may direct that specified activities must not be carried out until the contraventions of the law that give rise to the risk of serious personal injury are remedied. Moreover, the prohibition notice can have immediate effect, allowing inspectors to order a stop to business activities if required. It is a criminal offence to contravene any requirements within either an improvement or prohibition notice.[108] HSE inspectors are complemented by local authority inspectors. These have a wider remit, with particular responsibility for controlling food hygiene, but they enjoy similar powers to issue improvement and prohibition notices. HSE inspectors mainly focus on high hazard sites, such as manufacturing, while local authority inspectors deal with low hazard sites.

Clearly, there are limits to the capacity of the HSE to inspect all workplaces. An important source of information for the HSE is the mandatory reporting of injuries, diseases and dangerous occurrences. The reporting system, known by its acronym 'RIDDOR' requires the HSE to be notified of a range of events falling within the Regulations, including any accidents and injuries that cause the person to be incapacitated for more than seven days.[109] While this is a valuable source of information on health and safety at work, the HSE estimates that only 58% of reportable injuries are actually reported.[110]

The HSE evidently enjoys significant powers of inspection and enforcement. There is, though, a lively debate about the manner in which these powers are exercised. During the past decade, there has been a discernable shift in the regulatory model pursued by the HSE, moving away from a reliance on enforcement powers (such as prosecutions) and instead focusing on partnership with businesses to promote voluntary compliance.[111] This was initiated by the HSE in 2000, but reinforced by a government review of regulation published in 2005, referred to as the 'Hampton

---

106. Section 20 HSWA.
107. Section 21 HSWA.
108. Section 33(g) HSWA.
109. The Reporting of Injuries, Diseases and Dangerous Occurrences Regulations 1995, SI 1995/3163 (as amended).
110. HSE, *The Health and Safety Executive: Annual reports and Accounts 2009/10*, 12 HC20 (The Stationery Office 2010). Lord Young's review has proposed raising the threshold for reporting incidents to those leading to seven days of incapacity, cf. HM Government, *Common Sense Common Safety – A Report by Lord Young of Graffham to the Prime Minister following a Whitehall-Wide Review of the Operation of Health and Safety Laws and the Growth of Compensation Culture* 30 (Cabinet Office 2010).
111. S. Tombs & D. Whyte, *Regulatory Surrender: Death, Injury and the Non-Enforcement of Law*, 29 (Inst. Empl. Rights 2010).

report'.[112] This sought to reduce regulatory burdens on business by targeting those areas where the risk is highest; the HSE committed itself in response to the report to reduce inspections by 30%.[113] The strategy can also be attributed to the resources available to the HSE. Its annual budget peaked in 2002 at GBP 258 million, but has declined since then.[114] Its net operating cost in 2010/2011 was around GBP 175 million.[115] It is not surprising, therefore, to discover than the number of inspectors has declined during the past decade; the Field Operations Division had 962 inspectors in 2003,[116] but the TUC estimates that there are around 700 now.[117] Tombs and Whyte have assembled a detailed statistical analysis of inspections in the period 1999/2000–2008/2009.[118] Workplace inspections conducted by the HSE have declined by 69%. Looking at RIDDOR reported injuries, there have been major falls in the proportion of such incidents that are investigated by the HSE. For example, only 33.6% of injuries leading to amputations are investigated and 25.4% of injuries resulting in major burns. The number of improvement notices issued fell by 30%, and similarly the number of prohibition notices fell by 26%. The most recent figures show that the HSE issued 7,137 improvement notices and 3,835 prohibition notices in 2010/2011; this was a 23% increase in the number of improvement notices issued in 2009/2010.[119] Prosecutions, the most serious weapon in the HSE's arsenal, fell by 48%. Most of this decline occurred in the period to 2005, since when the annual number of prosecutions has been fairly stable.[120]

The data confirms a change in the regulatory model, which the HSE terms a 'risk-based approach to complaints investigation'.[121] In defence of this approach, the past decade has witnessed a downward trend in the rate of fatal and major injuries. The HSE estimates that this fell from 130 per 100,000 employees in 1999/2000 to 106 in 2008/2009, an 18% decline.[122] Reports of work-related ill-health are estimated to have declined from 2,190 per 100,000 workers in 1999/2000 to 1,810 in 2008/2009, a 17%

112. P. Hampton, *Reducing Administrative Burdens: Effective Inspection and Enforcement* (HMSO 2005).
113. S. Tombs & D. Whyte, *Regulatory Surrender: Death, Injury and the Non-Enforcement of Law*, 35 (Inst. Empl. Rights 2010).
114. P. James & D Walters, *Regulating Health and Safety at Work: An Agenda for Change?* 56 (Inst. Empl. Rights 2005).
115. HSE, *The Health and Safety Executive: Annual Reports and Accounts 2011/12*, HC204, 64 (Stationery Office, 2012).
116. P. James & D. Walters, *Regulating Health and Safety at Work: An Agenda for Change?* 55 (Inst. Empl. Rights 2005).
117. Trades Union Congress, *The Case for Health and Safety*, 2, http://www.tuc.org.uk/workplace/tuc-18421-f0.cfm, 30 Sep. 2010, 8.
118. S. Tombs & D. Whyte, *Health and Safety Gone Mad?* (Inst. Empl. Rights Briefing, Aug. 2010): http://www.ier.org.uk/sites/ier.org.uk/files/H&S%20gone%20mad%20briefing.pdf, 2 Mar. 2011.
119. HSE, *Annual Statistics Report 2010/11*, 12 (HSE 2011). Local authority inspectors issued 5,620 improvement notices and 1,600 prohibition notices.
120. The HSE brought 878 prosecutions in 2010/11, of which 85% resulted in convictions: HSE, *Annual Statistics Report 2010/11*, 11 (HSE 2011).
121. HSE, *The Health and Safety Executive: Annual Reports and Accounts 2009/10*, HC20, 24 (Stationery Office 2010).
122. HSE, *The Health and Safety Executive: Annual Reports and Accounts 2009/10*, HC20, 12 (Stationery Office 2010).

decline. Interpreting the data is difficult. James and Walters draw attention to the longer-term role played by shifts in the type of work within the economy, in particular the decline in manufacturing and the rise in the service sector.[123] This means that part of the decline in the rate of injuries may be due to a reduction in physically dangerous forms of work, rather than an actual improvement in the quality of health and safety. Moreover, the HSE data are based on RIDDOR and it is accepted that over 40% of reportable incidents are not in fact reported. Notwithstanding these notes of caution, it seems fair to acknowledge that the shift in the regulatory model has not prevented a continuing trend of decrease in occupational injuries. Yet it is simultaneously notable that the decline in the rate of inspections far outstrips any comparable decline in the rate of injuries. The HSE also acknowledges that the picture is less encouraging if the focus is on work-related ill-health, such as stress and musculoskeletal disorders.[124] In autumn 2010, the government announced deep cuts in public spending and this included a 35% cut in the HSE budget cut in the period to 2014/2015.[125] The government has announced that workplace inspections will be reduced by one-third.[126] This is in contrast to the conclusions of a House of Commons inquiry in 2008, which recommended an increase in inspections in key sectors of poor performance.[127]

[D] Promotion

The mandate of the HSE includes responsibilities for promoting occupational health and safety. It conducts research in this area and operates an information and advisory service.[128] Moreover, it is empowered to issue 'Approved Codes of Practice', subject to the consent of the relevant minister.[129] These aim to provide practical guidance on the law. Although not legally-binding, they are admissible as evidence in criminal proceedings under the HSWA.[130] The HSE operates an extensive website with a wide range of information on all manner of occupational health issues, including basic information available in thirty languages.[131] The construction industry is a key focus of the HSE given its higher rate of occupational injuries. In 2007, one in three construction sites inspected by the HSE were found to put the lives of workers at risk.[132] The HSE is conducting various campaigns to address these issues. For example, it distributed

123. P. James & D. Walters, *Regulating Health and Safety at Work: An Agenda for Change?*, 37 (Inst. Empl. Rights 2005).
124. Health and Safety Commission (HSC), *Performance Report 2007*, 3 (HSC 2007).
125. HSE, *The Health and Safety Executive Annual Report and Accounts 2010/11* HC1066, 20 (Stationery Office 2011).
126. Department of Work and Pensions, *Good Health and Safety, Good for Everyone*, http://www.dwp.gov.uk/policy/health-and-safety, 21 Mar. 2011.
127. House of Commons Work and Pensions Committee, *The Role of the Health and Safety Commission and the Health and Safety Executive in Regulating Workplace Health and Safety*, Third Report of Session 2007–2008, vol. 1, HC 246, 29 (Stationery Office, 2008).
128. Section 11(2) HSWA.
129. Section 16 HSWA.
130. Section 17 HSWA.
131. http://www.hse.gov.uk/index.htm, 3 Mar. 2011.
132. Health and Safety Commission (HSC), *Performance Report 2007*, 13 (HSC 2007).

110,000 pocket information cards aimed at migrant construction workers to make them aware of health and safety issues.[133] In 2010, it targeted inspections on the refurbishment branch of the construction industry where there was a disproportionately high accident record.[134]

§11.03 PERSONAL SCOPE OF APPLICATION AND DUTIES OF THE PARTIES

[A] Employers

Section 2 HSWA sets out a series of general duties for employers. These are: (a) the provision of a safe plant and system of work; (b) making arrangements for the safe use of substances; (c) the provision of information, training and supervision for employees; (d) keeping the place of work safe; (e) providing a safe working environment and making adequate arrangements for the welfare of employees. These are combined with those duties arising from the myriad of Regulations adopted under the Act, especially the MHSW Regulations. It should be noted that the HSWA excludes employment 'as a domestic servant in a private household'.[135]

There are various legislative provisions that extend the duties of the employer beyond those who are employees. Section 3(1) HSWA provides:

> It shall be the duty of every employer to conduct his undertaking in such a way as to ensure, so far as is reasonably practicable, that persons not in his employment who may be affected thereby are not thereby exposed to risks to their health or safety.

This is important because the concept of 'employee' in British law does not cover all those who provide work for an employer. In particular, casual workers and agency workers often fall outside the definition of 'employee'. Section 3(1) provides a mechanism for ensuring that the employer's duties will extend to cover non-standard forms of work. A similar situation is found in the MHSW Regulations, where the duty to conduct risk assessments extends also to those not employed by the employer.[136] These Regulations include specific provisions on the health and safety of temporary workers, such as their access to health surveillance provided by the employer.[137] Nevertheless, there remain certain provisions where the duties of the employer are limited to those categorized as employees. For example, the duty on employers to

133. HSE, *The Health and Safety Executive: Annual Reports and Accounts 2009/10*, HC20, 18 (Stationery Office 2010).
134. HSE, *The Health and Safety Executive: Annual Reports and Accounts 2009/10*, HC20, 18 (Stationery Office 2010).
135. Section 51 HSWA.
136. Regulation 3(1), MHSW Regs.
137. Regulation 15, MHSW Regs. This is designed to comply with Directive 91/383 supplementing the measures to encourage improvements in the safety and health at work of workers with a fixed-duration employment relationship or a temporary employment relationship [1991] OJ L206/19.

supply information on health and safety risks, preventative measures and procedures, only extends to 'employees'.[138]

The common law case-law on negligence has recognized that the liability of the employer can include matters that are 'reasonably incidental' to the performance of work.[139] This can include entering or leaving the workplace, or being on part of the employer's premises, even though the employee is not required to be in that place.[140] The employer may be both directly liable for her own actions and vicariously liable for the actions of her employees. In a significant development of the law, the Court of Appeal recently held that vicarious liability may also arise in respect of those who are working in 'a position akin to employment'.[141] The case concerned potential liability of a Catholic Diocese for sexual abuse in a children's home committed by a priest. While the priest was not an employee of the Diocese, the relationship under which he operated was held to be sufficiently akin to employment in order to give rise to vicarious liability. This decision may be significant in the future for claims of vicarious liability relating to the actions of atypical workers.

Vicarious liability extends to the actions of employees that take place in 'the course of employment'. There is a large body of case-law surrounding when an employee's actions may be regarded as being in the course of employment, despite not being the proper manner for performing work. Langstaff summarizes the position as the following: 'the employer is liable for tortious acts which amount to an unauthorized mode of performing authorized acts but not for acts which are not sufficiently connected with the authorized act so as to be a mode of doing it'.[142] Ultimately, this is a question of fact and the law reports are replete with fine distinctions between situations that fall inside or outside of this sphere.[143] Some general guidance was provided by the House of Lords in *Smith v. Stages*.[144] This case concerned two workers who were required to complete a job at a site far from their normal place of work. They travelled there in a private car. Although the employer had agreed to pay them for sleeping time before returning home, they did not take a rest and a serious accident occurred on the return journey. Given that they were being paid during the time spent travelling, the House of Lords held that this fell within the course of employment, even though the mode of travelling was at the discretion of the workers.[145] As a consequence, the employer could be vicariously liable for the negligence of the employee driving the car. More generally, Lord Lowry drew a distinction between 'the duty to

138. Regulation 10, MHSW Regs.
139. Justice Langstaff, *The Employer's Duty of Care*, in *Munkman on Employer's Liability*, 137 (14th ed., B. Cotter & D. Bennett eds., LexisNexis 2006).
140. Justice Langstaff, *The Employer's Duty of Care*', in *Munkman on Employer's Liability*, 138 (14th ed., B. Cotter & D. Bennett eds., LexisNexis 2006).
141. Paragraph 70, *JGE v. The Trustees of Portsmouth Roman Catholic Diocesan Trust* [2012] EWCA Civ 938.
142. Justice Langstaff, *The Employer's Duty of Care*, in *Munkman on Employer's Liability* 145 (14th ed., B. Cotter & D. Bennett eds., LexisNexis 2006).
143. One source of litigation is accidents and injuries that occur outside the workplace but in the course of work-related events, such as organized excursions: D. Brodie, *Away Days and Employers' Liability: Reynolds v. Strutt and Parker*, 41 Indus. L.J. 93 (2012).
144. [1989] AC 928 (HL).
145. *Ibid.*, 939.

turn up for one's work with the concept of already being "on duty" while travelling to it.'[146] In his view, the employee travelling between her home and her place of work (and returning back again) is normally not acting within the course of employment, so the employer has no liability for such journeys. Nevertheless, there could be exceptions to this general principle, such as where an employee is called from home to attend an emergency; in this circumstance, the journey from home to the place of work could be regarded as being in the employer's time.

[B] Employees

Section 7 HSWA creates two duties for an employee: (a) to take reasonable care for her own health and safety, as well as that of others who may be affected by her acts and omissions; (b) to cooperate with the employer in complying with any statutory duty or requirement. This is supplemented by a further duty on the employee not to 'intentionally or recklessly interfere with or misuse anything provided in the interests of health, safety or welfare'.[147] Duties are also placed on employees within the Regulations adopted under the HSWA. For example, the MHSW Regulations require employees to use equipment and substances in accordance with training and instructions provided to them.[148] Employees must also inform either their employer or an employee with responsibility for health and safety of: (a) any work situation presenting a serious and immediate danger; and (b) any shortcoming in the employer's arrangements for protecting health and safety.[149] The statutory duties of employees can be relevant in a civil law action for damages in relation to whether there has been contributory negligence on the part of the employee.[150] If this is found to be the case, then the court may reduce any damages awarded to the claimant.[151]

[C] Self-employed

Section 3(2) HSWA places every self-employed person under a duty 'to conduct his undertaking in such a way as to ensure, so far as is reasonably practicable, that he and other persons (not being his employees) who may be affected thereby are not thereby exposed to risks to their health or safety.' Further duties on the self-employed can be found in the Regulations adopted under the HSWA. For example, the Provision and Use of Work Equipment Regulations 1998 specify that the requirements imposed on employers also apply to self-employed persons 'in respect of work equipment he uses at work'.[152] One of the most significant obligations on self-employed persons is found

146. *Ibid.*, 956.
147. Section 8 HSWA.
148. Regulation 14, MHSW Regs.
149. Regulation 14, MHSW Regs.
150. I. Smith et al., *Health and Safety: The Modern Legal Framework* 31 (2nd ed., Butterworths 2000).
151. See further, W.V.H. Rogers, *Winfield and Jolowicz on Tort* 375–377 (18th ed., Sweet & Maxwell 2010).
152. Regulation 3(3), SI 1998/2306.

within the MHSW Regulations, which require the self-employed person to conduct risk assessments in respect of her own personal safety as well as that of any others who may be affected.[153] The Löfstedt Review raised concerns that this was unduly onerous and the government has announced that it will seek to change the law to 'remove health and safety burdens from the self-employed in low-risk occupations, whose activities represent no risk to other people.'[154]

The regulation of the self-employed is a key issue in relation to the construction industry, where it is estimated that one-third are self-employed.[155]

[D] Sub-contractors

One issue that proves challenging for health and safety law is the question of sub-contractors; that is, where the employees of one employer are working on the premises of another. This is particularly relevant to certain forms of employment, such as construction. Specific provision for this is made in the MHSW Regulations. If an employee of one employer is working on the premises of another employer, or of a self-employed person, then the host employer/self-employed person has a number of obligations. For example, the person working in the host undertaking must be provided with 'appropriate instructions and comprehensible information regarding any risks to that person's health and safety'.[156] These provisions also apply if the situation is one of a self-employed person working in the premises of another employer or self-employed person.

In respect of the tort of negligence, the courts have been willing to attach liability to a person who was not the employer of the claimant, but who is deemed to have control at the relevant time when the damage occurred.[157] The case-law in this area is often concerned with accidents occurring on construction sites where there may be multiple contractors working simultaneously on the same site, some being self-employed and others being sub-contracted labour from another employer. While the attribution of liability depends on the specific facts, the case-law includes examples of situations where the courts impute a duty on the principal contractor or the owner of the property to protect the health and safety of those working on the site.[158] This corresponds with the approach in the Construction (Design and Management) Regulations 2007, where duties are not attached to the status of being an 'employee', but are instead linked to being a contractor, or working under the control of a contractor.[159] The broader statutory language is evidently designed to ensure that the precarious

153. Regulatioin 3(2), MHSW Regs.
154. Department for Work and Pensions, *The Government Response to the Löfstedt Report*, http://www.dwp.gov.uk/docs/lofstedt-report-response.pdf, Nov. 2011, 9.
155. D. Walters et al., *The Role and Effectiveness of Safety Representatives in Influencing Workplace Health and Safety*, Research Report 363, 81 (HSE 2005).
156. Regulation 12(3), MHSW Regs.
157. G. Pitt, *Employment Law* 441 (7th ed., Sweet & Maxwell 2009).
158. Justice Langstaff, *The Employer's Duty of Care*, in *Munkman on Employer's Liability*, 140 (14th ed., B. Cotter & D. Bennett eds., LexisNexis 2006).
159. Regulation 25(1).

working arrangements common in the construction industry do not circumvent the scope of health and safety law.

§11.04 WORKERS' REPRESENTATION

There is a general duty in section 2(6) HSWA to consult with representatives appointed by recognized trade unions with a view to 'promoting and developing measures to ensure the health and safety at work of employees'. The detailed arrangements for the implementation of this duty are found within the Safety Representatives and Safety Committee Regulations 1977.[160] Originally, this was the only route for worker participation in the area of health and safety. The main shortcoming was that it was limited to those workplaces where there was a recognized trade union. Prior to 1999, there was no statutory process whereby employers could be compelled to recognize a trade union. So worker participation in the area of health and safety was completely dependent on the employer's discretionary decision regarding whether or not to recognize a trade union. Unsurprisingly, this is especially problematic in workplaces where there is low union density, such as construction, as well as areas where precarious work dominates (e.g., temporary or agency work).[161]

The linkage of worker participation rights to trade union recognition was an approach familiar to other aspects of British collective labour law, but it was called into question by two decisions of the Court of Justice.[162] The cases concerned the duty to consult workers' representatives under EU law relating to collective redundancies and the transfer of undertakings. The Court held that it was not consistent with the obligations of EU law to limit such consultation to situations where the employer had recognized a trade union. Although the case did not concern the UK's implementation of EU health and safety legislation, it seemed very likely that the Court would adopt a similar approach in respect of other aspects of worker participation law.[163] This stimulated the government to introduce an alternative pathway for worker participation in workplaces where there was no recognized trade union. These rules are found in the Health and Safety (Consultation with Employees) Regulations 1996.[164] As there are important differences between the two sets of Regulations on worker participation (1977 and 1996), it is necessary to consider each in turn.

[A] Worker Participation where there Is a Recognized Trade Union

The 1977 Regulations permit recognized trade unions to appoint 'Safety Representatives'. These representatives are given a variety of powers. First of all, there are

160. SI 1977/500.
161. P. James & D. Walters, *Regulating Health and Safety at Work: An Agenda for Change?* 105–106 (Inst. Empl. Rights 2005).
162. Case C-382/92, *Commission v. UK* [1994] ECR I-2435; Case C-383/92, *Commission v. UK* [1994] ECR I-2479.
163. See further, V. Howes, *Workers' Involvement in Health and Safety Management and Beyond: The UK Case*, 23 Intl. J. Comp. Lab. L. Indus. Rel. 245, 256 (2007).
164. SI 1996/1513.

investigatory powers, such as the investigation of potential hazards, dangerous occurrences, the causes of accidents, and complaints by an employee.[165] Second, the representative has the power to inspect the workplace, including where there has been a notifiable accident or dangerous occurrence.[166] The representative is entitled to inspect and take a copy of any relevant document.[167] Third, the representative has the function of making representations to the employer on behalf of the workforce in relation to health and safety, and this extends to representing the employees in consultations with the inspectors of the HSE.[168] The employer is under a duty to consult 'in good time' safety representatives on a wide range of matters pertaining to health and safety, such as the introduction of new technologies at work.[169] Moreover, the employer must provide the representatives with paid time off to perform their functions and to be trained, as well as providing them with reasonable facilities and assistance.[170] If at least two safety representatives request it, the employer must establish a Safety Committee for the purpose of consultations with Safety Representatives.[171] The TUC estimate that there are currently 150,000 trained health and safety representatives, but that these only cover around half of the workforce.[172] It has been suggested that the serious injury rate in workplaces with trade union safety representatives is half of that found in similar workplaces without such representatives.[173] While there is some debate over the reliability of such figures, studies have supported the basic premise that 'systematic and consultative management' is linked to better health and safety outcomes.[174]

[B] Workplaces without a Recognized Trade Union

The 1996 Regulations place employers under an equivalent duty to consult their employees on health and safety matters, but the employer can choose between direct consultation with all employees or consulting 'representatives of employee safety' who have been elected for that purpose by the employees. Moreover, the Regulations permit an employer who has been consulting elected employee representatives to decide to change to direct consultation.[175] The Regulations say little about what direct consultation with employees would entail, except that the employer must provide the employees with such information 'as is necessary to enable them to participate fully

165. Regulation 4(1)(a)–(b), Safety Representatives and Safety Committees Regulations 1977.
166. Regulations 5 and 6.
167. Regulation 7.
168. Regulation 4(1).
169. Regulation 4A.
170. Regulation 4(2) and 4A(2). See further, F. Wright, *Law of Health and Safety at Work* 189–190 (Sweet & Maxwell 1997).
171. Regulation 9.
172. Trades Union Congress, *The Case for Health and Safety*, 2, http://www.tuc.org.uk/workplace/tuc-18421-f0.cfm, 30 Sep. 2010, 12.
173. Trades Union Congress, *The Case for Health and Safety*, 2, http://www.tuc.org.uk/workplace/tuc-18421-f0.cfm, 30 Sep. 2010, 6.
174. D. Walters et al., *The Role and Effectiveness of Safety Representatives in Influencing Workplace Health and Safety*, Research Report 363, xii (HSE 2005).
175. Regulation 4(4), Health and Safety (Consultation with Employees) (HSCE) Regs 1996.

and effectively in the consultation'.[176] If the elected representatives route is chosen by the employer, then the representatives are empowered to make representations to the employer and HSE inspectors.[177] The representatives must be provided with paid time off for carrying out their functions and they must be provided with training *organized by the employer*.[178] This is in contrast to the 1977 Regulations where the employers must provide paid time off for training, but there is no duty on the employer to ensure that training takes place. This would seem to reflect the likelihood that a trade union representative will have access to training organized by the trade union.

Strikingly, the other functions of (trade union) Safety Representatives are not extended to elected representatives of employee safety. They have no powers to investigate or inspect the workplace. They have no powers to inspect documents and there is no provision for employee representatives to require the employer to establish a Safety Committee. Breach of the 1996 Regulations will not expose the employer to any civil liability, except in respect of the duty to provide paid time off for representatives of employee safety. Once again this is a departure from the 1977 Regulations where there is no restriction of civil liability.

[C] Worker Participation in Health and Safety: Time for Reform?

Unsurprisingly, the dichotomy contained in the two sets of Regulations has led to calls for reform. Research has suggested that the possibility for elected employee representatives in the 1996 Regulations is rarely used in practice,[179] while a majority of employers rely on direct consultation.[180] Research in the construction industry revealed that direct consultation, in the absence of trade unions, was often so informal as to fall short of an accepted concept of what consultation should entail.[181] Yet the HSE is reluctant to exercise enforcement powers in relation to worker participation. Between 1997 and 2008, five improvement notices were issued under the 1977 Regulations and forty-two under the 1996 Regulations.[182] There was no evidence of any prosecutions ever having been brought in relation to breaches of the consultation duties.

The current legal framework seems out of step with contemporary trends in the labour market, in particular, the steep decline in trade union membership since the 1977 Regulations were adopted. In the private sector, only 15.1% of workers are

176. Regulation 5(1).
177. Regulation 6.
178. Regulation 7.
179. P. James & D. Walters, *Regulating Health and Safety at Work: An Agenda for Change?* 106 (Inst. Empl. Rights 2005).
180. S. Deakin & G. Morris, *Labour Law* 824 (5th ed., Hart Publg. 2009).
181. D. Walters et al., *The Role and Effectiveness of Safety Representatives in Influencing Workplace Health and Safety*, Research Report 363, 100 (HSE 2005).
182. House of Commons Work and Pensions Committee, *The Role of the Health and Safety Commission and the Health and Safety Executive in Regulating Workplace Health and Safety*, Third Report of Session 2007–2008, vol. 1, HC 246, 56 (The Stationery Office 2008).

members of trade unions.[183] This means that many workers are only covered by the lower standards in the 1996 Regulations. A straightforward reform would, for example, be the levelling-up of the functions of representatives of employee safety with those of (trade union) Safety Representatives, while the House of Commons Work and Pensions Committee has proposed removing the power of the employer to opt for direct consultation.[184] Nevertheless, when reform was considered by the HSE in 2003, it concluded that legislative change was not appropriate and that it would pursue voluntary methods for promoting worker participation.[185] In the HSE's current strategy document, it reasserts the importance of worker involvement[186] and it is supporting initiatives towards this end, such as part-funding the training of representatives in non-unionized workplaces.[187] Yet research suggests that a strong legislative foundation is a key prerequisite to effective worker participation in health and safety,[188] so it must be questioned how much progress can be made in the absence of a willingness to consider legislative reform as part of the overall package of measures in this field.

§11.05 CONSEQUENCES ON THE PARTIES AND REMEDIES IN CASE OF VIOLATION OF HEALTH AND SAFETY PROVISIONS

[A] Enforcement by the Health and Safety Executive

As discussed earlier, the HSE enjoys powers to inspect workplaces and to investigate possible breaches of health and safety legislation. Its ultimate sanction is to bring a prosecution under the HSWA.[189] Section 33 HSWA creates a range of criminal offences related to the operation of the legislation. Some of the offences concern the substance of the law; for example, it is an offence 'to fail to discharge a duty' which has been imposed by the HSWA or to contravene any requirement in both the HSWA and the Regulations adopted thereto. In addition, there are offences that are linked to the implementation of the law, such as obstructing a health and safety inspector.

The penalties vary according to the specific offence, however, most of the offences created by the HSWA attract the same penalty. Most offences under the Act

183. J. Achur, *Trade Union Membership 2009*, at 1 (Department for Business, Innovation and Skills 2010). By coincidence, this is also the rate of union membership in the construction sector.
184. House of Commons Work and Pensions Committee, *The Role of the Health and Safety Commission and the Health and Safety Executive in Regulating Workplace Health and Safety*, Third Report of Session 2007–2008, vol. 1, HC 246, 55 (Stationery Office 2008).
185. House of Commons Work and Pensions Committee, *The Role of the Health and Safety Commission and the Health and Safety Executive in Regulating Workplace Health and Safety*, Third Report of Session 2007–2008, vol. 1, HC 246, 119 (Stationery Office, 2008).
186. HSE, *The Health and Safety of Great Britain – Be Part of the Solution*, http://www.hse.gov.uk/strategy/strategy09.htm#a08a, accessed on 9 Mar. 2011.
187. Department of Work and Pensions, *The Government Response to the Rita Donaghy Report: One Death Is Too Many – Inquiry into the Underlying Causes of Construction Fatal Accidents*, 22 (Stationery Office 2010).
188. D. Walters et al., *The Role and Effectiveness of Safety Representatives in Influencing Workplace Health and Safety*, Research Report 363, xi (HSE 2005).
189. In Scotland, prosecutions are brought by the Procurator Fiscal.

can be brought either summarily or on indictment.[190] If there is a summary conviction, then the penalty is a maximum of twelve months' imprisonment, a fine not exceeding GBP 20,000 (approximately EUR 25,000), or both.[191] If there is a conviction on indictment, then the penalty is a maximum of two years' imprisonment, a fine (subject to no prior limit), or both. In 2010/2011, there were 912 offences prosecuted in Britain resulting in 774 (85%) convictions.[192] The average penalty was GBP 24,005 (approximately EUR 30,000) per breach of the legislation.[193] Statistics from 2002–2007 show that 30%–40% of convictions are in the construction industry.[194] There have been occasional high profile cases where substantial fines have been imposed. Notably, the gas company, Transco, was fined GBP 15 million for an incident where four people were killed,[195] and Network Rail was fined GBP 3 million for a railway accident where seven people died.[196] Imprisonment is unusual, with normally only three to four cases per year.[197] Barrett reports that 'the vast majority of prosecutions are brought against an organization in its role of employer',[198] while the TUC notes that prosecutions are rarely in respect of occupational *disease* rather than *injury*.[199]

There has been a long-running debate over the liability of senior officers in companies, such as the Board of Directors, for health and safety violations. This stems from the difficulty of identifying a specific individual who was responsible for the failure in systems of work, notwithstanding clear evidence that the senior management collectively failed to give this sufficient priority within the organization.[200] If the organization is found to have committed a breach of the legislation, then there is the possibility in the HSWA for individual senior officers to be prosecuted if the offence 'is proved to have been committed with the consent or connivance of, or to have been attributable to any neglect on the part of, any director, manager, secretary or other similar officer of the body corporate ...'.[201] Following an amendment of the legislation in 2008, this offence now carries the possibility of up to two years' imprisonment. Furthermore, the Corporate Manslaughter and Corporate Homicide Act 2007 created a new offence for organizations if 'the way in which its activities are managed or organized: (a) causes a person's death, and (b) amounts to a gross breach of a relevant

190. A summary trial is before a Magistrates' Court; trial on indictment is before a Crown Court where there is a jury. Normally, more severe penalties can only be imposed by the latter.
191. Schedule 3A HSWA.
192. HSE, *Annual Statistics Report 2010/11*, 12 (HSE 2011).
193. HSE, *Annual Statistics Report 2010/11* (HSE 2011).
194. House of Commons Work and Pensions Committee, *The Role of the Health and Safety Commission and the Health and Safety Executive in Regulating Workplace Health and Safety*, Third Report of Session 2007–2008, vol. 1, HC 246, 42 (Stationery Office 2008).
195. M. Ford & J. Clarke, *Redgrave's Health and Safety* 1.22 (5th ed., LexisNexis Butterworths 2007).
196. BBC, *Potters Bar Crash: Network Rail Fined £3m*, http://www.bbc.co.uk/news/uk-13389147, 13 May 2011.
197. B. Barrett, *The Health and Safety (Offences) Act 2008: The Cost of Behaving Dangerously at the Workplace*, 38 Indus. L.J. 77 (2009).
198. B. Barrett, *Liability for Safety Offences: Is the Law Still Fatally Flawed?* 37 Indus. L.J. 100, 110 (2008).
199. Trades Union Congress, *The Case for Health and Safety*, 2, http://www.tuc.org.uk/workplace/tuc-18421-f0.cfm, 30 Sep. 2010, 9.
200. B. Barrett, *Liability for Safety Offences: Is the Law Still Fatally Flawed?* 37 Indus. L.J. 108 (2008).
201. Section 37(1) HSWA.

duty of care owed by the organization to the deceased.'[202] Section 1(3) clarifies the intended focus of the Act as it applies where the role of 'senior management is a substantial element in the breach'. Barrett identifies the significant overlap between the HSWA and the 2007 Act; death due to a breach of health and safety by an organization could already be prosecuted under the HSWA and, like the 2007 Act, there is no limit on the fine that can be imposed if tried on indictment. She concludes that the main impact of the 2007 Act could be the 'greater stigma' attached to a conviction for corporate manslaughter.[203] The first conviction under the Act was obtained in February 2011, when a company was convicted for operating a dangerous system of work following the death of a geologist when earth collapsed on top of him.[204] A fine of GBP 385,000 (approximately EUR 488,000) was imposed on a firm with an annual turnover of GBP 333,000; the firm was permitted a ten year period during which to pay the fine.[205]

[B] Enforcement by Individuals

As discussed earlier, individuals have the possibility to seek a remedy via civil proceedings, which may take the form of the torts of breach of statutory duty or negligence, or an action in contract law for breach of the employer's implied duty of care. While space does not permit a detailed examination of the remedies available in tort and contract law, the most common remedy will be damages. These are not subject to a prior maximum limit and the objective is to place the claimant in the position she would have been in, had the wrong not been committed.[206]

In order to protect victims of health and safety violations, the Employers' Liability (Compulsory Insurance) Act 1969 requires all employers to be insured in respect of personal injury claims by employees.[207] Moreover, the Employers' Liability (Defective Equipment) Act 1969 imposes strict liability on employers in respect of damage caused to employees by defective equipment used at work, even if the defect was due to the actions of the third party. It is notable that both of these Acts only apply in relation to those employed under a 'contract of service'; this means that those with more precarious working arrangements, such as casual workers, are unlikely to be covered. Workers in the construction industry, where there is a high rate of casual work and self-employment, may find themselves not covered by these laws. James and Walters note the general difficulty that construction workers can face in bringing personal

202. Section 1(1).
203. B. Barrett, *Liability for Safety Offences: Is the Law Still Fatally Flawed?* 37 Indus. L.J. 117 (2008).
204. Crown Prosecution Service, *Cotswold Geotechnical Holdings Convicted of First Corporate Manslaughter Charge under New Act*, http://www.cps.gov.uk/news/press_releases/107_11/, 8 Aug. 2011.
205. http://www.shoosmiths.co.uk/news/3439.asp, 8 Aug. 2011.
206. See further, W.V.H. Rogers, *Winfield and Jolowicz on Tort*, 1025 (18th ed., Sweet & Maxwell 2010).
207. Employers can take out a policy with any authorized insurance provider (a register is maintained by the Financial Services Authority). The insurance policy must be for at least liabilities of GBP 5 million. See further: HSE, *Employers' Liability (Compulsory Insurance) Act 1969: A Guide for Employers*, http://www.hse.gov.uk/pubns/hse40.pdf, 23 Aug. 2012.

injury claims where industrial injuries and diseases were caused over time and through working for a range of different employers.[208]

[C] Reparation

While an individual might be successful in claiming damages for an injury or disease sustained at work, this will typically follow a lengthy process of litigation and it will not assist in dealing with the immediate consequences for the individual. This section explores several ways in which law provides support for individuals experiencing work-related ill-health. It looks first at social security, followed by the protection of sick workers and workers with disabilities.

[1] Social Security and Work-Related Ill-Health

In the short term, the most important social security benefit for those incapable of working is statutory sick pay.[209] This is payable after four consecutive days of incapacity for a maximum of twenty-eight weeks. The current rate is GBP 85.85 per week (around EUR 109). Statutory sick pay is paid by the employer, but in order for the employee to qualify, s/he must have average weekly earnings of at least GBP 107 (around EUR 136) during the previous eight weeks. This is the threshold at which point the employee begins to pay National Insurance Contributions. This benefit is available to those who are 'employed earners', defined as 'a person who is gainfully employed in Great Britain ... under a contract of service'.[210] One of the complexities of British law is that the definition of 'employed earner' does not necessarily coincide with the definition of 'employee', which is found in employment rights legislation. The definition of 'employed earner' used in relation to tax and social security is broader and may capture some engaged in non-standard employment, such as casual work or agency work, even though these persons might not be employees for the purposes of employment rights.[211] Those who are self-employed will not, though, be eligible for statutory sick pay. Once an employed earner has exhausted their entitlement to statutory sick pay, or if s/he was not eligible for this benefit, then s/he may be entitled to the 'Employment and Support Allowance'. This is a longer-term social security benefit which is payable both to those who have made National Insurance Contributions and to those who have not. The level of the payment is adjusted to reflect individual circumstances.[212]

---

208. P. James & D Walters, *Regulating Health and Safety at Work: An Agenda for Change?*132 (Inst. Empl. Rights 2005).
209. Part XI, Social Security Contributions and Benefits Act 1992. For full details, see *Statutory Sick Pay*, at http://www.direct.gov.uk/en/MoneyTaxAndBenefits/BenefitsTaxCreditsAndOtherSupport/index.htm, 8 Aug. 2011. I am very grateful to Julie McClelland for providing helpful information on this section.
210. Section 2(1)(a) Social Security Contributions and Benefits Act 1992.
211. S. Deakin & G. Morris, *Labour Law*, 124–125 (5th ed., Hart Publishing 2009).
212. Department of Work and Pensions, *Employment and Support Allowance: Help If You Are Ill or Disabled*, http://www.dwp.gov.uk/docs/dwp1001.pdf, 8 Aug. 2011.

There is also a specific set of social security benefits for work-related ill-health. The Industrial Injuries Disablement Benefit can be provided to 'employed earners' who have an occupational accident, or who experience disease or deafness from their work;[213] it does not extend to those who are self-employed. The injury or disease should result from events 'in the course of employment',[214] but this is designed to capture a broad range of situations. The fact that the employed earner was acting contrary to the instructions of the employer, or any legal requirement, will not prevent eligibility for the benefit so long as 'the act is done for the purposes of and in connection with the employer's trade or business'.[215] The level of the benefit provided depends on the degree of disablement; those deemed to be 100% disabled and over the age of 18 receive GBP 158.10 per week (around EUR 200).[216] There are a number of additional benefits that can be provided depending on the specific circumstances of the individual. For example, those who are 100% disabled and need daily care may be entitled to a 'Constant Attendance Allowance', as well as an 'Exceptionally Severe Disablement Allowance'.[217]

[2] The Legal Protection of Sick Workers and Workers with Disabilities

A key long-term issue for workers who experience injury or illness is the ability to return to work. British law does not impose a general duty on the employer to rehabilitate those experiencing work-related ill-health. Nevertheless, this may arise indirectly under the disability provisions of the Equality Act 2010. The first issue is to establish whether the worker's situation meets the threshold to be treated as a disability for the purposes of the Equality Act. The definition of disability can be found in section 6(1):

> A person (P) has a disability if – (a) P has a physical or mental impairment, and (b) the impairment has a substantial and long-term adverse effect on P's ability to carry out normal day-to-day activities.

Further detail on the meaning of disability can be found in Schedule 1 of the Act. This clarifies, for example, that the effect of an impairment is 'long-term' if it is likely to last more than twelve months.[218] A large volume of litigation has centred on the preliminary issue of whether or not the individual is disabled, often in relation to whether their impairment has a 'substantial' adverse effect on 'day-to-day' activities. Section 212(1) defines 'substantial' as 'more than minor or trivial', so this indicates the need for a broad interpretation.

213. Part V, Social Security Contributions and Benefits Act 1992.
214. Section 94(3), Part V, Social Security Contributions and Benefits Act 1992.
215. Section 98, Part V, Social Security Contributions and Benefits Act 1992.
216. http://www.direct.gov.uk/en/MoneyTaxAndBenefits/BenefitsTaxCreditsAndOtherSupport/Disabledpeople/DG_10018839, 23 Aug. 2012.
217. http://www.direct.gov.uk/en/MoneyTaxAndBenefits/BenefitsTaxCreditsAndOtherSupport/Disabledpeople/DG_10018839, 23 Aug. 2012.
218. Paragraph 2(1), Schedule 1, Equality Act 2010. See also, Equality Act 2010 (Disability) Regulations, SI 2010/2128.

If a person is disabled for the purposes of the Equality Act, then there is a duty on employers to make reasonable adjustments; failure to comply with this duty constitutes unlawful discrimination.[219] There are three limbs to the duty to make reasonable adjustments, the first of which is:

> a requirement, where a provision, criterion or practice of A's [the employer] puts a disabled person at a substantial disadvantage in relation to a relevant matter in comparison with persons who are not disabled, to take such steps as it is reasonable to have to take to avoid the disadvantage.[220]

The second duty is to make adjustments to physical features of the workplace that create substantial disadvantage and the third is to provide auxiliary aids, where failure to do so would result in substantial disadvantage.[221] The potential value of the duty can be seen in the leading case of *Archibald v. Fife Council*.[222] In this case, a road sweeper was left unable to walk following an operation. The House of Lords held that the duty to make reasonable adjustments required the Council to permit her to transfer to another role in the organization that she was capable of performing. While the Council had allowed her to apply for other positions, she had to compete against other internal candidates and she was repeatedly unsuccessful. The extent of the duty to make reasonable adjustments was such that they should not have required her to undergo competitive assessments in seeking to make a job transfer. Baroness Hale recognized that the duty 'necessarily entails an element of more favourable treatment'.[223]

Those workers whose injuries or illnesses constitute a disability will therefore be able to draw upon the duty to make reasonable adjustments as an instrument to assist in returning to work, even if modifications are needed to their original job. The Equality Act has a broad application, so it extends beyond those in a 'standard' employment relationship. For example, the duty to make reasonable adjustments covers contract workers supplied by another employer, and partners within a business.[224] No such obligations arise in respect of a worker who is sick, but who does not fall within the definition of disability; for example, a worker who experiences an injury from which she is expected to recover in less than twelve months. In relation to those who are employees, some protection may be derived from unfair dismissal legislation. Section 98 of the Employment Rights Act 1996 treats 'capability' as a potentially fair reason for dismissal, and this is defined as 'skill, aptitude, health or any other physical or mental quality'. It may, therefore, be legitimate for an employer to dismiss someone on grounds of ill-health. Nevertheless, for such a dismissal to be regarded as fair, the tribunals and courts have emphasized that a fair procedure needs to be followed, which will normally include dialogue with the employee, as well as a consideration of all relevant factors, such as the length of the absence.[225] In *McAdie v. Royal Bank of*

219. Section 21(2) Equality Act.
220. Section 20(3).
221. Section 20(4)–(5).
222. [2004] IRLR 651 (HL).
223. [2004] IRLR 651 (HL), para. 47.
224. Sections 41 and 44 Equality Act.
225. S. Deakin & G. Morris, *Labour Law* 460 (5th ed., Hart Publg. 2009).

Scotland, a bank worker sustained a psychiatric illness following harassment at work. The Court of Appeal approved the approach of the Employment Appeals Tribunal, which suggested that the obligations on an employer prior to dismissal were greater where the ill-health was caused by the employer's actions: 'it may, for example, be necessary in such a case to "go the extra mile" in finding alternative employment for such an employee, or to put up with a longer period of sickness absence than would otherwise be reasonable.'[226] On the facts of the case, however, the employee was on indefinite sick leave with little prospect of return, so dismissal was held to be a fair course of action. In such cases, the employee's remedy lies in a personal injury claim rather than unfair dismissal.

§11.06 TAKING STOCK OF UK LAW ON OCCUPATIONAL HEALTH AND SAFETY

UK health and safety law has several key strands. On the preventative side, there are the Health and Safety at Work Act 1974 and the accompanying Statutory Regulations. These seek to advance health and safety by imposing duties on key actors, such as employers, employees and the self-employed. In terms of substantive content, they are heavily shaped by EU legislation and this has been the main driving force in UK occupational health and safety law since the late 1980s. The HSE plays a promotional role insofar as it seeks to stimulate voluntary compliance with these duties, but it also enjoys preventative powers in the form of workplace inspections and the possibility to issue improvement and prohibition notices, or ultimately to initiate prosecutions. Worker participation is part of this preventative infrastructure, although it appears constrained by changes in industrial relations, as well as the weakness of the statutory provisions. In addition to prevention, the legal framework also includes a range of tools for responding to breaches of the legislation. The HSE has an important role to play here in bringing criminal prosecutions, but individuals can also tackle breaches of health and safety law through civil proceedings, often based in tort or contract law. While civil litigation is essentially a reactionary process, arising in response to violations of the law, undoubtedly it plays an indirect role in encouraging the prevention of injuries and ill-health at work. Employers' awareness of the risk of being sued for damages by individual employees is certainly an incentive for compliance that runs parallel to the impact of the HSE as a regulator.

At the beginning of this essay, the polemical debate today around health and safety was highlighted. This is premised on concerns that there are excessive obligations on businesses arising from the combination of statutory regulation on the one hand, and civil proceedings in tort and contract on the other. The Prime Minister, David Cameron, has declared that 'businesses are drowned in red tape, confusion and the fear of being sued for even minor accidents'.[227] Given the strength of the rhetoric, especially

226. [2007] IRLR 895, para. 37.

227. HM Government, *Common Sense Common Safety – A Report by Lord Young of Graffham to the Prime Minister following a Whitehall-Wide Review of the Operation of Health and Safety Laws and the Growth of Compensation Culture*, 5 (Cabinet Office 2010).

in the popular media, it can be difficult to form an objective picture of the reality of health and safety law. Stepping back from the politics of this issue, a range of statistics confirm that there has been progress in improving health and safety in the workplace during the past ten to fifteen years, with discernible declines in the rate of fatal and non-fatal injuries. Moreover, the performance of the UK compares relatively well to other EU Member States. The rate of fatal injuries at work is one of the lowest in the EU,[228] although the rate of serious accidents at work is simply in line with the EU-27 average.[229] Statistics in the past decade suggest that the UK has the lowest level of reported work-related stress in the EU,[230] and amongst the lowest levels of musculoskeletal disorders.[231] Such figures have to be treated with some caution; reporting systems may not be entirely equivalent. Nevertheless, the general picture that emerges from the statistical portrait might be taken as an indication that the regulatory mix in the UK is enjoying some success in achieving its objectives and that the bureaucratic burdens, insofar as they exist, may be a necessary means of delivering this record. Behind the headlines, the government itself appears to recognize that the problem, if there is one, does not lie with the legal framework. Lord Young's review proposes some alterations in the legal requirements, but these do not call into question the basic approach in the law. Indeed, his criticisms are mainly directed at health and safety *practice* and a misunderstanding of what the law demands from employers. Likewise, the Löfstedt Review concluded that 'in general, there is no case for radically altering current health and safety legislation'.[232] In any case, the close proximity of the British legal framework to the requirements of EU legislation means that there is relatively little space for the UK to dilute its laws without running the risk of contravening EU law.

Like the government, trade union criticisms are not primarily directed at the underpinning legal framework, although they have made modest proposals for it to be strengthened.[233] Their main concern also lies with health and safety *practice*, but in this instance the approach to enforcement on the part of the HSE. During the past decade, the HSE has experimented with a significant reduction in inspections and its use of 'hard' enforcement powers, such as prosecutions. So far, that has not prevented continuing declines in the rate of fatal and non-fatal injuries at work. It is, though, impossible to divine whether retaining the previous level of inspections and prosecutions would have resulted in even greater declines in injury rates. Moreover, progress

228. http://www.hse.gov.uk/statistics/european/index.htm ,accessed on 8 Aug. 2011.
229. http://www.hse.gov.uk/statistics/european/index.htm accessed on, 8 Aug. 2011.
230. Figures are from 2005: European Agency for Safety and Health at Work, *OSH in Figures: Stress at Work – Facts and Figures*, 20 (Office for Official Publications of the European Communities 2009).
231. Figures are from 2000 to 2005: European Agency for Safety and Health at Work, *OSH in Figures: Work-Related Musculoskeletal Disorders in the EU – Facts and Figures* 32–33 (Publications Office of the European Union 2010).
232. R. Löfstedt, *Reclaiming Health and Safety for All: An Independent Review of Health and Safety Legislation*, Cmnd 8219, 1 (Stationery Office 2011).
233. For example placing company directors under a specific duty to ensure health and safety: Trades Union Congress, *The Case for Health and Safety*, 2, http://www.tuc.org.uk/workplace/tuc-18421-f0.cfm, 30 Sep. 2010, 12.

in relation to tackling work-related ill-health appears to be more gradual. When questioned about this by the House of Commons Work and Pensions Committee, the then Chief Executive of the HSE accepted that there was a 'tipping point' where reductions in enforcement activity would be such that it would 'cease to have a deterrent effect'.[234] Recent evidence from the construction industry suggests that there is an ongoing need for enforcement. The HSE reported that in 2009/2010, it visited 2,014 refurbishment sites and needed to issue 691 enforcement notices.[235] In 2009, a construction industry 'blacklist' was uncovered. This elaborate system involved a private company collecting data on trade union activists within the construction sector, as well as those who raised health and safety concerns. Forty-four construction companies subscribed to this database and apparently used it to avoid recruitment of those on the list.[236] While this example spurred the previous government to prohibit blacklisting,[237] it provides a salutary reminder that many employers do not have an altruistic commitment to health and safety and there is a need for enforcement to remain an active element in the regulatory toolbox. Given the extent to which enforcement has already been scaled back in the past decade, it would appear highly risky to proceed with yet further reductions. Nevertheless, the deep cuts to the HSE's budget during the next three years appear to make such reductions an inevitability.

The combination of a deregulatory government agenda and reduced public spending on health and safety poses challenges for those seeking to ensure further improvements in occupational health. One avenue that might offer a way through this impasse is to place greater priority on worker participation. A range of research confirms a correlation between stronger worker participation and better health and safety performance.[238] Moreover, worker participation offers an alternative regulatory strategy to the external oversight of the HSE. If the latter is going to lack the resources to revitalize its enforcement activities, then worker participation may hold an attraction as means of promoting inspection and monitoring from *within* the organization. The barriers to enhancing worker participation are, though, significant. In particular, the low level of unionization in the private sector imposes a need to ensure structures for worker participation that deal with both union and non-union workplaces. The current legal framework goes some way to addressing the former, but remains weak and fragile in relation to the latter. This is compounded by the apparent reluctance of the HSE to use its enforcement powers in relation to the legal requirements on worker

234. House of Commons Work and Pensions Committee, *The Role of the Health and Safety Commission and the Health and Safety Executive in Regulating Workplace Health and Safety*, Third Report of Session 2007–2008, vol. 1, HC 246, 26 (Stationery Office 2008).
235. HSE, *The Health and Safety Executive: Annual Reports and Accounts 2009/10*, HC20, 18 (Stationery Office 2010).
236. House of Commons Work and Pensions Committee, *Workplace Health and Safety: Follow-Up Report*, Fourth Report of Session 2008–2009, paras 55–60.
237. The Employment Relations Act 1999 (Blacklists) Regulations 2010, SI 2010/493.
238. House of Commons Work and Pensions Committee, *The Role of the Health and Safety Commission and the Health and Safety Executive in Regulating Workplace Health and Safety*, Third Report of Session 2007–2008, vol. 1, HC 246, 51 (Stationery Office 2008); D. Walters et al., *The Role and Effectiveness of Safety Representatives in Influencing Workplace Health and Safety*, Research Report 363, x (HSE 2005).

participation. Moreover, the proliferation of non-standard forms of work, such as casual and agency work, poses a challenge to traditional forms of worker participation, which tend to focus on employees.

In conclusion, the outlook for health and safety law, and its enforcement, seems uncertain in the light of current political and economic pressures. Yet weakening the statutory infrastructure for health and safety might have the perverse effect of compelling workers to rely more on the protection derived via personal injury litigation based on tort law. While this can deliver compensation after the event, it is costly and time-consuming for both workers and employers, as well as feeding popular perceptions about an excessively litigious culture. A preventative approach, such as that based on worker participation and external inspection and enforcement, remains a more constructive model to pursue.

BIBLIOGRAPHY

Achur, J. *Trade Union Membership 2009*. London: Department for Business, Innovation and Skills, 2010.

Barrett, B. 'Liability for Safety Offences: Is the Law Still Fatally Flawed?'. *Industrial Law Journal* 37 (2008): 100–118.

Barrett, B. 'The Health and Safety (Offences) Act 2008: The Cost of Behaving Dangerously at the Workplace'. *Industrial Law Journal* 38 (2009): 73–79.

BBC. 'Potters Bar crash: Network Rail Fined £3m'. <http://www.bbc.co.uk/news/uk-13389147>, 13 May 2011.

Brodie, D. 'Away Days and Employers' Liability: *Reynolds v. Strutt and Parker*'. *Industrial Law Journal* 41 (2012): 93–97.

Cotter, B & D. Bennett (eds). *Munkman on Employer's Liability*. 14th edn. London: LexisNexis, 2006.

Crown Prosecution Service. 'Cotswold Geotechnical Holdings Convicted of First Corporate Manslaughter Charge under New Act'. <http://www.cps.gov.uk/news/press_releases/107_11/>, 8 August 2011.

Deakin, S. & G. Morris. *Labour Law*. 5th edn. Oxford: Hart Publishing, 2009.

Department of Work and Pensions. *The Government Response to the Rita Donaghy Report: One Death Is Too Many – Inquiry into the Underlying Causes of Construction Fatal Accidents*. London: The Stationery Office, 2010.

Department of Work and Pensions. 'Good Health and Safety, Good for Everyone'. <http://www.dwp.gov.uk/policy/health-and-safety>, 21 March 2011.

Department of Work and Pensions. 'Employment and Support Allowance: Help If You Are Ill or Disabled'. <http://www.dwp.gov.uk/docs/dwp1001.pdf>, accessed on 8 August 2011.

Department for Work and Pensions. 'The Government Response to the Löfstedt Report'. <http://www.dwp.gov.uk/docs/lofstedt-report-response.pdf>, November 2011.

European Agency for Safety and Health at Work. *OSH in Figures: Stress at Work – Facts and Figures*. Luxembourg: Office for Official Publications of the European Communities, 2009.

European Agency for Safety and Health at Work. *OSH in Figures: Work-Related Musculoskeletal Disorders in the EU – Facts and Figures*. Luxembourg: Publications Office of the European Union, 2010.

Ford, M. & J. Clarke. *Redgrave's Health and Safety*. 5th edn. London: LexisNexis Butterworths, 2007.

Hampton, P. *Reducing Administrative Burdens: Effective Inspection and Enforcement*. London: HMSO, 2005.

HM Government. *The Coalition: Our Programme for Government*. London: Cabinet Office, 2010.

HM Government. *Common Sense Common Safety – A Report by Lord Young of Graffham to the Prime Minister following a Whitehall-Wide Review of the Operation of Health and Safety Laws and the Growth of Compensation Culture*. London: Cabinet Office, 2010.

Health and Safety Commission. *Performance Report 2007*. Suffolk: HSC, 2007.

Health and Safety Executive. *The Health and Safety Executive: Annual Reports and Accounts 2009/10*. HC20. London: The Stationery Office, 2010.

HSE. *The Health and Safety Executive Annual Report and Accounts 2010/11*. HC1066. London: The Stationery Office, 2011.

HSE. *The Health and Safety of Great Britain – Be Part of the Solution*. http://www.hse.gov.uk/strategy/strategy09.pdf, accessed on 9 Mar. 2011.

HSE. *The Health and Safety Executive: Annual Reports and Accounts 2011/12*. HC204. London: The Stationery Office, 2012.

HSE. 'Employers' Liability (Compulsory Insurance) Act 1969: A Guide for Employers'. <http://www.hse.gov.uk/pubns/hse40.pdf>, 23 Aug. 2012.

House of Commons Work and Pensions Committee. 'The Role of the Health and Safety Commission and the Health and Safety Executive in Regulating Workplace Health and Safety'. Third Report of Session 2007–2008. Vol. 1. HC 246. London: The Stationery Office, 2008.

House of Commons Work and Pensions Committee. 'Workplace Health and Safety: Follow-Up Report'. Fourth Report of Session 2008–2009.

Howes, V. 'Workers' Involvement in Health and Safety Management and Beyond: The UK Case'. *International Journal of Comparative Labour Law and Industrial Relations* 23 (2007): 245–265.

Howes, V. 'Duties and Liabilities under the Health and Safety at Work Act 1974: A Step Forward?'. *Industrial Law Journal* 38 (2009): 306–317.

James, P. & D. Walters. *Regulating Health and Safety at Work: An Agenda for Change?*. London: Institute for Employment Rights, 2005.

Langstaff, J. 'The Employer's Duty of Care'. In *Munkman on Employer's Liability*, edited by B. Cotter & D. Bennett. 14th edn. London: LexisNexis, 2006.

Pitt, G. *Employment Law*. 7th edn. London: Sweet & Maxwell, 2009.

Löfstedt, R. *Reclaiming Health and Safety for All: An Independent Review of Health and Safety Legislation*. Cmnd 8219. London: The Stationery Office, 2011.

Robens, Lord. *Safety and Health at Work: Report of the Committee 197–1972*. Cmnd 5034. London: HMSO, 1972.

Rogers, W. *Winfield and Jolowicz on Tort*. 18th edn. London: Sweet & Maxwell, 2010.

Smith, I., et al. *Health and Safety: The Modern Legal Framework*. 2nd edn. London: Butterworths, 2000.

Tombs, S. & D. Whyte. *Regulatory Surrender: Death, Injury and the Non-enforcement of Law*. Liverpool: Institute for Employment Rights, 2010.

Tombs, S. & D. Whyte. 'Health and Safety Gone Mad?'. Institute for Employment Rights Briefing, August 2010. http://www.ier.org.uk/sites/ier.org.uk/files/H&S%20gone%20mad%20briefing.pdf, 2 March 2011.

Trades Union Congress. 'The Case for Health and Safety'. http://www.tuc.org.uk/workplace/tuc-18421-f0.cfm, 30 September 2010.

Walters, D., et al. 'The Role and Effectiveness of Safety Representatives in Influencing Workplace Health and Safety'. Research Report 363. Suffolk: HSE, 2005.

Wilby, D. 'The General Principles of Negligence'. In *Munkman on Employer's Liability*, edited by B. Cotter & D. Bennett. 14th edn. London: LexisNexis, 2006.

Wright, F. *Law of Health and Safety at Work*. London: Sweet & Maxwell, 1997.

CHAPTER 12
Occupational Health and Safety: A Comparative Perspective

Edoardo Ales

§12.01 OCCUPATIONAL HEALTH AND SAFETY LAW: THE BIG PICTURE

In all the countries analysed in this book, the protection of occupational health and safety has a longstanding tradition, dating back, in many cases, to the second half of the nineteenth century. That was the time in which regulations on the exercise of industry (like in Austria, Germany and Hungary) were adopted obliging the entrepreneur to guarantee a healthy and safe functioning of the productive activity, working conditions included. Moreover, since the beginning of the Twentieth century several regulations were adopted in all the countries in order to reduce the danger produced to occupational health and safety by the excessive working time for young workers and women, especially as far as night-work was concerned. After that, general regulations on maximum daily and/or weekly working hours were introduced too, confirming the relevance of such provisions in the health and safety domain.

In the same period, a general safety obligation in any employment relationship was recognized within Civil Codes (like in Germany, Italy and The Netherlands), general employment legislation (Belgium), specific occupational accident regulations (Spain) or by case law (implied duties of the employer: United Kingdom).

More recently, although in a few cases (Italy and Spain), such recognition has been backed up by constitutional provisions affirming a general right to health.

From a practical point of view, the safety obligation has been, at first, realized through specific, detailed, rule and protection oriented regulations. According to these, in case of activities regarded as risky, the employer was – and in some case still is – obliged to provide the worker with individual tools protecting him or her from harmful events that can derive from the exposure to a source of danger. According to this approach the employer is not obliged to intervene on the source of danger in order to eliminate or reduce it.

From the 1980s on, there has been a move from that specific, detailed, rule and protection oriented regulation to a more general, goal and prevention oriented framework legislation. This has meant the establishment of procedural rules that the employer has to respect in order to reach the final goal of a healthy and safe working environment. In such a perspective, a major role is played by the so-called risk assessment that has to be carried out by the entrepreneur in each working activity in order to eliminate or, where impossible, to reduce dangers and risks. Therefore individual protection tools have to be seen as a last resort, since the main idea is to intervene drastically on the sources of danger. However, detailed prevention rules still play a relevant role.

A further crucial development refers to the definition of the healthy and safe workplace as an environment in which the physical, psychical and psychosocial well-being of the worker is guaranteed (what we can call the comprehensive approach to health and safety). This has led to emphasize the concept of human dignity at work (the Dutch and Hungarian 'humanization' of working conditions) and to establish a direct link between occupational health and safety, on the one hand, and the fight against harassment at work, on the other.

From the point of view of the harmful consequences produced on the worker by an unhealthy and unsafe work environment, it has to be stressed that, with the exception of the United Kingdom, all the countries have provided, at least since the beginning of the Nineteenth century, a mandatory social insurance coverage which exonerate the employer from civil liability in case of occupational accidents or diseases occurred to the worker. To this end, the employer was and still is obliged to pay contribution to the social insurance which will, at its turn, provide benefits in cash and in kind to the injured or ill employee.

To cope with the disincentive that such an exoneration is likely to produce in terms of employers' commitment on the prevention of occupational accidents and diseases, the solution of establishing a link between the level of contribution and the rate of accidents and diseases in the company has been adopted in many countries. Moreover social insurance bodies can act in regress against the employer if the harmful event has been caused by his or her faulty or intentional violation of health and safety regulations.

In many countries, above all in case of heavy fault of the employer (see, for instance, the *faute inexcusable* in France and in Belgium) linked to the patent violation of health and safety prescriptions, the worker can award in court a supplementary compensation for the part of the damage which exceeds the amount of the social insurance benefit. In Spain, the law provides that when the accident occurs as a result of insufficient safety measures, the employer is obliged to pay in addition between 30% and 50% of the benefit. In these cases, the benefit is divided into two parts: one being the ordinary benefit supplied by the social security system, and the other being the extra benefit. In order for the extra benefit to be authorized there must be a strict causal relationship between the lack of safety measures and the accident. The decision to grant the extra benefit lies with the social security authorities.

In the same perspective, the worker can act against the employer for kind of damages not covered by the social insurance or against the same social insurance in order to get an ill-health covered and indemnified. In Germany, bipartite bodies established at sectoral level which manage in a self-administration regime the social insurance against occupational accident and diseases (*Berufsgenossenschaften*) are obliged by the law to issue accident prevention rules (*Unfallverhütungsvorschriften*). Such rules are binding for employers and employees due to the fact that *Berufsgenossenschaften* are established by the law.

Last but not least, a link has been established between health and safety regulations on the one hand and sickness and disability law on the other, since the latter, in case the harmful event has produced a reduction of the work ability, imposes on the employer a duty to reintegrate the employee in the same or in another suitable position within the company or even, in certain cases, within the group.

§12.02 NATURE AND SCOPE OF THE HEALTH AND SAFETY OBLIGATIONS

Health and safety obligations are established by the law, so, prima facie, one may say that they are public law obligations. Nevertheless, they are usually fed into the employment relationship as general and/or specific contractual obligations of the parties. For instance, according to the German *Transformationslehre*, echoed by the Swiss doctrine of the *Rezeptionsklausel*,[1] duties provided under public law can be transformed into (or regarded as) private law/contractual obligations if they can constitute the content of the contract. In the same perspective, according to Italian law, the contract obliges the parties not only to what is explicitly prescribed within it but also to all the consequences the law derives from its conclusion (Article 1374 Civil Code – principle of integration of the contract).

By consequence: (a) the worker can bring action in contract against the employer for the fulfilment of the safety obligation(s) before an harmful event occurs; (b) the harmful event gives rise to a right to compensation in contract; (c) being also legal in nature, the action against the violation of safety obligation(s) can be brought by the worker also in tort; (d) compensation can be awarded also in tort.

The adoption of the goal oriented framework legislation summarily described in the above, has posed the problem of the scope of the safety obligation for the employer. Indeed, being such framework legislation mainly procedural in nature, it is up to the employer to find out the technical solution which suits best to each situation where the safety obligation has to be fulfilled. Therefore, with the exception of the residual but

1. 'Bundesgesetz betreffend die Ergänzung des Schweizerischen Zivilgesetzbuches' (Fünfter Teil: Obligationenrecht) Zweite Abteilung: Die einzelnen Vertragsverhältnisse; Zehnter Titel: Der Arbeitsvertrag; Erster Abschnitt: Der Einzelarbeitsvertrag; Art. 342 I. Vorbehalt und zivilrechtliche Wirkungen des öffentlichen Rechts: '2 Wird durch Vorschriften des Bundes oder der Kantone über die Arbeit und die Berufsbildung dem Arbeitgeber oder dem Arbeitnehmer eine öffentlich-rechtliche Verpflichtung auferlegt, so steht der andern Vertragspartei ein zivilrechtlicher Anspruch auf Erfüllung zu, wenn die Verpflichtung Inhalt des Einzelarbeitsvertrages sein könnte.'

still relevant cases in which specific prevention and protection measures are imposed by the law, one may say that the entrepreneur enjoys a discretionary power in the fulfilment of the safety obligation. Such a power can be traced back to the fact that the entrepreneur is responsible for the work organization. In such a perspective the safety obligation cannot be seen as an obligation of means: it is rather regarded as a general and, sometimes, unlimited obligation of result (like in France, Italy, Spain and The Netherlands) to keep the worker healthy and safe.

This is confirmed by the fact that even if labour and health authorities or bilateral (social insurance) bodies have verified the respect, in a certain moment, of general and detailed health and safety regulations, the entrepreneur's liability for damages derived to the worker from an harmful, work-related event is not excluded per se. This depends both from the fact that the fulfilment of the safety obligation is dynamic in nature, both from the consideration that an harmful event has in any case occurred and a damage has derived to the worker, a damage that is not always covered or covered to its full extent by the social insurance.

The question is, then, if the entrepreneur bears a no-fault liability or is admitted to prove that what has been done in terms of prevention and protection within the company excludes his or her liability. In all the countries analysed in this book liability is linked to fault (or intent of course). At its turn, fault is recognized in case the entrepreneur has violated precise prevention and protection rules. However, the real point is to find out which is the level of prevention and protection that has to be realized by the entrepreneur in order to see his or her liability excluded in case of lack of precise prevention or protection rules. This is crucial also in the perspective of criminal liability which may derive from the mere violation of precise prevention and protection rules, from a violation of those rules that has produced a harmful event to the worker and from the simple fact that a harmful event has occurred to the worker also in absence of violation of specific rules.

Indeed, the lack of specific binding rules comparable to the German *Unfallverhütungsvorschriften*, to the Swedish Provisions issued by the Work Environment Authority or to the British Regulations adopted under the auspices of the Health and Safety at Work Act, does not mean that the are no instruments available to help the entrepreneur in the proper fulfilment of the safety obligation(s). This is the case, for instance, of the Technical Notes issued by the Spanish National Institute of Occupational Safety and Health, the self-regulations adopted by employers and employees at sectoral level (The Netherlands: the so-called ARBO Catalogues), the private standards (The Netherlands) or the Workplace Health and Safety Management Systems certified by bilateral bodies (Italy).

The adoption of all those instruments may succour the entrepreneur when s/he is called to prove in court the proper fulfilment of the safety obligation. As a matter of fact, case law plays a crucial and ultimate role in defining the scope of the safety obligation(s) although judicial approaches highly differ from country to country.

In the United Kingdom, for instance, judges have adopted a restrictive approach to the safety obligation accepting that it is relevant in assessing the scope of the entrepreneur's liability to take into account the likelihood of a particular risk materializing (the so-called reasonable foreseeability of the harmful event) and that it has to be determined whether the entrepreneur had taken all reasonably practicable steps to avoid the risk (the so-called reasonably practicable test of the preventive or protective measure). However, it must be borne in mind that the burden of proof when asserting that a certain step was *not* reasonably practicable lies on the defendant, i.e., the entrepreneur. As a matter of fact, section 40 HSWA, adopted under the influence of the above mentioned case law, is an example of a reversed burden of proof in the criminal law: 'it shall be for the accused to prove that it was not practicable or not reasonably practicable to do more than was in fact done to satisfy the duty or requirement'.

On the contrary, in Italy, judges have adopted an extensive approach to the safety obligation according to which the employer has to prove that s/he has used the diligence required in the specific case, i.e., by implementing the most advanced technology available at sectoral level (in some decision even outside the relevant sector) and using his or her experience in the field. However, it has to be borne in mind that the burden of proof when assessing that more advanced technologies were available to the employer lies on the claimant, i.e., the worker.

It goes without saying that the employer may not be able to cope alone with such a wide and multifaceted obligation. For this reason, the goal oriented framework legislation has multiplied the number of figures and bodies who shall assist the single employer in the proper fulfilment of the safety obligations. All those, however, do not bear any civil, criminal or administrative liability, since they act in support of the entrepreneur and not on his or her behalf (see below). A different case is that one in which the entrepreneur formally delegates managers or executive employees to the fulfilment of (part of) the safety obligation(s) by assigning them organizational powers and economic means adequate to the task (see below).

§12.03 BASIC NOTIONS

In order to get into the structure of the health and safety discourse it is useful to draw attention to some basic notions, first of all the notion of 'danger' in connection with that one of 'risk'. Indeed, with the exception of Italy, in all the countries analysed in this book as well as in EU Law these two notions overlap, creating a super-concept of 'risk' which includes the concept of 'danger'. Only the Italian legislator makes a distinction which, in our view, is worth to be emphasized. According to Italian Law, 'danger' has to be regarded as the capacity that a given entity has to cause a worker harm, whereas 'risk' pinpoints the probability that an harmful event caused by that danger occurs to the worker.

Therefore, in theory, the evaluation of danger should precede the evaluation of risk. In practice, also those evaluations overlap. Indeed legislators, the Italian included, when establishing entrepreneur's duties, insist on the evaluation of risk assuming that a risk can be eliminated or reduced by means of protection tools without the need of eliminating the relevant source of danger. In such a perspective, the overlapping of the notions as well as of the evaluations of danger and risk is likely to put into question the notion of safety which emphasizes the preventative approach and which regards protection as last resort to be used only exceptionally.

Furthermore, as we will see immediately after, such an overlapping approach to danger and risk, already hardly acceptable in case of physical risks, becomes even more questionable when referred to psychical and psychosocial risks in which the source of danger is represented by a bad work organization (excessive working time, hardness at work, monotonous and solitary work) or by a human behaviour (moral or sexual harassment).

Indeed, by making reference to the notion of health provided by the World Health Organization (WHO) as 'a state of complete physical, psychical and social well-being', legislators have widen the scenario of the safety obligation from the narrow perspective of the tangible harmful event which produces a physical ill-health to the worker to a more comprehensive view in which also psychical and psychosocial damages which can derive from a tangible harmful event or psychical and psychosocial risks which originate in a bad work organization or in a human misbehaviour are taken into account.

Such a dramatic change is also likely to put into question the very scope of the social insurance protection traditionally limited to occupational accident and diseases. These are defined as such by the fact that an harmful event linked to (in case of occupational accident) or caused by (in case of occupational disease) the working activity has produced the loss or a significant reduction of the working capacity, leading to a permanent (partial or total) or temporary (total) inability, which, at its turn, has produced a reduction or a withdrawn of the earning capacity of the worker, damaging him or her from an economic point of view. It is rather clear that in the view of the new comprehensive approach described in the above many factors of danger and many harmful events will not be covered by the traditional definition of occupational accident and disease. Therefore, alternative solutions have to be found either inside the same social insurance system, by enlarging its scope of application to situations in which the psychophysical and psychosocial well-being of the worker is at stake, or outside it, by recognizing the compensation of damages not related to the loss or the reduction of the working as well as the earning capacity (damages to health, reputation, social life, etc.).

In such a double labour and social insurance law perspective, case law above all but also legislators and the doctrine have played a decisive role in updating the legal discourse by looking at already existing phenomena in terms of new factors of danger (risk) and of possible sources of harmful events.

This has been the case of musculoskeletal disorders deriving from an incorrect posture to which the worker is obliged during his or her work. Musculoskeletal disorders have been admitted by case law to the coverage of the social insurance as occupational diseases or to damage compensation. In the same perspective Italian case law has recently recognized the right to compensation also of damages not related to the loss or the reduction of the working and earning capacity. As a consequence, the Italian legislator has decided to extend the scope of application of the social insurance to the psychical and psychosocial consequences of occupational accidents or diseases, thus going beyond the principle that only the worker whose working and earnings capacity has been reduced or withdrawn can be entitled to social insurance benefits.

As far as non-physical risks are concerned, the most recent health and safety debate has focused on the concept of psychosocial risk. Indeed what happened with 'risk' – a super-concept which includes the concepts of 'danger' and 'risk', has happened also with 'psychosocial risks' – a super-concept which includes the concepts of 'psychophysical risk' and 'psychosocial risk'. As a matter of fact, the comparative analysis shows that a distinction is made in legislation and case law, between risk related to the work organization which are likely to produce negative consequences on the psychophysical sphere of the worker and risks related to behaviour coming either from the employer or from other employees which are likely to produce negative consequences on the emotional and relational sphere of the worker.

Therefore, in our view, the first should be defined as 'psychophysical' and the second as 'psychosocial risks'.

Practically speaking, according to this distinction, excessive working time, hardness at work (the French *pénibilité au travail*), monotonous and solitary work, night-work fall within the scope of 'psychophysical risk' whereas moral harassment (also called mobbing in Austria, Italy and The Netherlands, for instance or victimization in Sweden) or sexual harassment fall within the scope of 'psychosocial risk'.

Indeed, in many countries, the legislature and the judiciary have adopted a very pragmatic approach towards 'psychophysical' and 'psychosocial' risks by regarding as unlawful, often in the name of health and safety, the above mentioned organizational solutions and individual or collective behaviour. By consequence, further to the penal and administrative sanctions which usually assist legislative provisions, such misconducts constitute themselves a breach of the safety obligation and as such give rise to the right to compensation in favour of the affected worker.

Such a pragmatic approach to non-physical risks has made also possible to get around the unsolved and perhaps unsolvable problem of the recognition of work-related stress as psychophysical consequence of a bad work organization or of unlawful behaviour. Indeed, since no medical evidence proves the link among the work organization or the behaviour, the unease of the worker and his or her physical or psychical ill-health, stress is mainly regarded (also by the European Social Dialogue) as the product of a subjectively distorted perception of the work organization or of the behaviour by the worker. In such a perspective, the employer can hardly be held liable for the breach of the safety obligation since the reasons of worker's ill-health lie mainly in the worker him or herself.

The fact that already existing phenomena are now regarded as new factors of risk (or sources of danger) has obliged (legislators, above all) to define their very notion in order to make them subject to sanction and/or to compensation. Some definitions are worth to be reported from the French legislation which seems to be one of the more advanced in the view of protecting workers from psychophysical and psychosocial risks.

Let us start with moral harassment (*harcèlement moral*). According to Article L 1152-1 of the French Labour Code, 'workers shall not bear repeated behaviour aiming at, or resulting in, a deterioration of their working conditions capable to alter their rights or dignity, their physical or psychical health or to jeopardize their professional future'. Case law plays a crucial role in interpreting such a broad concept, since it is difficult sometimes to distinguish moral harassment from a highly demanding work organization. However, the large number of cases brought to court under this provision shows quite clearly that a lot of workers in France feel like facing moral harassment. The recent increase of the number of suicides at work is attesting the reality of the problem. According to Article 222-33-22 of the French Penal Code, 'the fact of harassing someone through repeated behaviour aiming at, or resulting in, a deterioration of his or her working conditions capable to alter his or her rights or dignity, his or her physical or psychical health or to jeopardize his or her professional future is punished with two years of imprisonment and a 30.000 euro fine'. Compensation of damages is admitted too.

A second, interesting example is that one of sexual harassment. Before 2012, the French legislation did not provide a detailed definition of it, just prohibiting harassing behaviour aimed at obtaining sexual favours for the harasser or for a third person (former Article L 1153-1 LC). Nevertheless, sexual harassment was punished with one year of imprisonment and EUR 15,000 of fine (former Article 222-33 Penal Code). In May 2012 the French Constitutional Court declared unconstitutional Article 222-33 Penal Code since it allowed the punishment of sexual harassment without defining it precisely, thus infringing the principle of legality (no crime without precise law). Therefore, after that decision, sexual harassment, if so assessed by a court, might have only given rise to compensation under civil law.

The decision of the Constitutional Court has obliged the French legislator to amend both the above mentioned labour and penal provisions (by Act n. 2012-954 of 6 August 2012), introducing a detailed definition of sexual harassment regarded as 'any repeated proposal or behaviour related to sex which either endangers workers' dignity due to its humiliating or degrading character or create for the workers an intimidating, hostile or offensive situation'. Sexual harassment is prohibited, as well as facts which can be assimilated to it like 'forms of heavy pressure, also isolated, exercised with the real or apparent goal of obtaining an act of sexual nature in favour of the harasser or of a third person' (new Article l 1153-1 LC).

Sexual harassment or sexual harassment-like facts are now punished with two years of imprisonment and EUR 30,000 fine (new Article 222-33 Penal Code). Compensation of damages is admitted too.

A third, crucial example is that one of the above mentioned *pénibilité au travail* that can be translated as 'hardness at work'. According to the French legislator, within the framework of the safety obligation, the employer has to adopt any measure necessary to protect the physical and psychical health of the worker, including actions aimed to prevent professional risks and *hardness at work* (Article L 4121-1 LC). Therefore hardness at work is something different from a traditional professional risk. Indeed, according to Article L 4121-3-1 LC, hardness at work is a condition which derives from the exposure to a sum of professional risks specified by decree which can be traced back to strong physical constraints, to an aggressive physical environment, or to certain rhythms of work likely to produce durable, identifiable and irreversible effects for the health of the worker. The employer has to take written note of the periods in which the worker has faced such a condition.

According to Articles L 138-29 and L 138-30 Social Security Code, if more of the half of the employees is facing hardness at work (Article D 138-26 Social Security Code), companies employing at least fifty workers have to reach an agreement or to adopt an action plan aimed to the prevention of hardness. The agreement or the action plan have to provide for the reduction of hardness and the adaptation of the workplace to the worker (Article D 138-27 and D 138-28 Social Security Code). The agreement or the action plan have to be renewed every three years. A significant administrative fine is provided in case the employer does not sign an agreement or does not adopt an action plan.

§12.04 PERSONAL SCOPE OF APPLICATION OF THE HEALTH AND SAFETY LEGISLATION

Comparative analysis shows that, even though still referring to it formally, national legislators have gone beyond the notion of 'employer' in order to indicate who is bound by the health and safety legislation. In such a broader perspective, one may say that *anyone who organizes a productive activity and/or exercises managerial prerogatives* falls under the scope of the health and safety legislation in the sense that s/he will be held responsible for the application of the prescribed preventive and protective measures within the general framework of the safety obligation.

Practically speaking, this means: (a) arrangement of a healthy and safe work environment for anyone who works within his or her premises even outside a traditional employment relationship (like temporary workers, self-employed, students in laboratories, interns, inmates, volunteers); (b) control of the respect of the health and safety regulations by anyone who works within his or her premises even outside a traditional employment relationship; (c) control of the application of the health and safety legislation by contractors and subcontractors, being them companies or self-employed workers; (d) assessment and coordination of the prevention of risk of interference deriving from the simultaneous presence in the same site of various contractors and subcontractors performing different activities.

Therefore, leaving aside the formal references still made by national legislations to the 'employer', the expression 'anyone who organizes a productive activity and/or exercises the managerial prerogatives' seems to be very much closer to 'entrepreneur'. The same can be said with reference to the expression 'anyone who works within the premises of somebody who organizes a productive activity even outside a traditional employment relationship', which seems to be very much closer to 'worker' than to 'employee'. For sure, 'entrepreneur' and 'worker' are more far-reaching than 'employer' and 'employee' in the domain of occupational health and safety law.

The reference to 'anyone who organizes a productive activity and/or exercises the managerial prerogatives' explains also the fact that further to the entrepreneur, being it a legal or natural person, which is the prime responsible for health and safety, also the so-called hierarchical structure can be bound by and liable for the application of the health and safety legislation.

In such a perspective, a paramount example comes from Belgian legislation. According to it, persons such as foremen, managers, supervisors, team leaders, heads of departments who have the authority to give orders and assignments are all members of the hierarchical structure. So the hierarchical structure consists of all persons who have authority over one or more persons or a department. Notwithstanding the fact that there is no legal definition of the term, which is only mentioned in the Royal decree on well-being at work, the legislator has assigned a number of specific tasks to the hierarchical structure. As a matter of fact, because of its specific position, the hierarchical structure is able to play a role in the making of the well-being policy, the planning, the implementation and assessment of the prevention measures.

Therefore, even if the entrepreneur is the prime responsible for the well-being of the workers in the workplace, the hierarchal structure is liable for the implementation of the company's well-being policy as defined by the entrepreneur. In other words the hierarchical structure has to take care of the daily management of the well-being of the workers.

Apart from the case in which the legislator directly assigns specific functions to the hierarchical structure, the allocation of tasks and related responsibilities remains with the entrepreneur who has the possibility and sometimes the obligation (like in Sweden) to delegate functions, duties and powers to members of the hierarchical structure. Delegation shall be in written and shall be clear and precise as for the definition of all its elements. In short, one may say that the delegated person in order to be held liable for the implementation of the health and safety regulations shall have enough authority, resources and knowledge. The delegated person has to accept the delegation explicitly, most of the times in writing.

Controversial are the consequences that the delegation may produce in terms of residual criminal and civil liability of the delegating person (the entrepreneur, above all). To avoid that the claimant does not get the compensation in cases in which the amount of damage to be paid is so high that the delegated person is not able to cope with it whilst the company would be, German legislation affirms that the civil liability remains with the entrepreneur also in case of delegation (§ 278 BGB). Another solution is that one adopted by the Italian legislator which provides that delegation does not, of itself, exclude the entrepreneur's duty to ensure that the transferred functions are carried out correctly. In Austria, the so-called liable authorized representative will be held responsible for not having taken appropriate and sufficient measures according to the provisions of the ASchG. However, the employer will still be held liable in case of wilful inducement or non-prevention of infringement.

Exceptions to the application of the health and safety legislation are provided for (parts of) the public sector in relation to the peculiarity of the tasks performed by the workers i.e., activities in which risk is an essential and inevitable component (policeman, fireman, soldiers, etc.). Apart from these activities, the general trend seems to be that one of applying the legislation on health and safety also to the public sector without distinctions linked to the nature of the employment relationship (being it private or public).

Another relevant trend is that one of providing higher standards of protection in favour of 'especially sensitive worker' i.e., to quote the Spanish legislation, 'any worker whose personal characteristics lead to a condition of physical, mental or sensory impairment, which makes him or her particularly vulnerable to occupational risks'. Thus the subjective condition of the worker takes precedence over the conditions or circumstances of the job itself which has to be adapted to the 'especially sensitive worker'. These are, generally speaking, young, protected women and disable. In such cases the general safety obligation of the employer involves the adoption of measures with a more specific and intense character.

An opposite trend has to be highlighted with reference to domestic workers and home-workers. Generally speaking they are excluded from the scope of application of the health and safety legislation (with the exception of The Netherlands as far as home-workers are concerned). In Spain the householder is generically obliged to ensure that the work of domestics is carried out under suitable health and safety conditions and such obligation seems not to be meaningless. The respect of health and safety regulations is sometimes provided by the law (in Austria, for instance) or by collective agreement (in Italy, for instance) in case of that particular form of homework called telework, above all when it is performed within premises which belong to the company although placed outside its headquarters. This is in line with the Framework Agreement on Telework signed by the European Social Partners on the 16 July 2002.

Last but not least, one has to stress the fact that also anyone who finds him or herself within or nearby a productive site without being a 'worker', like costumers, visitors, people walking by and people who live nearby, falls under the scope of the health and safety legislation. This means that while designing preventive and protective measures, the entrepreneur has to take into account the fact that the above mentioned persons may be exposed to dangers and risks present at the workplace. Therefore, in many countries civil and criminal liability of the entrepreneur are extended also to non workers and even to persons who occasionally or permanently find themselves near the productive site. In such cases, occupational health and safety law intertwines with penal environmental law.

§12.05 SPECIFIC DUTIES AND PREROGATIVES OF THE PARTIES INVOLVED

As we have pointed out in the above, in order to make the safety obligation effective, specific duties have been imposed on the parties involved i.e., the entrepreneur and the worker. It goes without saying that in setting such duties national legislators have been deeply influenced by EU Law. As for the general framework, this has happened for risks evaluation, the establishment of the protection and prevention service, the duty to inform, train and consult workers or their representatives on health and safety issues. Nevertheless, in some cases, duties imposed by national legislators go beyond what is required by EU health and safety law also because, being the latter provided through directives, the usual room has been left to national legislators in its implementation. However, the purpose of this chapter is neither to describe nor to check the proper implementation of EU health and safety law. Its purpose is to draw a comparative picture of national health and safety legislations. Therefore, dealing with the specific duties imposed to entrepreneurs and workers, we will provide some examples of provisions which either go beyond the mere reproduction of EU law or are not related to it at all.

[A] The Entrepreneur

Starting from the entrepreneur, it will be interesting to highlight some duties which confirm the systematic approach that has been adopted by national legislator to the fulfilment of the safety obligation.

[1] The Obligation to Carry Out a Systematic Work Environment Management

A first example is the obligation to carry out a Systematic Work Environment Management (hereafter also SWEM) which is one of the major changes introduced by the Swedish legislator in 1991 to the Work Environment Act. That obligation has been specified in the Provision called 'Systematic Work Environment Management' (AFS 2001:1). The idea – which comes from Norway, is that employers must approach work environment issues as part of their daily responsibilities. This includes designing a work environment policy, making risk assessments, allocating tasks with regard to health and safety, and drawing up and following a work environment action plan including follow-up routines. All of this should be set out in writing. Employers shall also investigate all injuries occurred at work and constantly monitor possible hazards in the work environment and take measures to neutralize them. When measures cannot be taken immediately, the employer must draw up an action plan including a prepared scheme of job adaptation and rehabilitation, in compliance with the Social Insurance Code and the Work Environment Act. The employer shall make an annual follow-up inspection to observe and record the effects of SWEM. It is the employer's duty to organize job adaptation and reintegration as part of the SWEM.

A kind of SWEM is implicitly or explicitly (see Italy for instance) provided in all the countries analysed in this book. A common problem is that one of the scope of application which is often limited to medium-large size companies.

[2] The Establishment of the Prevention and Protection Service

A second example of the above mentioned systematic approach to health and safety is that one of the establishment of the prevention and protection service provided by EU Law but also known before in some of the countries analysed in this book. This was the case of Sweden, where the Fifties and the Sixties saw the conclusion of collective agreements proposing the setting up the Occupational Health Service (hereafter OHS). However, the status of the OHS has been a problematic issue in Sweden. It started as an integrated service in big companies in which the company doctor and safety engineer worked together. At the present, the Work Environment Act includes a section dealing with OHS, requiring entrepreneurs to provide it as appropriate according to working conditions. The OHS shall be 'an independent expert resource' in the field of 'working environment and rehabilitation'. It must 'work for the prevention and elimination of health risks at workplaces and shall have the competence to identify and describe connections between the working environment, organization, productivity and health' (Chapter 3, section 2c WEA).

The already mentioned Provision 2001:1 concerning SWEM includes a section on OHS. Nevertheless, the Work Environment Authority has found it difficult to enforce this section. The problematic side of it is that a company can simply pay for OHS in order to comply with the law, but is not obliged by the law to follow the recommendations of the OHS. Entrepreneurs can choose someone else to help them if needed. In such a problematic perspective, collective agreements may play a crucial role in making the OHS effective. Nowadays OHS consists of specially educated occupational health doctors, nurses, physiotherapists, safety engineers, and psychologists or persons trained in the behavioural sciences.

The concern about the expertise of persons involved in the systematic management of the safety obligation is shared by all the legislators of the countries analysed in this book. For instance, in The Netherlands the Arbo Act has introduced a system of certification, not required by EU Law. The Act provides that in case the entrepreneur is making use of an external service - a so-called Arbo service - only certified institutions may be used to carry out the various tasks required by occupational health and safety law. Although the entrepreneur is no longer obliged to contract such a service - and s/he can carry out these tasks by him or herself usually by one or more of his or her employees with specific expertise in the field of health and safety, for some specific tasks, listed in the law, the entrepreneur has to rely on certified experts (individuals or services). The workers' representatives played a crucial role in this field. As a matter of fact, the entrepreneur needs their consent to the decision on how to organizing these tasks. If s/he fails to get it, s/he has to contract a certified Arbo service. In the same perspective, the instrument of certification has to be mentioned. By way of this, the quality of the agents involved in the health and safety policy and practice of companies is checked and supposed to be secured. An independent body assesses whether the requirements for certification are fulfilled. It also periodically supervises the action of the various bodies involved. The minister appoints the certification body and the Council for Accreditation checks the reliability and the independency of the certification body and its expertise.

[3] The Duty of Establishing Joint Health and Safety Bodies

The reference to workers' representative led us to stress the relevance of another duty of the entrepreneur - that one of establishing joint health and safety bodies, a solution adopted only in some of the countries analysed in this book.

In Austria, for instance, according to § 88 ASchG, in companies with more than 100 employees the entrepreneur is obliged to set up an Occupational Safety Committee (hereafter OSC). Members of the committee are the entrepreneur or his or her representative, the person liable for compliance with health and safety provisions, health and safety advisors, occupational health doctors, safety representatives and one competent delegate of the works council. The committee has to meet at least twice a year, though it may also come together when one third of its members insist on it. If the entrepreneur runs more than one work site, a Central Occupational Safety Committee, which has to meet at least once a year, has to be set up. OSC main duty is to guarantee information and coordination among its members and the groups they represent, as well as to work for better employment conditions and improved protection of workers' health and safety.

A paramount example of systematic work environment co-management is offered by the French *Comité d'Hygiène, de Sécurité et des Conditions de Travail* (hereafter CHSCT) which shall be set up in companies employing at least fifty workers (Article L 4611-1 LC). The members of the CHSCT are chosen by the members of the works council and the *délégués du personnel*, according to their skills on health and safety issues. Any competent employee of the plant may be appointed. The entrepreneur in its capacity of employer chairs this body as well as the works council.

First of all, the CHSCT is informed and consulted in case of any change having some influence on the organization of work and the working conditions. The CHSCT must analyse the modifications of working conditions to assess whether they are likely to result in health problems for the workers, as, for instance, an important modification in working time (especially in case of night-work or work in shifts). The consultation takes place approximately according to the same model used for the works council: the CHSCT must be informed in a satisfactory way, and is consulted. However, the entrepreneur is not obliged to follow its advice. In any case, the CHSCT has the right to visit regularly the premises in order to check the working conditions and to make inquiries. If the CHSCT identifies any risky situation for the workers, it has to give an alert to the entrepreneur. Then a procedure has to be started by the entrepreneur in order to check whether the danger is real. The CHSCT has no possibility to stop the work, even if is some danger exist. In such a case, it is up to the employee to make use of the '*droit de retrait*' (see below).

Joint health and safety committees shall be established also according to Belgian (see below), Spanish and Swedish law in companies employing more that fifty workers. In Germany, in plants with more than twenty employees, a Health and safety committee (*Arbeitsschutzausschuss*) has to be established according to § 11 ASiG. It is composed by the entrepreneur or his or her representative, two members of the works council, the occupational health doctor, the health and safety officers and the safety advisor. The Health and safety committee has the task to discuss health and safety issues and shall meet four times the year.

[4] Reintegration and Rehabilitation Programmes for Ill Workers

Last but not least, in the view of the systematic work environment management, the entrepreneur is obliged to provide reintegration and rehabilitation programmes for ill workers. This is a common feature of all the legal systems analysed in this book even though developed to different extents.

A very interesting example comes from Germany where § 84 section 2 SGB IX states the employer's duty to organize the so-called undertaking reintegration management (*betriebliches Eingliederungsmanagement*) in the event of inability to work due to illness. This provision is built on an older provision for prevention of dismissals of disabled persons laid down in § 84 sect. 1 SGB IX. Following the idea that it is easier to look for solutions for keeping the position of a disable than to search for a new job in case s/he get dismissed, the German legislator has established a general procedure of workplace reintegration management in the case of inability to work due to illness. If a worker is unable to work for more than six weeks during one year the employer has to clear the possibilities to get over the inability unless the worker refuses. Such a clearing procedure includes the worker and the employer, the workers representatives and, in case of already disable persons, their representatives. If necessary the occupational health doctor shall be involved. The reintegration management shall focus on any measure that could get over the inability including benefits and facilities. Social insurance bodies shall be involved if social security benefits in cash or in kind are to be considered.

In order to make such procedure effective, the *Bundesarbeitsgericht* has decided that employers who do not follow it will be penalized with respect to the burden of proof. As a matter of fact, usually an employee who cannot fulfil his or her obligations anymore may be dismissed by the employer unless there is another position which fits to his or her residual working capacities. This has to be proved by the employee. In case the employer has ignored the reintegration management procedure, s/he may not claim that the employee is no more able to perform his or her tasks. Then, it is up to the employer to prove that there is no suitable position left for the employee in the company as a whole.

However, the *Bundesarbeitsgericht* has decided that if the employer has arranged a reintegration procedure that meets the legal criteria and if s/he has taken, without success, measures for getting over the inability, the employee may not claim for other measures under the *ultima ratio* principle as provided by dismissal law unless these measures have shown to be feasible after the conclusion of the reintegration procedure. According to some German scholars, this case law is highly disputable because the *Bundesarbeitsgericht* creates legal rules turning the reintegration management procedure into the opposite of its legal aim i.e., an instrument for facilitating dismissals for health reasons.

Another interesting example of reintegration management comes from The Netherlands where the employer, on the basis of provisions of the Civil Code and Sickness and Disability Law, has to make serious and effective efforts to reintegrate ill and disabled employees in their job or at least in a suitable job within the company or even within other companies. The point is that for the fulfilment of these multiple tasks the employer has to make use of specialized bodies or persons. These can be special health and safety service providers or so-called 'prevention' persons who have special expertise in the field. Usually the 'prevention' persons are employed in the company but they have (or should have) an independent role, independent from the employer;

that means in this respect not subordinated to him or her. In case such a person fails to be present in the company a service from outside can be used. As already stressed in the above, external services need to be certified. Therefore, the provider will be assessed on reliability, independency and expertise. Next to these general legal requirements, certification norms have been developed by the parties involved in order to adjust those requirements to the needs and peculiarities of the branch.

A last, paramount example of rehabilitation management comes from Sweden where according to the Work Environment Act (Chapter 2, section 1 and Chapter 3, section 3), the employer has a general duty to adapt work to the employees and to have an organizations in place both for this purpose and for effecting rehabilitation. These duties are specified in the Act itself and in the Provision *Work Adjustment and Rehabilitation* (AFS 1994:1). The Act places several demands on the employer in this area, stipulating that 'Working conditions shall be adapted to people's differing physical and mental aptitudes' (Chapter 2, section 1).

This means that the employer must adapt work environments to the average worker, male or female, according to age and level of experience. The Act stipulates that every workplace must have 'a suitably organised scheme of job adaption and rehabilitation ... ' (Chapter 3, section 2a). Such schemes must fulfil the requirements of the WE Act and of Chapter 30 of the Code of Social Insurance. The employer must also consider an 'employee's specific aptitude for a given job by modifying working conditions and taking other appropriate measures if necessary.'

In planning a job an employer must also take into consideration the fact that 'individual persons have differing aptitudes for the tasks assigned them' (Chapter 3 section 3). This section prescribes the making of individual adjustments by employers. For example, chairs and desks are to be adjusted for the general comfort of employees — though, on the other hand, employers are not responsible for adapting every work situation to every individual. It is the duty of the SC to discuss the organization of job adaptation and rehabilitation (Chapter 6, section 9).

The Provision *Work Adaption and Rehabilitation*, requires both physical and mental job adaptation and rehabilitation. The employer must have set objectives for this and be constantly aware of changing needs in relation to it. Individual job adaptation may involve such things as choice of tasks, use of technical equipment, adjustment and layout of the work setting, various kinds of work aids, changes in organization, changes in work tasks and procedures, and individual support and coaching.

Employees must be informed how these forms of adaptation are to be carried out. The responsibility for organization for adaptation and rehabilitation can be assigned to a particular individual. This person must have the competence and be given the authority, resources, and knowledge needed for the job. Adaptation and rehabilitation are to be carried out in close contact with the persons involved and with their SD. The principle is that there shall be cooperation between the Social Insurance Office and other authorities involved. These might include a hospital, social agency, or OHS. The employer is to conduct an annual review of the organization and make any appropriate changes. The intended objectives, the allocation of tasks, and the routines for performing these tasks, along with the results of an annual revision, must be set forth in written documents. Special attention is to be given to employees with any kind of disability. Efforts to combat alcohol and drug abuse come under the rehabilitation scheme. If the employer does not comply with such demands, the Work Environment Authority may issue an injunction.

[B] The Worker

The fact that the position of the worker is completely different from that one of the entrepreneur who owns the production means and who exercises the managerial prerogatives does not mean that workers have no health and safety obligations. First of all, workers have to comply with the health and safety measures put in place by the entrepreneur. Employees in particular have the obligation of diligence, and must ensure their own safety and that of other workers who may be affected by their activity. The level of diligence is related to the level of qualification of the employee and to his or her personal conditions. Workers must make appropriate use of instruments and resources associated with their job. However, to do so, the entrepreneur must supply them with the correct instructions and information. Workers are also under the obligation of notifying their superiors, health and safety supervisors or other workers of any possible risk they discover while discharging their duties.

In this view, an interesting standpoint is that one of the employee who decides to lodge a criminal complaint against the company assuming that the management is not taking due care of health and safety conditions. This was the basis for the *Heinisch* case decided in 2011 by the European Court of Human Rights against Germany. Although the *Bundesverfassungsgericht* has declared the right to complain protected by the Constitution, the *Bundesarbeitsgericht* has adopted a negative aptitude towards whistle-blowing, advocating that, according to the weighing of the interests principle (*Interessenausgleichprinzip*), such criminal complaints may constitute an infringement of contractual obligations and, therefore, a compelling ground for dismissal without notice. In *Heinisch,* however:

> being mindful of the importance of the right to freedom of expression on matters of general interest, of the right of employees to report illegal conduct and wrongdoing at their place of work, the duties and responsibilities of employees towards their employers and the right of employers to manage their staff, and having weighed up the other various interests involved in the present case, the Court comes to the conclusion that the interference with the applicant's right to freedom of expression, in particular her right to impart information, was not 'necessary in a democratic society'.

Therefore the Court 'considers that (...) the domestic courts failed to strike a fair balance between the need to protect the employer's reputation and rights on the one hand and the need to protect the applicant's right to freedom of expression on the other. There has accordingly been a violation of Article 10 of the Convention.' (points 93–95).

A totally different case is that one of the employees refuse to perform their tasks and interrupt their activity leaving the workplace when they discover a serious and imminent risk to their life or health (the Spanish *ius resistentiae* and the French *droit de retraite*, among the others). They cannot be penalized for taking this action. If a serious and imminent risk occurs, it is because the employer has failed to comply with his or her obligation, or because an unforeseeable situation has arisen. As a matter of fact, in the field of health and safety, the principle of good faith is twofold: on the one hand, employees must not interrupt their work without a serious or justified cause, and on the other employers, when making their a posteriori evaluation of the employee's conduct must limit themselves to consider the circumstances at that particular time and place. The personal situation of the employee must also be taken into account.

Failure on the part of an employee to comply with the above mentioned obligations entitles employers to impose sanctions which shall be proportionate to the extent of the breach. The idea is clearly to exclude civil liability of the employee towards the employer and substitute it with disciplinary liability. Nevertheless, it has to be stressed that workers' misconduct may be taken into account when it comes to decide damage compensation in case of occupational accidents or diseases. As a matter of fact, the employer may claim for contributory negligence (in The Netherlands and in the United Kingdom, for instance) which may lead to a denial or a reduction of the compensation if the worker has failed to contribute to the prevention or to reduce the damages s/he might suffer or is suffering.

§12.06 WORKERS' PARTICIPATION

Workers' participation is a key element of health and safety legislation. Although boosted by EU Law, it already existed (to different extent) in all the countries analysed in this book. As far as participation is concerned, legislators have focused on the company level, even if in some countries (Italy and The Netherlands, for instance) trade unions acting as contracting parties or within joint bilateral bodies may exercise a strong influence on the making and the application of health and safety regulations.

In the lack of any indication from EU Law going beyond the generic and equivocal reference to 'balanced' participation, workers' involvement in the health and safety domain has been structured according to the different models of workers representation and participation already existing at national level. Therefore, the same problematic questions which have been risen by workers' representation and participation in general refer also to workers' representation and participation in the health and safety domain.

These are for instance: (a) the establishment of a representative body as an employees' right, as an employees' duty (like in Austria) or as an entrepreneur's duty (like in Hungary); (b) the nature of workers' representation (unionized or not); (c) the size of the undertaking as a determinant of the presence and the quantity of workers' representatives; (d) the level at which representative bodies has to be established; (e) the protection of workers' representatives and their duty of confidentiality; (f) the role of workers' direct involvement.

The fact that workers' representation and participation in the health and safety domain is modelled on the general rules governing representation and participation at national level does not exclude the presence of some distinctive features worth to be highlighted as typical in that domain.

As a first specific feature of the health and safety domain, it should be stressed that workers participation and even co-determination rights have the only function to promote compliance and to improve the health and safety standard. Therefore, these rights are workers' protection rights which neither limit entrepreneurs' liability nor make workers' representatives jointly and severally liable for the violation of health and safety rules.

Another distinctive feature is that participation rights are recognized to workers' representatives specialized in health and safety matters. However, this cannot be considered an absolute distinctive feature since in some countries the same rights are recognized to general representation bodies. Therefore, the nature and the function of the subject entitled to participation rights influence in decisive way also the very concept of 'participation' in the health and safety domain which finds its specific definition in each country according to the national approach to workers' involvement in general.

This means that common characters are even harder to find in the specific health and safety domain than in the general perspective of representation and participation. This is the reason why with reference to the ten countries analysed in this book we have sorted out three groups plus two unique experiences that cannot be classified according the grouping criteria.

[A] General Representation

A first grouping criterion is that one of the recognition of representation and participation rights in the health and safety domain to general representation bodies. This is the case of Germany and The Netherlands with works councils. Nevertheless, although usually regarded as systems inspired to the principle of co-determination the powers of works council are not of the same intensity. As we will see below, the same can be said about works council in Austria and Spain.

In Germany, according to § 80 no. 1 of the *Betriebsverfassungsgesetz* (hereafter BetrVG), the works council, established in plants employing five or more workers, has the general task to supervise employer's compliance with statutes, regulations and collective agreements. This includes any statute, regulation and agreement on occupational health and safety. Therefore, the works council is a very important player in the field of health and safety supervision.

Besides the supervision and advisory function, the works council is entitled to co-determination rights. Further to the involvement in the appointment of the occupational health doctor, the health and safety officer and the safety advisor, § 87 sect. 1 no. 7 BetrVG regulates a co-determination right on rules of prevention of occupational accidents and diseases and on health protection in the framework of legal rules and regulations. The co-determination right exists unless a statutory rule or a collective agreement is in place. If the entrepreneur and works council do not agree on a solution, an arbitration committee (*Einigungsstelle*) will take the decision.

The right to co-determination is given only in domains in which the employer has a discretionary power (no co-determination without determination) and it does not go beyond the compulsory level of protection. However, in any case of adoption of health and safety measures, the employer must involve the works council. This concerns in particular security instructions. On the contrary, if the employer wants to modify or change the workplace, the work process or the work environment, the works council has no possibility to oppose since it as no real co-determination rights in this field.

Nevertheless, the works council and the entrepreneur must bargain an agreement of balance of interest (*Interessenausgleich*) in specific situations regulated in §§ 111, 112, 112a BetrVG. This means that the employer may freely decide to alter or even withdraw the intended measures. In these cases, however, the works council has a co-determination right on measures for economic compensation aimed at coping with the disadvantages deriving to the workers from the employer's decision (the so-called *Sozialplan*). Furthermore, if those modifications or changes are in open contradiction to acquired ergonomic knowledge and if they have a specific negative impact on the workers, the works council has the right to claim for prevention, lowering or compensation. If the works council and the entrepreneur do not agree on such measures the above mentioned arbitration committee will take a decision.

In The Netherlands works council's functions, rights and duties are governed by the Works Council Act (hereafter WCA). The WCA itself refers in several provisions to the health and safety domain. The role of the works councils in this field has been strengthened over time. Initially the works council has been entitled to the right to be informed and consulted on health and safety issues as part of the right to give advice to the employer according to Article 25 WCA.

However, the most important provision in the WCA is Article 27 which recognizes to the works council *a right to consent on health and safety issues*. According to that, the employer needs the consent of the works council when taking decisions on the establishing, the changing or the withdrawal of rules or measures concerning the health and safety of his or her employees. Consequently, the potential scope of Article 27 WCA can be wide. Actually it will depend upon the role and acting of the works council. Although the WCA does not refer explicitly to the Arbo Act, it is obvious that the scope of Article 27 can be determined taking into account the various regulations on health and safety. However, it has to be kept in mind that since the works council operates at company level, its influence on entrepreneur's decisions is restricted to that level.

The link between Article 27 WCA and the health and safety regulations (Arbo Act and Catalogues) indicates the possible issues subject to the consent of the works council. So if the employer wants to specify the various obligations stemming from the health and safety regulations, the consent of the works council is required. The restriction is that the regulations concerned must have a general nature. That means the regulation must be generally applicable and not be restricted to a single case or employee.

The main subjects for which the works council can exercise its right of consent include: (a) the ruling on the inventory and evaluation of risks; (b) rulings on the prevention or restriction of heavy accidents due to dangerous substances; (c) the establishment and the way in which prevention services operate; (d) the choice of the external service deploying the health and safety activities to be performed.

Since Article 27 WCA contains also an obligation for the employer to pursue an effective policy on the prevention of ill-health of his or her employees as well as reintegration of workers who are partly disabled, the works council can exercise its right to consent on the regulation the employer intends to make in these fields. The works council can decide itself the degree of its involvement. It even can extend its power if it is able (and powerful enough) to widen its involvement by concluding an agreement with the employer entailing extra rights on specific issues related to health and safety.

Last but not least, Article 28 WCA attributes to the works council the task to encourage the actual application of all health and safety measures and regulations taken by the employer. This Article provides the works council with the possibility to check whether the employer actually applies his or her regulations and measures.

[B] Specialized Representation

A second grouping criterion is that one of the establishment of representation bodies or the appointment of representatives specialized in the health and safety domain. This happens in Belgium with the *Comités pour la Prévention et la Protection au travail,* in France with the *Comité d'Hygiène, de Sécurité et des Conditions de Travail* (although where both cannot be established because the size of the undertaking falls below fifty employees their powers will be exercised by the general representatives, respectively the *delegation syndicale* and the *délégués du personnel*) and in Hungary with workers' safety representatives.

The prerogatives of the French *Comité d'Hygiène, de Sécurité et des Conditions de Travail* have been already highlighted in the above, whilst the role of Safety Representatives is deemed by Hungarian legal doctrine to be only formal. Therefore, we can focus on the Belgian case which presents some interesting distinctive features in the relationship between specialized and general workers' participation.

In Belgium, the *Comités pour la Prévention et la Protection au travail* (hereafter the Commitee), exercises prerogatives which transcend the traditional ambit of information and consultation. In this respect, the nature of the prerogatives of the Committee are essentially different from those of the Belgian works council (*conseil d'entreprise*) which only needs to be informed and consulted in social, economic and financial matters. The overall image of the Belgian works councils established in undertakings employing at least 100 workers, is that of a passive body. They need to be informed and consulted, but have scarcely any margin of manoeuvre to take initiatives, let alone decisions.

Article 65 of the *Loi relative au bien être des travailleurs* takes a different stance. It states that the Committee has a general competence to examine and to submit proposals related to means and to contribute actively to everything that is done to improve the 'well-being' of the workers. Thus, the Committee is actually regarded as a pro-active agent involved in the employer's policy or strategy in the health and safety domain. The Royal Decree of 3 May 1999 (hereafter RD) confirms this impression, by highlighting the right of the Committee to give advice and to submit proposals. In other words, the Committee is not only a body which is informed and consulted when the employer is obliged to do so. It can also take initiatives in its own right and formulate proposals (Article 2 RD).

As far as consultation is concerned, it relates specifically to all the measures the employer intends to take with regard to occupational health and safety. The RD stresses that consultation must take place prior to the adoption of those measures, since the Committee needs to give a preliminary advice (*un avis préalable*). In some rare cases, the Committee has involvement rights which are close to decisional powers. Since the Committee is a mixed body, composed of the management and the elected workers' representatives, the exercise of such decision rights can be seen as a form of co-determination.

The Committee also has a role in the implementation of the health and safety policy. Article 5 RD refers to the elaboration and the implementation of means to propagate and to the adoption of measures related to the assistance of newly recruited workers (*accueil des travailleurs*). The same competence is recognized with regard to the information and training in the field of prevention and protection.

[C] Second Level Specialized Representation Which Stems from General Trade Unions Representation, with Alternative Route in Absence of Trade Union Representation within the Undertaking

A third grouping criterion is that one of the presence of second level specialized representation which stems from general trade unions representation, with alternative route in absence of trade union representation within the undertaking. This happens in Italy, Sweden and United Kingdom

In Italy, Workers Safety Representative (hereafter WSR) shall be elected or designated by the workers in the context of the trades union representation in the undertaking. In their absence, the WSR shall be elected by the workers from amongst themselves.

As regards the powers of the WSR the Italian path is twofold in its nature: basically a-conflictive, with reference to the powers regarding training, information and consultation, and conflictive when, in the absence of agreement between the employer and the WSR, the latter can take action directly. Under the 2008 legislative framework, therefore, consultation with the WSR is characterized by its timeliness and priority with regard to the most important of the employer's decision-making processes concerning the protection of workplace safety. Whilst it cannot be deemed binding on the employer, it is an essential pre-requisite, enabling WSR to participate in the entrepreneur's decision-making in a technical sense, making proposals and checking that the safety measures are implemented.

In such a perspective, Legislative Decree no. 81/2008 has added the principle that consultation must take place regarding two employer's decisions of primary importance: designation of the occupational health physician and, more importantly, the organization of training in general. If consultation takes place regarding the most important health and safety topics and encompasses the very great majority of employer's decisions potentially affecting the management of health and safety, the problematic side to the provision therefore lies in the fact that the opinion of the WSR is not binding. More precisely, where the views on the occupational safety measures that have been or are to be adopted are not identical, the naturally conflictive tendency of Italian industrial relations therefore reappears.

In terms of the concrete enforcement measures, should there be no election or designation of the WSR either in undertaking with fewer than fifteen workers or in those with more than fifteen workers, the legislator expressly provides (in Articles 47(8) and 48(1) of Legislative Decree no. 81/2008) for the subrogating intervention of the territorial workers' safety representative, establishing that the latter shall exercise the powers of the WSR.

In Sweden, a Safety Delegate (hereafter SD) must be appointed wherever there are five or more persons employed. It is the responsibility of the local trade union having signed a collective agreement with the employer to make this appointment. In the absence of such an agreement, the employees themselves may appoint the SD. Where there is more than one SD, a Senior SD must be appointed from among them and will be responsible for co-ordinating their activities.

Regional SD (hereafter RSD) are appointed by trade unions. At least one member of the union must be present among the employees on a work site for the RSD to have access thereto. The RSD have the same rights as other SD and are expected to help employees appoint SD and organize safety-oriented activities. Today, many RSD help employers in drawing up programmes of Systematic Work Environment Management.

SD represent employees in matters concerning their work environment. They have the right to participate in the planning of new or alteration of existing premises, equipment, and working methods and techniques, as well as in the organization of work and the use of chemical and other substances. An SD takes part in the preparation of action plans, that being a function of Systematic Work Environment Management (section 4 Work Environment Act).

Entrepreneurs must inform SD whenever changes are planned which will affect the work environment. SD have the right to inspect all relevant documents and to be given any other information relevant to their concern with the work environment (section 6 WEA).

SD have a right to demand that entrepreneurs conduct an investigation, and if necessary a correction, any time they deem the work environment unsatisfactory. If an entrepreneur does not comply with such a demand, the SD has the right to appeal to the Work Environment Authority, which is then required to conduct an inspection and, if appropriate, take legal action and issue a prohibition or injunction. If a SD is not satisfy with the resulting decision, the Senior SD for the workplace can appeal the issue to an administrative court (Chapter 9 section 3).

Last but not least, SD have the authority to order a work stoppage in situations where there is an immediate and serious danger to life or health and no help from the entrepreneur forthcoming. The SD can order work to stop if considerations of health and safety in solitary work so demand. Authority to call for a work stoppage also exists in cases where an entrepreneur has failed to respect a prohibition issued by the Work Environment Authority. SD cannot be held liable for any damage that occurs as a result of action taken by them under the terms of the law.

In the United Kingdom, according to the Safety Representatives and Safety Committee Regulations 1977 recognized trade union may appoint Safety Representatives (hereafter SR) amongst the employees. SR are given a variety of powers. First of all, there are investigatory powers, such as the investigation of potential hazards, dangerous occurrences, the causes of accidents, and complaints by an employee. Second, SR has the power to inspect the workplace, including where there has been a notifiable accident or dangerous occurrence. SR is entitled to inspect and take a copy of any relevant document. Third, SR has the function of making representations to the employer on behalf of the workforce in relation to health and safety, and this extends to representing the employees in consultations with the inspectors of the Health and Safety Executive. The employer is under a duty to consult 'in good time' safety representatives on a wide range of matters pertaining to health and safety, such as the introduction of new technologies at work. Moreover, the employer must provide the representatives with paid time off to perform their functions and to be trained, as well as providing them with reasonable facilities and assistance. If at least two safety representatives request it, the employer must establish a Safety Committee for the purpose of consultations with SR.

In workplaces without a recognized trade union, the Health and Safety (Consultation with Employees) Regulations 1996 place entrepreneurs under an equivalent duty to consult their employees on health and safety matters, but the entrepreneur can choose between direct consultation with all employees or consulting 'representatives of employee safety' (hereafter RES) who have been elected for that purpose by the employees. Moreover, the Regulations permit an entrepreneur who has been consulting RES to decide to change to direct consultation.

The 1996 Regulations say little about what direct consultation with employees would entail, except that the employer must provide the employees with such information 'as is necessary to enable them to participate fully and effectively in the consultation'. If the elected representatives route is chosen by the employer, then RES are empowered to make representations to the employer and Health and Safety Executive inspectors. The RES must be provided with paid time off for carrying out their functions and they must be provided with training organized by the employer. This is in contrast to the 1977 Regulations where the employers must provide paid time off for training, but there is no duty on the employer to ensure that training takes place. This would seem to reflect the likelihood that a trade union SR will have access to training organized by the trade union.

Strikingly, the other functions of (trade union) SR are not extended to elected RES. They have no powers to investigate or inspect the workplace. They have no powers to inspect documents and there is no provision for employee representatives to require the employer to establish a Safety Committee. Breach of the 1996 Regulations will not expose the employer to any civil liability, except in respect of the duty to provide paid time off for RES. Once again, this is a departure from the 1977 Regulations where there is no restriction of civil liability.

[D] General Non Unionized Representation from Which Stems a Specialized Representation: Spain

The first case of a unique experience that cannot be classified according the grouping criteria used in the above is Spain, where a specialized representation stems from a general non unionized representation. As a matter of fact, in Spain the legal system of representation in the health and safety domain bypasses the trade unions and is organized through the works council (or Staff Representative in the case of companies with less than fifty workers). The works council choose which of its members will have specific health and safety functions, and these are called Health and Safety Representatives (hereafter HSR). This means that HSR must be members of the works council (Article 35 *Ley de Prevención de Riesgos Laborales* - LPRL).

HSR are charged with overseeing and ensuring compliance with health and safety rules; receiving information from the employer regarding health and safety plans; and encouraging workers to cooperate with health and safety measures. Generally speaking, they receive information from the employer, and the latter consults them on health and safety matters. HSR receive general information, but this does not release the employer from the obligation to inform each worker individually about the risks involved in their specific job.

The system of workers' representation provided by the law can be derogated by collective bargaining. Nevertheless, empirical analysis has shown that it is applied in most of the cases. Collective agreements mainly bring in some small differences, mostly related to the way in which HSR are appointed.

The works council or the HSR if the council cannot be convened, may exceptionally take over the powers of the employer and order activity to be suspended if a serious or imminent risk arises that affects all or a particular group of workers (Article 21.3 LPRL). The decision to interrupt production can be adopted when the employer does not take these measures in a situation of imminent risk, or does not allow others to take them. This decision, which is a temporary expropriation of the managerial powers of the entrepreneur, must be taken by the majority of the works council, or by the HSR if the works council cannot be convened. It must be immediately communicated to the relevant labour authority which, within twenty-four hours, must ratify or overrule the decision. The extent and imminence of the risk must be weighed up on the basis of the exact circumstances and in the context of the moment at which the decision was taken. Workers' representative may not be penalized for any decision taken in good faith and workers may freely decide whether or not to follow the order to interrupt their activity. Workers who interrupt their activity retain the entitlement to receive their salary.

[E] The Co-existence of General (Non Unionized) and Specialized (but Employer Appointed) Representation: Austria

The second case of a unique experience that cannot be classified according to the grouping criteria used in the above is Austria where general non unionized (works council) and specialized but employer appointed (Safety Representatives) representation co-exist.

In companies employing at least five employees, it is the employees' duty to establish a works council, although infringements of such a duty are not sanctioned (§ 40/1 Arbeitsverfassungsgesetz – ArbVG). The main obligation of the works council, which represents employees on health and safety issues, is to control and to survey the implementation of the provisions of the *ArbeitnehmerInnenschutzgesetz* – ASchG as well as the respective regulations.

Yet, the works council's obligations with reference to health and safety at work are not provided in the ASchG but in the ArbVG.

According to § 89 Z 3 ArbVG, which provides the works council's duty to monitor the implementation and compliance with health and safety provisions, the works council is allowed to inspect the work site. The employer, on the other hand, is obliged to inform the works council immediately about any occupational accident as well as on any trial related to issues concerning the employees. § 92a/1 ArbVG stipulates the works council's right to be heard by the employer concerning all health and safety matters as well as the employer's obligation to consult the works council on these issues. According to § 92a/2 ArbVG, the employer is obliged to grant the works council access to health and safety-documents, to place documents and findings concerning e.g., labour organization or specific working materials at the works council's disposal, and to provide information about regulatory requirements, orders and authorizations as well as to hear the works council on e.g., health and safety related issues and measures or on first aid measures.

With regard to the appointment or the dismissal of safety advisors and occupational health doctors, the employer is obliged to consult the works council in advance. If the employer does not comply with this obligation, the act of appointment or dismissal is void (§ 92a/3 ArbVG). Only if the intended appointment or dismissal is dealt with in the Health and Safety Committee can the consultation with the works council be omitted.

Last but not least, the works council can delegate its responsibilities concerning occupational health and safety to the Safety Representatives (§ 92a/4 ArbVG). The safety representatives as well as the employer have to be informed about the delegation; the information is pre-requisite for the delegation to produce its effects.

Safety Representatives (hereafter SR) shall be appointed by the entrepreneur in companies employing more than ten employees (§ 10 ASchG). In principle, the appointment of any SR has to be approved by the works council. Only if no works council is established all employees have to be informed in writing about the appointment. When at least one third of the employees object the entrepreneur's choice (again in writing) the entrepreneur is obliged to select another person. Generally speaking, SR are appointed for four years, although there is the possibility of prior revocation. It is crucial to note that the appointment of SR does not diminish the employer's responsibility for the implementation of health and safety measures. Any transfer of responsibilities towards safety representatives is void.

According to Austrian legal doctrine, SR although appointed by the entrepreneur *are representatives of the employees*. As a result, the entrepreneur has to comply with his or her obligations of information, instruction etc. towards them and not towards the employees. The SR main duties are: (a) to provide information, support and advice for the workers and workers representatives as provided in collective labour law; (b) to represent workers; (c) to advise/support employers in implementing health and safety provisions; (d) to take care of the actual implementation of health and safety measures; (e) to inform the employer about any potential risks; (f) to cooperate with occupational health and safety advisors and occupational health doctors.

Although employed by the company, SR are not bound to instructions issued by the entrepreneur and they have the right to intervene by the entrepreneur as well as by the competent authorities concerning all matters of occupational health and safety. Furthermore, they have access to all health and safety related documents, accident records and test results.

§12.07 COMPLIANCE AND CONTROL

It is clear from the aforesaid that entrepreneurs' compliance with health and safety obligation is primarily controlled by workers' representatives whose powers go well beyond participation in the health and safety management. An extreme example of this is the power to stop work in case of imminent danger which is recognized to them by Dutch, Spanish and Swedish law. The single worker too takes part in the compliance and control system through the duty to alert the entrepreneur of danger and risks that may occur at the workplace. Workers may also lodge a claim against the employer in labour courts to get the safety obligation fulfilled. Following *Heinisch* lodging a criminal complaint against the company cannot produce any contractual consequences for the worker.

Nevertheless, in all the countries analysed in this book a crucial role in the compliance and control system is played by public authorities, due to the (partly) public law nature of the safety obligation. It is not a case that the role of public authorities is particularly developed in Austria where the safety obligation is regarded as a purely public law one.

[A] Labour Inspection

A first example of public authority dealing with compliance and control is the Labour Inspection which is usually part of the ministry for labour and social affairs.

Generally speaking labour inspectorates have the following rights and duties: (a) assist and offer consultation to entrepreneurs as well as employees in all matters of occupational health and safety; (b) enter and inspect plants, workplaces and construction sites at any time with or without prior notice; (c) interview entrepreneurs and employees (or their representatives) as well as ask entrepreneurs for written information; (d) inspect documents referring to occupational safety or employment conditions; (e) take pictures and take measurements also by appointing experts in specific fields; (f) take samples and arrange for analyses; (g) obtain information on material and machines from producers and distributors; (h) by all means attend to complaints without disclosing the source; (i) issue order and injunctions even in the view of suspending the activity; (l) request the competent authorities to enforce measures for the protection of employees; (m) impose sanctions or propose competent authorities to do so.

Classical form of labour inspection acting also in the health and safety domain are to be found in Austria, Belgium, France, Germany, Spain and The Netherlands. In Italy, the Labour Inspectorate is only marginally competent in that domain which is covered by Local Health Authorities inspectors. This results in an undesirable overlapping of competences.

[B] Specific and Dedicated Systems of Compliance and Control

In contrast, Hungary, Sweden and United Kingdom have developed specific and dedicated systems of compliance and control in the health and safety domain which go beyond the classical structure of labour inspection, although guaranteeing the same inspection powers.

[1] Hungary

In Hungary, the structure of the authorities responsible for the enforcement and supervision of health and safety regulations has been changed recently according to the centralization principle. In 2011, with effect from 1 January 2012 the main authority has been restructured and renamed from National Supervisory Authority for Occupational health and safety (*Országos Munkavédelmi és Munkaügyi Főfelügyelőség*) to National Authority for Labour (*Nemzeti Munkaügyi Hivatal*, hereafter NAL). The NAL has integrated three bodies working earlier in separate organizations, namely the Authority for Employment, the National Supervisory Authority for Occupational health and safety as well as the National Institute of Vocational Training and Adult Education. Within the single authority two separate directorates have been established, namely the Directorate of Occupational health and safety and Labour Issues as well as the National Employment Service. The first one is the major body responsible for occupational health and safety in every branch and for every profession. Only the mining sector has a separate authority maintaining its competence.

NAL is the major central body being responsible for occupational health and safety in Hungary and the national leader of the supervisory activities in this field. NAL supports the work of the regional bodies by guidelines and recommendations and operates the uniform electronic system with registered data. It organizes training, education and examination of the supervisors. At national level, NAL takes part in the elaboration of new acts and the national policy, the annual assessment of the situation of national occupational health and safety, the preparation of the annual report and the information of people. Furthermore, it takes part in the activities of European and international organizations and cooperates with the organizations of other countries.

Parallel to the restructuring at national level, also the organization at regional level has been changed. The Regional supervisory bodies of occupational health and safety have lost their autonomy and recently constitute a part of the Regional governmental agencies next to several other authorities. The regional supervisory bodies for occupational health and safety (*Munkavédelmi felügyelőség*) carry out the general tasks of the authority. They are the authorities at first instance and perform *inter alia* the supervisory activities.

Chapter VII, sections 81-86 of the Labour and Safety Act set common rules for the authorities performing supervisory activities in this field. They carry out their supervisory activities based on the supervisory guidelines published by the head of the authority. The guide indicates the primary fields of supervision in the year concerned, the concrete activities, the objectives to be achieved and the main activities, branches or occupations, which shall be controlled.

The supervisory activities include the following: (a) the monitoring of compliance with the duties and obligations of employers and employees in connection with occupational safety and health; (b) enforcement of the requirements related to the establishment of workplaces, operation of work equipment, technological processes and materials applied, and to personal protective equipment; (c) investigation, reporting and registration of occupational accidents, occupational diseases, and cases of extreme exposure, as well as measures taken with a view to prevention.

The authority can impose fines against the entrepreneur failing to fulfil the requirements of occupational health and safety law, and thereby seriously endangering workers' life or health. The law provides for an extensive list of cases which causes particularly serious endangerment of the life or health of workers.

The occupational safety and health authority shall maintain an official register with a view to be able to verify employers' compliance with occupational safety regulations in connection with proceedings before other bodies conducted in accordance with specific other legislation (e.g., providing budgetary subsidies for the employment of workers with some degree of incapacity). The authority publishes on its official website the name, registered office and tax number of employers sanctioned for breach of occupational safety regulations, including the description of the infringement, the amount of the fine imposed, the date of the resolution on the infringement and the date when it becomes operative.

[2] Sweden

In Sweden, in 2001 Labour Inspection and the National Board of Occupational Safety and Health (NBOSH) were merged into a single body we have already mentioned - the Work Environment Authority (hereafter WE Authority). One aim of this was bringing implementation by the two organizations into line with each other.

The WE Authority's duty is to ensure compliance with the WE Act and the Working Time Act — and, in certain respects with the Tobacco Act and the Environmental Code and the Provisions issued by the WE Authority itself.

The WE Authority is run by an Authority's Director General, as chair. S/he shall install a special committee for cooperation. This committee shall consist of representatives from the trade unions and the employers' confederations. When issuing new

Provisions, the WE Authority shall again cooperate with the Social Partners. There shall also be a special committee for transparency. This committee consisted today of three members of the Parliament and one work psychologist.

The WE Authority issues Provisions and manages the three regional Inspection Offices. Under these offices are ten Inspection Districts with work environment inspectors responsible for supervision by inspection. The WE Authority issues prohibitions and injunctions.

The WE Authority is invested with legal powers to assure that this Provision is respected, but it can only exercise them preventatively: e.g., by demanding injunctions against victimization and by advising employers concerning strategies to prevent it. The Authority does not deal with individual cases and therefore has no power to determine compensation for damages in such cases.

The WE Authority can make inspections on self-employed persons' premises, if there is a need for it or if the self- employed person has asked for it.

As supervision of the Working Time Act is the responsibility of the WE Authority, it can issue both positive commands and prohibitions. Non-compliance with any of these can lead to a contingent fine (see below). Non-compliance with the Act, whether intentional or because of negligence, can result in the imposition of a personal fine. The same penalty is incurred if an entrepreneur gives the WE Authority false information.

Last but not least, WE Authority can call for a work stoppage in cases where an entrepreneur has failed to respect a prohibition issued by it.

[3] United Kingdom

In the United Kingdom, under the Health and Safety at Work Act 1974 (as amended) (HSWA) the Health and Safety Executive (hereafter HSE) has been set up in order to support the Government's strategic aims and current targets for health and safety at work. Its main aim is to secure the health, safety and welfare of people at work and protect others from risks to health and safety from work activity. HSE operates as one of the Department of Work and Pensions's (DWP) Non Departmental Public Bodies (NDPBs). The Secretary of State has the principal responsibility for HSE. The DWP Minister with responsibility for health and safety will account for HSE's business in Parliament.

More specifically HSE main statutory duties are: (a) to propose and set necessary standards for health and safety performance; (b) to secure compliance with those standards; (c) to carry out research and publish the results and provide an information and advisory service; (d) to provide a Minister of the Crown on request with information and expert advice.

HSE is the primary delivery agent for DWP's strategic objective of improving health and safety outcomes. HSE's aims should be to: (a) continue to deliver its mission of preventing death, injury and ill-health to those at work and those affected by work activities; and (b) deliver any targets agreed for occupational health and safety.

In such a perspective, HSE shall: (a) protect the health and safety of workers and minimize risks from work to members of the public; and (b) ensure that the major hazard industries (such as nuclear, petrochemicals and offshore oil and gas) manage and control the risks around their work to a high standard which enhances assurance and allows these industries to operate with a high degree of public acceptance.

Therefore, HSE seeks to: (a) influence people and organizations – duty holders and stakeholders – to embrace high standards of health and safety; (b) promote the benefits of employers and workers working together to manage health and safety sensibly; (c) investigate incidents, enquire into citizens' complaints and enforce the law.

[C] Social Security Institutions

An important role in the control and compliance system is also played by social insurance institutions active in the occupational accident and disease domain.

A paramount example comes from Germany, where a dual system of safeguarding and promoting occupational health and safety is in place including tasks for public authorities as well as for the autonomous bodies of the occupational accident insurance (*Berufsgenossenschaften*). The latter has a longstanding tradition, dating back to the end of the nineteenth century. In that period the German government started to deal with social security of the working class in order to weaken the Socialist opposition by solving the problems of their potential supporters, the workers. In that context the Occupational Accident Insurance (hereafter the Insurance) was created. It took the place of the former *Reichshaftpflichtgesetz* which had established a strict liability in tort for factory owners. Departing from this idea of liability, the Insurance provides that the employer has to pay contributions being therefore not liable for damages caused by occupational accidents to his or her employees.

Nevertheless, the replacement of liability does not exempt the employer from fulfilment of the security obligation. The employer is liable for health and safety from the point of view of contract law as well as from the point of view of public law. On the other hand it is a substantial interest of the Insurance to prevent occupational accidents. Therefore, the *Sozialgesetzbuch VII* – SGB VII which deals with the Insurance, stipulates in § 14 the priority of prevention. For this purpose the Insurance has the power to adopt *Unfallverhütungsvorschriften* – UVV according to § 15 SGB VII. These are binding to the employers and employees.

Complementarily, Insurances have a monitoring and advisory duty. With reference to industrial diseases they have furthermore the duty to use any possible mean to cope with dangers of appearance or re-appearance or worsening of an illness. If this is not possible they have to try to convince the employee to stop the unhealthy work, according to § 3 section 1 of the *Berufskrankheitenverordnung* – BKV. This task of the Insurance is complemented by the health promotion duty of the health insurance (*Krankenkasse*) in the plant as described in § 20a SGB V. *Krankenkassen* shall cooperate with the Insurance and support its prevention activities.

In such a legal framework the Insurance bodies are obliged to employ inspectors in order to comply with their supervisory tasks (§ 18 SGB VII). These inspectors –

although employed by the self-administrated Insurance bodies – have public law powers in the health and safety domain (§ 19 SGB VII).

[D] Bilateral Bodies Established by Collective Agreement

Last but not least, bilateral bodies established by collective agreement may play an active role in the compliance and control system. This is the case of Italy where a fundamental role in occupational health and safety management is assigned to Joint Committees (hereafter JC). These are identified by Article 2(ee) of Legislative Decree no. 81/2008 as bodies founded on the initiative of one or more of the entrepreneurs/ workers comparatively most representative organization at national level.

It is the defining framework contained in Article 2(ee) of Legislative Decree no. 81/2008 that states this, specifying that the said bodies are privileged places for programming training activities and formulating and consolidating good preventive practices, developing actions connected with health and safety at work, assisting undertakings in fulfilling their health and safety duties and every other activity or function assigned them by law or by the collective agreements of reference.

Over and above the essential function of resolving disputes concerning application of the rights to representation, information and training, in accordance with all that was previously indicated by Article 20 of Legislative Decree no. 626/1994, the legislator therefore accords the JC the fundamental role for adopting measures actively geared to completing the participatory framework outlined by the legislation. The task of orienting and promoting training assumes a particular importance.

In completion of their specific functions, the same JC are also given the fundamental task of supporting companies in the identification of technical and organizational solutions aimed at guaranteeing and improving the protection of health and safety at work (Article 51(3) of Legislative Decree no. 81/2008). To such end, provided that they have staff with the relevant specific technical competences, the JC are entrusted with a particularly important task since they are given the power to carry out on-site inspections in workplaces falling within the territory and sector of production for which they have competence. In such a sense, therefore, JC support companies in identifying specific shortcomings in health and safety protection and promote the interventions needed to make workplaces safe.

Lastly, the JCs' role and functions have been enhanced by Article 30 of Legislative Decree no. 106/2009. This amended Article 51 of Legislative Decree no. 81/2008 by inserting paragraph 3-*bis* and 3-*ter* which attribute to (truth to tell) not better specified 'technically competent joint commissions' (to be set up within the joint committees) the important functions of:

> certifying performance of the activities and services supporting undertakings' systems, including 'asseverazione' (i.e. certification by professional experts) of the adoption and effective implementation of the models of safety organization and management referred to in Article 30, of which the supervisory bodies can take account for the purposes of programming their activities.

Therefore, Article 51(3)(*ter*) of Legislative Decree no. 81/2008 attributes important functions to technically competent joint commissions, to be set up within JC. These are the functions of certifying performance of the activities and services supporting companies' systems, including *asseverazione* of the adoption and effective implementation of the models of safety organization and management referred to in Article 30, of which the supervisory bodies (i.e., Local Health Authorities and, to a lesser extent, Labour inspectors) can take account for the purposes of planning their inspection activities.

Such fact guarantees a double advantage to entrepreneurs. On the one hand, they can profit from a checking that, if not less rigorous, is at least programmed by the supervisory bodies since it is performed periodically under the compliance auditing procedures carried out by the JC. On the other, they have the advantage that the *asseverazione* regarding adoption of a particular model of business management and organization exempts legal entities, companies and associations (including the unincorporated associations referred to in Legislative Decree no. 231 dated 8 June 2001) from administrative liability.

Confirming that the interpretation of the provisions on *asseverazione* set out above is the only one consistent with the national system of accident prevention (in terms of compatibility both with domestic constitutional law and with the EU provisions contained in Framework Directive), the European Commission opened infringement proceedings against the Italian government on 30 September 2011. The government has been asked to explain, *inter alia*, the effects of an *asseverazione* in relation to an employer's supervisory duty.

More specifically, the European Commission is challenging the Italian government regarding the incompatibility of a preventive finding of correct supervision on the part of the employer and those whom s/he delegates with the provisions of Article 5(3) of the Framework Directive 89/391/EC, which states: '*The workers' obligations in the field of safety and health at work shall not affect the principle of the responsibility of the employer.*'

Except that, in the Italian legislative and case law context, the so-called 'principle of employer's reliability' has a very limited role. This by virtue of the consideration (constantly affirmed by the constitutional case law) that such principle is not operative in situations where there exists a guarantee (such as the employer's guarantee to the company's employees) which continues even in the face of a formal scrutiny of compliance with the legislative provisions conducted by the administrative supervisory bodies, given that such compliance relates to a particular, 'frozen' moment.

That is to say that the *asseverazione* of a theoretically effective model of business management and organization does not free employers a priori from their potential liability in the case of subsequent accidents. Indeed, if an accident occurs, the existence of a direct connection between the employer's act or omission, even *in vigilando*, and the event occurring must be assessed by the court. This by virtue of the responsibility for safety that the legal order generally attributes to employers in principle and, more particularly, under Articles 32 and 41(2) of the Constitution and 2087 of the Civil Code. Thus entrepreneurs have to show at trial that their protective measures complied with the legal provisions, without prejudice to the court's exclusive competence to assess whether a direct connection existed between the fault and the accident at the moment when the damage occurred.

This in spite of provisions (such as the one being contested by the European Commission contained in Article 16(3) of Legislative Decree no. 81/2008) which, rather apodictically, provide for discharge of the employer's duty to supervise those to whom s/he delegates duties in those cases where the company's model of health and safety management has been adopted and effectively implemented.

That said, *asseverazione* of the adoption and effective implementation of the models of safety organization and management referred to in Article 30 of Legislative Decree no. 81/2008 does afford a partial alleviation of entrepreneurial liability in the said legislative context.

Indeed, since it constitutes a formal scrutiny of the compliance and efficiency of an undertaking's management and organizational model, it ought to facilitate the 'division of competences' between entrepreneurs and the employees they have deputed to carry out specific actions to protect the workplace.

§12.08 CONSEQUENCES OF VIOLATION OF HEALTH AND SAFETY PROVISIONS

[A] General Overview

All the parties involved, i.e., the entrepreneur and the worker shall, within the scope of their respective liabilities, bear the consequences of the violation of the health and safety provisions. The direct consequence of the violation – if detected – are sanctions.

Being the health and safety obligation, at the same time, public and private in nature, violations are not only relevant as breaches of law, but also as breaches of (the employment) contract or of the *neminem laedere* principle (action in tort). Therefore, sanctions are provided which are typical of labour law (disciplinary sanctions against the worker) and of public law, being the latter administrative or criminal. Sanctions provided under public law may affect both the entrepreneur and the worker. However, before any harmful event happens, the worker may lodge a civil (labour law) claim or a criminal complaint against the employer who, in the worker's opinion, does not fulfil his or her safety obligation.

In the same perspective, the violation of health and safety provisions is sanctioned in itself under administrative law even if it has not produced a harmful event. In such cases, administrative (pecuniary) sanctions may be avoided or reduced if the entrepreneur, following the prescriptions coming from a supervisory authority, restores a situation of legality. In Sweden, compliance to prescription may be 'stimulated' by contingent fines which have to be paid by the entrepreneur only in case s/he does not respect the prescription. However, if the prescription is not respected and a harmful event occurs, this may increase the sanction and will be taken into account in the assessment of entrepreneur's civil liability.

The mere state of danger produced by the violation of health and safety obligations does not exclude the application of civil (labour) law remedies such as workers' refusal to perform their tasks. However, the stop to production can be called by workers' representatives and by the supervisory authority in case of imminent danger or of persistent violation of health and safety prescriptions.

[B] Civil Liability & Social Insurance

At the beginning of this chapter, drawing the big picture of health and safety law, we have stressed the relevance of civil liability as a consequence of the violation of the health and safety provisions which has produced material or immaterial damages to the worker or to a third party. We have also stressed the link between civil liability and social insurance in the sense that the duty to insure the employee and to pay contribution to the latter exonerates the employer from the former in case the harmful event is classified as occupational accident or disease.

The fact that the employee was acting contrary to the instructions of the employer, or any legal requirement, will not prevent eligibility for the benefits provided by the social insurance so long as the act is done for the purposes of and in connection with the employer's trade or business. The social insurance may act in regress against the employer if the harmful event has been produced by the violation of health and safety provisions. Exoneration is excluded or differential damages compensation admitted if the harmful event has been caused by employer's intent or inexcusable fault.

That of inexcusable fault is a very interesting concept elaborated by French case law (but also present in Belgian and Spanish case law, as we have seen in the above) in order to cope with the problem of the unsatisfactory level of social insurance benefits in case of occupational accident or disease being the recourse to civil liability precluded.

In the French case law, the definition of the concept varies depending upon the author of the fault.

If the author is the employer, there is an inexcusable fault when s/he was or should have been aware of the danger existing for the worker and has not taken appropriate protective measures. According to case law it is not necessary that the fault of the employer has been determining as regards the damage. It is enough that his or her fault has been necessary to cause the damage. In addition, the employer may be held liable even if s/he has taken measures to eliminate the danger which have not impeded the harmful event to take place. By recognizing the employer's inexcusable fault, the judge may decide an increase of the social insurance benefit. The social security institutions, which shall pay the increased benefit to the worker will act in regress against the employer for the reimbursement of the additional payment.

If the worker is recognized as the author of an inexcusable fault, the judge may reduce the amount of the social insurance benefit. Case law defines the inexcusable fault of the worker as a deliberate fault, of an exceptional seriousness, creating without any relevant reason a danger, which the worker should have been aware of. However, the interpretation of the inexcusable fault of the employee is quite restrictive whilst case law provides now a wide interpretation of the inexcusable fault of the employer which was in the past restrictively defined.

An interesting relation between civil liability and social insurance has to be found in Hungary where a specialized insurance against occupational accidents and diseases does not exist. Yet the coverage of such risks is jointly guaranteed by the health insurance and by the pension insurance according to the extent of the incapacity to work. Moreover, Hungary presents a peculiar system of civil liability which can be activated in case of damages not covered by the combination of the health and the pension insurance.

According to the New Labour Code of 2012 as per which employers have *full and objective liability* for all damages suffered by the workers in connection with the employment relationship. Such a strict system of employer's liability has its roots in the socialist regime and was also present in the Old Labour Code of 1992. The employee has to prove the damage, its amount and the connection between the damage and the employment relationship. The employer's full and objective liability does not mean that s/he cannot shift the amount of compensation partly to the person who has caused the damage. Thus, the New Labour Code provided the possibility to allocate the compensation for damages in case of worker's negligence. The employer shall not compensate for the part of the damage, which was caused by the worker's contributory negligence.

Apart from contributory negligence, according to the old Labour Code, the employer could relieve from liability only in if s/he was able to prove that the damage was caused by an unavoidable event outside his or her field of operations or solely by the unavoidable conduct of the aggrieved party. The term 'employer's field of operation' was interpreted in a very broad way by the court. The opinion Nr. 29 of the Labour Committee of the Supreme Court stated that a reason causing damages shall only be considered falling outside 'the employer's field of operation', if there is no connection between the cause of the damage and the employer's activity. The employer's field of operation was not limited to the employer's premises especially since in many cases the employee performs his or her work usually outside them. The employer was also objectively liable, if the cause fall outside his or her field of operation, but it could have been addressed by the employer him or herself.

The new Labour Code has restructured and slightly limited the employer's liability. The main rule or full and objective liability remains unchanged. However, the grounds of exemption are redefined. The term used earlier 'field of operation' has been replaced by the expression 'field of control'. The aim of this change was to reduce the employer's liability elaborated in an extremely broad way by the case law of the labour courts. The legislator intended to exclude the employer's liability in cases, when the employer could not have had even an indirect effect on the cause of the damage.

Pursuant to the official explanation of the New Labour Code, the employer's 'field of control' includes all those circumstances (devices, energy, workers), on the operation of which the employer has influence. The employer has two opportunities to exempt from liability. In the first case if a damage has been caused by an external circumstance beyond his or her control which the employer did not have to reckon with and s/he was not expected to avoid the emergence of the circumstance giving rise thereto or to eliminate the same. In the second case, if the damage has been caused exclusively by the aggrieved party and his or her behaviour was not evitable.

As under the Old Labour Code, the employer shall compensate the employee for the totality of the damages occurred; however, two new rules have been introduced to reduce the amount of compensation.

First of all, no compensation shall be provided for damages, in connection with which the employer proves that their emergence was not foreseeable at the time they were caused. The foreseeability rule has been overtaken from the regulation of the Civil Code into the New Labour Code. No compensation shall be provided for the part of the damage which was caused through the employee's culpable conduct or which arose from the fact that the employee failed to meet his or her obligation of damage alleviation.

Second, the court may release the employer from the compensation for damages under special circumstances. The court shall consider particularly the financial situation of both parties, the gravity of the infringement and the consequences of paying compensation. This rule enables the labour courts to reduce the amount of compensation on ground of equity. According to the official explanation of the New Labour Code, this rule can be applied in case of small employers (particularly physical person), where a high amount of compensation would endanger their existence as an employer or – in case of physical persons – even their subsistence.

The other rules on compensation for damages have not been significantly modified in the New Labour Code. The employer has to reimburse all the lost income of the worker received in or outside of the employment relationship. Not only the lost wages but also the value of any regular benefits shall be taken into consideration to which the employee was entitled over and above his or her wages on the basis of his or her employment. The employer has to reimburse the worker for any other regular income outside the employment lost by virtue of the injury. No compensation shall be provided for the value of the benefits which, by virtue of their purpose, are only due in case of the performance of work, or for amounts received as a cost allowance. Compensation may also be awarded in form of monthly benefit. A regular benefit shall be awarded if it serves to support or to supplement the maintenance of the worker or his or her relative entitled to maintenance vis-à-vis the worker.

[C] Civil Liability without Social Insurance: United Kingdom

In the United Kingdom, where the employer is not obliged to insure workers against occupational accident and diseases, civil liability comes back at the centre stage as far as damages compensation is concerned. Moreover, the absence of a compulsory social insurance coverage means also the lack of insurance based inability or invalidity benefits paid to the worker from the moment the accident or the disease occurs. Therefore, a major role is played by the protection of sick workers and workers with disabilities. As a matter of fact, while an individual might be successful in claiming damages for an injury or disease sustained at work, this will typically follow a lengthy process of litigation and it will not assist in dealing with the immediate consequences for him or her.

Civil proceedings may take the form of the torts of breach of statutory duty or negligence, or an action in contract law for breach of the employer's implied duty of care. The most common remedy will be damages. These are not subject to a prior maximum limit and the objective is to place the claimant in the position s/he would have been in, had the wrong not been committed.

In such a perspective, in order to protect victims of health and safety violations, the Employers' Liability (Compulsory Insurance) Act 1969 requires all employers to be insured in respect of personal injury claims by employees. Moreover, the Employers' Liability (Defective Equipment) Act 1969 imposes strict liability on employers in respect of damage caused to employees by defective equipment used at work, even if the defect was due to the actions of the third party. It is notable that both of these Acts only apply in relation to those employed under a 'contract of service'; this means that those with more precarious working arrangements, such as casual workers, are unlikely to be covered. Workers in the construction industry, where there is a high rate of casual work and self-employment, may find themselves not covered by these laws. Generally speaking, workers can face difficulties in bringing personal injury claims where occupational ill-health was caused over time and through working for a range of different employers.

[D] Administrative and Criminal

As breaches of law, violations of health and safety provisions are sanctioned by administrative and/or criminal law provisions. Criminal law applies in case of violations which produce death or serious ill-health to the worker. As far as the applicable criminal law provisions are concerned further to the norms provided by general penal codes for death, injuries or procured disease, national legislators (the Austrian excluded) either: (a) provide for specific norms related to the health and safety domain within general penal codes (like in France, Hungary, Italy and Spain) or (b) have issued a criminal social code (like in Belgium) or (c) integrate criminal law provisions directly into health and safety regulations (like in Germany, Sweden, the Netherlands and United Kingdom).

The use of criminal law provisions gives rise to some well-known general problems. First of all, the respect of the principle of legality (no crime without precise law); second, the problem of the clear designation of the potential punishable offender which has been solved by referring to any person who has authority, resources and knowledge to take decisions in the health and safety domain (not necessarily the entrepreneur but also a delegated person); third, the problem of the criminal liability of legal persons, which has been approached by imposing fines on companies further to the penalty applied to directly liable physical persons; fourth, the reluctance of public prosecutors to open criminal proceedings against the entrepreneur in case of violations which have not produced the death or serious injuries to the affected person.

All the above mentioned difficulties have led legislators to opt, where socially sustainable and advisable, for a transformation of penalties into administrative sanctions with the advantage that the latter can be imposed by administrative authorities or courts without involving the criminal justice system.

In some cases, the same violation can be regarded as a breach of law punishable by an administrative sanction and as a crime punishable by a penalty. Therefore the problem is to know whether the heaviest sanction prevails according to the *ne bis in idem* principle or both sanctions have to be applied according to the *tot crimina tot poenae* principle. Both solutions are alternatively adopted, for instance the former in Spain and the latter in Italy.

§12.09 CONCLUSION

The conclusion of this chapter – that can only be synthetic, is that in the countries analysed in this book health and safety law has faced and is still facing an incremental development, taking stock from the past while looking at the challenges of the present. It is a permanent process in which 'additions' more than 'substitutions' contribute to the making of a dynamic and effective regulatory framework able to cope with the scientific and technological advance.

Paramount examples of this can be found in the relationship between collective prevention and individual protection, goal oriented legislation and rule oriented legislation, psychical, psychophysical, psychosocial risks on the one hand and physical risk on the other, workers' (representatives) participation and workers' (representatives) control.

In most of the cases, both 'additions' and 'substitutions' have been due to what we can call the 'EU turn' represented by the preventative and participative approach introduced by the Framework Directive at the end of the Eighties. As a consequence, in the countries analysed in this book the procedural part of health and safety law has been widely harmonized. On the contrary, the substantive part of it – that one related to the scope of the safety obligation, to the consequences of its violation and to workers' representation, remains national.

In such a perspective, the dynamic nature of the health and safety regulatory framework has compelled civil liability and social insurance law to admit the possibility to compensate or indemnify damages which go beyond the mere loss or reduction of work and earning capacity, touching the psychosocial sphere of the human being.

STUDIES IN EMPLOYMENT AND SOCIAL POLICY

1. W. Beck, L. van der Maesen & A. Walker (eds), *The Social Quality of Europe*, 1997 (ISBN 90-411-0456-9).
2. R. Blanpain, M. Colucci, C. Engels, F. Hendrickx, L. Salas & E. De Smyter, *Institutional Changes and European Social Policies after the Treaty of Amsterdam*, 1998 (ISBN 90-411-1018-6).
3. V. Lo: *Law and Industrial Relations: China and Japan after World War II*, 1998 (ISBN 90-411-1075-5).
4. A. Den Exter & H. Hermans (eds), *The Right to Health Care in Several European Countries*, 1998 (ISBN 90-411-1087-9).
5. M. Biagi (ed.), *Job Creation and Labour Law from Protection towards Pro-Action*, 2000 (ISBN 90-411-1432-7).
6. W. Beck, L. van der Maesen, F. Thomése & A. Walker (eds), *Social Quality: A Vision for Europe*, 2000 (ISBN 90-411-1523-4).
7. F. Pennings, *Introduction to European Social Law*, 3rd ed., 2001 (ISBN 90-411-1628-1).
8. J. Murray, *Transnational Labour Regulation: The ILO and EC Compared*, 2001 (ISBN 90-411-1583-8).
9. R. Blanpain & C. Engels (eds), *The ILO and Social Challenges of the 21st Century*, 2001 (ISBN 90-411-1572-2).
10. M. Biagi (ed.), *Towards a European Model of Industrial Relations? Building on the First Report of the European Commission*, 2001 (ISBN 90-411-1653-2).
11. J. Clasen (ed.), *What Future for Social Security? Debates and Reforms in National and Cross-National Perspective*, 2001 (ISBN 90-411-1671-0).
12. A. Numhauser-Henning (ed.), *Legal Perspectives on Equal Treatment and Non-Discrimination*, 2001 (ISBN 90-411-1665-6).
13. R. Blanpain (ed.), *Labour Law, Human Rights and Social Justice* 2001 (ISBN 90-411-1697-4).
14. M.-C. Kuo, H.F. Zacher & H.-S. Chan (eds), *Reform and Perspectives on Social Insurance: Lessons from the East and West*, 2002 (ISBN 90-411-1819-5).
15. P. Foubert, *The Legal Protection of the Pregnant Worker in the European Community*, 2002 (ISBN 90-411-1842-X).
16. M. Biagi (ed.), *Quality of Work and Employee Involvement in Europe*, 2002 (ISBN 90-411-1885-3).
17. F. Pennings, *Dutch Social Security Law in an International Context*, 2002 (ISBN 90-411-1887-X).

18. T. Carney & G. Ramia, *From Rights to Management: Contract, New Public Management and Employment Services*, 2002 (ISBN 90-411-1889-6).
19. R. Blanpain & M. Colucci, *European Labour and Social Security Law, Glossary*, 2002 (ISBN 90-411-1905-1).
20. I.U. Zeytinoglu (ed.), *Flexible Work Arrangements: Conceptualizations and International Experiences*, 2002 (ISBN 90-411-1947-7).
21. J. Berghman, A. Nagelkerke, M. Boos, R. Doeschot & G. Vonk (eds), *Social Security in Transition*, 2002 (ISBN 90-411-1969-8).
22. R. Blanpain, *The Legal Status of Sportsmen and Sportswomen under International, European and Belgian National and Regional Law*, 2003 (ISBN 90-411-1980-9).
23. R. Blanpain & M. Weiss (eds), *Changing Industrial Relations & Modernisation of Labour Law, Liber Amicorum in Honour of Professor Marco Biagi*, 2003 (ISBN 90-411-2008-4).
24. J. Malmberg (ed.), *Effective Enforcement of EC Labour Law*, 2003 (ISBN 90-411-2160-9).
25. M. De Vos (ed.), *A Decade Beyond Maastricht: The European Social Dialogue Revisited*, 2003 (ISBN 90-411-2163-3).
26. M. Sewerynski (ed.), *Collective Agreements and Individual Contracts of Employment*, 2003 (ISBN 90-411-2190-0).
27. R. Blanpain & M. Van Gestel, *Use and Monitoring of E-Mail, Intranet and Internet Facilities at Work: Law and Practice*, 2004 (ISBN 90-411-22-66-4).
28. A. C. Neal (ed.), *The Changing Face of European Labour Law and Social Policy*, 2004 (ISBN 90-411-2312-1).
29. E. Sol & M. Westerveld (eds), *Contractualism in Employment Services: A New Form of Welfare State Governance*, 2005 (ISBN 90-411-2405-5).
30. F. Pennings (ed.), *Between Soft and Hard Law – The Impact of International Social Security Standards on National Social Security Law*, 2006 (ISBN 978-90-411-2491-3).
31. L. Dickens & A. C. Neal (eds), *The Changing Institutional Face of British Employment Relations*, 2006 (ISBN 978-90-411-2541-5).
32. G. Sebardt, *Redundancy and the Swedish Model in an International Context*, 2006 (ISBN 978-90-411-2503-3).
33. A.M. Swiatkowski, *Charter of Social Rights of the Council of Europe*, 2007 (ISBN 978-90-411-2608-5).
34. M. Sargeant (ed.), *The Law on Age Discrimination in the EU*, 2008 (ISBN 978-90-411-2522-4).
35. G. Di Domenico & S. Spattini (eds), *New European Approaches to Long-Term Unemployment: What Role for Public Employment Services and What Market for Private Stakeholders?*, 2008 (ISBN 978-90-411-2614-6.)
36. C. Welz, *The European Social Dialogue under Articles 138 and 139 of the EC Treaty: Actors, Processes, Outcomes*, 2008 (ISBN 978-90-411-2744-0).
37. M. Rönnmar (ed.), *EU Industrial Relations v. National Industrial Relations: Comparative and Interdisciplinary Perspectives*, 2008 (ISBN 978-90-411-2770-9).

38. F. Pennings, Y. Konijn & A. Veldman (eds), *Social Responsibility in Labour Relations: European and Comparative Perspectives*, 2008 (ISBN 978-90-411-2783-9).
39. F. Pennings & C. Bosse (eds), *The Protection of Working Relationships: A Comparative Study*, 2011 (ISBN 978-90-411-3289-5).
40. S. Devetzi & S. Stendahl (eds), *Too Sick to Work?: Social Security Reference in Europe for Persons with Reduced Earnings Capacity*, 2011 (ISBN 978-90-411-3426-4).
41. U. Becker, F. Pennings & T. Dijkhoff (eds), *International Standard-Setting and Innovations in Social Security*, 2013 (ISBN 978-90-411-3233-8).
42. E. Ales (ed.), *Health and Safety at Work: European and Comparative Perspective*, 2013 (ISBN 978-90-411-4661-8).

Lightning Source UK Ltd.
Milton Keynes UK
UKHW02n1528190818
327414UK00003BA/132/P

9 789041 146618